INTERNATIONAL UNION OF CRYSTALLOGRAPHY
BOOK SERIES

IUCr BOOK SERIES COMMITTEE

IUCr Monographs on Crystallography
1 *Accurate molecular structures*
   A. Domenicano, I. Hargittai, editors
2 *P.P. Ewald and his dynamical theory of X-ray diffraction*
   D.W.J. Cruickshank, H.J. Juretschke, N. Kato, editors
3 *Electron diffraction techniques, Vol. 1*
   J.M. Cowley, editor
4 *Electron diffraction techniques, Vol. 2*
   J.M. Cowley, editor
5 *The Rietveld method*
   R.A. Young, editor
6 *Introduction to crystallographic statistics*
   U. Shmueli, G.H. Weiss
7 *Crystallographic instrumentation*
   L.A. Aslanov, G.V. Fetisov, J.A.K. Howard
8 *Direct phasing in crystallography*
   C. Giacovazzo
9 *The weak hydrogen bond*
   G.R. Desiraju, T. Steiner
10 *Defect and microstructure analysis by diffraction*
   R.L. Snyder, J. Fiala and H.J. Bunge
11 *Dynamical theory of X-ray diffraction*
   A. Authier
12 *The chemical bond in inorganic chemistry*
   I.D. Brown
13 *Structure determination from powder diffraction data*
   W.I.F. David, K. Shankland, L.B. McCusker, Ch. Baerlocher, editors
14 *Polymorphism in molecular crystals*
   J. Bernstein
15 *Crystallography of modular materials*
   G. Ferraris, E. Makovicky, S. Merlino

16   *Diffuse X-ray scattering and models of disorder*
     T.R. Welberry

17   *Crystallography of the polymethylene chain: an inquiry into the structure of waxes*
     D.L. Dorset

18   *Crystalline molecular complexes and compounds: structure and principles*
     F. H. Herbstein

19   *Molecular aggregation: structure analysis and molecular simulation of crystals and liquids*
     A. Gavezzotti

20   *Aperiodic crystals: from modulated phases to quasicrystals*
     T. Janssen, G. Chapuis, M. de Boissieu

21   *Incommensurate crystallography*
     S. van Smaalen

22   *Structural crystallography of inorganic oxysalts*
     S.V. Krivovichev

23   *The nature of the hydrogen bond: outline of a comprehensive hydrogen bond theory*
     G. Gilli, P. Gilli

24   *Macromolecular crystallization and crystal perfection*
     N.E. Chayen, J.R. Helliwell, E.H. Snell

25   *Neutron protein crystallography: hydrogen, protons, and hydration in bio-macromolecules*
     N. Niimura, A. Podjarny

IUCr Texts on Crystallography

1   *The solid state*
     A. Guinier, R. Julien

4   *X-ray charge densities and chemical bonding*
     P. Coppens

8   *Crystal structure refinement: a crystallographer's guide to SHELXL*
     P. Müller, editor

9   *Theories and techniques of crystal structure determination*
     U. Shmueli

10   *Advanced structural inorganic chemistry*
     Wai-Kee Li, Gong-Du Zhou, Thomas Mak

11   *Diffuse scattering and defect structure simulations: a cook book using the program DISCUS*
     R. B. Neder, T. Proffen

12   *The basics of crystallography and diffraction, third edition*
     C. Hammond

13   *Crystal structure analysis: principles and practice, second edition*
     W. Clegg, editor

14   *Crystal structure analysis: a primer, third edition*
     J.P. Glusker, K.N. Trueblood

15   *Fundamentals of crystallography, third edition*
     C. Giacovazzo, editor

16   *Electron crystallography: electron microscopy and electron diffraction*
     X. Zou, S. Hovmöller, P. Oleynikov

17   *Symmetry in crystallography: understanding the International Tables*
     P.G. Radaelli

18   *Symmetry relationships between crystal structures: applications of crystallographic group theory in crystal chemistry*
     U. Müller

19   *Small angle X-ray and neutron scattering from solutions of biological macromolecules*
     D.I. Svergun, M.H.J. Koch, P.A. Timmins, R.P. May

20   *Phasing in crystallography: a modern perspective*
     C. Giacovazzo

Small Angle X-Ray and Neutron Scattering from Solutions of Biological Macromolecules

Dmitri I. Svergun

European Molecular Biology Laboratory, Hamburg, Germany

Michel H. J. Koch

University of Leuven, Leuven, Belgium

Peter A. Timmins

Institut Laue-Langevin, Grenoble, France

Roland P. May

Institut Laue-Langevin, Grenoble, France

OXFORD

UNIVERSITY PRESS

OXFORD
UNIVERSITY PRESS

Great Clarendon Street, Oxford, OX2 6DP,
United Kingdom

Oxford University Press is a department of the University of Oxford.
It furthers the University's objective of excellence in research, scholarship,
and education by publishing worldwide. Oxford is a registered trade mark of
Oxford University Press in the UK and in certain other countries

First published 2013
First published in paperback 2020

Impression: 1

Published in the United States of America by Oxford University Press
198 Madison Avenue, New York, NY 10016, United States of America

British Library Cataloguing in Publication Data
Data available

Library of Congress Cataloging in Publication Data
Data available

ISBN 978–0–19–963953–3 (hbk.)
ISBN 978–0–19–885421–0 (pbk.)

Printed and bound by
CPI Group (UK) Ltd, Croydon, CR0 4YY

Preface

Small angle scattering (SAS) has rapidly spread as a tool in many areas of science and technology, and its applications have become increasingly specialised. The limited aim of the present book is to provide an introduction for the growing community wishing to use the technique for the characterisation of biological macromolecules—mainly proteins, nucleoproteins and their complexes—in solution. In this area, recent decades have seen tremendous progress not only in X-ray (SAXS) and neutron (SANS) instrumentation but also in modelling approaches to obtain low-resolution (about 1–2 nm) structures from the scattering data. We have tried to reflect this progress in the book, which covers the basics of SAXS/SANS, instrumentation, data collection, modelling techniques and complementary use with other methods. The book is oriented towards biochemists, biophysicists and researchers using related structural methods (macromolecular crystallography, nuclear magnetic resonance, electron microscopy). It should be useful not only for scientists who have already used SAS but also for novices with little or no experience in solution scattering.

Part I of the book presents the major theoretical concepts of small angle scattering of X-rays and neutrons by macromolecular solutions required to understand the basics of the technique (Chapter 1). SAS instrumentation on laboratory and synchrotron X-ray sources, and, for SANS, on reactors and spallation sources, is reviewed in Chapter 2. Specific features of SAXS and SANS experiments, sample requirements, measurement protocols and data reduction procedures are described in Chapter 3.

Part II describes the methods for interpreting SAXS and SANS data in terms of structural models of biological macromolecules. Chapter 4 is devoted to the analysis of data from dilute monodisperse solutions containing identical particles, for which not only overall parameters but also low-resolution three-dimensional models can be reconstructed. Modern approaches to construct structural models *ab initio* or by hybrid modelling are considered. Chapter 5 treats polydisperse systems (such as oligomeric mixtures or flexible macromolecules) and systems of interacting particles. In these cases the aim of using SAS is not to determine a structure, but to quantitatively characterise the composition of mixtures or to analyse the interactions between particles in solution.

Part III presents practical applications of SAXS and SANS to study the structure of biological macromolecules in solution. These are grouped in sections covering different types of experiments and, accordingly, different methodological approaches. In Chapter 6, studies of the static structure of individual macromolecules and complexes, and of the equilibrium composition of mixtures are considered. The characterisation of the kinetics of biological processes in time-resolved experiments is explained in Chapter 7. Interacting systems, where SAS helps to determine the factors influencing interparticle interactions are covered in Chapter 8. Finally, Chapter 9 is devoted to the joint use of SAXS and SANS with other structural, biophysical and computational techniques to illustrate the advantages of multipronged approaches in modern structural biology and to describe the recent trends in biological SAS applications.

As the book is intended for readers without extensive theoretical background, the concepts are, wherever possible, explained in simple terms. For those willing to learn more rigorous treatments, major mathematical tools are presented in the Appendices, including Fourier transformations, spherical harmonics and calculation of interaction potentials. These Appendices also contain a selection of useful Web links online at the time of publication of this book.

Several aspects of SAXS/SANS in the life sciences are not covered, such as fibres, semi-crystalline, lipid or colloidal systems and gels, which often occur in living systems and in applications in the food industry and in pharmaceutical technology. We also did not enter into detailed aspects of instrumentation such as collimation and detectors, as it was assumed that the readers will use commercial instruments or the instruments available at large facilities, where users have no influence on these aspects.

We thank the numerous colleagues, old masters and inquisitive coworkers, from whom we learned what we know about SAS, and in particular also M. Agamalian, C. Baldock, P. Bernado, W. Bouwman, W. Bras, O. Byron, M. Capel, J. Carrascosa, P. Chacon, C. Dewhurst, D. Durand, S. Egelhaaf, O. Glatter, I. Grillo, P. Høghøj, T. Irving, S. King, J. Kuriyan, P. Lindner, L. Makowski, P. Meersman, P. Panine, J.S. Pedersen, J. Pérez, S. Perkins, A. Round, T. Narayanan, D. Niessing, V. Ramakrishnan, J. Tainer, N. Terrill, J. Trewhella, P. Vachette, B. Vestergaard, M. Wilmanns and G. Zaccai, who read parts of the text or kindly provided information and illustrations. Dmitri Svergun thanks the BioSAXS group of the EMBL, Hamburg, and the SAXS group of the Institute of Crystallography, Moscow, for the invaluable contributions to the methods development, and D. Petoukhova for technical assistance in the preparation of the book.

No textbook is free of errors, we take full responsibility for any that are present, and we would be very much appreciative if they were reported to us.

Contents

Introduction   **1**
References   9

PART I THEORY AND EXPERIMENT

1   Basics of small angle scattering   **13**
1.1   Elastic scattering of X-rays and neutrons   13
1.2   Scattering by macromolecular solutions   18
1.3   Resolution and contrast   21
1.4   Absorption and anomalous X-ray scattering   25
References   26

2   X-ray and neutron scattering instruments   **27**
2.1   Characteristics of sources   27
2.2   X-ray sources   29
2.3   Neutron sources   37
2.4   SAXS instruments   43
2.5   SANS instruments   52
2.6   Special instruments   57
References   62

3   Experimental practice and data processing   **65**
3.1   Sample requirements for SAXS and SANS   65
3.2   Experimental protocols and instrumental corrections   68
3.3   Special features of SAXS: radiation damage   72
3.4   Special features of SANS: the use of H_2O/D_2O mixtures   75
3.5   Random and systematic errors   79
3.6   Basic structural information and data quality   81
3.7   Calibration to absolute scale and molecular mass   85
References   88

PART II DATA ANALYSIS METHODS

4   Monodisperse systems   **93**
4.1   Overall parameters of particles   93
4.2   Multipole representation of SAS intensity   108
4.3   Shannon sampling   110
4.4   *Ab initio* shape analysis   112
4.5   Computation of scattering patterns from atomic models   125

4.6 Hybrid methods   128
4.7 Labelling and triangulation   134
References   145

5   Polydisperse and interacting systems   **152**
5.1 Size, shape and conformation polydispersity   152
5.2 Size distribution functions   153
5.3 Shape polydispersity and oligomeric mixtures   155
5.4 Conformational polydispersity and flexible systems   159
5.5 Interacting systems and structure factor   163
References   166

PART III BIOLOGICAL APPLICATIONS
OF SOLUTION SAS

6   Static structural studies   **171**
6.1 Applications of *ab initio* shape determination   171
6.2 Quaternary structure analysis of proteins and complexes   181
6.3 Equilibrium mixtures and oligomeric composition   192
6.4 Membrane proteins and lipoproteins   198
6.5 Flexible systems   205
References   212

7   Kinetic and perturbation studies   **220**
7.1 Dynamics and kinetics   221
7.2 Perturbation methods   223
7.3 Temperature scans and T-jumps   230
7.4 High-pressure experiments   233
7.5 Stopped-flow and continuous-flow mixing   237
7.6 Light-triggered processes   247
References   250

8   Analysis of interparticle interactions   **259**
8.1 Basic physical chemistry of interactions   259
8.2 Experimental SAS studies of protein–protein interactions   265
8.3 Structure factor calculations for proteins   266
8.4 Interactions in nucleic acids   280
References   281

9   SAS in multidisciplinary studies   **286**
9.1 Automation and high-throughput SAS   286
9.2 Joint use with high-resolution methods   291
9.3 SAS and low-resolution crystallography   298
9.4 Complementary biophysical methods   303
9.5 Bioinformatics and model validation   309
References   314

Conclusions and future prospects   **320**

Appendices   **323**
Appendix 1: Basic physics and mathematics of wave phenomena   323
Appendix 2: Spherical harmonics and their applications for SAS   338
Appendix 3: Interactions between spherical molecules   343
Appendix 4: Web resources   349

Abbreviations list   **353**

Index of computer programs   **355**

Index   **357**

Introduction

Small angle scattering (SAS) of X-rays (SAXS) or neutrons (SANS) is a powerful method for analysing the structure and structural changes of biological macromolecules in solution. Since the beginnings in the 1930s, methods for extracting structural information from SAS patterns of non-crystalline samples in physics, materials science and biology have made much progress (Guinier and Fournet 1955, Glatter and Kratky 1982, Feigin and Svergun 1987). Advances in instrumentation and computational methods, which largely paralleled the development of new X-ray and neutron sources starting around the 1970s, led to exciting applications in structural biology. These applications and the wish to characterise the many new proteins generated by the various '-ics' initiatives have led to a renewed interest in the method on the part of biochemists and to its more systematic application.

Like other methods in structural biology such as X-ray crystallography and nuclear magnetic resonance (NMR), SAS requires milligramme amounts of highly purified material, preferably in a monodisperse solution. While sample requirements are similar for the three techniques (except that no additional crystallisation step is required for solution methods) a distinct advantage of SAS is, however, the speed of both data collection and processing. On modern neutron and synchrotron radiation sources, SAS data can be collected, processed and analysed in minutes or seconds, allowing an almost immediate characterisation of the sample using a few overall parameters and even online generation of low-resolution particle shapes. SAS can also be used for rapid screening of samples in various solvent/additives conditions to study intermolecular interactions, including, for example, identification and optimisation of crystallisation conditions.

Scattering is a process whereby a beam of radiation or particles is deviated from its initial trajectory by the inhomogeneities in the medium which it traverses. A scattering experiment is thus conceptually very simple and requires only a source, a sample and a detector, as illustrated in Fig. 1. Solution scattering does not require any special sample processing like crystallisation or cryocooling. Isotopic labelling can be useful in neutron scattering but is usually not indispensable. The sample is usually an aqueous solution of biological macromolecules, and the scattering of interest is that arising from the inhomogeneities due to the solute particles. Both X-rays and neutrons interact with all atoms

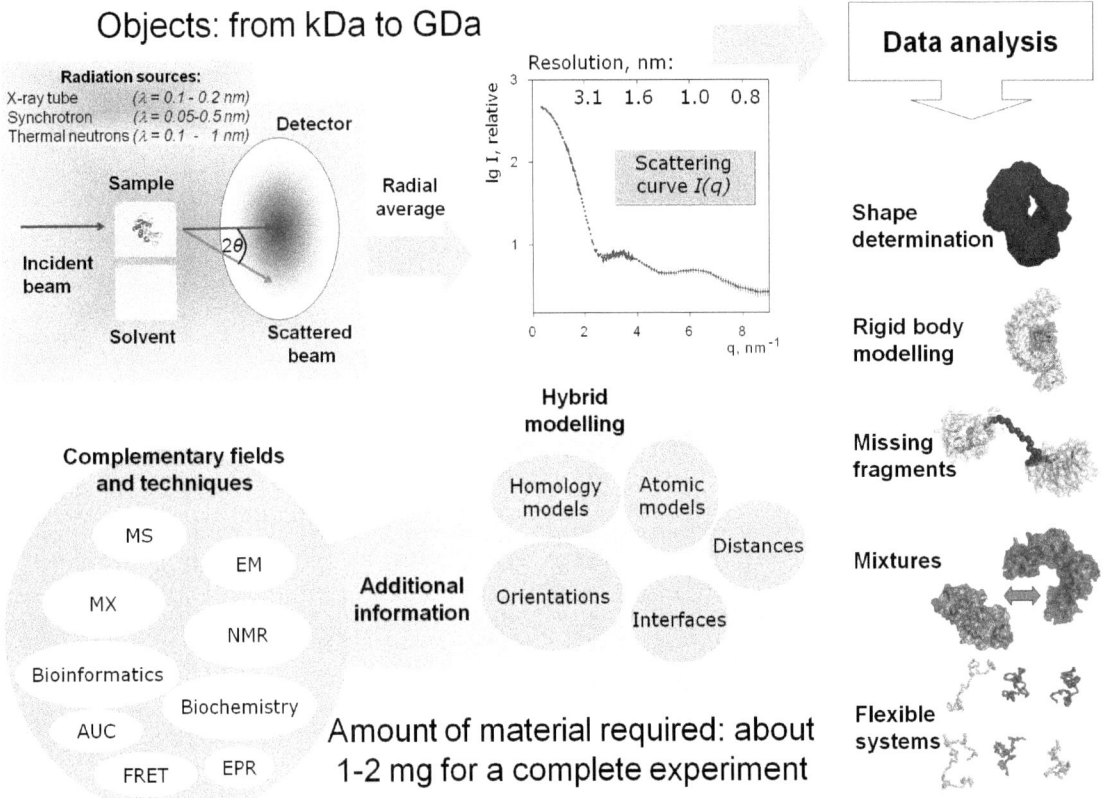

Fig. 1 General scheme of an SAS experiment; structural questions addressed by the technique and its synergistic use with other methods. The nominal resolution of the data in the scattering pattern is given as $d = 2\pi/q$. Abbreviations: MS, mass spectroscopy; MX, macromolecular crystallography; AUC, analytical ultracentrifugation; FRET, fluorescence resonance energy transfer; EM, electron microscopy; NMR, nuclear magnetic resonance; EPR, electron paramagnetic resonance. Adapted, with permission, from Svergun (2010), © De Gruyter.

in the irradiated volume, making them the sources of scattered radiation. X-rays interact largely with electrons, whereas neutrons interact with the nuclei and spin. In many instances, it is convenient to describe the object not with discrete atoms, but as a continuous distribution of scattering density (electron density for X-rays and nuclear/spin density for neutrons).

We shall deal exclusively with elastic scattering of X-rays and neutrons, where the energy and wavelength (λ) of the incident and scattered radiation are identical. The scattering pattern obtained after background subtraction, instrumental corrections and normalisation, $I_{total}(q)$, results from the interference of the scattered waves. The scattering from aqueous solutions of biological macromolecules is isotropic. and thus solely depends on the variable $q = (4\pi \sin \theta)/\lambda$, the momentum transfer, where 2θ is the scattering angle between the incident beam and the direction of observation.

A measurement of the scattering of a solution of macromolecules must always be accompanied by that of the corresponding solvent, and the latter pattern is subtracted from the former to eliminate the solvent scattering and the instrumental background scattering. The net signal from the macromolecules depends on the number of molecules in the illuminated volume (that is, on the solute concentration) and on the excess scattering density (often also called the contrast), which is the difference between the scattering density of the solute and that of the solvent. Clearly, the higher the concentration and the contrast, the stronger the useful signal. In most structural studies, dilute solutions with solute concentrations below 1% must be used to avoid aggregation, and as the contrast of biomolecules in aqueous solutions is usually small (especially for X-rays) the useful signal is rather weak. SAS instruments must therefore be optimised carefully to minimise the background.

   The importance of low-background instruments for SAS was appreciated very early on, and the technique really took off only in the 1930s, when André Guinier was asked for his PhD to construct 'a camera with monochromatic primary radiation and with minimum parasitic radiation', to look at scattering between Bragg reflections. The device is now called a Guinier camera. He writes:

"In the ordinary cameras utilised at that time, there was a strong parasitic scattering in the small angle region, which prevents any correct measurement. My camera had not this defect. That is why I studied especially small angle scattering. I observed that small angle scattering does not exist when the sample is homogeneous. It is strong when it contains fine grains (10 to 100 nm). I produced next the means of determining the grain size from the scattering curve. It was the beginning of X-rays small angle scattering which, some years later, was followed by neutron small angle scattering." (Guinier 1999)

The construction of large scale facilities—high flux reactors and spallation sources for neutrons and synchrotron-radiation X-ray sources—was a major step forward for SAS, as the new sources brought large improvements in signal-to-background ratio. This made SAS techniques increasingly attractive not only for soft condensed-matter physics and materials science, but especially for the study of non-crystalline biological samples which give weak signals. One area that has particularly gained from these developments as well as from new computational approaches to the interpretation of scattering patterns is the study of biological macromolecules in solution. It thus seemed timely to devote a monograph to these applications, which would also take into account the profound changes in the user community, where biochemists are rapidly replacing biophysicists and physical chemists. In this introductory chapter we present the very basic concept of the technique and list its major biological applications. Because of differences in background between generations of SAS users, some readers may find Appendix 1 useful to refresh their knowledge about the basics of physics and mathematics of waves before proceeding to Chapter 1, which describes the scattering processes in more detail.

The net SAS intensity after solvent subtraction may under some assumptions be expressed as a product of two terms, $I_{total}(q) = I(q) \times S(q)$. The form factor, $I(q)$, arises from the scattering from individual particles in solution and contains the information about their structure. The structure factor, $S(q)$, is due to interference of scattered waves emitted by different particles, and contains information about the interparticle interactions (about the structure of the solution). Dilute solutions of monodisperse macromolecules with concentrations in the µM–mM range are usually employed, to avoid interparticle interference effects and approach the limit of an ideal (non-interacting) system and to be able to analyse $I(q)$, assuming that $S(q) = 1$. SAS is, however, also useful, and actively used, to study interactions between macromolecules in solution based on the analysis of the structure factor $S(q)$.

Three important cases are distinguished in structural SAS studies: (i) monodisperse systems, where all particles are identical, (ii) polydisperse system, when they have the same chemical composition but differ in size and/or shape, and (iii) mixtures, where they may differ in size, shape and chemical composition. For monodisperse solutions, the net intensity $I(q)$ is proportional to the scattering from a single particle averaged over all its orientations. This averaging process precludes going back from the scattering pattern to the three-dimensional scattering density as can be done with crystals. Consequently, with SAS one obtains only the distribution $p(r)$ of distances (r) within the particle, averaged over all orientations, which is, obviously, a one-dimensional function.

This suffices to determine the molar mass M and overall geometrical parameters of the particles—for example, their radius of gyration R_g, which is a measure of their extent, their maximum dimension (D_{max}) and their hydrated volume (V_p). Advances in computational methods have made it possible to not only extract these simple parameters, but also to obtain reliable three-dimensional structures from scattering data. This has been verified in a number of cases by the comparison with crystal structures determined later (see, for example, Schmidt *et al.* 1995 and Riboldi-Tunnicliffe *et al.* 2001 or Armbruster *et al.* 2004 and Drory *et al.* 2004). Low-resolution (1–2 nm) models can be obtained *ab initio* or through refinement of available high-resolution structures and/or homology models. While the former analysis provides a low-resolution shape of the molecule in question and often adds insight to the biological problem at hand, the latter combination of SAS and complementary data is a powerful method for modelling the organisation of macromolecular complexes, especially where there are conformational changes induced by, for example, a change of pH, binding of cofactors or ions (see, for example, King *et al.* 2005). In addition to modelling, SAS is routinely used for the validation of structural models obtained by other methods, the quantitative analysis of oligomeric states and the estimation of fractions of the solute volume of components in mixtures and/or polydisperse systems.

For polydisperse systems or mixtures consisting of K different types of non-interacting particles, the measured scattering pattern can be written as a linear combination of the scattering intensity of the K types of particles ($i_k(q)$), where the coefficients $v_k > 0$ represent their fractions of the solute volume:

$$I(q) = \sum_{k=1}^{K} v_k i_k(q) \tag{1}$$

In this case, the overall parameters reflect the average values over the ensemble, but, of course, shapes of individual components cannot be reconstructed given only the experimental scattering from the mixture. However, if the scattering intensities from the components are known *a priori*, their fractions of the solute volume can be determined from eq. (1). A typical application is the characterisation of equilibrium mixtures of oligomers, but much more complicated systems can also be analysed. In particular, SAS can be readily employed for the analysis of flexible systems including multi-domain proteins with flexible linkers or intrinsically disordered proteins (IDPs).

Although the fundamental physical phenomena underlying the scattering of X-rays and neutrons are very different, the theory and most approaches for data analysis are similar. The major difference is that X-ray scattering and absorption are sensitive to the electron density, and hence particularly to heavier atoms, as they contain more electrons. Neutron scattering is due to the nuclear (and sometimes spin) scattering density, which is not simply related to the atomic number. In SANS, samples that strongly absorb X-rays (due to high salt concentrations, for example) can be measured, and the macromolecules usually do not suffer from radiation damage. SANS is sensitive to the isotopic composition of the sample, and, most importantly, there is a large difference in the scattering properties of hydrogen (H) and deuterium (D). Both have one electron and would thus in principle be indistinguishable with X-rays if isotope effects did not also result in small differences in electron density (Henderson 1999). Contrast variation by H/D substitution, using water/heavy water mixtures and/or partially or perdeuterated biological material, yields precious additional information about the structure of macromolecular complexes. The disadvantages of SANS, for which no laboratory sources are available, are that more material is usually required than for SAXS and that so-called incoherent scattering effects create a strong background scattering. As a rule of thumb, the use of SANS is recommended, when additional information can be obtained by H/D exchange in the solvent—for example, in the case of nucleoprotein complexes or membrane proteins, or in multiprotein complexes where individual subunits can be distinguished by deuterium labelling. SAXS and SANS are complementary methods, and their joint use is often of great advantage.

SAS allows one to address a broad variety of structural questions, as schematically illustrated in Fig. 1. It can be employed as a stand-alone tool to obtain overall characteristics and low-resolution models, but is often more useful in combination with other structural, hydrodynamic, computational and biochemical methods. SAS can be used in a very broad range of macromolecular sizes, from small proteins and even peptides to large macromolecular machines and sub-cellular aggregates. Importantly, the method is not only useful for static structural modelling but also for rapid analysis of the response to changes in experimental conditions (pH, temperature, pressure, ionic strength, ligand binding, and so on). It is also possible to follow the time course of processes such as folding/unfolding and assembly/dissociation over several orders of magnitude in time, but such experiments usually require a large amount of material.

SAS modelling has a long history, to which several groups have contributed using different techniques (for early references see, for example, Koch *et al.* 2003). The application of spherical harmonics to this problem marked an important step forward (Stuhrmann 1970), but initially received little attention, and the results of neutron contrast variation yielding the shape of the 50S ribosomal subunit (Stuhrmann *et al.* 1977) were largely met with disbelief, even within the SAS community. In another example, an electron density map of bacteriophage T7 constructed in axially symmetric approximation from the SAXS data back in 1982 (Svergun *et al.* 1982) depicted not only the overall shape of the virus but also intriguing structural features such as an internal core and a cylindrical protrusion (Fig. 2, left panel). These results could

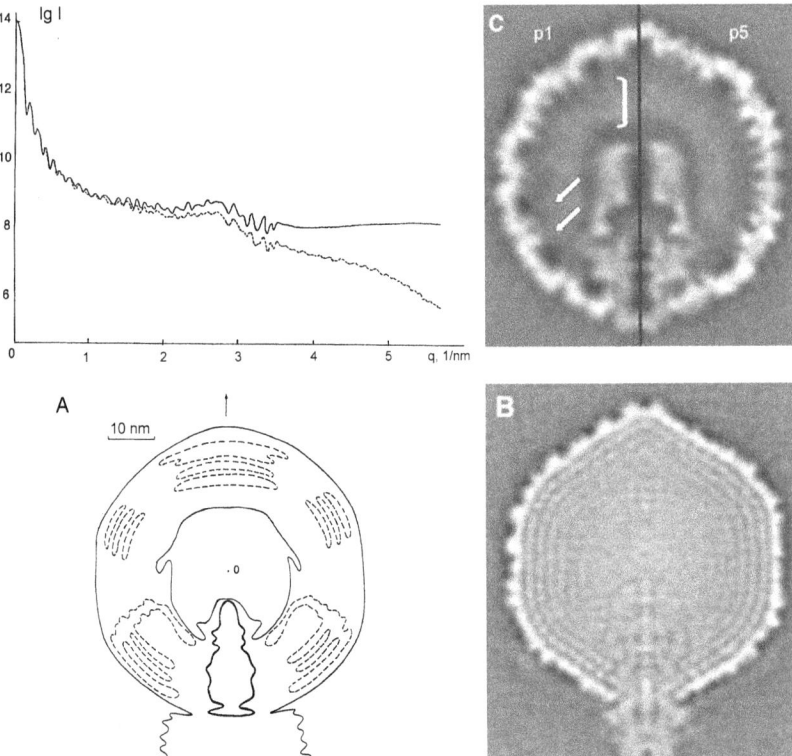

Fig. 2 Structural models of bacterial virus T7. (A): the model constructed *ab initio* by (Svergun *et al.* 1982) from the SAXS data. The processed experimental curve (top panel) is displayed as solid line, the scattering from the model as dashed line. Bottom panel: cross-section of the electron density map containing the rotation axis. Adapted, with permission, from Svergun *et al.* (1982), © International Union of Crystallography. (B) and (C) are cross-sections of the cryo-EM 3D of the mature virion and of the DNA-free prohead, respectively. Adapted, with permission, from Agirrezabala *et al.* (2005) © Macmillan Publishers Ltd., 2005.

not be further validated by other methods at that time, and remained barely noticed. They were, however, fully confirmed by cryo-EM (Agir-rezabala *et al.* 2005)—but only 25 years later—as illustrated in Fig. 2, right panel.

In subsequent years, the synchrotron radiation and SANS results from studies of the quaternary structure of tubulin assemblies (Bordas *et al.* 1983), components of the immune system (Perkins 1985) or the higher-order structure of chromatin (for a review, see Koch 1989) illustrated that SAS could usefully complement and sometimes challenge the results of EM, but also offered the possibility of time-resolved measurements.

Further, a number of results like those of rigid-body modelling of the SAXS patterns of the allosteric enzyme aspartate transcarbamoylase (ATCase) based on the available crystallographic models helped to convince macromolecular crystallographers that SAS could also be useful to them. Indeed, these results showed that crystal and solution structures, especially of large assemblies such as the R-state of ATCase, could differ significantly (Svergun *et al.* 1997).

It the meantime it has also become generally accepted that crystal structures provide only a snapshot of the macromolecules and that a more detailed description of conformational changes—sometimes too subtle to detect by SAS—is required to understand function. As the emphasis in structural biology shifts towards larger systems with a higher biological relevance, structural studies are now rarely limited to a single technique, and one increasingly attempts to obtain consistent hybrid models based on different methods. In this context, SAS is used mostly in combination with other structural, biophysical and biochemical methods. Moreover, SAXS and SANS are also often employed together, making use of the complementarity of the two techniques.

An example is given by the analysis of a complex between two proteins: *Bacillus subtilis* histidine kinase KinA and a DNA-damage checkpoint inhibitor Sda. (Whitten *et al.* 2007). This complex was prepared with perdeuterated Sda, allowing one to distinguish between the two proteins using neutron scattering. The measured SANS profiles in solutions with different concentrations of D_2O and also the SAXS data (Fig. 3, left panel) were simultaneously fitted by a model constructed from the available high-resolution structures of the two proteins. This rigid-body model revealed two Sda molecules bound to the base of the dimerisation domain of the KinA dimer (Fig. 3, right panel). Hybrid approaches synergistically using complementary methods have become the most powerful and reliable instruments for interpreting data from macromolecular solutions. Thanks largely to this analysis strategy there has been a rapid growth in the number of publications containing results of SAS on biological macromolecules, as illustrated in Fig. 4.

The examples in the following chapters give an overview of the capabilities of SAS in the study of solutions of biological macromolecules, as well as a sample of the variety of approaches used by different

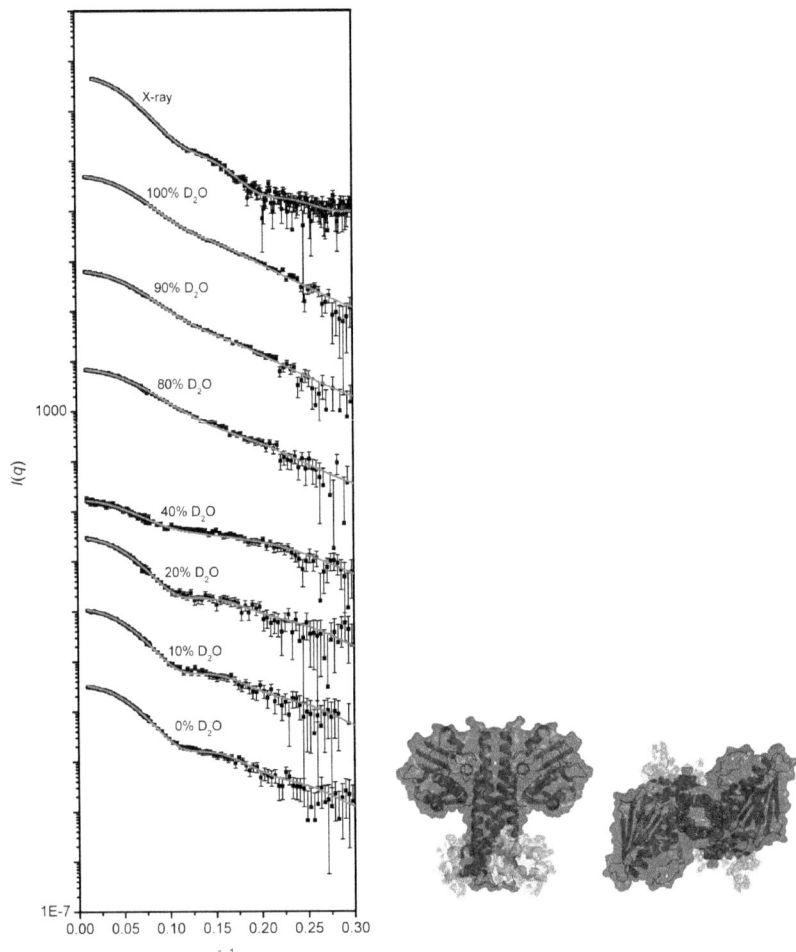

Fig. 3 Structural modelling of the KinA-Sda complex. Left panel: the experimental X-ray and neutron scattering data (dots with error bars) and the scattering patterns computed from the best model constructed from the high-resolution structures of the two proteins. Right panel: best-fit model of the KinA dimer (dark grey surface and skeleton) complexed by two Sda molecules (light grey surface and skeleton) obtained from rigid-body modelling. The right view is rotated counterclockwise by 90° around the horizontal axis. Adapted from Whitten *et al.* (2007), with permission from Elsevier, © 2007.

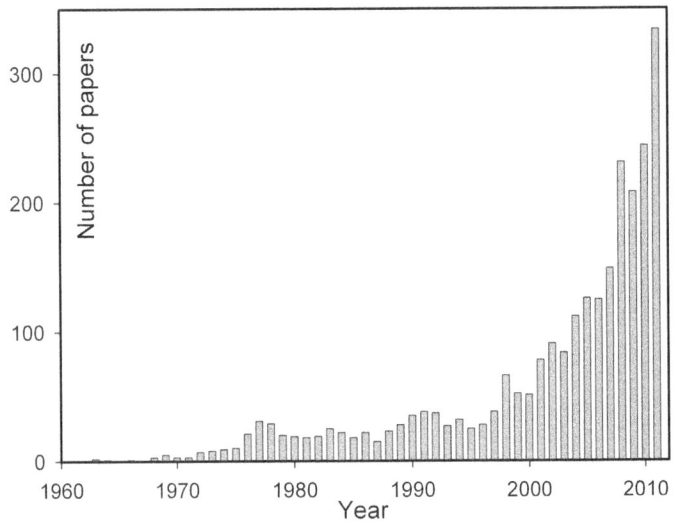

Fig. 4 Evolution of the number of publications reporting SAS results on biological macromolecules during the last fifty years, based on a PubMed search.

groups. Before turning to these in Parts II and III, the basic concepts of SAS will be introduced in Part I. These should allow the reader to appreciate the progress made in the recent past, and enable them to develop an ability to critically assess the increasing number of articles reporting SAS results while hopefully avoiding the pitfalls of over-interpretation.

References

Agirrezabala, X., Martin-Benito, J., Caston, J. R., Miranda, R., Valpuesta, J. M. and Carrascosa, J. L. (2005). 'Maturation of phage T7 involves structural modification of both shell and inner core components', *EMBO J.* 24: 3820–9.

Armbruster, A., Svergun, D. I., Coskun, U., Juliano, S., Bailer, S. M. and Gruber, G. (2004). 'Structural analysis of the stalk subunit Vma5p of the yeast V-ATPase in solution', *FEBS Lett.* 570: 119–25.

Bordas, J., Mandelkow, E. M. and Mandelkow, E. (1983). 'Stages of tubulin assembly and disassembly studied by time-resolved synchrotron X-ray scattering', *J. Mol. Biol.* 164: 80–135.

Drory, O., Frolow, F. and Nelson, N. (2004). 'Crystal structure of yeast V-ATPase subunit C reveals its stator function', *EMBO Rep.* 5: 1148–52.

Feigin, L. A., and Svergun, D. I. (1987). *Structure Analysis by Small-Angle X-Ray and Neutron Scattering.* New York: Plenum Press.

Glatter, O., and Kratky, O. (1982). *Small-Angle X-Ray Scattering.* New York: Academic Press.

Guinier, A. (1999). 'As chance would have it...', *The Rigaku Journal* 16: 1–2, <http://www.rigaku.com/downloads/journal/Vol16.11.1999/guinier.pdf>

Guinier, A., and Fournet, G. (1955). *Small-Angle Scattering of X-Rays.* New York: Wiley.

Henderson, S. J. (1999). 'Isotope effects in solution small-angle X-ray scattering', *J. Appl. Crystallogr.* 32: 113–4.

King, W. A., Stone, D. B., Timmins, P. A. *et al.* (2005). 'Solution structure of the chicken skeletal muscle troponin complex via small-angle neutron and X-ray scattering', *J. Mol. Biol.* 345: 797–815.

Koch, M. H. J. (1989). 'The structure of chromatin and its condensation mechanism', in Heinemann, U. and Saenger, W. (eds.), *Protein-Nucleic Acid Interaction.* London: McMillan.

Koch, M. H. J., Vachette, P. and Svergun, D. I. (2003). 'Small-angle scattering: a view on the properties, structures and structural changes of biological macromolecules in solution', *Quart. Rev. Biophys.* 36: 147–227.

Perkins, S. J. (1985). 'Molecular modelling of human complement subcomponent C1q and its complex with C1r$_2$C1s$_2$ derived from neutron scattering curves and hydrodynamic properties', *Biochem. J.* 228: 13–26.

Riboldi-Tunnicliffe, A., König, B., Jessen, S. *et al.* (2001). 'Crystal structure of Mip, a prolylisomerase from Legionella pneumophila', *Nat. Struct. Biol.* 8: 779–83.

Schmidt, B., König, S., Svergun, D., Volkov, V., Fischer, G. and Koch, M. H. J. (1995). 'Small-angle X-ray solution scattering study on the dimerisation of the FKBP25mem from Legionella pneumophila', *FEBS Lett.* 372: 169–72.

Stuhrmann, H. B. (1970). 'Ein neues Verfahren zur Bestimmung der Oberflächenform und der inneren Struktur von geloesten globularen Proteinen aus Roentgenkleinwinkelmessungen', *Zeitschrift fuer Physikalische Chemie (Neue Folge)* 72: 177–98.

Stuhrmann, H. B., Koch, M. H. J., Parfait, R., Haas, J., Ibel, K. and Crichton, R. R. (1977). 'The shape of the 50S subunit of Escherichia Coli ribosomes', *P. Natl. Acad. Sci. USA* 74: 2316–20.

Svergun, D. I. (2010). 'Small-angle X-ray and neutron scattering as a tool for structural systems biology', *Biol Chem* 391: 737–43.

Svergun, D. I., Feigin, L. A. and Schedrin, B. M. (1982). 'Small-angle scattering: direct structure analysis', *Acta Crystallogr. A* 38: 827–35.

Svergun, D. I., Barberato, C., Koch, M. H. J., Fetler, L. and Vachette, P. (1997). 'Large differences are observed between the crystal and solution quaternary structures of allosteric aspartate transcarbamylase in the R state', *Proteins* 27: 110–17.

Whitten, A. E., Jacques, D. A., Hammouda, B. *et al.* (2007). 'The structure of the KinA-Sda complex suggests an allosteric mechanism of histidine kinase inhibition', *J. Mol. Biol.* 368: 407–20.

THEORY AND EXPERIMENT

Part
I

In this part we start with presenting the major theoretical concepts of small angle scattering of X-rays and neutrons by macromolecular solutions required to understand the basics of the technique (Chapter 1). The instrumentation for SAXS, including laboratory X-ray and synchrotron X-ray sources, and for SANS (instruments for the steady-state reactors and for the spallation sources) are reviewed in Chapter 2. Peculiarities of the SAXS and SANS experiments, sample requirements, measurement protocols and procedures for data reduction are described in Chapter 3.

Basics of small angle scattering

1.1 Elastic scattering of X-rays and
    neutrons   13

1.2 Scattering by macromolecular
    solutions   18

1.3 Resolution and contrast   21

1.4 Absorption and anomalous
    X-ray scattering   25

References   26

Scattering of X-ray photons or neutrons is due to their interactions with atoms in the sample, with the difference that X-rays interact with the electrons and neutrons with the nuclei. For structural studies, elastic scattering effects, where there is no energy exchange between the radiation and the atoms (that is, the wavelength λ of the scattered beam is equal to that of the incident beam), are exploited. Although the physical mechanisms of elastic X-ray and neutron scattering are different, they can be conveniently described by the same mathematical formalism. The basics of scattering are therefore presented simultaneously, pointing where necessary to the differences between the two types of radiation. This section will describe the basic theoretical aspects of X-ray and neutron scattering required to understand the principles of the technique as applied to solutions of biological macromolecules. We shall concentrate on elastic scattering phenomena and introduce the major concepts describing these processes for macromolecular solutions with a minimum of equations. (The mathematical concepts leading to a more rigorous treatment are presented in Appendix 1.)

1.1 Elastic scattering of X-rays and neutrons

In the following description, both X-ray photons and neutrons are considered to be radiation waves. There is a simple relationship between energy E and wavelength λ. In the case of X-rays, $E = hc/\lambda$, and X-ray photons with an energy E (in keV) have a wavelength $\lambda = 1.256/E$, where λ is expressed in nm. For neutrons with a (group) velocity v, $E = m_n v^2/2$, where m_n is the mass of the neutron. Following de Broglie's relationship, $E = h/2m_n\lambda^2$, one obtains $\lambda = h/mv$ or $\lambda[\text{nm}] = 396.6/v[\text{m/s}]$. For structural studies, relatively hard X-rays with energies around 10 keV (λ about 0.10–0.15 nm) and thermal neutrons with wavelengths λ around 0.10–1.0 nm are typically employed.

In a scattering experiment both the distances between the source of radiation and the sample and between the sample and the detector (L_2)

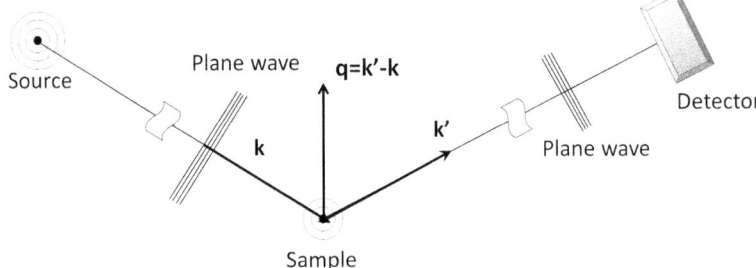

Fig. 1.1 Schematic representation of a scattering experiment in the Fraunhofer approximation.

of the scattered radiation are macroscopic quantities (measured typically in m). The characteristic size of the structural inhomogeneities in the sample (d, typically 10^0–10^3 nm) is much smaller than these distances, whereas the wavelength of the radiation λ is yet smaller (sub-nm range). This leads to a so-called Fresnel number $d^2/\lambda L_2 >> 1$, indicating that the incident radiation and the scattered radiation in any given direction ($\mathbf{k'}$) can be considered as plane waves (this is the case of a far-field diffraction or Fraunhofer approximation, see Fig. 1.1) (Willmott 2011). Overall, the scattering is, of course, divergent.

When an object is illuminated by a monochromatic plane wave with a wavevector with modulus $k = |\mathbf{k}| = 2\pi/\lambda$, atoms within the object interacting with the incident radiation become sources of spherical waves. For elastic scattering—that is, without energy transfer—the modulus of the scattered wave is $k' = |\mathbf{k'}| = k$. In this book we shall only consider the case of elastic scattering, which depends on the momentum transfer $\mathbf{q} = \mathbf{k'} - \mathbf{k}$ and is most relevant for structural studies. There are also inelastic scattering effects, where the energy of the scattered radiation differs from that of the incoming radiation. An example of inelastic X-ray scattering is Compton scattering (Cooper 2004), which originates from an elastic collision between a photon and an electron where the scattered photons always lose energy compared to the incoming ones. There is also so-called Raman scattering, where the energy and momentum of the photon are transferred to intrinsic excitations of the material, which provides information about the electron dynamics (Schülke 2007). For neutrons, the energy exchange occurs at the level of atoms or molecules, and inelastic neutron scattering allows one to study atomic and molecular motions as well as magnetic excitations (Fitter *et al.* 2005). In inelastic scattering experiments, both energy and momentum transfer are relevant, and the signal is recorded as a function of these two variables.

The elastic scattering process of X-rays and neutrons is described in the first Born approximation in the quantum theory of scattering (Gottfried 1966), where each point in the scatterer is assumed to be driven by the incident field and not by the total field. This kinematic scattering approximation is accurate if the scattered field is much weaker than the incident field. This is always the case for the neutron potential scattering where the interaction potentials are relatively weak

Table 1.1 X-ray and neutron scattering lengths of some elements.

Atom	H	D	C	N	O	P	S	Au
Atomic mass	1	2	12	14	16	30	32	197
Z electrons	1	1	6	7	8	15	16	79
$b_X, 10^{-12}$ cm	0.282	0.282	1.69	1.97	2.16	3.23	4.51	22.3
$b_N, 10^{-12}$ cm	−0.374	0.667	0.665	0.940	0.580	0.510	0.280	0.760

(see the neutron scattering lengths in Table 1.1). For X-ray studies of biological material mostly containing light atoms, the scattering lengths defined below are also sufficiently small to neglect multiple scattering and thus remain within the Born approximation.

Elastic scattering is due to interference of waves emitted by the individual scatterers within the coherently scattering volume, where the scatterers have correlated positions in space (for example, single protein in solution; see Fig, 1.2). The scattering process involves a transformation from the 'real' space of laboratory coordinates $\mathbf{r}$, where the structure of the scattering object is defined, to the 'reciprocal' space of scattering vectors $\mathbf{q}$, in which the scattered radiation is measured. In the Born approximation, this process is described by a Fourier transformation. Importantly, the Fourier transform involves a reciprocity between dimensions in real and reciprocal space implying that the smaller the 'real' size, the larger the corresponding 'reciprocal' size (Fig. 1.3). For solutions, the scattering is isotropic and the scattered intensity depends only on the modulus of the momentum transfer $|\mathbf{q}| = q = (4\pi \sin \theta)/\lambda$.

The strength of the interaction between the incident beam and the sample is best described as illustrated in Fig. 1.4 by the differential scattering cross-section, $d\sigma/d\Omega$, which is defined as the ratio between the scattered energy/(unit solid angle · unit time) and the incident energy/(unit area · unit time) (that is, the incident intensity I_0). Obviously this ratio has the dimensions of an area.

When N scatterers located at positions $\mathbf{r}_i$ (with $1 \leq i \leq N$) scatter coherently the total scattering amplitude is the sum of the waves scattered by the individual centres, which are described by a factor (b_i), the scattering length, and therefore has units of length: cm. The total scattered amplitude $(A(\mathbf{q}))$ from such an ensemble depends on the nature of the

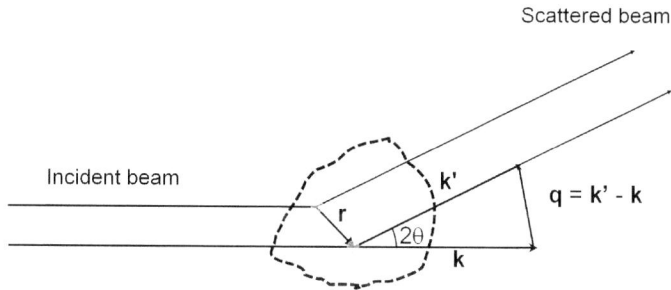

Fig. 1.2 A scheme of elastic scattering. The incident beam is scattered through the angle 2θ by two scattering centres separated by a distance $\mathbf{r}$. The waves emitted by atoms at different positions in the sample interfere because of the resulting phase shift. Note that $|\mathbf{q}| = q = 2 \cdot (2\pi/\lambda) \sin \theta = (4\pi \sin \theta)\lambda$.

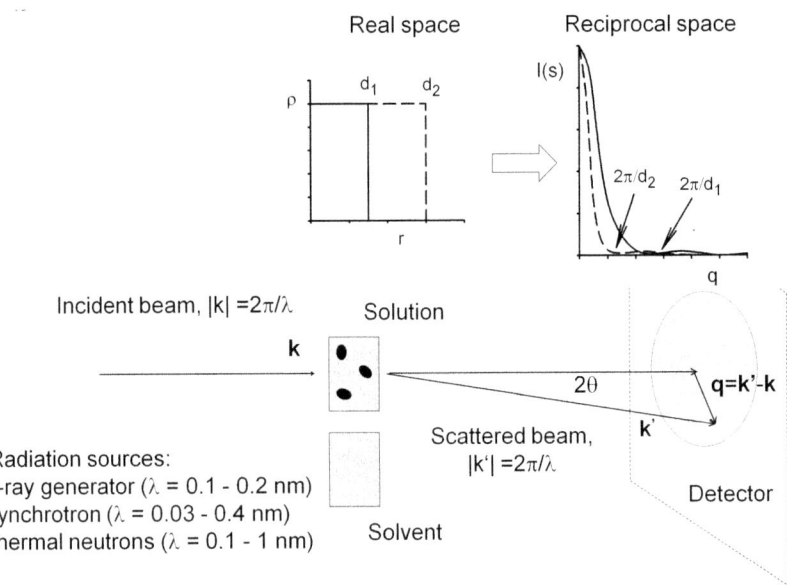

Fig. 1.3 Schematic representation of a small angle scattering experiment and the Fourier transformation from real to reciprocal space. Adapted, with permission, from Svergun and Koch (2003), © IOP Publishing Ltd, 2003.

Radiation sources:
X-ray generator (λ = 0.1 - 0.2 nm)
Synchrotron (λ = 0.03 - 0.4 nm)
Thermal neutrons (λ = 0.1 - 1 nm)

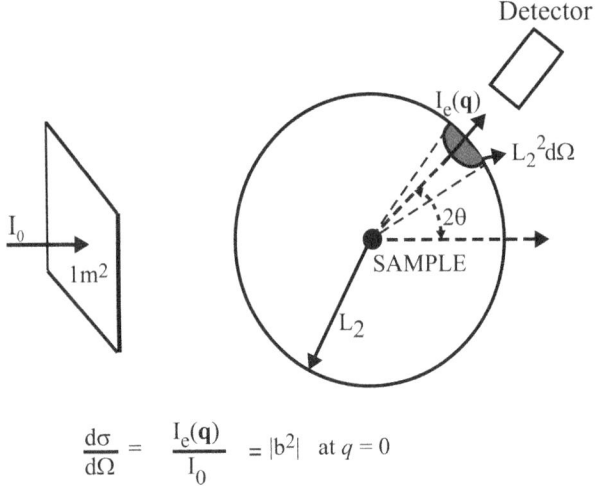

Fig. 1.4 Definition of the scattering cross-section. $I_e(\mathbf{q})$ is the energy scattered/(unit solid angle $\times$ s) in the direction 2θ, whereas I_0 is the incident energy/(unit area $\times$ s). The ratio of these two quantities has the dimensions of an area. L_2 is the distance between the sample and the detector. The area associated with the solid angle $d\Omega$ is $L_2^2 d\Omega$, and the intensity measured by the detector is thus $I_{\text{meas}}(\mathbf{q}) = \frac{I_e(\mathbf{q})}{L_2^2} = \frac{I_0}{L_2^2}\frac{d\sigma}{d\Omega}(\mathbf{q})$.

$$\frac{d\sigma}{d\Omega} = \frac{I_e(\mathbf{q})}{I_0} = |b^2| \quad \text{at } q = 0$$

individual scatterers (that is, their scattering lengths) and on the phase factors ($\exp(i\mathbf{q}\cdot\mathbf{r}_i)$) defined by their positions:

$$A(\mathbf{q}) = \sum_{i=1}^{N} b_i \exp(i\mathbf{q}\cdot\mathbf{r}_i) \tag{1.1}$$

The scattering amplitude is thus the Fourier transform (see Appendix 1) of the distribution of scatterers.

If the object consisting of these N scatterers can take any orientation (for example, molecules in a gas or in a solution) the phase factor is given by the following expression, where the brackets represent the spherical average:

$$< \exp(i\mathbf{q} \cdot \mathbf{r}) > = \frac{\sin(qr)}{qr} \tag{1.2}$$

Although an object may be represented by a discrete distribution of scatterers, it is often more practical to use a continuous distribution of the scattering length density $\rho(r)$, defined as the total scattering length of the atoms per unit volume, and below, both approaches will be used, depending on the application.

The above concepts hold for both X-ray and neutron scattering by matter, but the scattering length density distributions are different and depend on the physics of the interaction of the two types of radiation with the object. For hard X-rays, it is safe to assume that the radiation interacts only with the electrons, as they have a much smaller mass than protons. For a single electron $b = r_0 = 2.82 \times 10^{-13}$ cm, the Thomson or classical electron radius. Hence, for an atom with radial electron density $\rho(r)$ placed at the origin:

$$b_x(q) = r_0 4\pi \int \rho(r) r^2 \frac{\sin(qr)}{qr} dr \tag{1.3}$$

Since $\lim(\sin(x)/x)_{x \to 0} = 1$, the atomic scattering length in the forward direction $(q = 0)$ is $b_x(0) = Zr_0$, where Z is the atomic number.

Clearly, X-rays interact more strongly with heavier atoms than with lighter ones (see Table 1.1). The X-ray scattering length does not depend on the wavelength unless the photon energy is close to an absorption edge of the atom. In the latter case there is resonant or anomalous scattering, as described in Section 1.4. In X-ray scattering one traditionally uses the scattering factor f_x, which is the ratio of the amplitude scattered by an object to that scattered by a single electron in the same conditions (that is, $f_x(0) = Z$). Atomic scattering factors along with other useful information are now available on the Web from numerous online sources (for example, Chantler *et al.* 2003).

In contrast to X-rays, which are scattered by charged particles (electrons), neutrons interact with the nuclear potential and with the spin, and the neutron scattering length consists of two terms $b_n = b_p + b_s$. In the structural studies of biological materials the potential interactions are most important. Unlike the situation with X-rays, b_p does not increase with the atomic number but is sensitive to the isotopic composition. Table 1.1 reveals two major differences between the X-ray and neutron scattering lengths. First, neutrons are more sensitive to lighter atoms than X-rays and there is no regular difference between the light and heavy atoms. The neutron amplitudes, contrary to the X-ray ones, may be negative (which means that the scattered wave is in phase with the incoming wave), due to the contribution of the so-called nuclear resonance scattering (Bacon 1975). This property of neutron scattering is extensively used in neutron crystallography to localise hydrogen atoms in crystals (Shu *et al.* 2000). Secondly, there is a very large difference between the scattering length of hydrogen, which is negative (-0.374×10^{-12} cm), and that of deuterium, which is positive (0.667×10^{-12} cm). This difference represents one of the most important advantages of neutrons in many fields of structural research, and provides an effective tool for selective labelling and contrast variation in neutron scattering and diffraction. Note that due to the small (10^{-13} cm) size of the nuclei the neutron scattering lengths of atoms can be considered to be constants (that is, q-independent values).

The spin scattering term can be expressed as $b_s = 2B\,\boldsymbol{IS}$, where B is the spin scattering length of an atom, and $\boldsymbol{S}$ and $\boldsymbol{I}$ are spin operators of the neutron and of the nucleus, respectively (Abragam and Goldman 1982). If the spins of the atoms in the sample and/or of the incoming neutrons are not ordered, spin interactions yield only incoherent scattering (B is therefore called the incoherent scattering length of the atom). As we saw, coherent scattering describes interference between waves produced by multiple scatterers with correlated positions in the sample, such that the scattering varies strongly with the angle (or momentum transfer). In incoherent scattering there is no correlation between the waves emitted by different scatterers, which thus do not interfere. For neutrons this incoherent scattering yields a flat background. For many atoms the incoherent scattering length significantly exceeds the coherent one. The most striking example is given by hydrogen, for which $b_p = -0.374 \times 10^{-12}$ cm, whereas $2B = 5.824 \times 10^{-12}$ cm. For the spin values of $I = \pm 1/2$, b_s ranges between $\pm 2.912 \times 10^{-12}$ cm; that is, the incoherent scattering exceeds the coherent one by a factor of 80. The strong incoherent scattering leads to severe experimental difficulties in neutron scattering experiments by degrading the signal-to-noise ratio (see Section 3.4). With polarised neutrons the spin-dependent scattering can, however, be used as a probe to study the magnetic structure (Abragam and Goldman 1982) and, when using polarised samples (where the nuclear spins have been aligned), the spin scattering becomes coherent and can be used to very effectively vary the contrast of the sample (Stuhrmann *et al.* 1986), see Section 6.7. For X-rays there is no direct analogue to the incoherent neutron scattering. The already-mentioned Compton X-ray scattering (Cooper 2004) is also unrelated to the atomic positions, but is inelastic, and its amplitude approaches zero for low momentum transfers, which means that it can be neglected in SAXS.

1.2 Scattering by macromolecular solutions

The scattering from assemblies of atoms is most conveniently described in terms of their scattering length density distribution $\rho(\mathbf{r})$. Experiments on macromolecules in solution require separate measurements of the scattering from the solution and the solvent (Fig. 1.4). The scattering pattern of the solvent is subtracted from that of the solution to yield the net scattering from the particles. Assuming that the solvent is a featureless matrix with a constant scattering density ρ_s, the difference between the scattering length from a single particle relative to that of the equivalent solvent volume, is the excess scattering length density $\Delta\rho(\mathbf{r}) = \rho(\mathbf{r}) - \rho_s$. It is related to the scattering amplitude of the particles by a Fourier transformation:

$$A(\mathbf{q}) = \Im[\rho(\mathbf{r})] = \int_V \Delta\rho(\mathbf{r})\exp(i\mathbf{q}\mathbf{r})d\mathbf{r} \tag{1.4}$$

where the integration is performed over the particle volume (V). Note that the amplitude $A(\mathbf{q})$ is, generally speaking, a complex number (see Appendix 1). One cannot directly measure the amplitude but only the intensity, which is a product of the amplitude and its complex conjugate, $I(\mathbf{q}) = A(\mathbf{q})A^*(\mathbf{q})$, and corresponds to the scattered energy; that is, for monochromatic radiation to the number of photons or neutrons scattered per unit area and unit time in the given direction 2θ. From the definition of the scattering cross-section in Fig. 1.4 it follows that the intensity measured by the detector is:

$$I_{meas}(\mathbf{q}) = \frac{I_e(\mathbf{q})}{L_2^2} = \frac{I_0}{L_2^2}\frac{d\sigma}{d\Omega}(\mathbf{q}) \tag{1.5}$$

For monochromatic radiation the number of photons/neutrons counted in a detector pixel of area A during the time Δt, $N_{meas}(\mathbf{q})$ is therefore

$$N_{meas}(\mathbf{q}) = I_e(\mathbf{q})\frac{A}{L_2^2}\Delta t = I_0\frac{A}{L_2^2}\frac{d\sigma}{d\Omega}(\mathbf{q})\Delta t \tag{1.5a}$$

As the measured intensity and the scattering cross-section $d\sigma/d\Omega$ are proportional, we shall below—not entirely accurately—refer to the scattering cross-section as scattering intensity, remembering that it corresponds to the fraction of photons or neutrons scattered from the incident beam per unit solid angle and unit time in the direction 2θ.

If one now considers an ensemble of identical particles, the total scattering will depend on their distribution in the sample. The total scattering density in the object (that is, of the total irradiated volume) will be a convolution of the particle density distribution $\Delta\rho(\mathbf{r})$ and a function $d(\mathbf{r})$ describing the positions and orientations of the particles within the volume: $\Delta\rho_{total}(\mathbf{r}) = \Delta\rho(\mathbf{r})*d(\mathbf{r})$.

Following the properties of Fourier transforms (see Appendix 1), the scattering amplitude of the ensemble will be a product of the transforms of the two terms, $A_{total}(\mathbf{q}) = \Im[\Delta\rho(\mathbf{r})] \times \Im[d(\mathbf{r})] = A(\mathbf{q})F(\mathbf{q})$. The measured intensity will (on certain conditions) also be a product, $I_{total}(\mathbf{q}) = I(\mathbf{q})S(\mathbf{q})$—the first term depending on the structure of the particles, the second on their distribution (Fig. 1.5).

Two major limiting cases should be considered. In an ideal single crystal, all particles in the sample have defined and correlated orientations and are regularly distributed in space, so that the scattering amplitudes of the individual particles have to be summed to account

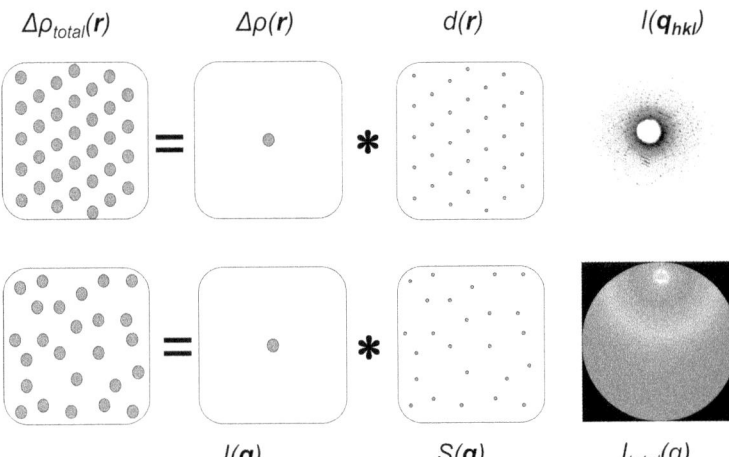

$\Delta\rho_{total}(\boldsymbol{r})$ $\Delta\rho(\boldsymbol{r})$ $d(\boldsymbol{r})$ $I(\boldsymbol{q}_{hkl})$

$I(\boldsymbol{q})$ $S(\boldsymbol{q})$ $I_{total}(q)$

Fig. 1.5 Formation of scattering pattern from a crystal (top row) and from a solution (bottom row). The right panels display typical experimental patterns on a detector (in this case using X-rays): top, a diffraction pattern from a crystal; bottom, scattering pattern from a solution (note that in the latter case the beam centre is offset).

for all interparticle interferences. This means that the crystal volume irradiated by coherent radiation also scatters coherently. As a result, the total scattered intensity is redistributed along specific directions defined by the reciprocal lattice, and the discrete three-dimensional function $I(\boldsymbol{q}_{hkl})$ measured corresponds to the density distribution in a single unit cell of the crystal (Giacovazzo *et al.* 1992). These images are usually called diffraction (and not scattering) patterns (Fig. 1.5, top row). If the particles are randomly distributed and their positions and orientations are uncorrelated, their scattering intensities rather than their amplitudes are summed. In this case, the coherently scattering volume is that of a single particle and there is no interference between the waves emitted by the individual molecules. Consequently, the intensity from the entire ensemble is a continuous isotropic function proportional to the intensity scattered by a single particle averaged over all orientations $I(q) = <\boldsymbol{I}(\boldsymbol{q})>_\Omega$. Dilute solutions of monodisperse non-interacting biological macromolecules under specific solvent conditions correspond to this second limiting case, which will mostly be considered below. If the particles in solution are randomly oriented but also interact (non-ideal semi-dilute solutions), local correlations between the neighbouring particles must be taken into account. The scattering intensity from the ensemble will still be isotropic as long as the molecules remain randomly oriented, and for spherical particles can be written as a product $I_{total}(q) = I(q) \times S(q)$. In the literature the particle scattering $I(q)$ and the interference term $S(q)$ are called 'form factor' and 'structure factor', respectively. In biological applications, SAXS/SANS is used to analyse the structure of macromolecules in solution (based on the particle scattering) as well as the interactions based on the interference term. Separation of the two terms for semi-dilute solutions is possible by using measurements at different concentrations or/and for different solvent conditions (pH, ionic strength, and so on). For systems of particles differing in size and/or shape, the total scattering intensity will be given by the weight average of the scattering from the different types of particle.

The description above gives a simplified picture valid for the X-ray or neutron sources with a limited coherence (most existing sources are like this). For highly coherent incident radiation, sometimes available at SR sources, non-isotropic scattering may be observed even for isotropic systems (so-called speckle) (Thurn-Albrecht *et al.* 1996). Highly coherent X-ray sources (for example, X-ray lasers) are described in Chapter 2.

1.3 Resolution and contrast

The Fourier transformation of the box function in Fig. 1.4 illustrates that most of the intensity scattered by an object of linear size d is concentrated in the range of momentum transfer up to $q = 2\pi/d$. It is therefore assumed that if the scattering pattern is measured in reciprocal space up to q_{max} it provides information about the real space object with a resolution $\Delta = 2\pi/q$. For single crystals, due to the redistribution of the diffracted intensity into reflections, the data can be recorded to high resolution ($d \sim \lambda$). For spherically averaged scattering patterns from solutions, $I(q)$ usually decays rapidly as a function of momentum transfer, and only low-resolution patterns ($d >> \lambda$) can be obtained. It is thus clear that solution scattering cannot provide information about the atomic positions but only about the overall structure of macromolecules in solution. It should also be noted that the information content in the SAS data is severely limited due to the random positions of particles in solution, and the estimate $2\pi/q$ gives only a nominal resolution of the scattering pattern, rather than the actual resolution of the models, which can be determined from these data. Although it is in principle technically possible to record the scattering from macromolecular solutions up to reasonably high resolution (even up to 0.2 nm, so-called wide-angle X-ray scattering, WAXS), it is not possible to uniquely reconstruct structural models to this resolution even from the high-quality WAXS data. The fact that models built from SAS are not unique is one of the most severe problems of the technique with which we shall be confronted in all sections dealing with the structural interpretation of the experimental data.

As we have seen in Section 1.1, small-angle scattering arises from fluctuations in electron density (for X-rays) or scattering length density (for neutrons) in solution. The difference between the average electron density or neutron scattering length density of a macromolecule in solution and that of its surrounding solvent $\Delta\rho = <\Delta\rho(\mathbf{r})> = <\rho(\mathbf{r})> - \rho_s$ is called the contrast. This is another important characteristic of the sample, and Fig. 1.6 schematically illustrates the concept of contrast in solution scattering studies. The scattering density of the particle can be represented as $\rho(\mathbf{r}) = \Delta\rho \times \rho_C(\mathbf{r}) + \rho_F(\mathbf{r})$, where $\rho_C(\mathbf{r})$ is the shape function equal to 1 inside the particle and 0 outside, whereas

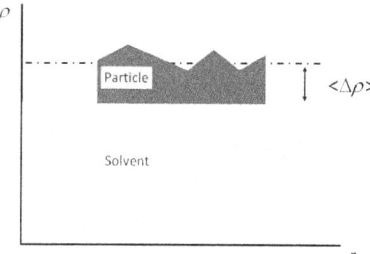

Fig. 1.6 Schematic representation of contrast in SAS.

$\rho_F(\mathbf{r}) = \rho(\mathbf{r}) - <\rho(\mathbf{r})>$ represents the fluctuations of the scattering density around its average value (Stuhrmann and Kirste 1965). Inserting this expression into eq. (1.4), the scattered amplitude contains two terms $A(\mathbf{q}) = \Delta\rho \times A_C(\mathbf{q}) + A_F(\mathbf{q})$, so that the average intensity can be written in terms of three basic scattering functions:

$$I(q) = (\Delta\rho)^2 I_C(q) + 2\Delta\rho I_{CF}(q) + I_F(q) \qquad (1.6)$$

where $I_C(q)$, $I_F(q)$ and $I_{CF}(q)$ are the scattering from the particle shape, fluctuations and the cross-term, respectively (Stuhrmann and Kirste 1965). This equation is of general value, and the contributions from the overall shape and internal structure of particles can be separated using measurements in solutions with different solvent densities (different $\Delta\rho$). This technique is called 'contrast variation'.

In a SAS experiment the environment is usually water or an aqueous solution, which may contain salts or small molecules. In the case of X-rays the electron density of an aqueous solution may depend very strongly on these solutes. The electron density of pure H_2O for example is 3.34×10^2 e nm^{-3}, whereas for a 1 M solution of NaCl it is 3.50×10^2 e nm^{-3}. As the electron density of an average protein is 4.200×10^2 e nm^{-3}, its contrast in pure water is therefore 0.854×10^2 e nm^{-3}, whilst in 1M NaCl it is only 0.701×10^2 e nm^{-3}. An even more striking example is that of sucrose. A 50% (w/v) sucrose solution has an electron density of 4.00×10^2 e nm^{-3}, giving a protein contrast of only 0.2×10^2 e nm^{-3}. The X-ray contrast of a macromolecule in solution can therefore be manipulated by changing the electron density of the environment (see Fig. 1.7). This property was indeed exploited in the early days of protein crystallography (see Section 9.4) to determine the molecular envelope of the haemoglobin molecule (Bragg and Perutz 1952). Later it was used in SAXS to determine the molecular envelope of tomato bushy stunt virus (Harrison 1969). In principle,

Fig. 1.7 Possibilities of contrast variation with X-rays and with neutrons. The grey areas depict the ranges of solvent density reachable by addition of small molecules in X-rays (left) and with H/D exchange for neutrons (right). Note that the scattering length densities of proteins and nucleic acids slightly increase with D_2O content due to H/D substitution of exchangeable hydrogens. The percentages of D_2O above 100% can of course not be reached physically, but are given for information (the contrast estimates of macromolecular components are often reported as percentages of D_2O).

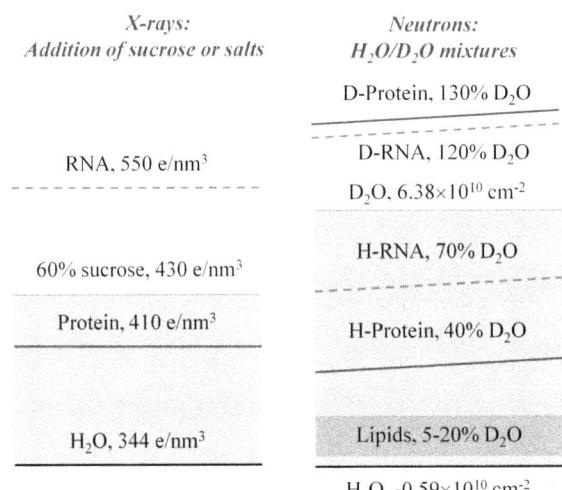

X-ray contrast could also be manipulated by heavy-atom labelling of the macromolecule, but such experiments are rather difficult and have rarely been tried with success. In general, contrast variation with X-rays has been limited for reasons of practical difficulty and by the fact that manipulating the contrast by addition of salts or small molecules may influence the conformation of the macromolecular system or its hydration shell under study. Electron densities and neutron scattering length densities of macromolecules and selected salt/small molecule solutions are shown in Table 1.2. These values are calculated using published molalities and mass densities for NaCl and sucrose, and partial specific volumes of 0.73 cm^{3}g^{-1} for protein and 0.53 cm^{3}g^{-1} for DNA.

Contrast variation is most effective and reliable in neutron scattering, as described in more detail in Section 3.3. Briefly, however, the main advantage of contrast variation in SANS is that it is carried out simply by exchanging hydrogen for its heavy isotope deuterium either in the aqueous environment or in the dissolved macromolecule— modifications which have very little or no influence on molecular shape or interactions. The coherent scattering length of hydrogen is -3.74×10^{-12} cm, whilst that of deuterium is 6.67×10^{-12} cm, leading to coherent scattering length densities of -0.562×10^{-14} cm^{-2} for H$_2$O and 6.33×10^{-14} cm^{-2} for D$_2$O. Thus the contrast can be changed by partially or fully exchanging hydrogen for deuterium in the aqueous environment or in the dissolved macromolecule.

Table 1.2 lists the coherent neutron scattering length densities for a number of biological macromolecules as a function of the deuterium content of the aqueous environment. It is important to realise that the scattering length density of the macromolecule changes with the H/D content of the aqueous environment due to the exchange of labile protons with those of the water. Labile protons are usually assumed to be those attached to electronegative atoms such as oxygen, nitrogen or

Table 1.2 X-ray and neutron scattering length densities of biological components.

	X-rays*		Neutrons		
	ρ (electrons nm^{-3})	Matching solvent	ρ in H$_2$O (10^{10} cm^{-2})	ρ in D$_2$O, (10^{10} cm^{-2})	Matching % D$_2$O
H$_2$O	334	—	−0.56	—	—
D$_2$O	334	—	—	6.33	—
50% sucrose	400	—	1.2	—	—
0.1 M NaCl	335	—	−0.55	—	—
1.0 M NaCl	343	—	−0.48	—	—
2.0 M NaCl	352	—	−0.40	—	—
Lipids	300	—	0.3	0.3	$\approx$ 5–20%**
Proteins	420	65% sucrose	1.8	3.1	$\approx$ 40%
D-proteins	420	65% sucrose	6.6	8.0	—
Nucleic acids	550	—	3.7	4.8	$\approx$ 70%
D-nucleic acids	550	—	6.6	7.7	—

* For X-rays, the scattering length densities are often expressed in terms of electron density; that is, the number of electrons per nm^3: 1 electron nm^{-3} = 2.82 cm × 10^8 cm.

** The match point for lipids in SANS depends very strongly on the nature of the headgroup and in particular the number of labile (exchangeable) protons.

sulphur. The extent to which the protons in proteins actually undergo exchange depends on their accessibility to the solvent. Surface atoms exchange in seconds or minutes, whereas buried protons attached to the polypeptide backbone may not exchange in months or years. A reasonable rule of thumb is that ≈75% of the protons exchange within 24 hours. For nucleic acids or carbohydrates where a large part of the molecule is accessible to water almost all protons exchange within 24 hours. The exchange of macromolecular protons with the solvent has a number of important consequences:

1. The scattering length density of a macromolecule in a D_2O-containing solvent can be calculated exactly only if the number of exchanged protons (and the protein partial specific volume) is known.
2. Exchange is a kinetic process, and it is important to ensure that the H/D content of a molecule is as constant as possible for the duration of a scattering experiment. As described above, the rates of exchange of different protons can vary very widely within and between different proteins. This should therefore, if possible, be explored by, for example, NMR.
3. In a SANS experiment, it is important that solution and buffer have the same H/D content. To ensure that this is so, the sample should ideally be prepared by dialysis to equilibrium.

In addition to changing the scattering length density of the buffer, specific deuteration is a very effective method for highlighting selected structural fragments in complex particles, and extremely valuable information can be obtained by selective labelling, where a portion of the structure is specifically perdeuterated. This can be done by growing cells in deuterated medium. Under these conditions both the exchangeable hydrogens (such as those of NH, NH_2, NH_3, OH and SH groups) and the non-exchangeable carbon-bound H-atoms are replaced by deuterons. Perdeuteration of proteins allows one to effectively create contrast, for example, between different proteins in protein–protein complexes. Figure 1.7 presents the scattering densities of deuterated and protonated proteins and nucleic acids as a function of the deuterium content of the buffer in which they are dissolved. For proteins the scattering length densities depend on the amino acid content, but they do not vary much (Zaccai and Jacrot 1983) and the values given in Table 1.2 are averages. The same is true for nucleic acids, but for lipids and carbohydrates the values depend strongly on the particular chemical composition. Specific examples are therefore given in the table, but the values should in all cases be calculated using the known composition and partial specific volumes.

It is evident from Fig. 1.7 that for all except perdeuterated protein and DNA molecules there is an H_2O/D_2O composition at which the macromolecule has the same scattering length density as its aqueous environment. This is known as the contrast match or isopicnic point, at which the average contrast is zero. This does not mean that there is no scattering—only that the scattering at $q = 0$ is zero. At higher angles, residual scattering will be present due to density fluctuations within the macromolecule. The calculation of precise match points is important and will be described in more detail in Chapter 3. Note that although contrast may be highest in either H_2O or D_2O, depending on the macromolecular composition/deuteration, in practice the most accurate measurements are often made in high-D_2O-content buffers, as this minimises the spin incoherent scattering, which contributes only to background.

1.4 Absorption and anomalous X-ray scattering

The interaction of X-rays and neutrons with a sample is a complex process where elastic scattering of a photon or neutron is only one of the possible outcomes. In practice, one of the most important effects is absorption, where the energy of radiation is absorbed by the material and the beam is attenuated. For a sample with thickness t (cm) the intensity of the transmitted beam (I) is related to that of the monochromatic incident beam (I_0) with energy E by

$$I(E) = I_0(E)e^{-\mu(E)t} \qquad (1.7)$$

where $\mu(E)[\text{cm}^{-1}]$ is the total linear attenuation. For absorption calculations, it is more convenient to use the mass absorption $\mu/\rho(\text{cm}^2\ \text{g}^{-1})$, where ρ is the mass density of the material, which is additive in terms of the mass fraction of each of the elements in the sample.

The values of μ/ρ for X-rays as a function of wavelength for all elements can be found at <http://physics.nist.gov/PhysRefData/XrayMassCoef/> or in the International Tables for Crystallography. In the X-ray range, two phenomena contribute to the total absorption: photoelectric absorption and scattering ($\mu = \tau + \sigma$). Photoelectric absorption is due to the excitation of electrons to higher quantum levels, which results in fluorescence or ejection of a photoelectron with excess kinetic energy. For low molecular mass elements at wavelengths used for SAS on biological systems photoelectric absorption dominates, so that $\mu \cong \tau$. Note that the optimal thickness of the sample in terms of scattering is given by $t_{\text{opt}} = 1/\mu(\text{E})$. Much thicker and much thinner samples should be avoided if possible.

For purely elastic X-ray scattering without energy exchange between the photons and the atoms, the atomic scattering length does not depend on the wavelength and is simply proportional to the number of electrons, Z. Therefore, collecting the scattering patterns at different λ using the same detector (for example, at the same set of the scattering angles 2θ) just leads to a linear transformation of the angular axis following the definition of the momentum transfer ($q = (4\pi \sin \theta)/\lambda$). The situation changes if the photon energy is close to an absorption edge of

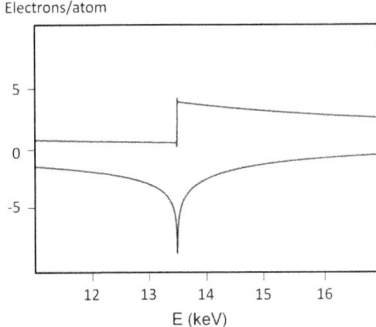

Electrons/atom

Fig. 1.8 Real and imaginary correction to the scattering factor for bromine in the vicinity of its K-absorption edge.

the atom. In the general case, the X-ray scattering factor can be written as a complex function

$$f(\lambda) = f_0 + f'(\lambda) + if''(\lambda) \tag{1.8}$$

where at small angles $f_0 = Z$, whereas the correction factors $f'(\lambda)$ and $f''(\lambda)$ become significant close to an absorption edge of an atom and may change its scattering length up to about 50% (Schaaffs 1957) (see the example in Fig. 1.8). This so-called anomalous scattering allows one to observe the changes in the scattering or diffraction pattern due to the contribution of specific atoms in the sample, and therefore obtain additional information about the structure. This X-ray counterpart of the contrast variation due to H/D exchange in neutron scattering is used for experimental phase determination in crystallography (Hendrickson and Ogata 1997), and also in some SAS applications to highlight specific atoms in biological macromolecules (Stuhrmann 1981).

References

Abragam, A. and Goldman, M. (1982). *Nuclear Magnetism: Order and Disorder*. Oxford and New York: Clarendon Press.

Bacon, G. E. (1975). *Neutron Diffraction*. Oxford: Clarendon Press.

Bragg, W. L. and Perutz, M. F. (1952). 'The external form of the haemoglobin molecule. I', *Acta Crystallogr.* 5: 277–83.

Chantler, C. T., Olsen, K., Dragoset, R. A., Kishore, A. R., Kotochigova, S. A., and Zucker, D. S. (2003). *X-Ray Form Factor, Attenuation and Scattering Tables*, 2.0 edn. Gaithersburg, MD: National Institute of Standards and Technology.

Cooper, M. (2004). *X-ray Compton Scattering*. Oxford and New York: Oxford University Press.

Fitter, J., Gutberlet, T., and Katsaras, J. (2005). *Neutron Scattering in Biology*. New York: Springer.

Giacovazzo, C. *et al.* (1992). *Fundamentals of Crystallography*. New York: Oxford University Press.

Gottfried, K. (1966). *Quantum Mechanics*. New York: W. A. Benjamin.

Harrison, S. C. (1969). 'Structure of tomato bushy stunt virus. I. The spherically averaged electron density', *J. Mol. Biol.* 42: 457–83.

Hendrickson, W. A. and Ogata, C. M. (1997). 'Phase determination from multiwavelength anomalous diffraction measurements', in Carter, C. W. and Sweet, R. M. (eds.), *Method. Enzymol.* New York: Academic Press.

Schaaffs, W. (1957). *Röntgenstrahlen*. Berlin: Springer.

Schülke, W. (2007). *Electron Dynamics by Inelastic X-Ray Scattering*. Oxford and New York: Oxford University Press.

Shu, F., Ramakrishnan, V., and Schoenborn, B. P. (2000). 'Enhanced visibility of hydrogen atoms by neutron crystallography on fully deuterated myoglobin', *P. Natl. Acad. Sci. USA* 97: 3872–7.

Stuhrmann, H. B. (1981). 'Anomalous small angle scattering', *Quart. Rev. Biophys.* 14: 433–60.

Stuhrmann, H. B., and Kirste, R. G. (1965). 'Elimination der intrapartikulaeren Untergrundstreuung bei der Röntgenkleinwinkelstreuung an kompakten Teilchen (Proteinen)', *Zeitschr. Physik. Chem. Neue Folge* 46: 247–50.

Stuhrmann, H. B., Scharpf, O., Krumpolc, M., Niinikoski, T. O., Rieubland, M., and Rijllart, A. (1986). 'Dynamic nuclear polarisation of biological matter', *Eur. Biophys. J.* 14: 1–6.

Svergun, D. I., and Koch, M. H. J. (2003). 'Small-angle scattering studies of biological macromolecules in solution', *Rep. Progr. Phys.* 66: 1735–82.

Thurn-Albrecht, T., Steffen, W., Patkowski, A. *et al.* (1996). 'Photon correlation spectroscopy of colloidal palladium using a coherent X-ray beam', *Phys. Rev. Lett.* 77: 5437–40.

Willmott, P. (2011). *An Introduction to Synchrotron Radiation: Techniques and Applications*. Chichester: Wiley.

Zaccai, G., and Jacrot, B. (1983). 'Small-angle neutron scattering', *Annu. Rev. Biophys. Bioeng.* 12: 139–57.

X-ray and neutron scattering instruments

2

2.1 Characteristics of sources   27
2.2 X-ray sources   29
2.3 Neutron sources   37
2.4 SAXS instruments   43
2.5 SANS instruments   52
2.6 Special instruments   57
References   62

The theory of scattering, including small-angle scattering, for neutrons and X-rays is very similar, but neutron and X-ray sources are very different and their various characteristics determine their suitability for small-angle scattering studies.

Conventional X-ray sources for SAXS are compact devices available to individual laboratories. In contrast, with the exception of table-top synchrotrons, SR and neutron sources are available only at a few large central facilities. Only the basic principles of the different types of sources are explained below. For details and pictures of specific facilities the reader is referred to the worldwide directory of SR sources and free electron lasers, which provides links to the websites of the different facilities at <http://www.lightsources.org>, and to a similar directory for neutron sources available at <http://www.ncnr.nist.gov/nsources.html>.

2.1 Characteristics of sources

2.1.1 Spectral brightness

Even if accurate calculations may require the knowledge of the three-dimensional characteristics of the source, one usually makes the simplifying assumption that radiation sources are planar and have the shape of a Gaussian ellipse. As illustrated in Fig. 2.1A, the properties of such a source are defined by its spectral brightness ($SB = n(x, y, \theta, \psi, E, t)$), which corresponds to the rate at which radiation with energy (E) in the range $E - \Delta E/2 \leq E \leq E + \Delta E/2$ is emitted by a surface element (dx, dz) centred at (x, z) into an element of solid angle $(d\theta, d\psi)$ in the direction (θ, ψ). The SB is usually expressed as the number of particles (p) with a 0.1% energy band-pass emitted per unit area of the source into a unit solid angle in the direction (θ, ψ) per unit time and given in $p/(\text{s} \cdot \text{mm}^2 \cdot \text{mrad}^2 \cdot 0.1\% \, \Delta E/E)$. The properties of any source are completely defined by its SB, its time structure and its degree of coherence. As the SB is conserved along the path of a beam and real (non-ideal) optical elements can only degrade it, its value sets an upper limit to the

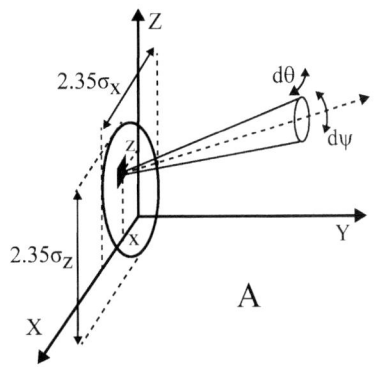

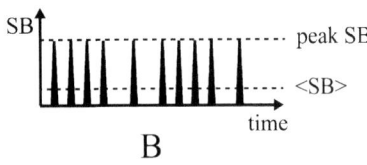

Fig. 2.1 A: The spectral brightness $SB = n(x, z, \theta, \psi, E, t)$ is the rate at which radiation (or particles) with an energy range $E - \Delta E/2 \leq E \leq E + \Delta E/2$ is emitted by a source element centred at (x,z) into an element of solid angle $(d\theta, d\psi)$ in the direction (θ, ψ). It is generally expressed as the number of particles/unit time · unit area of the source · unit solid angle of emission ·0.1% band pass $(\Delta E/E)$. The source is usually assumed to have a Gaussian shape with standard deviations σ_X and σ_Z and full width at half maximum of 2.35σ in the X and Z directions. Adapted from Koch *et al.* (1983), with permission from Elsevier, © 1983. B: Example of the time structure of a pulsed source—here a repetition of non-equally spaced pulses. <SB> is the average spectral brightness.

quality of optical instruments such as SAS cameras that can be built on a given source. This also implies that the *SB* of the source cannot be taken as a measure of instrument performance.

The *SB* is particularly well suited to compare naturally collimated sources such as lasers and SR sources, but less so for isotropic sources like X-ray tubes or neutron sources. In those cases one often prefers to consider an 'effective' source. For a conventional X-ray source this might be the beam at the exit of the mirror, for example.

In practice, a number of other measures are therefore also used, such as the radiant intensity (energy/unit solid angle) *N*, obtained by integrating the *SB* over the source area:

$$N(\vartheta, \psi, E, t) = \int_{\text{Source}} n \, dx \, dz \, (\text{p s}^{-1} \, \text{mrad}^{-2} \, \text{per } 0.1\% \Delta E/E) \quad (2.1)$$

or the spectral flux obtained by integrating the radiant intensity over the solid angle Ω:

$$\Phi_s(E, t) = \int_{\Omega} N \, d\theta \, d\psi \, (\text{p s}^{-1} \, \text{per } 0.1\% \Delta E/E) \quad (2.2)$$

or the total flux:

$$\Phi_T(t) = \int_0^{\infty} \frac{1}{E} \, \Phi_s dE \, (\text{p s}^{-1}) \quad (2.3)$$

As explained in Appendix 1, the longitudinal coherence properties of the beam depend solely on the choice of wavelength and band pass, which may both be limited by the source (for X-ray tubes, for example). In contrast, the transverse coherence length depends not only on the source size but also on the dimensions of the instrument. The necessary condition to observe useful scattering from particles in solution is that the coherence lengths, longitudinal and transverse, should be larger than the largest dimension of the particles.

The time structure of pulsed sources refers to the shape, duration and separation of the pulses, as illustrated in Fig. 2.1B. To compare pulsed and continuous sources one distinguishes between peak brightness and average brightness. For conventional SAXS the quantity that matters most is the average brightness, but the detectors must of course be able to cope with the peak brightness, which can be several orders of magnitude higher than the average brightness when the pulses are short and the repetition rate is low.

2.1.2 Conversion of energy to wavelength

There is a simple relationship between energy (E) and wavelength (λ). For X-rays, $E = hc/\lambda$ and

$$E[\text{keV}] = 123.9 \, \lambda[\text{nm}] \quad (2.4)$$

whereas for neutrons with a speed v, $E = m_n v^2/2 = h/2m_n \lambda^2$ and

$$E[\text{meV}] = 0.8182/(\lambda\,[\text{nm}])^2 \qquad (2.5)$$

It is easy to calculate from this that a neutron with a wavelength of 0.15 nm has about 2×10^5 times less energy than a photon of the same wavelength, which also explains why there is practically no radiation damage with cold neutrons, contrary to X-rays.

Distributions that are dependent on energy can be easily transformed to their equivalent as a function of wavelength or time of flight, and *vice versa*. For any variable z that is a function of energy ($z = f(E)$), the distribution $D(E)$ is related to its equivalent as a function of z, $F(z)$, by the Jacobian of the transformation $|dE/dz|$, where the vertical bars indicate the absolute value:

$$F(z) = D(E)|dE/dz| = D(E)/|(df/dE)| \qquad (2.6)$$

If $z = KE^n$, for example, $df/dz = E/nz = 1/|(df/dE)|$. To convert $D(E)$ to $F(\lambda)$ for X-rays, note that $\lambda = f(E) = 123.9\,E^{-1}$ and $|df/dE| = 123.9\,E^{-2} = \lambda/E$. Hence, $F(\lambda) = D(E) \cdot E/\lambda$.

To convert from E to λ for neutrons using the expression above, $\lambda = f(E) = 0.904\,E^{-1/2}$ and $|df/dE| = \lambda/2E$ and $F(\lambda) = D(E) \cdot 2E/\lambda$.

On spallation neutron sources the energy is often measured by the time of flight; that is, the time (τ in μs) taken by the neutrons to travel the distance (L) from the source to the detector. The term is also used for the inverse of the velocity (in $\mu\text{s} \cdot \text{m}^{-1}$). A look at the units will avoid any confusion. To convert from E to time of flight (τ): $E = m_n v^2/2 = m_n L^2 \tau^{-2}/2$, $\tau = f(E) = (m_n L^2/E)^{1/2}$ and $|df/dE| = \tau/2E$, hence $F(\tau) = D(E) \cdot 2E/\tau$.

The inverse transformation from $F(z)$ to $D(E)$ is done by multiplying by $|dz/dE| = 1/|dE/dz|$. An example of such a transformation is shown in Fig. 2.12.

2.2 X-ray sources

X-ray sources used in SAXS are based on the acceleration of electrons or positrons. The two main types are X-ray tubes used in laboratory instruments, which are continuous sources, and SR sources, usually bending magnets and insertion devices on electron (or positron) storage rings, which are pulsed sources.

2.2.1 Laboratory sources

In laboratory sources, electrons produced by heating a cathode are accelerated by a high-voltage gradient (10–90 kV) towards an anode consisting of a metal such as copper (Cu), molybdenum (Mo), silver (Ag) or tungsten (W). The electrons collide with atoms in the target and are scattered. Most of their energy is lost as heat, but a

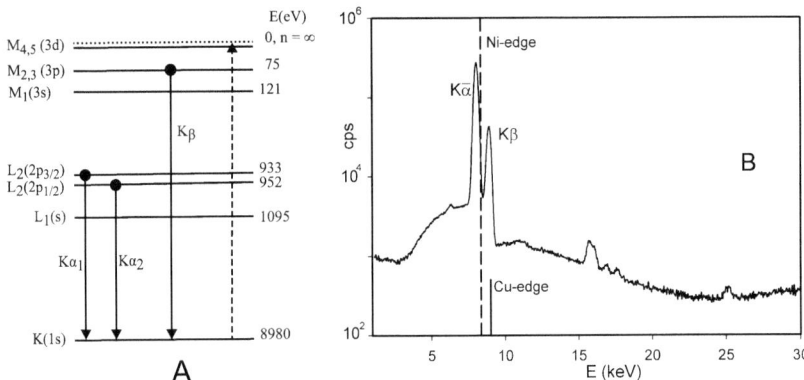

Fig. 2.2 A: X-ray emission spectrum of copper, illustrating the origin of the characteristic lines when a copper target is hit by electrons with energies above the critical energy. B: Experimental spectrum (cps: counts/s) of a standard X-ray tube with a Cu target and a beryllium window at an excitation voltage of 50 kV. (Courtesy P. Panine, Xenocs, Grenoble). A Ni filter absorbs most of the radiation above the Ni edge. The position of the CuK absorption edge is also indicated.

very small fraction ($\ll$1%) is dissipated by emission of a continuous spectrum of X-rays: the Bremsstrahlung, the German word for 'braking' radiation. Above a critical acceleration voltage some electrons have sufficient energy to eject the electrons from the inner shell of the target atoms, thereby creating a vacancy. The critical voltage is defined by the K-absorption edge of the target, which corresponds to the minimum energy that the incoming electrons must have to ionise the K-shell. As illustrated for copper in Fig. 2.2A, when an electron from a higher shell fills the vacancy—by dropping from the L- or M-shell to the K-shell, for example—an X-ray photon of well defined energy is emitted (2p $\rightarrow$ 1s for K$\bar{\alpha}$ or 3p $\rightarrow$ 1s for Kβ-radiation). Depending on the tube voltage, these sharp characteristic emission lines are 50–100 times more intense than the continuous Bremsstrahlung on which they are superimposed.

In the emission spectrum of Cu, both the Kα and Kβ lines are doublets, which are not separated in practice, with average wavelengths of 0.1542 nm and 0.1386 nm, respectively. The K edge in the absorption spectrum corresponding to the ejection of a 1s electron to the continuum is at 0.1380 nm (8.98 keV). The L-line of Cu at 1.33 nm (0.9 keV) is absorbed by the beryllium (Be) window of the tube, as Be has a mass attenuation coefficient that decays exponentially from 604 cm^2 g^{-1} at 1 keV to 0.64 cm^2 g^{-1} at 10 keV. This is also the cause of the rapid drop in intensity below 7 keV in Fig. 2.2B.

For SAXS from biological macromolecules one generally uses a Cu target. A Mo target—the next possible choice in terms of wavelength, with a characteristic energy of 17.5 keV ($\lambda = 0.071$ nm)—would be less useful because q_{min} becomes larger and the resolution of the detector must be higher as the q-range is compressed over a smaller angular range.

In order to be useful for SAXS, the radiation of the main characteristic line (Cu Kα) must be selected from the spectrum, while the other lines and the Bremsstrahlung must be eliminated by the optical system. On older installations this is done by inserting a Ni-filter which has an absorption edge at 0.1488 nm (8.33 keV) and effectively suppresses the Kβ line and harder radiation at the cost of an approximately 40% reduction in intensity of the Kα line.

The coherence length of a Cu source calculated from its band pass, approximately $(\lambda K\alpha_1 - \lambda K\alpha_2)/\lambda K\bar{\alpha} \approx 2.5 \times 10^{-3}$, is of the order of 50 nm, which is also the value calculated for the transverse coherence length at 1 m from a typical $1\,\mathrm{mm}^2$ source.

A great variety of X-ray tubes is being produced in particular for medical and imaging applications, and only a few common design features can be sketched out here. On standard sealed X-ray tubes the cathode is usually elongated and the anode is a fixed target, which is viewed at a small take-off angle to obtain a foreshortened image of the footprint of the electron beam on the target as illustrated in Fig. 2.3A. These tubes often have four side windows—two with a nearly square source and two with a line source depending on the viewing angle on the target. In the past, line sources were frequently used with standard tubes to obtain a sufficient photon flux for SAXS on isotropic samples. As each point on the line is effectively an independent point source, this leads to convolution (that is, the experimental scattering pattern consists of a superposition of scattering patterns shifted along the line), as illustrated in Fig. 2.3E. The resultant distortions in the scattering pattern are often difficult to correct, and this approach is therefore rapidly being abandoned in favour of instruments with point collimation.

Where higher power is required, rotating anodes consisting of a disk cooled with liquid metal, for example, are used as illustrated in Fig 2.3B. These rotate at several thousand rpm to increase the track of the focal spot and the heat capacity, which in turn reduces the temperature rise of the anode. These devices, however, require significantly more maintenance than sealed tubes.

Modern microfocus tubes (Fig. 2.3C) are equipped with electrostatic or magnetic lenses that focus the electrons into a small spot (10–50 μm diameter) or a thin line ($10 \times 200\,\mathrm{\mu m}^2$) to reduce the power load on the

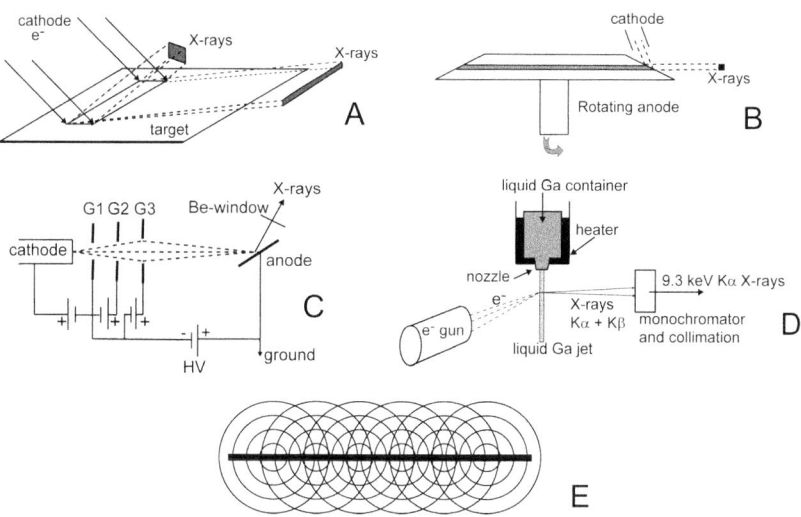

Fig. 2.3 A: Sketch of a standard X-ray tube with a line source or a smaller rectangular source depending on the viewing direction on the target. B: Rotating anode target used in high-power X-ray tubes. C: Microfocus tube with electrostatic focusing. In the electron gun the grid G1 controls the flow of electrons that are extracted from the cathode by G2 and focused in a small spot on the anode by G3. D: Principle of a liquid gallium jet X-ray source. E: Effect of a line source requiring corrections for the convolution to obtain the correct experimental scattering pattern.

anode. These tubes, which usually have a single window, and a fixed or rotating target are air- or water-cooled. The final size and shape of the beam at the exit window depend on the viewing direction and the take-off angle, but the beam diameters are of the order of 20–50 μm. Sealed microfocus tubes have a brightness that is 3–5 times higher than that of standard tubes, whereas another factor of ten is gained with microfocus rotating-anode tubes.

The most recent laboratory X-ray sources are based on jets of a liquid metal such as gallium (Fig. 2.3D), which provides 9.24 keV X-rays (Kα) with a flux that is at least an order of magnitude higher than that of rotating anodes (Otendal *et al.* 2008). Liquid gallium at about 50° C above its melting point (29.8° C) is injected under high pressure, as a jet of about 30 μm diameter, into a vacuum chamber. This jet is irradiated by a focused 50 kV electron gun giving a 10 μm FWHM X-ray spot. Optimisation of parameters like the jet speed and allowing complete vapourisation of the jet could yield a peak *SB* around 10^{12} photons/(s · mm^2 · mrad^2 · 0.1%$\Delta E/E$), comparable to that of bending magnets on SR sources.

2.2.2 Synchrotron radiation sources

SR sources are installed on circular accelerators or rings where a slowly exponentially decaying current consisting of regularly spaced short pulses or bunches of electrons (or positrons) can be kept circulating for many hours at constant particle energy, hence the name storage ring. On some rings operating in the so-called top-up mode the current is kept nearly constant by injecting electrons at short intervals to compensate for the losses.

In the bends of the ring, the electrons with energy E are kept on an orbit with radius of curvature R_c by the magnetic field (B, in Tesla) of the dipole bending magnets as determined by the relationship $R_c(m) = 3.335\ E(\text{GeV})/B(T)$, which essentially defines the geometry of the ring. The curved part of the orbit can be regarded as a short section where the acceleration first increases and then decreases again, thereby creating a pseudo-oscillatory movement. Fig. 2.4A illustrates that in a frame moving along x at the same velocity $v \ll c$ as the electron, the latter emits the classical doughnut-shaped dipolar radiation pattern due to the acceleration by the field of the bending magnet and the radiation is linearly polarised in the plane of the orbit. When the electrons reach a velocity close to that of light (c) their emission pattern in the frame of the observer is distorted by the Lorentz transformation and the Doppler effect, and appears as an intense cone of light along the instantaneous velocity vector. This cone has a very small vertical width and a horizontal width $\approx \gamma^{-1}$, where $\gamma = E/m_e c^2$ is the ratio between the total energy of the relativistic electrons and their rest energy given by the product of their mass (m_e) and the square of the velocity of light in vacuum (that is, for electrons, $\gamma = 1957 \cdot E$ (in GeV)).

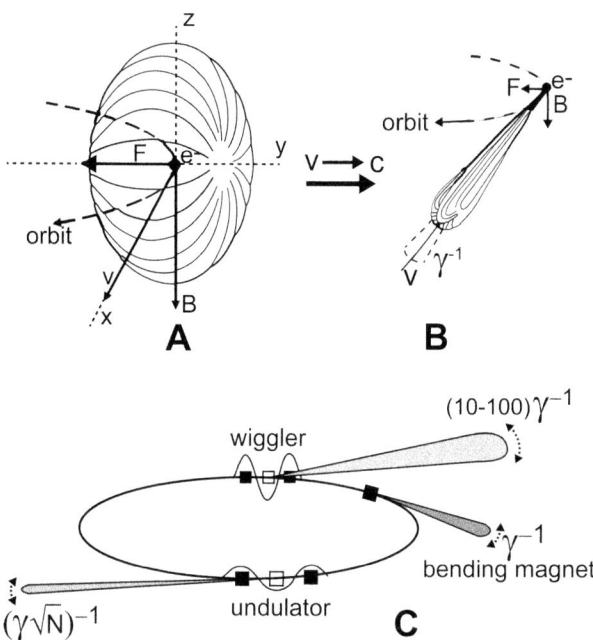

Fig. 2.4 A: An electron with a velocity (v) well below that of light ($v \ll c$) accelerated (by a magnetic field, for example) emits the classical monochromatic doughnut-shaped radiation pattern with linear polarisation along the direction of acceleration. B: When v approaches c the emission pattern in the laboratory frame of relativistic electrons travelling through a bending magnet is distorted by the Lorentz transformation and the Doppler effect, and appears as a narrow cone of polychromatic radiation. C: Different types of SR sources in a storage ring. Note the large difference in the horizontal spread of the beams from bending magnets and insertion devices. Adapted from Koch *et al.* (1983) with permission from Elsevier, © 1983.

As the electrons move along a periodic circular orbit with radius R the narrow cone of light sweeps like a searchlight around the orbit. A fixed detector only detects pulses of radiation during the short time intervals ($T = L/c$, where L is the spatial length of the bunches) when the velocity vector of the electrons points in its direction. If there is a single bunch, these short pulses are separated by an interval $T_0 = 2\pi R/c$. Modern rings usually have different filling patterns that can be optimised for special applications. On the PETRA III (DESY, Hamburg) ring, for example, there are 960 bunches of 40 ps separated by 8 ns. The properties of Fourier transforms imply that such a periodic train of light pulses, similar to the one in Fig. 2.1B, results from a broad distribution of frequencies or energies up to a critical energy E_c (keV) $= 0.6650 \, E^2$ (GeV) $B(T)$. Above E_c the emission rapidly falls off, as illustrated in Fig. 2.5A. Note that E_c, which defines an equipartition of the spectrum in terms of energy, does not correspond to the maximum of emission and that the linear–log graphs in Fig. 2.5A differ significantly in shape from the more common log–log representations of the *SB* or flux. For the bending magnets in PETRA III ($E = 6$ GeV, $B = 0.87$ T), the critical energy is 20.8 keV. The main properties of SR, high-intensity, linear polarisation in the plane of the orbit, pulsed time structure and broad range of energy or wavelength make storage rings ideal sources for UV–X-ray research.

The insertion devices (IDs)—wigglers and undulators—consist of a linear periodic structure of N dipole magnets with alternating polarity, with period λ_0, installed in the straight section of a storage ring. Along the axis of such devices the electrons follow a near sinusoidal path and

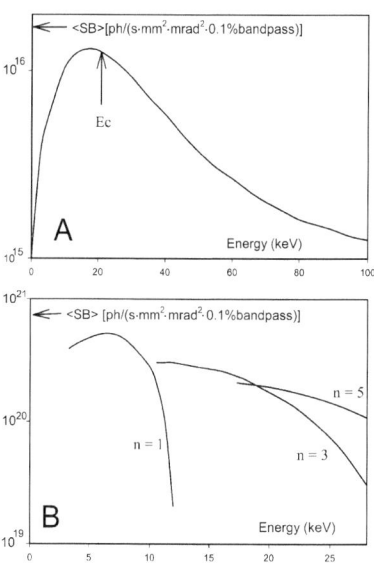

Fig. 2.5 A: SB of a bending magnet with a field of 0.87 T at the PETRA III ring (6 GeV) assuming a stored current of 100 mA. E_c is the critical energy (20.8 keV). B: On-axis *SB* for the first three odd harmonics of a standard undulator ($B = 0.81 \, T$) at PETRA III for 100 mA stored current. Compare the *SB* values and width of the first harmonic (about 3.45 keV) with that of the spectrum of a bending magnet on the left.

the characteristics of the radiation which they emit depend on the field-strength parameter $K = 0.934 \, B[T] \, \lambda_0$ [cm] and on the angle θ between the axis of the ID and the direction of observation. The main differences between the properties of the two types of IDs are due to the difference in the strength of the magnetic field, which is weak in undulators ($K \ll 1$) and strong in wigglers ($K \gg 1$).

In an undulator the amplitude of the deviations of the electron trajectory from the axis of the ID are small and hence there is coherent superposition of the radiation from the N sources, which leads to an increase in brightness by a factor of up to N^2. In a frame moving along the axis of the undulator at the same velocity as a single electron travelling along an ideal sinusoidal path, the electron is accelerated by an electromagnetic field defined by an oscillating magnetic field with period λ_0/γ because of the Lorentz contraction and a corresponding oscillating electric field perpendicular to it. Consequently, the electron emits a sharp line with a wavelength λ_0/γ, which is detected in the laboratory frame as having a wavelength of $\lambda_0/2\gamma^2$ because of the Doppler effect, which also narrows the width of the emission cone. In practice, the electrons in a bunch have a certain energy spread and the resulting emission spectrum has a width $\Delta E/E \approx 1/nN$, corresponding to the envelope of all the slightly displaced lines, as illustrated in Fig. 2.5B. As the motion of the electrons becomes more complex at increasing K-values, higher harmonics also appear. On-axis ($\theta = 0$) there are only odd harmonics ($n = 1, 3, 5 \ldots$) whereas off-axis there are also even harmonics. The horizontal width of the emission cone ($\approx (\gamma N^{1/2})^{-1}$) is much smaller than that of bending magnets or wigglers. The wavelength (λ_n) and energy (E_n) of the harmonics (n) are given by:

$$\lambda_n [\text{cm}] = \frac{\lambda_0 [\text{cm}]}{2n\gamma^2}(1 + 0.5K^2 + \gamma^2\theta^2) \tag{2.7}$$

$$E_n [\text{KeV}] = \frac{0.95 \cdot n \cdot (E[\text{GeV}])^2}{(1 + 0.5K^2 + \gamma^2\theta^2)\lambda_0[\text{cm}]} \tag{2.8}$$

With wigglers ($K \gg 1$) the amplitude of displacement of the electron from the axis is much larger and there is therefore incoherent superposition of the radiation from the N magnets with an increase in brightness by a factor of N. The number of harmonics becomes very large so that the shape of the emission spectrum is very similar to that of the bending magnet in Fig. 2.5A. The critical energy is given by the same expression as for the bending magnet, but the horizontal width of the emission cone is much larger ($\approx 10 - 100\gamma^{-1}$).

On low-energy rings, special short wigglers (wavelength shifters), with a high field (4–5 T) in the centre and weaker fields at both ends to obtain a straight average trajectory of the electrons, are used to shift the emission spectrum to higher energies. Contrary to the situation in bending magnets, the field and hence also the critical energy are not fixed by the geometry of the ring. Moreover, like other IDs the devices, which usually require superconducting magnets to achieve high fields, can be switched off or removed without altering the effective trajectory of the electrons or the overall performance of the ring.

As explained in Section 2.4.2, the use of SR for SAXS requires more complex optical systems than laboratory instruments, because of the necessity to use monochromators to select an appropriate wavelength band and mirrors to eliminate higher harmonics and/or focus the beam.

The longitudinal and transverse coherence lengths at synchrotron-radiation sources are several orders of magnitude larger than on conventional sources. The high flux makes it possible to use monochromators with very narrow band passes, and the low beam divergences make it possible to build very long instruments without excessive loss of flux.

On third-generation sources, coherence lengths of the order of 50 μm in the horizontal direction and 500 μm in the vertical direction can be obtained at 50–100 m from the source. This should not affect conventional SAXS measurements but may be useful in the study of intermolecular interactions, although it may be difficult to conserve coherence throughout the optical system.

The high brightness of IDs simplifies the design of optical system, but heat load on the optical elements becomes a problem and optical components must be cryogenically cooled. The high SB presents the possibility of producing high-intensity beams with which high-precision data can be obtained in short times for time-resolved measurements, whereas tunability makes anomalous scattering possible. The consequence of the natural collimation of the beams is that all SAXS instruments at SR facilities are based on point collimation. As illustrated by the comparison of different types of sources in Fig. 2.6, the SB of third-generation sources is so high that measurements are no longer limited by the available intensity, but radiation damage becomes a serious concern.

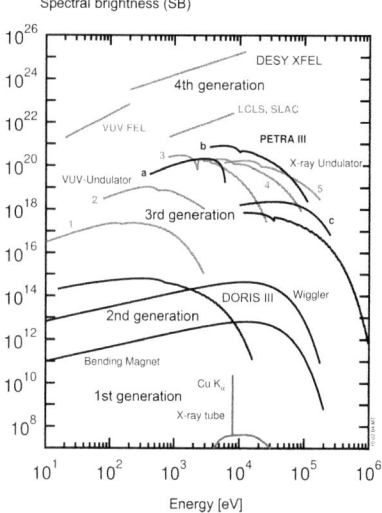

Spectral brightness (SB)

Fig. 2.6 Average spectral brightness of SR and free-electron laser (FEL) photon sources available or planned at DESY, Hamburg, compared with the actual performance of other third-generation storage rings and the LCLS (SLAC, Stanford). Labels are as follows: 1. BESSY II U125: 2. ALS U5: 3. DIAMOND U46: 4. ESRF ID16: 5. SPring-8 BL46. PETRA III: a. soft-X-ray undulator (4 m, high-β): b. standard $K_{max} \approx 2.2$ undulator (5 m, high-β): c. hard X-ray wiggler ($K_{max} \approx 7,5m$, high-β). Adapted, with permission, from the PETRA-3 Technical Design Report, DESY, Hamburg (2004).

2.2.3 Coherent sources

As explained in Section 2.2.2, in an undulator the electrons travel along a sinusoidal path and spontaneous emission yields a narrow cone of radiation. The width of this cone is of the same order as the amplitude of deflection of the electrons from their average direction of propagation. Because the longitudinal velocity of the electrons (v_l) is smaller than the velocity of light, the electrons lag behind the electromagnetic field which they emit. The condition for interference of the radiation emitted in successive periods of an undulator with period λ_0 is that on axis the difference in transit times would be equal to one wavelength of the emitted radiation (that is, at every period the electrons lag behind the electromagnetic field they emit by one wavelength).

$$\lambda/c = \lambda_0(1/v_l - \lambda/c) = \lambda[cm] = \frac{\lambda_0[cm]}{2\gamma^2 c}(1 + 0.5K^2) \qquad (2.9)$$

The electrons interact with this electromagnetic field, and depending on the phase relationship between their oscillation and the field they are accelerated or decelerated, which leads to the SASE (Self Amplified Spontaneous Emission) phenomenon on which all existing X-FELs (X-ray Free Electron Lasers) and those under construction are based (Margaritondo and Ribic 2011). SASE leads to micro-bunching, as illustrated in Fig. 2.7. As it progresses through the undulator, the bunch with initially randomly distributed electrons is sliced in its direction of propagation into increasingly sharply defined and regularly spaced layers of high and low electron density with thickness $\lambda/2$. As all electrons in such a high-density layer radiate in phase, coherent emission increases, which in turn leads to stronger micro-bunching. Close to the entry of the undulator the radiated power is proportional to the number of electrons in the bunch, and it grows exponentially along the length of the undulator to a value proportional to the square of this number, after which, emission saturates. The output of the undulator is a 50–100 fs X-ray pulse consisting of a train of sharp spikes of about 1 fs with randomly varying intensity.

Such devices are called free electron lasers (FELs), because the electrons are not bound to nuclei, and hence restricted to specific transitions, as is the case in the lasing medium of conventional lasers. Their radiation has many properties in common with that of conventional lasers: it is linearly polarised and highly collimated with a high transverse coherence. Due to the random nature of the production of electrons at the cathode of the injector, the wavelength spread in the X-ray beam ($\Delta\lambda/\lambda \approx 0.2\%$) corresponds to a longitudinal coherence length that is much shorter than the length of the electron bunch or than the longitudinal coherence lengths of conventional lasers.

To implement an X-FEL based on the SASE principle one needs an injector with a very low emittance (small spatial and momentum spread of the electrons), a high-energy linear accelerator sending bunches with a high charge density through a long undulator with a very accurately defined magnetic field. The LCLS (Linear Coherent Light Source in Stanford, USA), for example, consists of thirty-three undulator sections each 3.4 m long with 224 magnets of alternating polarity through which a 14-GeV electron beam with 250 pC ($\approx 1.5 \times 10^6\,e^-$) bunches produced by the linear accelerator passes. The angular deviations from a straight line in the alignment of the magnets in this undulator do not exceed $1\,\mu$rad/m.

The peak SB of fourth-generation sources such as the LCLS is about nine orders of magnitude above those of undulators at third-generation sources (ESRF, APS, SPring8), whereas the average SB is similar.

Present X-FELs are single-pass machines with relatively low repetition rates ($\sim$100 Hz) where the electron beam is dumped after passing through the undulator while the light beam continues its straight trajectory. The highly coherent very intense short pulses make it possible to acquire the scattering pattern of an object before it is destroyed by Coulomb explosion. As the structure of successive pulses is different and the samples are destroyed after each pulse, experiments require a very

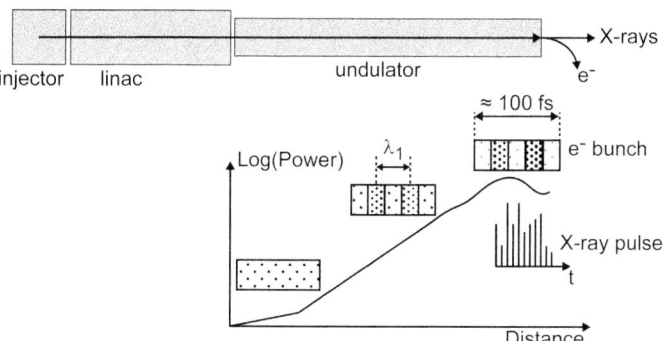

Fig. 2.7 Schematic representation of an X-ray free electron laser (X-FEL) based on the SASE principle and of the power gain along the length of the undulator. The electrons, which are initially randomly distributed in the bunch, become progressively ordered (microbunching) by interaction with the increasingly coherent radiation which they emit as they travel through the undulator. The output X-ray pulse is a train of equally spaced very intense short spikes with random amplitudes.

different approach from conventional measurements, including novel instrumentation and data evaluation methods. Details about recent experiments which ultimately aim at obtaining the three-dimensional structure of single molecules can be found in the literature.

2.3 Neutron sources

High-power neutron sources are based on reactors where neutrons are produced by a nuclear fission chain reaction (usually of ^{235}U), or on proton accelerators where the impact of high-energy protons (0.6–3 GeV) on a heavy metal target (for example, tantalum, liquid mercury, tungsten) produces excited nuclear states. These decay by spallation, emitting neutrons, γ-rays and neutrinos without initiating any chain reaction. Both types of source can be continuous or pulsed. OPAL (ANSTO, near Sydney, Australia), ILL (Institut Laue-Langevin, Grenoble, France), BERII (HZB; Helmholtz Zentrum, Berlin, Germany) and FRM2 (Garching, Germany) are continuous reactor sources, whereas the IBR2 (Dubna, Russia) is a mechanically pulsed reactor source. Similarly, SINQ (Villigen, Switzerland) is a continuous spallation source, whereas ISIS (UK), LANSCE (Los Alamos, USA), SNS (Oak Ridge, USA), J-PARC (Tokai, Japan) and CSNS (PR China) are pulsed spallation sources. A new pulsed spallation source (ESS; European Spallation Source) is planned to be built in Lund, Sweden, and should be operational around 2020.

As illustrated in Fig. 2.8, reactor sources have reached technological limits mainly related to heat dissipation, and their performance has therefore not greatly increased during the last decades. Pulsed spallation sources offer significantly higher instantaneous fluxes in short pulses, which are advantageous for many spectroscopic applications. Until recently their average flux was, however, lower than that of reactors or of the continuous spallation source (SINQ), but the flux of J-PARC, SNS and the future ESS may equal the effective flux of the most powerful reactors for SANS studies. Spallation sources produce less radioactive waste than reactors, and are safer to operate as they

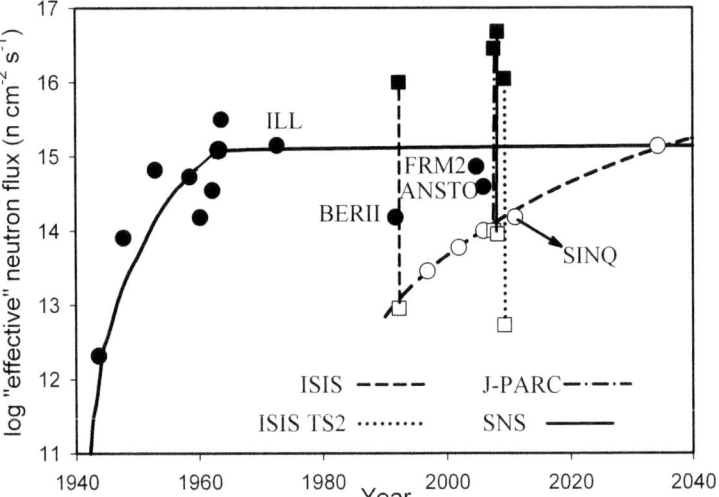

Fig. 2.8 Evolution of the 'effective' neutron flux of existing neutron sources. The 'effective' flux is well defined for reactors, but for pulsed spallation sources it is based on simulations. For the pulsed spallation sources (ISIS, J-PARC and SNS) the black squares represent the instantaneous flux and the white one the average flux. Adapted, with permission, from Clausen (2008), © Springer, 2008.

can be easily switched off. Indeed, like all accelerator-based machines the challenge is to keep them running. Their design is also more flexible than that of reactor sources. As the order-of-magnitude increases promised by inertial fusion sources, where H/D pellets are fused under the influence of high-power lasers, are still many years away, it can be expected that in the foreseeable future neutron research will be based mainly on spallation sources, and that like at reactor sources further gains will have to come largely from better optical components and instrument design.

2.3.1 Reactor-based sources

The basic layout of a reactor-based neutron source is shown in Fig. 2.9A. In the core of the reactor fission of a ^{235}U nucleus produces two nuclei with a total kinetic energy close to 200 MeV as well as γ-rays ($\sim$7 MeV), and on average 2.5 neutrons with an energy of 2 MeV. This energy is too high to sustain the chain reaction, and the neutrons must therefore first be slowed down by collisions in a moderator and partly reflected into the core—a process which takes 1–2 ms. To keep the power of the reactor constant a control rod is progressively retracted from the core during a cycle of operation to compensate for the loss of fuel. At the end of a cycle, which typically lasts four to eight weeks, the fuel elements must be replaced. A compact core and efficient cooling of the reactor are factors determining the performance of reactor-based sources. Safety is, of course, a primary concern at all installations, which are therefore also equipped with redundant safety systems such as shutdown rods of neutron-absorbing material which can be moved into the moderator or the core to stop the chain reaction. As the energy of the neutrons is most efficiently reduced by collisions with light elements such as hydrogen, which has nearly the same mass as the neutron, one often uses

H_2O, D_2O, graphite or liquid or solid methane as moderators. Some re-actors are additionally equipped with a beryllium reflector between the reactor core and the moderator to reflect neutrons into the core.

The maximum neutron flux in the vicinity of the core is around $10^{14}-10^{15}$ neutrons cm$^{-2} \cdot$ s^{-1}. Very fast neutrons and γ-rays must be absorbed by thick shielding consisting usually of steel and boron-loaded concrete. Neutrons with different wavelength distributions can be obtained by inserting moderators at different temperatures. Hot sources usually consist of a block of graphite heated by the reactor radiation to about 2700 K, whereas the cold sources, which produce the long-wavelength neutrons (0.45–2 nm) needed for SANS, consist of small volumes of liquid H_2 or D_2 at about 10–20 K.

After thermalisation in the moderator one obtains approximately a Maxwell–Boltzmann distribution of velocities. The most probable speed is $v_{\mathrm{mp}} = (2k_B/m_n)^{1/2} T^{1/2} = 128.4 \, T^{1/2}$, where T is the absolute temperature. At 323 K, the usual temperature of a water moderator, v_{mp}

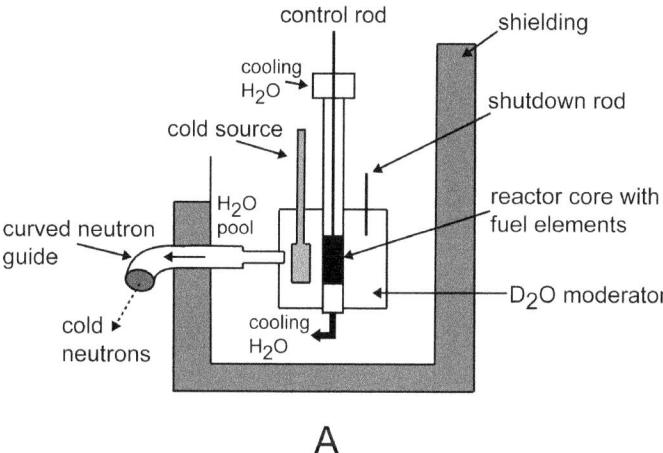

A

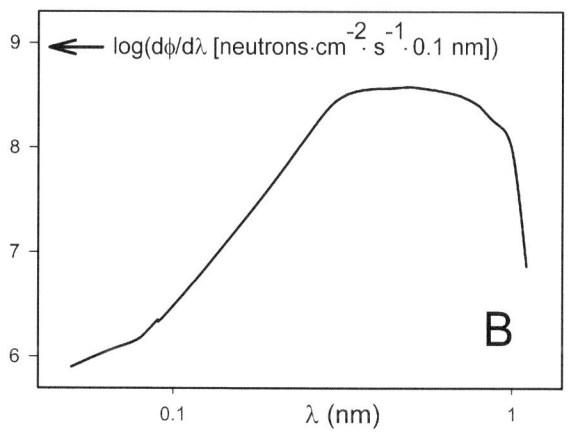

B

Fig. 2.9 A: Schematic representation of a reactor-based neutron source. B: Wavelength distribution of the H15 beam on the vertical cold source at the ILL (guide with 27 km radius of curvature).

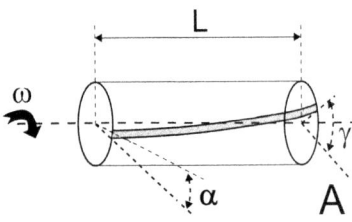

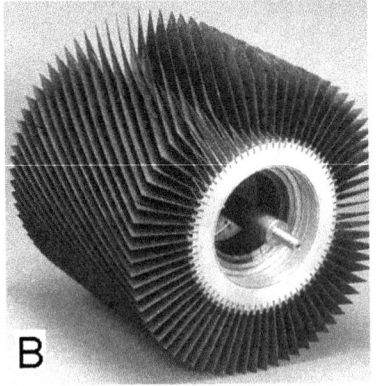

Fig. 2.10 A: Schematic diagram of a velocity selector with a single helical groove with angular width α. The condition for neutrons with velocity v to traverse the groove without hitting the walls is $L/v = \varphi/\omega$, where φ is the angle between the entry and exit points of the groove, and ω is the speed of rotation. B: Compact velocity selector with optimised blade system of the GENF facility at the GKSS. (Courtesy Prof A. Schreyer.)

is thus about 2307 m/s, and at 20 K, the temperature of a cold source, $v_{\mathrm{mp}} = 574$ m/s. Using de Broglie's relationship, which links the speed of the neutrons and their wavelength ($\lambda = h/m_n v$), one finds that this corresponds to wavelengths of about 0.17 nm and 0.69 nm respectively. A typical spectrum of a cold source is shown in Fig. 2.9B.

The neutrons are transported to the experiments using guides consisting of rectangular tubes with borofloat glass, nickel-coated glass or supermirrors (multilayers) which totally reflect the neutrons above a critical angle $\theta_c = \kappa\lambda$, where κ is a constant depending on the reflecting material and λ the wavelength of the neutron. To prevent fast (short-wavelength) neutrons or γ-rays from reaching the instruments and detectors, curved neutron guides are used. These eliminate the fast neutrons because their angle of incidence becomes higher than the critical angle in the bend.

The next step is to obtain a monochromatic beam. Since the velocities of thermal neutrons are of the order of 2000 m/s one can use mechanical velocity selectors for the purpose. As illustrated in Fig. 2.10A, velocity selectors are cylindrical devices with helical grooves separated by neutron-absorbing walls. The rotation speed determines the velocity of the neutrons which can traverse the device, and the ratio of the width of the grooves to the length of the cylinder determines the band pass ($\Delta\lambda/\lambda$), which is usually around 5–10%, but sometimes higher on lower-power reactors. To achieve a high resolution there must be a corresponding number of slots, and as the transmission diminishes with the ratio of the wall surface to the total surface, the walls should be as thin as possible. There is also a loss of diverging neutrons due to the fact that there cannot be any neutron guide over the length of the selector; therefore, the selector should be as short as possible, but rotate at higher speed (up to 28,000 rpm). This leads to designs with curved blades, as shown in Fig. 2.10B.

Note that as neutrons have mass over long distances they will follow a parabolic trajectory just like an artillery shell. In SANS this effect can be compensated for by using suitably designed prisms (Forgan and Cubitt 1998), oscillating collimation as on the LQD instrument at LANSCE, or less satisfactorily in the data-processing software. The effect, which increases with the square of the length of the neutron flight path, becomes important for sample-detector distances well over 10 m and longer wavelengths (>1 nm) that are rarely used in biology.

2.3.2 Spallation sources

As already mentioned, spallation sources can be continuous or pulsed. This depends on whether a proton cyclotron, where a continuous proton current (continuous wave; CW) circulates, or a synchrotron, where the protons are bunched, is used to accelerate the protons to energies of 0.6–3 GeV before they are sent to a heavy-metal target (Pb in zirconium alloy rods with a Pb blanket at SINQ, tantalum-cladded tungsten plates at ISIS or liquid mercury at SNS and J-PARC). The number of neutrons produced by the spallation reaction following the impact of a proton on a target atom depends on the energy of the proton and the nature of the target. On more recent sources liquid targets such as mercury or PbBi tend to be used because they are less damaged by radiation than solid targets, and as they have a higher atomic number than tungsten they yield a higher number of neutrons. Liquid targets also make it easier to dissipate the large amount of heat (1–2 MW) that is produced and to cope with the effects of the shock waves resulting from the impact of the high-energy proton pulses, but difficulties may arise from cavitation. The energy of the neutrons emanating from the target is higher than that in a reactor, so that there is even more need for efficient moderators.

A very schematic representation of the different types of spallation sources is shown in Fig. 2.11. As in the previous cases, a number of important technical installations such as cryogenic plants for super-conducting RF cavities, cooling water systems and radioactive waste handling facilities are not shown.

In the CW source at SINQ (Wagner 2002) a continuous proton current is emitted by the source into a pre-accelerator followed by two cyclotrons where the protons are progressively accelerated to 0.6 GeV. This proton beam is sent to the bottom of a cylindrical target so that the entire periphery of the target is available to install instruments on different types of moderators (such as hot and cold sources). From the point of view of the user a continuous spallation source is similar to a reactor source, the neutrons can be monochromated with velocity selectors and the same detectors and data-acquisition systems can be used.

Pulsed spallation sources consist of an ion source, which produces pulses of H^- ions (200 µs at ISIS) which are progressively shaped and accelerated by a sequence of different types of accelerator such as radio-frequency quadrupoles (RFQ) and drift-tube linear (DTL) accelerators, before being transferred to a linear accelerator (linac) consisting of a series of radio-frequency cavities providing further acceleration, and of

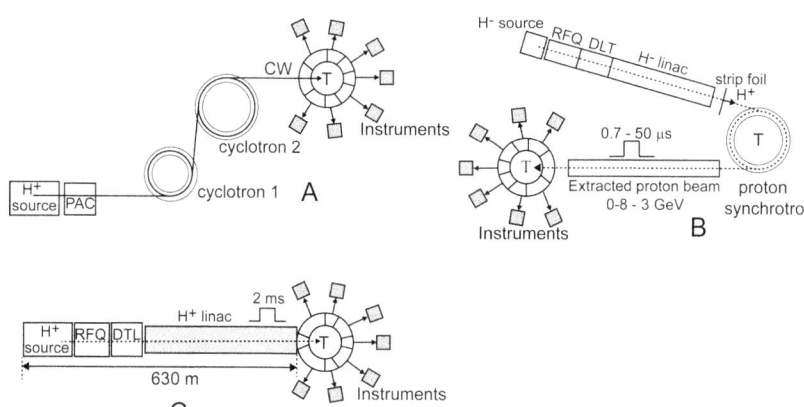

Fig. 2.11 Schematic representation of different types of spallation source. A: Continuous source (SINQ). B: Short-pulse source (for example, ISIS, SNS). C: Long-pulse source (ESS project). PAC: pre-accelerator, T: target, RFQ: radio-frequency quadrupole, DTL: Drift-Tube Linac.

beam-steering and focusing magnets. Between the linac and the synchrotron the H^- ions pass through a thin aluminium oxide or carbon foil, where their electrons are stripped off. After 1,000 to 10,000 turns in the synchrotron the proton pulses are compressed to 0.1–0.7 μs and accelerated to energies of 1–3 GeV, depending on the particular installation, before being sent to the target at frequencies of the order of 50–60 Hz.

On these short-pulse spallation sources the moderators must be thinner than on a CW or reactor source in order not to excessively increase the width of the neutron pulse at the exit of the moderator, which is in any case longer than the proton pulse (~400 ns for a ~100 ns proton pulse at ISIS). This is a major difference between continuous and pulsed sources. Indeed, a thin moderator implies that the resultant spectrum is undermoderated and has a large epithermal (>1 eV) component, as illustrated in Fig. 2.12. When the moderator is coupled to a premoderator or a reflector the neutron pulses tend to have long tails due to slow neutrons being reflected into the moderator. One way to avoid this and to decouple the moderator from the reflector for slow neutrons is to make a decoupled moderator by inserting a thin sheet of neutron-absorbing material (such as cadmium) between reflector and moderator. Some moderators also contain gadolinium poison plates, which have resonance absorptions for slow neutrons, for pulse shaping. The performances of the moderators are also influenced by specific features such as grooves in the face of the moderator viewed by the neutron guide, which can, depending on the wavelength, increase the flux by up to 50% compared to a flat face.

With short pulses the neutrons effectively leave the target simultaneously and, provided the incident and scattered neutrons are very well collimated and follow a path of known length, their energy can be determined from their time of flight (TOF) to the detector. Collimation is less of an issue on continuous sources where the neutrons are monochromated with velocity selectors. In TOF measurements all neutrons can in principle be used, which is more efficient than on continuous sources and allows an instrument to cover a much larger q-range in a single measurement, but at the cost of a more complex collimation and detection system. To have a useful beam and avoid detector saturation the prompt pulse of fast neutrons and γ-rays leaving the target immediately after the impact of the proton pulse must be eliminated. Moreover, the time interval (frame) during which neutrons are allowed to pass through the instrument must be accurately defined. This can be achieved with choppers—devices consisting of disks of neutron absorbers rotating at multiples or submultiples of the frequency of the source with which they are kept in phase. The prompt pulse can, for example, be eliminated with a high-speed T_0 chopper, whereas for neutrons in the appropriate energy/wavelength range the frame lengths are selected with slower bandwidth-limiting choppers with variable opening angles. These devices are also required to avoid

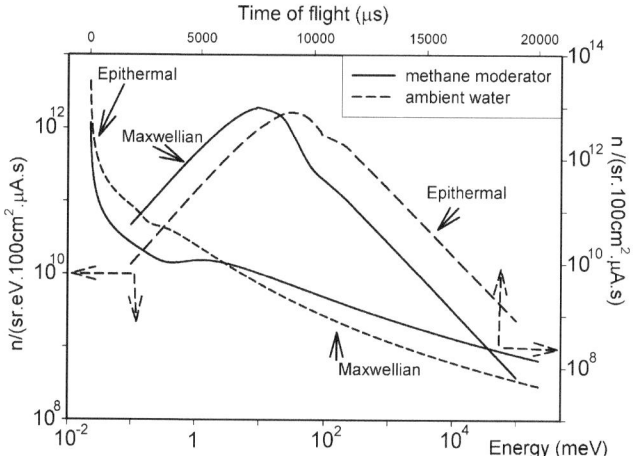

Fig. 2.12 Neutron flux at the exit of methane and ambient water moderators on Target Station 1 (TS1) at ISIS as a function of energy or time of flight, assuming a 12 m flight path. The distribution as a function of time of flight (τ) is obtained from that as a function of energy by multiplication by $2E/\tau$ (see Section 2.1.2).

frame overlap—a situation whereby the slow neutrons of one pulse are overtaken by the fast ones in the following pulse.

The proposal for the European Spallation Source (ESS)—a 600 m long machine to be built in Lund (Sweden) by 2020—is based on a multiple-proton source, a proton linac which accelerates 2-ms proton pulses to 1 GeV (300 kJ or 150 mA during 2 ms) onto a mercury or lead target with a frequency of 16.67 Hz (50 Hz/3). The aim is to achieve the same average flux as at the most powerful reactor source (ILL) while making more efficient use of the neutrons. The 300 kJ proton pulses preclude the use of a ring accelerator as in short-pulse machines. With its long pulses the ESS is a pseudo-continuous source. As already done in many experiments on steady-state sources, the shape, duration or energy distribution of the neutron pulses will be controlled using choppers and other instrumentation.

2.4 SAXS instruments

2.4.1 Laboratory SAXS instruments

Most of the pioneering work in SAXS was done on laboratory instruments, and the geometry of various cameras with slit, block or point collimation have been described in textbooks on SAXS (Kratky 1982, Holmes 1982). In recent years, progress with X-ray sources, multilayer optics and detectors have led to a new generation of instruments with point collimation.

Modern microfocus, or rotating anode and metallic jet sources make it possible to build cameras that are comparable in performance at a fixed wavelength of 0.154 nm with some of the early second-generation SR instruments. This is achieved by using magnifying optics to collect a large angle (4–8 mrad) from the divergent source and produce a lower exit divergence (0.5–1 mrad). Table 2.1 and Fig. 2.13 present an overview

Table 2.1 Characteristics of conventional SAXS instruments available from commercial suppliers.

Supplier	Collimation	Detector
Anton Paar SAXSess mc^2 www.anton-paar.com	Point or line collimation, Kratky camera	Imaging plate, Charge-coupled device (CCD), other
Bruker AXS Nanostar www.bruker-axs.com	Two Göbel mirrors, point collimation, three pinholes	Hi-Star multiwire gas proportional detector Våntec 2000 microstrip proportional counter
Rigaku www.rigaku.com	Point collimation, three pinholes Two-dimensional Kratky camera	Multiwire gas proportional detector Hybrid pixel photon-counting detector
Xenocs Xeuss www.xenocs.com	Microfocus source, Two-dimensional collimating optics Two sets of four-blade scatterless slits	Hybrid pixel photon-counting detector, imaging plate

of the features of SAXS cameras offered by some of the main suppliers. Most of these instruments are made for the characterisation of materials (such as polymers, nanomaterials, and so on), which scatter much more strongly than solutions of biological macromolecules. Some of these devices therefore also offer wide-angle X-ray scattering (WAXS) capabilities, which are useful in studying lipid systems, for example, but not indispensable for solutions of biological macromolecules. The instruments that are based on microfocus fixed target or rotating anode sources have q_{min} values of $(3-9) \times 10^{-2}$ nm^{-1} and fluxes at the sample of 10^8–10^9 photons s$^{-1} \cdot$ mm^{-2}, with a beam diameter of 30–1000 μm.

As an example, the Anton Paar camera utilises a block collimation (so-called Kratky collimator), which in principle guarantees the lowest q_{min} value and lowest background, but only in one half of the scattering

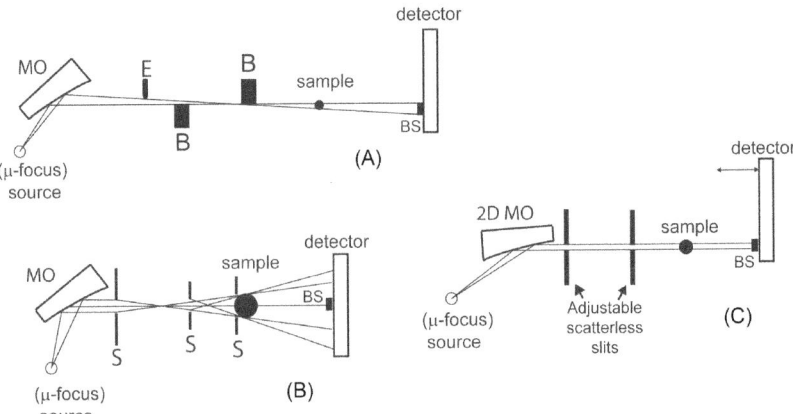

Fig. 2.13 Sketch of the geometry of some commercial laboratory instruments with block collimation (A) (MO: multilayer optics; B: block; BS: beamstop; E: edge; S: slits) pinhole collimation (B) and scatterless collimation (C).

pattern. The classical Kratky camera has a line focus (the beam is about 10 mm wide in the horizontal direction, whereas the scattering pattern is measured in the vertical direction). This yields higher intensity but leads to smearing of the scattering pattern; the smearing can be significantly reduced by horizontal slits at the expense of a lower flux. A recent development by Rigaku uses a two-dimensional Kratky camera, where the beam is focused onto the detector plane by a multilayer optics. For isotropic scattering this camera yields point collimation without loss of intensity.

The various types of camera differ significantly in the space required. Before acquiring any such device it is useful to familiarise oneself with some of the design considerations (Bergmann *et al.* 2000, Pedersen 2004) that have led to the choice of particular geometries. The most effective way to compare the X-ray cameras remains, however, to measure a set of well-characterised reference samples on the different instruments.

SAXS cameras can also be built using the Bremsstrahlung only with an energy-dispersive detector (Portale *et al.* 2007), but like their SR counterparts it is probable that because of radiation damage these devices will not become standard for biological applications.

The performance of laboratory instruments will still increase as new laboratory sources, including microfocus sources with liquid gallium jet anodes (Otendal *et al.* 2008) and compact synchrotrons based on the reverse Compton effect (Loewen 2003), as well as new optical components such as scatterless hybrid slits (Li *et al.* 2008), are being developed.

Perhaps the two most crucial parameters for the success of SAXS measurements—especially on modern SR sources, where beam sizes can be very small—are the positional and angular stability of the source and the mechanical stability of the instruments. If the necessary precautions are taken at installation stability is less of an issue on laboratory instruments, which are more compact and can be kept at constant power with variations in flux of only a few per cent over long periods.

2.4.2 Synchrotron SAXS instruments

The spectral brightness available to SR-SAXS stations is three to ten orders of magnitude higher than for laboratory sources, as illustrated in Fig. 2.6. This tremendous increase provides more precise data, shortens the exposure times for time-resolved studies, reduces the amount of material required and also paves the way for high throughput analysis. Most instruments provide tuneable wavelength for anomalous SAXS experiments. It is thus not surprising that although a number of laboratory X-ray SAXS cameras are available and actively used in biological research, most challenging projects rely on the use of SR.

SAXS beamlines are available at all major SR facilities in the world. Most of these are multipurpose beamlines used not only for solution scattering from biological macromolecules but also for studies on materials, surface scattering from films (grazing incidence SAXS), fibre diffraction, and so on. The requirements for low background and high stability for solution scattering experiments are rather stringent and not met by all SAXS stations. Below we shall briefly describe several SR beamlines, which are optimised for SAXS from solutions of biological macromolecules.

A typical optical layout of a synchrotron SAXS beamline features a monochromator (monocrystal or multilayer) to select the X-ray wavelength λ and mirror(s) to reject higher harmonics ($\lambda/2, \lambda/3\ldots$). These optical elements are also employed to focus the beam in both horizontal and vertical directions. In contrast to the situation in crystallography, the beam for the SAXS stations is not focused on the sample, but rather in the detector plane to avoid smearing effects. The parasitic scattering around the beam is reduced by using several pairs of slits made from highly absorbing material such as tungsten or tantalum.

An example of a high flux station dedicated to biological SAXS is given by the X33 beamline constructed in the beginning of 1980s at the EMBL Hamburg Outstation on a bending magnet of storage ring DORIS (Koch and Bordas 1983, Roessle *et al.* 2007). Although the DORIS ring was closed down in 2012, the instrument provides a good example of a bending magnet beamline (Fig. 2.14). The source size is $1.961 \times 0.512\,\text{mm}^2$ (FWHM) with a divergence of $0.438 \times 0.0238\,\text{mrad}^2$. A water-cooled bent triangular Si 111 monochromator focuses the beam horizontally with a fixed wavelength ($\lambda = 0.15\,\text{nm}$). A single 1000 mm rhodium (Rh)-coated gravimetric U-bending mirror (SESO, Aix-en-Provence, France) focuses the beam vertically and eliminates higher harmonics. At 4.5 mrad incident angle the Rh-coating provides about 90% reflectivity at $\lambda = 0.15\,\text{nm}$ and good harmonics suppression. Several slit systems (pairs of tungsten blades) are used for beam conditioning. The first water-cooled pair (primary slits, PS) defines the size of the white beam on the monochromator. The second pair, Secondary Slits 1 (SS1), controls the beam acceptance of the mirror, and the pair SS2 positioned just after the mirror cuts out the stray radiation. Downstream of SS2, the next pair SS3 forms the beam actually reaching the sample, and the sole purpose of SS4 in front of the sample is to cut the parasitic scattering from all previous optical elements.

The station provides a reasonably high flux (about 10^{11} photons s^{-1}), with a typical focused beam size on the detector of $1.5 \times 0.3\,\text{mm}^2$ (FWHM, H $\times$ V). A temperature-controlled sample cell (with flat mica or polycarbonate windows) is placed in vacuum to reduce the background. The scattered photons travelling through the evacuated detector tube are registered in both downward and upward directions by two photon-counting solid-state Pilatus detectors (Dectris, Villigen, Switzerland) (Kraft *et al.* 2009, Broennimann *et al.* 2006). The larger detector, Pilatus-1M, contains ten 100k modules arranged in two stripes, yielding an active area of $423.6 \times 70\,\text{mm}^2$ (2463×407 pixels), positioned at 2.6 m from the sample and allowing one to record the scattering patterns in

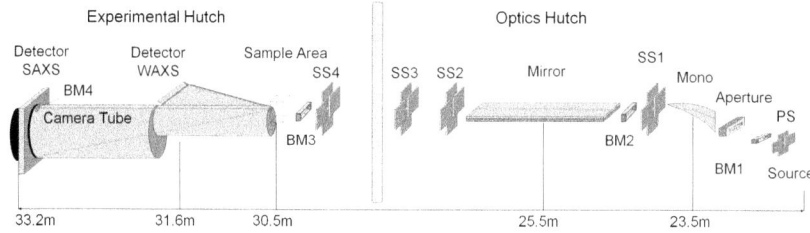

Fig. 2.14 Layout of the X33 SAXS beamline (bending magnet at DORIS, DESY, Hamburg). PS: primary slits; SS: secondary slits; Mono: triangular monochromator; BM: beam monitor.

the 2θ-range from about 0.1° to 8.5° ($0.07 \, \text{nm}^{-1} \leq q \leq 6.2 \, \text{nm}^{-1}$), corresponding to a resolution range of about 90 nm to 1 nm.

The WAXS Pilatus-300k detector has three modules arranged in a vertically oriented column, yielding a total active area of $254 \times 33.5 \, \text{mm}^2$ (1475×195 pixels). It is positioned at 0.75 m from the sample and collects photons scattered upwards at scattering angles between 5° and 25° ($3.7 \, \text{nm}^{-1} \leq q \leq 18 \, \text{nm}^{-1}$ or a resolution from 1.7 nm to 0.35 nm). The SAXS and WAXS data are collected simultaneously such that the station provides a broad range of resolution, from about 90 nm to about 0.35 nm, without need to change the setup. The solution scattering measurements are further facilitated by a dedicated robotic sample changer (Round *et al.* 2008), providing temperature controlled storage for up to 192 samples that can be measured automatically. The experiments are controlled by a beamline metaserver (BMS) allowing attended, unattended or remote modes of operation (the first remote SAXS experiment was performed from Singapore at X33 in May 2009). The BMS system controls not only hardware but also software units, such that the primary data processing (including normalisation, radial averaging and buffer subtraction) and analysis steps (calculation of overall parameters, assessment of data quality, *ab initio* shape determination) are performed in the background without necessity for user intervention. The users are provided with a convenient log-file summarising the obtained results, which, being opened by a Web browser, also contains hyperlinks to the processed data and constructed models.

The description of the X33 beamline presents the optical parameters and operational characteristics as well as the hardware and software elements facilitating data collection and analysis. This reflects the requirements of modern biological users, who, for most of the projects, expect an SR beamline to provide maximum automation and efficient use of the beam time. The trend towards automation, which revolutionised biological MX (McPhillips *et al.* 2002), has also reached the SAXS field, and many modern SR-SAXS beamlines offer robotic data collection and data-processing pipelines (David and Perez 2009, Classen *et al.* 2010, Pernot *et al.* 2010) (see more about automation and high throughput in Section 9.1).

Insertion devices provide much better beam characteristics than bending magnets. One of the first SAXS instruments suitable for solution work is the ID2 station located at a high-ß section (i.e. an undulator section with a low beam divergency) of the ESRF storage ring

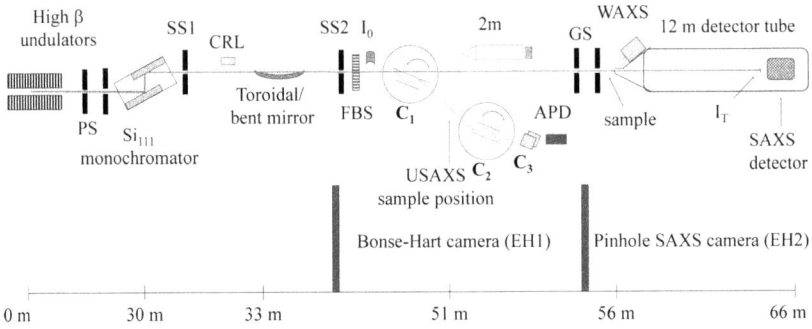

Fig. 2.15 Layout of the ID2 SAXS beamline (undulator, ESRF, Grenoble). PS: primary slits; SS: secondary slits; CRL: compound refractive lenses; FBS: fast beam shutter; APD: avalanche photodiode; GS: guard slits; I_T: transmitted intensity; EH: experimental hutch. (Courtesy T. Naranayan.)

(Grenoble, France), shown in Fig. 2.15 (Narayanan *et al.* 2001). The first optical element is a liquid-nitrogen-cooled Si-111 channel-cut monochromator, which can select wavelengths between $\lambda = 0.155$ nm and $\lambda = 0.073$ nm, with a band pass of $\Delta\lambda/\lambda = 2.10^{-4}$ at $\lambda = 0.1$ nm ($E = 12.4$ keV), and vertically displaces the beam by 30 mm, thus reducing the background radiation from the storage ring. The focusing in both directions is performed by an uncooled Rh-coated mirror with a fixed toroidal shape chosen to minimise the beam size at the end of the SAXS station (65 m). The standard size of the focused beam is $400 \times 200\,\mu\text{m}^2$ (FWHM, H $\times$ V) with a divergence of $40 \times 20\,\mu\text{rad}^2$. The beamline is optimised for an energy of 12.4 keV, yielding a maximum photon count of the monochromatic focused beam at the sample position of about 10^{14} photons s^{-1} for 100 mA of stored current in the ring.

ID2 has two separate experimental stations—one for combined SAXS/WAXS and the other for USAXS experiments using the Bonse–Hart geometry. This last instrument allows one to reach sub-micron particle sizes (down to $q_{\min}$ about 10^{-3} nm^{-1} (see Section 2.6.1). The SAXS detector in the SAXS/WAXS setup can be positioned as far as 10 m away from the sample such that the instruments can cover the range $0.01\,\text{nm}^{-1} \leq q \leq 40\,\text{nm}^{-1}$ at $\lambda = 0.1$ nm. ID2 is employed in a broad range of soft-matter, biological and nanoscience applications; for biological solution scattering studies, the SAXS/WAXS station is largely positioned as a time-resolved instrument.

The main SAXS detector is a fibre optics coupled FReLoN (Fast-Readout, Low-Noise) CCD based on a Kodak KAF-4320 image sensor, with an active area of $10 \times 10\,\text{cm}^2$, a dynamic range of 16 bits and a full frame rate of 3 frames s^{-1} (for 2048×2048 pixels). By binning 8×8 pixels, a rate up to 15 frames/s can be achieved, making the system suitable for millisecond-range time-resolved stroboscopic experiments. The WAXS detector—an AVIEX PCCD-4284—has an input area of $8.4 \times 4.2\,\text{cm}^2$, a dynamic range of 16 bits and a frame rate up to 6 frames/s for combined SAXS/WAXS experiments.

The use of optical elements with fixed geometry, like the toroidal mirror, at ID2, limits the possibilities for optimal focusing, and most recent SAXS instruments employ so-called Kirkpatrick–Baez (KB) mirrors. The original design consisted of two total-reflection mirrors focusing the beam in two orthogonal directions (Kirckpatrick and Baez 1948). These mirrors, fabricated, for example, from coated silicon substrates can be mechanically bent with high accuracy and flexibility into elliptical shapes, allowing one to move the focal point. Instead of mechanical bending, many modern beamlines employ bimorph systems, where the desired shape of the mirror is achieved by applying a voltage to sections of piezoelectric ceramics. In contrast to mechanical actuators, piezoelectric systems usually have more than a dozen electrodes, and tuning the mirror becomes itself a complicated inverse problem (Huang 2011). Still, automated focusing procedures for bimorph mirrors are available, and these mirrors, allowing rapid and flexible changes of the focusing conditions, are becoming increasingly popular.

The I22 beamline at Diamond (Oxford, UK) provides a typical example of a modern undulator beamline (Fig. 2.16). The beam is monochromated by a cryo-cooled fixed-exit double-crystal monochromator (DCM) and focused by a pair of bimorph KB mirrors, positioned approximately halfway between the source and the detector. Such a 1:1 geometry is characteristic for SAXS stations. In contrast to crystallographic beamlines, where the focusing elements are positioned closer to the sample to demagnify the beam, SAXS stations rather use 1:1 to ensure low beam divergence and thus better q- and energy resolution. I22 provides a $325 \times 110\,\mu m^2$ horizontal $\times$ vertical $(H \times V)$ beam delivering about 10^{13} photons/sec on the sample at a wavelength of 0.1 nm for the ring current of 300 mA. Initially, I22 employed gas-filled detectors (RAPID (2D) and HOTSAXS/HOTWAXS (1D)) allowing photon-counting with time resolutions down to $1\,\mu s$ (Berry *et al.* 2003), which were later replaced by a Pilatus 2M pixel detector. A stand-alone microfocusing station placed 5 m further away from the source than the standard sample (that is, at 53 m from the source) is also available and is attached to the main beamline conditioning optics by a transfer pipe. Compound beryllium refractive lenses (Snigirev *et al.* 1998) are used as secondary focusing optics yielding beam sizes on

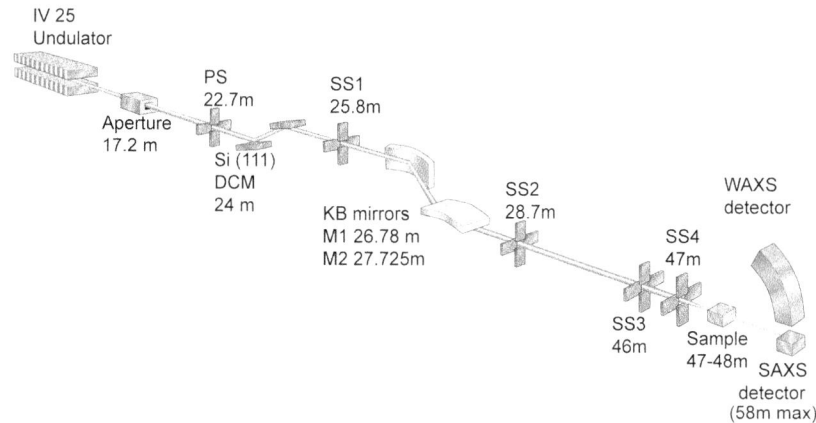

Fig. 2.16 Layout of the I22 SAXS undulator beamline (Diamond, Oxford). (Courtesy N. Terrill.)

the sample smaller than 10 μm. The microfocused beams are not used for typical-solution SAXS studies, but allow scanning experiments, for example, on tissues, single fibres or single particles trapped by optical tweezers.

Like ID2 at the ESRF, I22 at Diamond is a multi-purpose SAXS beamline serving the needs of biological and material science communities. Prompted by user requests, both facilities established, in addition, beamlines specifically dedicated to biological solution scattering. At the ESRF, beamline ID14-3 was converted from an MX to a SAXS beamline (Pernot *et al.* 2010), and later relocated to a bending magnet (BM29). At Diamond, construction of a dedicated high-throughput bending-magnet beamline HATSAXS started in 2011. A similar trend is observed at other facilities, thus, BioCAT, an undulator beamline at the Argonne National Laboratory (USA), initially intended to be essentially a fibre diffraction and absorption spectroscopy beamline (Fischetti *et al.* 2004), became most actively used for solution SAXS/WAXS applications. The wiggler beamline BL4-2 at the Stanford Synchrotron Radiation Laboratory (Stanford, USA), which takes advantage of SPEAR3, a third-generation storage ring is also optimised for biological solution SAXS (Smolsky *et al.* 2007). This beamline is equipped with a Si(111) DCM or a W/B_4C multilayer monochromator with 2% band pass providing higher flux and allowing time-resolved studies with millisecond time resolution. A Bonse–Hart USAXS device is also available on this instrument.

Another multi-purpose SAXS undulator beamline optimised and largely used for biological solution scattering is SWING at SOLEIL (Gif-sur-Yvette, France). The beamline offers a variable X-ray wavelength in the 5–16 keV energy range utilising a fixed-exit DCM (Si 111 at 20 m from the source) and a pair of fixed-incidence focusing KB mirrors at 22.5 m from the source. The beam size on the sample is $450 \times 40 \, \mu m^2 (H \times V)$ with a flux of 8×10^{12} photons s^{-1} at 7 keV and 8×10^{11} photons s^{-1} at 16 keV with 400 mA ring current. The sample is positioned at 30–32 m from the source, and the sample–detector distance can be varied between 0.5 and 8 m, such that the accessible angular range allows one to study macromolecules as large as 600 nm. SWING has different dedicated sample environments, which can be easily exchanged. Most importantly for solution scattering studies, the beamline offers a system for online purification of protein solutions using high-performance liquid chromatography (HPLC) (David and Perez 2009). The online purification ensures high sample monodispersity and also allows one to separate components in cases where a slow equilibrium between oligomers or protein complexes exists. The device, utilising a commercial modular apparatus (Agilent Technologies), can also be used as an automated sample dispenser bypassing the HPLC column. Although attempts at online purification were already reported earlier (Mathew *et al.* 2004), SWING provided the first system available for general user access. The advantages of online purification have been recognised rapidly by the users community, and such systems are now becoming standard on other instruments. The dedicated BioSAXS undulator beamline P12 of the EMBL at the PETRA-3 ring (DESY, Hamburg) not only provides a high-brightness beam and an automated data-collection and analysis system with a robotic sample changer similar to that described above for the X33 beamline. Similarly to the SWING beamline, P12 offers sample purification by size exclusion chromatography, which additionally is coupled with an online sample characterisation using a modular box (Malvern Instruments) to measure refractive index, static light scattering and UV absorption of the fractions.

Unique possibilities for time-resolved studies are provided by 'pink' beams, using the fundamental peak of undulator radiation (typically, with an energy peak-width of about 2%) as a quasi-monochromatic X-ray beam. This increases the flux by about two orders of magnitude compared to the DCM, and the higher harmonics of the undulator radiation can still be eliminated by the mirrors. Beamline BL40XU at SPring-8 (Himeji, Japan) is one of the first facilities utilising the 'pink beam' radiation for biological studies (Inoue *et al.* 2001). The X-ray source of BL40XU is a helical undulator, for which the energy of the fundamental radiation is concentrated in the core, while most of the higher harmonics are emitted off axis. So, by extracting the central part of the radiation cone, typically $15 \times 5\,\mu\mathrm{rad}^2$ (H $\times$ V), the fundamental can be used without loss of flux, while the higher harmonics are significantly reduced. By changing the undulator gap the energy of the fundamental radiation can be varied between 8 and 17 keV. The water-cooled horizontally and vertically focusing KB mirrors (silicon coated with Rh) have a glancing angle of 3 mrad for the first (horizontally focusing) mirror and 4 mrad for the second one to further suppress the remaining higher harmonics. The mirrors are located closer to the sample position to demagnify the beam size to about 1/4 of the source, yielding an X-ray focus size down to $250 \times 40\,\mu\mathrm{m}^2$ (H $\times$ V). At a ring current of 100 mA, the maximum flux achieved at 12 keV (1.8% band width) is about 10^{15} photons s^{-1} (that is, two orders of magnitude higher than that of a DCM-based instrument). The very intense beam is successfully employed for time-resolved studies on muscle, microbeam diffraction and the analysis of protein dynamics using X-ray photon correlation spectroscopy. Pilot experiments with such a beam, which burns paper, melts lead pieces and boils water, can help developing methodologies for fourth-generation sources (Iwamoto *et al.* 2005). Measurements on frozen-hydrated biological specimens performed at this beamline represent a good example of such developments. At the ESRF, the 'white beam' station ID09B is dedicated to the studies of ultrafast structural changes in condensed matter (Plech *et al.* 2002), including pump–probe experiments, where the reaction of the sample is triggered by a short laser pulse and the X-ray flash is the probe signal. This makes it possible to follow the kinetics of processes such as the allosteric transition in CO-haemoglobin in solution with nanosecond time resolution (Cammarata *et al.* 2008).

As already indicated, all major SR facilities offer SAXS beamlines, and the number of stations capable of reliably measuring biological solution scattering patterns is rapidly increasing. We have therefore not attempted to compile a full list of the available beamlines, but instead to describe the design of a few typical instruments illustrating the experimental possibilities of modern SR sources. The basic requirements, common to all biological SAXS stations, can perhaps be summarised by several important 'highs' and 'lows': high brightness, low background, high stability, low-beam divergence, high degree of automation, low-noise detector.

2.5 SANS instruments

The kind of instrument used for SANS depends on the type of neutron source on which it is located. Steady-state sources usually exploit a monochromatic beam with a bandwidth of 5–25% (FWHM). Pulsed sources exploit the white beam, which is analysed by TOF methods. It is possible to use TOF analysis on a steady-state source by pulsing the incident beam using choppers, and this technique is used on the D33 instrument at ILL.

2.5.1 SANS instruments on steady-state sources

Commonly, SANS cameras on reactors (and on the continuous spallation source SINQ) are pinhole cameras. As in a camera obscura, the image of a virtual neutron source in the form of the end of a neutron guide or an aperture with a typical size of several cm^2 is projected through an aperture (pinhole) at the sample position on a two-dimensional detector. SANS cameras are very large compared to conventional X-ray equipment, since due to the comparatively low flux—even at 'high-flux' reactors—large source and sample areas are necessary to collect enough scattered neutrons. Moreover, in order to separate the scattered neutrons from the direct beam at small angles, the primary and secondary flight paths as well as the detector dimensions must be large. The 'father' of all SANS instruments of this kind, D11 at the ILL, has dimensions of $2 \times 40\,m^2$ (collimation and neutron tank) (Ibel 1976). Until today, the instrument has undergone several modernisations, keeping it at the forefront of neutron instrumentation (Lindner and Schweins 2010).

A 'World Directory of SANS Instruments available for outside users', founded by Konrad Ibel, is now maintained by Ralf Schweins at the ILL (<http://www.ill.eu/instruments-support/instruments-groups/groups/lss/more/world-directory-of-sans-instruments/>).

A SANS instrument that covers a particularly large q-range, and is much used for biological studies, is D22 (<http://www.ill.eu/instruments-support/instruments-groups/instruments/d22>; May and Thomas 1986) at the ILL. A schematic drawing is shown in Fig. 2.17.

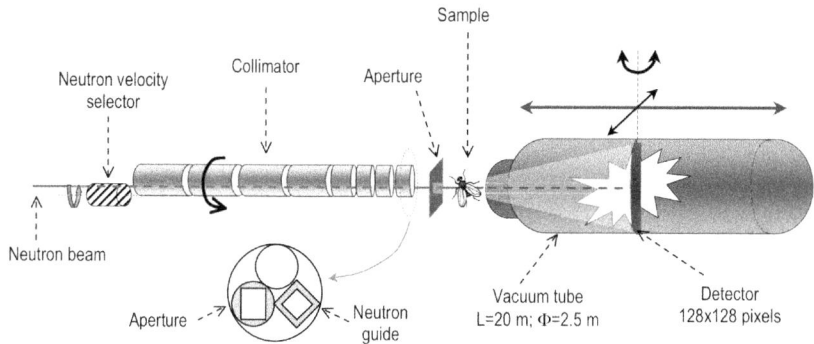

Fig. 2.17 Schematic view of the SANS camera D22 at the ILL. (Courtesy Isabelle Grillo, Institut Laue-Langevin; modified.)

D22 has an overall length of $2 \times 20\,\mathrm{m}^2$. Neutrons arriving through neutron guides from the horizontal cold source of the reactor traverse the Dornier velocity selector that selects a 10% wide band pass (FWHM) of wavelengths in the range of 0.45 to usually 2.5 nm. These neutrons enter a rotating-drum collimation system consisting of eight sections under vacuum. Each drum has three horizontal tubes, of which one contains a neutron guide, one is lined with boron-plastic and has an antiparasitic aperture at its end, and one is empty; only the drum directly after the selector (farthest away from the sample) contains a neutron polariser. The antiparasitic aperture consists of a sandwich of boron carbide facing the beam and a thin layer of lithium fluoride defining the beam. Every guide in place moves the virtual neutron source (the end of the respective neutron guide) closer to the sample. This provides eight 'collimation lengths' or rather free neutron flight paths between 1.4 and 17.6 m in front of the sample. The optimal sample size would be half the size of the neutron guide cross-section of $5.5 \times 4\,\mathrm{cm}^2$ in linear dimensions. In practice, biological sample cells are always smaller—typically 1–$2\,\mathrm{cm}^2$.

The sample containers—usually quartz cells—are placed in thermostated sample-changers and illuminated through a transparent window (such as sapphire) so that light can be used for sample adjustment. To reduce secondary γ radiation damage, the beam-defining aperture in front of the sample is a sandwich of B_4C and cadmium. The scattered beam enters the evacuated detector tank through a sapphire window located a few cm upstream from the cell window to limit air scattering.

The sample 'sees' the end of the neutron guide with a solid angle that decreases with the square of the collimation distance L_1. At the same time, the resolution improves as the size of the direct beam on the detector decreases. In order to match the image of the direct beam to the source dimension, the detector can be moved inside an evacuated tank of 20 m length and 2.5 m diameter. It can also be shifted laterally by 0.5 m and even rotated to reduce parallax effects in particular at short sample–detector distances, though this option is rarely used.

SANS detectors usually have sizes in the $0.5 \times 0.5\,\mathrm{m}^2$ to $1 \times 1\,\mathrm{m}^2$ range and resolutions of 0.5–1 cm. The detector of D22 (Van Esch *et al.* 2001) consists of 128 Reuter–Stokes 1D position-sensitive ^{3}He-filled stainless steel detector tubes of 0.8 mm diameter and 1 m active length. The position of every tube in the horizontal direction and the calculated vertical position along the tube result in a detector image with 16k pixels. Most other SANS detectors nowadays are multiwire ^{3}He chambers with one volume. The multi-tube detector has the advantage of increasing the maximum detector count rate by a factor of about 100 to several 10^6 s^{-1} (for 10% dead-time loss) at the expense of multiplying the necessary detector electronics.

The q-ranges that can be covered on steady-state SANS cameras reach from less than 10^{-2} to about 10 nm^{-1}. The smallest q-values require long detector distances and wavelengths, whereas large q-values can be reached with the shortest distances and wavelengths. In general, two or three instrument settings are required to cover the full necessary q-range, but using large detectors and an offset, one is sometimes sufficient for biological samples.

D33—a new SANS instrument at the ILL—bridges the gap towards cameras on spallation sources by providing a flexible time-of-flight option in parallel to the classical monochromatic mode on reactor-based sources (Dewhurst 2008). In TOF mode, the q_{max}/q_{min} ratio is extended up to >1000. Two detectors are employed simultaneously—a rear detector with 128×128 pixels for low q-values, and a front detector consisting of four panels of 32×128 pixels with a central opening letting the scattering pass to the rear detector. D33 is equipped with beam polarisation and ^{3}He spin analysis, making the instrument a first choice for studying magnetism. This will be facilitated by the distant position of the instrument so that strong magnets can be used.

2.5.2 SANS instruments on spallation sources

Cameras on the continuous spallation source (SINQ) are similar to those on reactor sources (Kohlbrecher and Wagner 2000), and the SANS II instrument is one that was simply transferred from the reactor source at Risø (Denmark) when the latter was closed. These instruments will thus not be discussed further here.

The design of instruments on short pulse spallation sources is more challenging than on steady-state sources because a number of opposing requirements must be taken into account among others in the design of the premoderator/moderator, especially where beams are shared by different instruments. The characteristics of the instruments given in Table 2.2 illustrate that, as a result, the instruments differ more than those on steady-state sources.

The schematic diagram of LOQ at ISIS in Fig. 2.18 illustrates a few of the possible solutions entering into the design of a successful instrument. Collimation on such a pinhole instrument is defined by the positions of two apertures, A_1 and A_2, and of the detector, which define the primary (L_1) and secondary (L_2) flight paths.

All instruments need a cold moderator to obtain the long wavelengths necessary to reach low q-values. Liquid-hydrogen moderators

Table 2.2 Characteristics of some TOF SANS instruments. LOQ, SANS2D and NIMROD are at ISIS (UK), and EQ SANS at the SNS (USA). NIMROD is not strictly a SANS instrument, but its parameters make it an ideal instrument for collecting medium–high q-data which are important in modelling.

Instrument	LOQ (Heenan *et al.* 1997)	SANS2D (Heenan 2006)	EQ SANS (Zhao *et al.* 2010)	NIMROD (Bowron *et al.* 2010)
Moderator	Decoupled 25 K liquid H_2	Grooved coupled solid CH_4 (26K) + liquid H_2 (18K)	coupled supercritical H_2	coupled solid CH_4 + liquid H_2 (17 K) + H_2O premoderator
Number of choppers	1	1	3	0
Incident wavelength (nm)	0.22–1.0 (25 Hz) 0.63–1.0 (50 Hz)	0.20–1.4 (10 Hz)	0.2–1.25 (30 Hz)	5×10^{-3}–1.0 (10 Hz)
q-range (nm^{-1})	6×10^{-2}–2.4, 1.5–14	2×10^{-2}–30.	4×10^{-2}–15	2×10^{-1}–500
Beam diameter at sample (mm)	2–20, typically 8	8–15	10–20	30×30 or smaller (square profile)
Typical time averaged flux (n cm^{-2} s^{-1})	$2\,10^5$ (50 Hz 200 µA)	$>10^6$ (10 Hz, 40 µA)	$\sim 10^7$–10^8	$>10^6$ (10 Hz, 40 µA)
Detector	^{3}He–CF$_4$ position sensitive proportional detector, 64×64 cm^2, resolution 5 mm	Two ^{3}He–CF$_4$ position-sensitive proportional detectors 100×100 cm^2, resolution 5 mm	~ 1 m^2 of ^{3}He tubes (Ø 8 mm, length 1041 mm)	^{6}LiF/ZnS(Ag) scintillation detectors

are usually preferred, though solid methane, which has a higher hydrogen density, is a more efficient moderator. Its radiochemistry results, however, in the formation of radicals which can both polymerise and lead to a sudden release of energy—a phenomenon known as burping, which also makes these moderators unusable on continuous or high-energy sources.

The main challenge for a TOF SANS instrument is to obtain a well-collimated and accurately timed beam with a very low background. To achieve this, the short-wavelength (<0.15–0.2 nm) and γ-ray components of the prompt pulse must be eliminated. Rather than a T_0 chopper or crystalline MgO filters (Thiyagarajan *et al.* 1997), one uses Soller super-mirror benders consisting of vertical parallel curved plates,

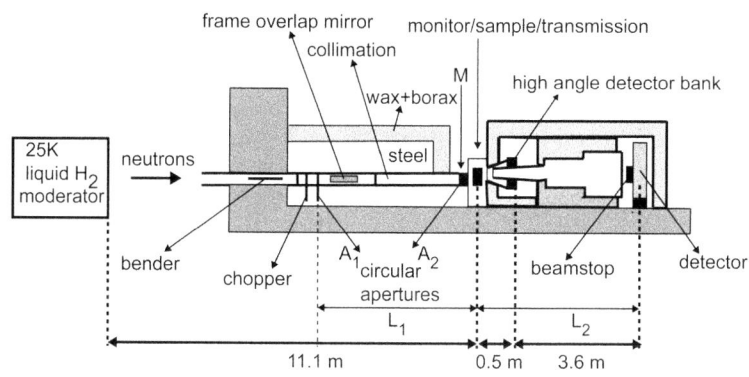

Fig. 2.18 Schematic representation of the LOQ instrument at ISIS. Adapted, with permission, from Heenan *et al.* (1997), © International Union of Crystallography.

which remove a direct line of sight between the moderator and the detector. The faster component of the delayed neutrons resulting from slow decay of long-lived nuclear states formed at the impact of the protons, which represent about 1% of the total, are also eliminated by the bender. These give a time-independent background with a wavelength distribution similar to that of the prompt pulse. A bender is, of course, not useful when the short wavelengths are also needed, such as on NIMROD on the ISIS second target station. Even with a bender, the prompt pulse results in short spikes in the neutron (flux versus τ) spectrum.

The longer-wavelength component of the delayed neutrons and the time-independent background due to neutrons which are slowly moderated by interactions with the shielding and collimation system are largely eliminated by a frame overlap mirror, which has a cut-off around 1.0–1.4 nm.

If L is the total flight path from the moderator to the detector, the time of flight (τ) is related to the wavelength by $\tau(\mathrm{ms}) = m_n \lambda L / h \approx 2.528\,\lambda(\mathrm{nm})L(\mathrm{m})$ or $\lambda \approx 0.4\tau/L$ or as the frequency of the proton pulses $f(\mathrm{Hz}) = 1/\tau(\mathrm{ms})$, $\lambda_{\max}(\mathrm{nm}) = 0.1h/(fm_nL)$. With a source operating at $f = 50\,\mathrm{Hz}\,(\tau = 20\,\mathrm{ms})$, assuming a 15 m flight path, the maximum usable wavelength would be $\lambda_{\max} = 8/15 = 0.53\,\mathrm{nm}$, which is much lower than on the cold source at a reactor. This implies that the total flight path should be as short as possible and/or the frequency should be lowered. A compromise must thus be found between data rate and q-range. There are two mechanisms to lower the frequency. One is to eliminate pulses with a single chopper placed after the bender as done on LOQ. The effective frequency can be lowered further by skipping frames using multiple phased choppers, as done on EQ SANS at SNS.

Absorption and Bragg reflections from aluminium windows or other material in the flight path further modify the spectrum and introduce glitches at different wavelengths. Together with limitations on the usable wavelength range described above, this leads to incident neutron spectra that are quite different and not as smooth as those in Fig. 2.12, as illustrated in Fig. 2.19.

The resolution of SANS instruments is determined essentially by geometrical factors, and with pinhole cameras the optimal count rates are obtained with $L_1 = L_2$ and $A_1 = 2A_2$, and $L_1 + L_2$ (Fig. 2.16) is as long as possible (May 1994). Besides the two defining apertures a number of additional apertures or vanes are required to eliminate unwanted background due to reflection and/or scattering along the primary and secondary flight path and any sample-dependent background. This is even more important on instruments using the short-wavelength neutrons such as NIMROD. Some instruments have a variable flight path and hence a variable q-resolution. On SANS2D at ISIS, for example, the position of the effective source is changed by inserting guide tubes between the bender and the sample and the detector is moved to a new position, whereas on EQ SANS only the secondary flight path is variable, whereas LOQ and NIMROD have a fixed flight path.

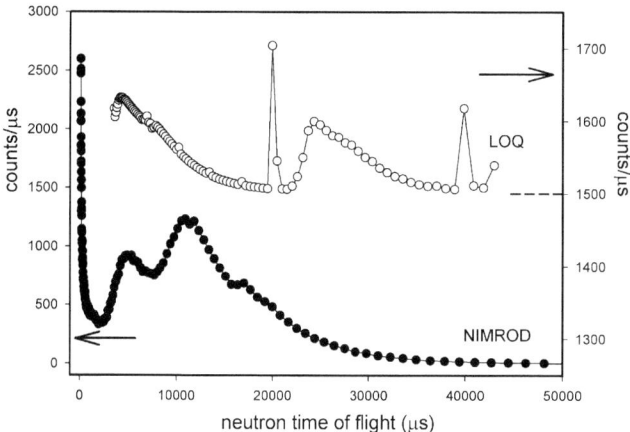

Fig. 2.19 Monitor spectrum of the incident beam on LOQ at ISIS covering two frames and on NIMROD. The spikes at 20 and 40 ms are due to the prompt neutron pulse. Other small discontinuities are due to Bragg reflection from the aluminium windows. This should be compared with the spectrum of the moderator in Fig. 2.12.

The detectors (position-sensitive proportional counters or arrays of linear gas tubes) are similar to those used on continuous sources, but the data acquisition system is more complex, as each detected neutron must be tagged with its corresponding time of flight. In contrast, the detection system of NIMROD, consisting of hundreds of scintillation detectors with photomultipliers, is unique to TOF instruments.

SANS instruments on pulsed sources do not reach as low q-values as instruments on reactor sources (5×10^{-3}–10^{-2} nm^{-1}). These q-values could be reached by further lowering the effective frequency of the source and lengthening the beam line, but this would result in low data rates. The main advantage of the TOF instruments lies in the fact that a large q-range can be obtained with a good resolution in a single measurement. This is particularly useful for systems like polyelectrolytes (for example, nucleic acids), which may still present distinct features at higher q-values.

The fact that there is no design that fits all the needs also suggests that there is potential for further developments at SANS TOF instruments.

2.6 Special instruments

Some SANS instruments that are not frequently used in biology, but may become interesting for some special applications, are described below. This holds for cameras that reach very low $q(<10^{-3}$ nm^{-1}), that is, double-crystal, reflecting-lenses and focusing-mirror cameras and Spin-Echo SANS that works in real space and resolves large structures up to 0.1 mm.

2.6.1 USAXS/USANS, Bonse–Hart double-crystal cameras

Ultra-small angle X-ray and neutron cameras extend the q-range into dimensions that are usually studied with visible light or electron

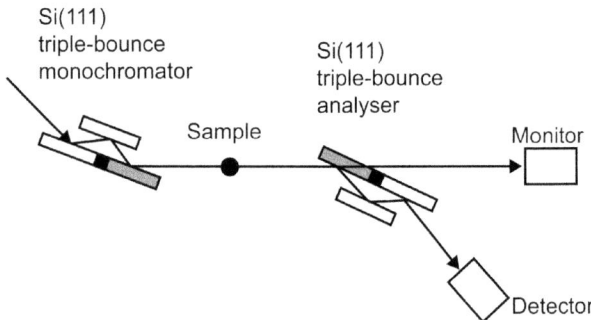

Fig. 2.20 Neutron Bonse–Hart setup. A cadmium insert in the long walls of the channel-cut monochromator and analyser crystals absorbs neutrons reflected within the white or grey part of the receiving crystal. This reduces the parasitic scattering of neutrons that would otherwise arrive on the detector. Adapted, with permission, from Agamalian and Koizumi (2011), © John Wiley and Sons, 2011.

microscopy. They are of limited use for biology. Instruments at SR sources and at reactors make use of the Bonse–Hart setup (Bonse and Hart 1965). The original device was designed for X-rays, and there are cameras of this type at several facilities—for example, at the ESRF, Grenoble, France (Sztucki *et al.* 2008), and at the Advanced Photon Source (APS) at Argonne, USA (Ilavsky *et al.* 2009).

As shown in Fig. 2.20, a sample is placed between two triple-bounce channel-cut monochromators. The first one selects a narrow wavelength band and the second scans the scattered intensity at very small angles. The resolution which can be reached depends on the intrinsic Darwin width of the diffraction by an ideal crystal, but there are also other contributions, due, for example, to the quality of the monochromator surfaces. Typically, a q-range of 10^{-3}–$1.0\,\mathrm{nm}^{-1}$ can be covered.

Neutron Bonse–Hart (Ultra-SANS) instruments ($10^{-4} < q < 10^{-1}$ nm^{-1}) exist at the ILL, Grenoble, France (Hainbuchner *et al.* 2000), at NIST NCNR (USA) (Barker *et al.* 2005) and elsewhere, but suffer from very low count rates because of the need to scan small angular steps with a small-footprint monochromatic beam. An important step in their development was the finding that one had to suppress reflections from the back surface and internal dynamical diffraction effects of the monochromators, because silicon is much more transparent for neutrons than for X-rays (Agamalian *et al.* 1997). This was first achieved by placing a Cd plate in a groove cut in the middle of the double-bounce face of the crystals.

The SNS (USA) is proposing to build a TOF-USANS instrument. This would allow different orders of Bragg reflection to be separated in time, optimising the neutron flux and resolution (<http://neutrons.ornl.gov/usans/>).

2.6.2 Focusing-mirror instruments

Whereas Bonse–Hart-type cameras work in one dimension, with the additional disadvantage of slit smearing, high-quality (and expensive) curved mirrors can be used to focus neutrons that are diverging in two dimensions from a small source through a sample onto a two-dimensional detector with a direct-beam spot size close to the source size (Alefeld *et al.* 1989).

This principle has been realised in a unique instrument (KWS-3) run by the Jülich Centre for Neutron Science (JCNS) outstation at the FRMII reactor in Garching, Germany (<http://www.fz-juelich.de/jcns/EN/Leistungen/Instruments2/Structures/KWS3/_node.html>). KWS-3 has a toroidal mirror 11 m distant from an aperture of $2 \times 2\,\mathrm{mm}^2$ that serves as source. A q-range $1.6 \times 10^{-3} < q < 3.5 \times 10^{-2}\,\mathrm{nm}^{-1}$ can be covered with the detector set at 10 m from the sample. With a second sample chamber at 1 m from the detector, the q-range can be extended to $3.5 \times 10^{-1}\,\mathrm{nm}^{-1}$. Large samples can be used with such a camera, because the resolution does not depend on the sample area.

2.6.3 Refracting lenses and other techniques

Arrangements of concave neutron-refracting lenses (such as MgF_2 crystals) are used at SINQ and NIST to focus long-wavelength (>1.5 nm) neutrons on the detector and thus improve the flux while simultaneously reducing $q_{\min}$ by a factor of two.

The VSANS instrument V16 at the Helmholtz Zentrum, Berlin, Germany, uses gravitational velocity selection through multi-pinhole grids or TOF, together with a high-resolution (2 mm) detector to provide a q-range from $10^{-3}\,\mathrm{nm}^{-1}$ to $8.5\,\mathrm{nm}^{-1}$.

A technique that uses multiple small-angle neutron scattering has been described (Grünzweig *et al.* 2007). Grids placed at the front-end of the collimator and close to the sample produce many overlapping USANS patterns with a resolution of $3 \times 10^{-3}\,\mathrm{nm}^{-1}$, which may be extended to $10^{-4}\,\mathrm{nm}^{-1}$.

2.6.4 Spin Echo SANS (SESANS)

This innovation (Rekveldt 1996), which builds on neutron spin echo techniques initially developed for spectroscopy (Mezei *et al.* 2003), uses the precession of polarised neutrons passing through magnetic fields to encode their (small) scattering angles. The design of SESANS instruments is still evolving, so that in contrast with the conventional SANS devices above 'standard' instruments do not yet exist.

All SESANS instruments are, however, based on a division of the space along the trajectory of the neutrons in two regions with exactly opposite magnetic properties. Further, one should remember that the total precession angle of the spin in a plane perpendicular to a magnetic field is proportional to the time that the neutron spends in the field. Hence, slower neutrons precess more than fast ones leading to a distribution of precession angles.

In the SESANS instrument shown in Fig. 2.21 the polarisation of the neutrons is manipulated by homogeneous magnetic fields (B_{eff}) inducing precession and magnetised foils which flip the polarisation (Rekveldt 1996). The neutrons are initially polarised in the z-direction

Fig. 2.21 Schematic representation of a side view in the x,z plane of a SESANS instrument (based on Krouglov *et al.* 2003b, Rekveldt *et al.* 2005). Magnets produce a homogeneous field B. The inclined bold arrows represent the local magnetic field in the plane of the π-flip foils placed between the magnet poles. Coils around the foils give a field (B_x) in the x direction so that the effective homogeneous field ($B_{\text{eff}} = B + B_x$) is nearly normal to the plane of the foil separating the space in each magnet in two equal regions—one with positive precession (dark grey) and the other with reverse precession (light grey). P: polarisation of the direct beam; S: sample; w: width of the homogeneous magnetic field in each device. The clock diagrams indicate the precession of the neutrons in the precession plane perpendicular to B_{eff} at different points along the trajectory of the beam. A field stepper placed between device II and III (not shown) guarantees a sharp inversion without depolarisation between the regions with opposite fields. The dashed arrow along x represents the direct neutron beam, and the full line the beam scattered over a small angle (2θ) by the sample. At the exit of IV a spin-echo condition is realised for all neutrons that have not been scattered. The polarisation is measured in the analyser. For details, see the text.

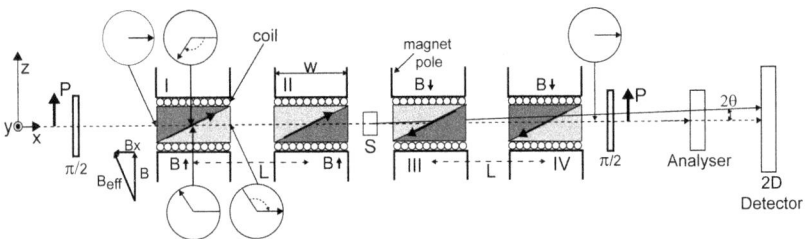

by a super-mirror, and after passing through an adiabatic $\pi/2$-flipper their polarisation is in the x,y plane (Kraan *et al.* 2003). They then enter the homogeneous magnetic field (B_{eff}) which is nearly perpendicular to the plane of a π-flip foil placed at the centre of the homogeneous field and inclined at a small angle $\theta_0 (\sim 5°)$ relative to the beam. The polarisation is flipped around the magnetic field in the plane of the foil, whereby the direction of precession changes and the distribution of precession angles as function of velocity is reversed. The foil thus divides the space between the magnet poles in triangular zones with opposite precession directions. The total precession at the exit of the magnet is proportional to the difference in time spent in the regions with positive and negative precession; that is, the difference in path length divided by the velocity of the neutrons, which is proportional to the inverse of their wavelength (λ). If the total width of the magnet (for example, magnet I in Fig. 2.21) is w, on axis the neutrons travel $w/2$ in each region and their total precession is zero (that is, the spin echo condition is realised). At a height h above the central axis they travel $(w/2 + h\cot\theta_0)$ in one region and $(w/2 - h\cot\theta_0)$ in the other. This difference in path length results in a precession $\varphi_1 = 2\gamma_L B_{\text{eff}}\lambda h\cot\theta_0$, where γ_L is the Larmor precession constant ($\gamma_L = 4.632 \times 10^{14}$ Tesla^{-1} m^{-2}). The precession angles thus linearly encode the height at which the neutrons have traversed the magnet.

When two such devices are installed in series with a distance L between their centres (such as I and II in Fig. 2.21) their effects cancel out when the neutron path is parallel to the axis joining the centres. If the incident neutron beam makes a (small) angle θ_i with this axis there is a difference in height at the entry points in I and II of $L\theta_i$, resulting in a precession $\varphi_{12} = 2\gamma_L B_{\text{eff}}\lambda L\theta_i\cot\theta_0$. This provides a way of aligning successive pairs of devices coaxially with the neutron beam.

If a sample is placed between two pairs of aligned devices (I, II and III, IV) at the exit of IV there will be a precession (φ) depending on the (small) scattering angle, the spin echo condition being realised for neutrons that are not scattered: $\varphi = 2\gamma_L\lambda B_{\text{eff}}L\theta\cot\theta_0 = Zq_z = Z2\pi\theta/\lambda$, where $Z = \gamma_L\lambda^2 B_{\text{eff}}L\theta\cot\theta_0/\pi$ is the so-called spin echo length. This quantity can be varied by changing the magnetic field B_{eff} and also the wavelength when monochromatic radiation is used, the precession length L or the inclination of the foils (θ_0). In front of the analyser a $\pi/2$-flip foil brings the polarisation parallel to that of the analyser. The polarisation is measured using the strong polarisation dependence of the neutron absorption cross-section of ^{3}He.

SESANS measurements yield the normalised polarisation $(P(Z)/P_0)$, where P_0 is the polarisation in absence of sample and Z is the real space coordinate along the polarisation direction of the analyser. For a sample with an average number of scattering events σ per unit length along its thickness D the normalised polarisation is in the general case related to the SESANS correlation function $G(Z)$ by (Krouglov *et al.* 2003b):

$$\frac{P(Z)}{P_0} = \exp(\sigma D(G(Z) - 1)) \tag{2.10}$$

The normalised polarisation is measured as a function of the magnetic field B_{eff}. This directly yields $G(Z)$, a real space function, in contrast with the conventional SANS intensity $I(q)$ in reciprocal space, which must be Fourier transformed to obtain the usual Debye correlation function or autocorrelation function $\gamma(r)$ (see Chapter 3). $G(Z)$ and $\gamma(r)$ are related by an Abel transform reflecting the fact that $G(Z)$ is a projection of the spherically symmetric autocorrelation function $\gamma(r)$ along x (Krouglov *et al.* 2003b). With $r = (x^2 + Z^2)^{1/2}$ this yields:

$$G(Z) = \int_z^\infty \frac{\gamma(r)r}{\sqrt{r^2 - Z^2}} dr \tag{2.11}$$

The range of spin echo length that can be covered ($5\,\text{nm} \leq Z \leq 20\,\mu\text{m}$) limits the method to the study of large objects. The smallest dimension that can be measured is defined by the accepted divergence, which is itself determined by apertures at the entrance and exit of the precession regions (Rekveldt *et al.* 2005). Unlike the situation in SANS, SESANS thus does not require a collimated beam. The larger divergence increases the usable intensity by orders of magnitude. This is definitely an advantage compared to USANS, and should make kinetic measurements with a time resolution of seconds possible. As the signal is superimposed on the direct beam, SESANS is applicable only when 5–10% of the direct beam is scattered. Moreover, since no distinction is made between different types of scattering the incoherent scattering of the sample should be minimised (for example, by working in D_2O in the case of aqueous systems).

The method is thus particularly useful for concentrated systems of large particles and colloidal systems, which often occur in pharmaceutical technology and food processing and are difficult to study by other methods. Another advantage of SESANS in this respect is that it is less affected by multiple scattering, which can be quite strong at small q-values in SANS from concentrated systems.

As in the case of SANS, contrast variation methods can be used to obtain additional structural information for solutes with a non-uniform scattering density (Li *et al.* 2012).

A recent proposal combining SESANS and SANS in a single instrument to cover spatial resolutions from 1 nm to 0.1 mm suggests that with further progress in instrumentation the method still has considerable potential for development (Bouwman *et al.* 2011).

The demonstration instrument at the low-power reactor of the Delft University of Technology (Reactor Institute Delft, Netherlands) already routinely measures across three orders of magnitude in length scale up to $20\,\mu\text{m}$ ($q \sim 3 \times 10^{-4}\,\text{nm}^{-1}$). SESANS can be used on both continuous and pulsed sources, though this requires different instruments (Bouwman *et al.* 2004) and TOF-SESANS instruments are being designed at ISIS (UK) (<http://www.isis.stfc.ac.uk/instruments/Larmor/>) and SNS (USA) (Zhao *et al.* 2010). The method has hitherto been applied to isotropic samples, but it could also be extended to two dimensions for anisotropic samples (Krouglov *et al.* 2003a).

References

Agamalian, M., and Koizumi, S. (2011). 'Ultra-small-angle neutron scattering', in Imae, T., Kanaya, T., Furusaka, M. and Torikai, N. (eds.) *Neutrons in Soft Matter*. Hoboken: John Wiley & Sons.

Agamalian, M., Wignall, G. D. and Triolo, R. (1997). 'Optimisation of a Bonse–Hart ultra-small angle neutron scattering facility by eliminating the rocking curve wings', *J. Appl. Crystallogr.* 30: 345-352.

Alefeld, B., Schwahn, D., and Springer, T. (1989). 'New developments of small-angle neutron scattering instruments with focusing', *Nucl. Instrum. Meth. A* 274: 210–6.

Barker, J. G., Glinka, C. J., Moyer, J. J., Kim, M.-H., Drews, A. R. and Agamalian, M. (2005). 'Design and performance of a thermal-neutron double-crystal diffractometer for USANS at NIST', *J. Appl. Crystallogr.* 38: 1004–11.

Bergmann, A., Orthaber, D., Scherf, G. and Glatter, O. (2000). 'Improvement of SAXS measurements on Kratky slit systems by Göbel mirrors and imaging-plate detectors', *J. Appl. Crystallogr.* 33: 869–75.

Berry, A., Helsby, W. I., Parker, B. T. *et al.* (2003). 'The RAPID2 X-ray detection system', *Nucl. Instrum. Meth. A* 513: 260–3.

Bonse, U., and Hart, M. (1965). 'Tailless X-ray single crystal reflection curves obtained by multiple reflection', *Appl. Phys. Lett.* 7: 238–40.

Bouwman, W. G., Duif, C. P., Plomp, J., Wiedenmann, A. and Gähler, R. (2011). 'Combined SANS–SESANS from 1 nm to 0.1 mm in one instrument', *Physica B* 406: 2357–60.

Bouwman, W. G., Stam, W., Krouglov, T. V. *et al.* (2004). 'SESANS with a monochromatic beam or with time-of-flight applied to colloidal systems', *Nucl. Instr. Meth. A.* 529: 16–21.

Bowron, D. T., Soper, A. K., Jones, K. *et al.* (2010). 'NIMROD: the Near and InterMediate Range Order Diffractometer of the ISIS second target station', *Rev. Sci. Instrum.* 81: 033905.

Broennimann, C., Eikenberry, E. F. and Henrich, B. (2006). 'The PILATUS 1M detector', *J. Synchrotron Radiat.* 13: 120–30.

Cammarata, M., Levantino, M., Schotte, F. *et al.* (2008). 'Tracking the structural dynamics of proteins in solution using time-resolved wide-angle X-ray scattering', *Nat. Methods* 5: 881–6.

Classen, S., Rodic, I., Holton, J., Hura, G. L., Hammel, M. and Tainer, J. A. (2010). 'Software for the high-throughput collection of SAXS data using an enhanced Blu-Ice/DCS control system', *J. Synchrotron Radiat.* 17: 774–81.

David, G., and Perez, J. (2009). 'Combined sampler robot and high-performance liquid chromatography: a fully automated system for biological small-angle X-ray scattering experiments at the Synchrotron SOLEIL SWING beamline', *J. Appl. Crystallogr.* 42: 892–900.

Dewhurst, C. D. (2008). 'D33: a third small-angle neutron scattering instrument at the Institut Laue Langevin', *Meas. Sci. Technol.* 19: 034007.

Fischetti, R., Stepanov, S., Rosenbaum, G. *et al.* (2004). 'The BioCAT undulator beamline 18ID: a facility for biological non-crystalline diffraction and X-ray absorption spectroscopy at the Advanced Photon Source', *J. Synchrotron Radiat.* 11: 399–405.

Forgan, E. M. and Cubitt, R. (1998). 'Cancellation of gravity for slow neutrons in small angle scattering experiments', *Neutron News* 9: 25–31.

Grünzweig, C., Hils, T., Mühlbauer, S. *et al.* (2007). 'Multiple small angle neutron scattering: A new two-dimensional ultrasmall angle neutron scattering technique', *Appl. Phys. Lett.* 91: 203504.

Hainbuchner, M., Villa, M., Kroupa, G. *et al.* (2000). 'The new high resolution ultra small-angle neutron scattering instrument at the high flux reactor in Grenoble.' *J. Appl. Crystallogr.* 33: 851–4.

Heenan, R. K., King, S. M., Turner, D. S. and Treadgold, J. R. (2006). 'SANS2D at the ISIS second target station,' *Proc ICANS-XVII*, 780–5.

Heenan, R. K., Penfold, J. and King, S. M. (1997). 'SANS at pulsed neutron sources: Present and future prospects', *J. Appl. Crystallogr.* 30: 1140–7.

Holmes, K. C. (1982). 'Instrumentation. Point collimation and synchrotron radiation', in Glatter, O. and Kratky, O. (ed.), *Small Angle X-Ray Scattering*. London: Academic Press.

Huang, R. (2011). 'The inverse problem of bimorph mirror tuning on a beamline', *J. Synchrotron Radiat.* 18: 930–7.

Ibel, K. (1976). 'The neutron small-angle camera D11 at the high-flux reactor, Grenoble', *J. Appl. Crystallogr.* 9: 296–309.

Ilavsky, J., Jemian, P. R., Allen, A. J., Zhang, F., Levine, L. E. and Long, G. G. (2009). 'Ultra-small-angle X-ray scattering at the Advanced Photon Source', *J. Appl. Crystallogr.* 42: 469–79.

Inoue, K., Oka, T., Suzuki, T. *et al.* (2001). 'Present status of high flux beamline (BL40XU) at SPring-8', *Nucl. Instrum. Meth. A* N467-468: 674–7.

Iwamoto, H., Inoue, K., Fujisawa, T. and Yagi, N. (2005). 'X-ray microdiffraction and conventional diffraction from frozen-hydrated biological specimens', *J. Synchrotron Radiat.* 12: 479–83.

Kirckpatrick, P., and Baez, A. V. (1948). 'Formation of optical images by X-rays', *J. Opt. Soc. Am.* 38: 766–74.

Koch, E. E., Eastman, D. E. and Farge, Y. (1983). 'Synchrotron Radiation: a powerful tool in science', in Eastman, D. E. and Farge, Y. (eds.) *Handbook on Synchrotron Radiation*. Amsterdam: North Holland Publishing.

Koch, M. H. J. and Bordas, J. (1983). 'X-ray diffraction and scattering on disordered systems using synchrotron radiation', *Nucl. Instrum. Meth. A* 208: 461–9.

Kohlbrecher, J., and Wagner, W. (2000). 'The new SANS instrument at the Swiss spallation source SINQ', *J. Appl. Crystallogr.* 33: 804–6.

Kraan, W. H., Grigoriev, S. V., Rekveldt, M. T., Fredrikze, H., De Vroege, C. F. and Plomp, J. (2003). 'Test of adiabatic spin flippers for application at pulsed neutron sources', *Nucl. Instrum. Meth. A* 510: 334–45.

Kraft, P., Bergamaschi, A., Broennimann, C. *et al.* (2009). 'Performance of single-photon-counting PILATUS detector modules', *J. Synchrotron Radiat.* 16: 368–75.

Kratky, O. (1982). 'Instrumentation: Experimental technique, slit collimation', in Glatter, O. and Kratky, O. (eds.), *Small Angle X-Ray Scattering*. London: Academic Press.

Krouglov, T. V., De Schepper, I. M., Bouwman, W. G. and Rekveldt, M. T. (2003a). 'Real-space interpretation of spin-echo small-angle neutron scattering', *J. Appl. Crystallogr.* 36: 117–24.

Krouglov, T. V., Kraan, W. H., Plomp, J., Rekveldt, M. T. and Bouwman, W. G. (2003b). 'Spin-echo small-angle neutron scattering to study particle aggregates', *J. Appl. Crystallogr.* 36: 816–9.

Li, X., Wu, B., Liu, Y. *et al.* (2012). 'Contrast variation in spin-echo small angle neutron scattering', *J. Phys.-Condens. Matter* 24: 064115.

Li, Y., Beck, R., Huang, T., Choi, M. C. and Divinagracia, M. (2008). 'Scatterless hybrid metal-single-crystal slit for small-angle X-ray scattering and high-resolution X-ray diffraction', *J. Appl. Crystallogr.* 41: 1134–9.

Lindner, P. and Schweins, R. (2010). 'The D11 small-angle scattering instrument: A new benchmark for SANS', *Neutron News* 21: 15–18.

Loewen, R. J. (2003). 'A compact light source: Design and technical feasibility study of a laser-electron storage ring X-ray source', SLAC Report 632, Stanford University, USA.

Margaritondo, G. and Ribic, P. R. (2011). 'A simplified description of X-ray free-electron lasers', *J. Synchrotron Radiat.* 18: 101–8.

Mathew, E., Mirza, A. and Menhart, N. (2004). 'Liquid-chromatography-coupled SAXS for accurate sizing of aggregating proteins', *J. Synchrotron Radiat.* 11: 314–8.

May, R. P. (1994). 'Geometrical optimisation of neutron small-angle scattering instruments', *J. Appl. Crystallogr.* 27: 298–301.

May, R. P., and Thomas, M. (1986). 'A new low-*q* scattering instrument at the 2nd cold source of the ILL', ILL Technical report 86MA07T. Grenoble, France: Institut Laue Langevin.

McPhillips, T. M., McPhillips, S. E., Chiu, H. J. *et al.* (2002). 'Blu-Ice and the distributed control system: software for data acquisition and instrument control at macromolecular crystallography beamlines', *J. Synchrotron Radiat.* 9: 401–6.

Mezei, F., Pappas, C. and Gutberlet, T. (eds.) (2003). *Neutron Spin Echo Spectroscopy: Basics, Trends and Applications*. Berlin: Springer.

Narayanan, T., Diat, O. and Boesecke, P. (2001). 'SAXS and USAXS on the high brilliance beamline at the ESRF', *Nucl. Instrum. Meth. A.* 346: 1005–9.

Otendal, M., Tuohimaa, T., Vogt, U. and Hertz, H. M. (2008). 'A 9 keV electron-impact liquid-gallium-jet X-ray source', *Rev. Sci. Instrum.* 79: 016102.

Pedersen, J. S. (2004). 'A flux- and background-optimised version of the NanoSTAR small-angle X-ray camera for solution scattering', *J. Appl. Crystallogr.* 37: 369–80.

Pernot, P., Theveneau, P., Giraud, T. *et al.* (2010). 'New beamline dedicated to solution scattering from biological macromolecules at the ESRF', *J. Phys. Conf. Series* 247: 012009.

Plech, A., Randler, R., Geis, A. and Wulff, M. (2002). 'Diffuse scattering from liquid solutions with white-beam undulator radiation for photoexcitation studies', *J. Synchrotron Radiat.* 9: 287–92.

Portale, G., Longo, A., D'ilario, L., Marinelli, A., Caminiti, R. and Rossi Albertini, V. (2007). 'Small-angle energy dispersive X-ray scattering using a laboratory-based diffractometer with a conventional source', *J. Appl. Crystallogr.* 40: 218–31.

Rekveldt, M. T. (1996). 'Novel SANS instrument using neutron spin echo', *Nucl. Instrum. Meth. B* 114: 366–70.

Rekveldt, M. T., and Plomp, J., Bouwman, W. G., Kraan, W. H., Grigoriev, S. V. and Blaauw, M. (2005). 'Spin-echo small angle neutron scattering in Delft', *Rev. Sci. Instrum.* 76: 033901–9.

Roessle, M. W., Klaering, R., Ristau, U. *et al.* (2007). 'Upgrade of the small-angle X-ray scattering beamline X33 at the European Molecular Biology Laboratory, Hamburg', *J. Appl. Crystallogr.* 40: s190–4.

Round, A. R., Franke, D., Moritz, S. *et al.* (2008). 'Automated sample-changing robot for solution scattering experiments at the EMBL Hamburg SAXS station X33', *J. Appl. Crystallogr.* 41: 913–7.

Smolsky, I. L., Liu, P., Niebuhr, M., Ito, K., Weiss, T. M. and Tsuruta, H. (2007). 'Biological small-angle X-ray scattering facility at the Stanford Synchrotron Radiation Laboratory', *J. Appl. Crystallogr.* 40: s453–8.

Snigirev, A., Kohn, V., Snigireva, I., Souvorov, A. and Lengeler, B. (1998). 'Focusing high-energy X-rays by compound refractive lenses', *Appl. Opt.* 37: 653–62.

Sztucki, M., Gorini, J., Vassalli, J.-P., Goirand, L., Van Vaerenbergh, P. and Narayanan, T. (2008). 'Optimisation of a Bonse–Hart instrument by suppressing surface parasitic scattering', *J. Synchrotron Radiat.* 15: 341–9.

Thiyagarajan, P., Epperson, J. E., Crawford, R. K., Carpenter, J. M., Klippert, T. E. and Wozniak, D. G. (1997). 'The time-of-flight small-angle neutron diffractometer (SAD) at IPNS, Argonne National Laboratory', *J. Appl. Crystallogr.* 30: 280–93.

Van Esch, P., Guérard, B., May, R. P., Sicard, A., Buffet, J. C. and Millier, F. (2001). 'High count rate detectors for small-angle neutron scattering', ILL Millennium Symposium. Grenoble, France: Institut Laue-Langevin.

Wagner, W. (2002). Target operation at the high-power neutron spallation source SINQ: safety and reliability issues. <http://cdsweb.cern.ch/record/688517/files/p15.pdf>

Zhao, J. K., Gao, C. Y. and Liu, D. (2010). 'The extended Q-range small-angle neutron scattering diffractometer at the SNS', *J. Appl. Crystallogr.* 43: 1068–77.

Experimental practice and data processing

3

3.1 Sample requirements for SAXS and SANS   65

3.2 Experimental protocols and instrumental corrections   68

3.3 Special features of SAXS: radiation damage   72

3.4 Special features of SANS: the use of H_2O/D_2O mixtures   75

3.5 Random and systematic errors   79

3.6 Basic structural information and data quality   81

3.7 Calibration to absolute scale and molecular mass   85

References   88

In this chapter we discuss the details of SAS measurements as well as the procedures for data reduction up to the point where data are processed to obtain structural information, to be described in the following chapter.

3.1 Sample requirements for SAXS and SANS

Most SAXS and SANS experiments—particularly those aimed at obtaining detailed structural information—require pure, monodisperse solutions. Exceptions are when solutions are inherently polydisperse or when concentrated solutions are being investigated to understand intermolecular interactions. Such systems are described in Chapter 5. For both SAXS and SANS, sample requirements are similar.

3.1.1 Monodispersity

Monodispersity is an essential prerequisite for structural modelling based on SAS data. As the scattering from particles is related to the square of their excess scattering length, aggregates contribute disproportionately to the signal and in most cases render the data uninterpretable. In practice, samples should be at least 95% monodisperse in order to obtain useful structural information. Proteins should, for instance, migrate as a single species on a native gel. The best way to check for monodispersity (absence of aggregates) is through multiangle light scattering (MALLS), but dynamic light scattering (DLS), gel filtration (size exclusion chromatography) or analytical ultracentrifugation (AUC) can also be used. Aggregation manifests itself in SAS as an increase in the scattering at the lowest q-values.

3.1.2 Concentrations

Concentrations are usually in the range 1–10 mg/mL. SAS theory (see Section 1.4) requires that solutions should be infinitely dilute to avoid interparticle interactions which perturb the scattering in characteristic ways. In order to obtain data from an infinitely dilute solution the best way is to measure a concentration series and to extrapolate the results to zero concentration. The ease with which this can be done may depend on the amount of material available as well as the X-ray or neutron beam-time available. However, for a protein of 50 kDa molecular mass a typical concentration series might consist of measurements at 10, 5, 2 and 1 mg/mL. If data are to be placed on an absolute scale, and for molecular mass calculations, it is essential to know the exact macromolecule concentration. For proteins the best way is via the absorbance at 280 nm (A_{280}) so long as the extinction coefficient is known accurately (by quantitative amino acid analysis, for example). To make a good A_{280} measurement it is important to have a good buffer measurement, particularly when using dithiothreitol (DTT), which when oxidised has a strong absorption. Errors in molecular mass determination are directly proportional to errors in sample concentration measurement (see Section 3.7)

3.1.3 Sample volumes

Sample volumes will depend on the instrumentation being used. SAXS measurements are often done in 1 or 2 mm diameter glass capillaries, and volumes as small as 3–4 μL can be used for a single exposure. However, if the sample is flowed in order to reduce the radiation damage (see Section 3.3) a number of exposures may be required, and it is usually advisable to have 20–30 μL of concentrated solution available. For measurements in flat cells (usually with mica, polycarbonate or Si_3N_4 windows) made in manual mode, considerably larger volumes (20–30 μL) are required for each measurement, leading also to greater total volumes to avoid possible radiation damage.

For SANS, samples are typically measured in high-precision path length quartz cells (for example, Hellma, Müllheim, Germany) of either rectangular or circular cross-section and 1–2 mm path length. The volume of solution exposed to the beam is usually of the order of 70–100 μL, but taking into account dead volume this will more realistically be ~150 μL for 1 mm path length cells. The same volume is required for each H_2O/D_2O content. In principle, a single sample could be dialysed several times against different H_2O/D_2O mixtures, but in practice this would become prohibitively time-consuming. As with SAXS, a concentration series should be measured to allow extrapolation to infinite dilution. Usually, this need be done only at the highest contrast, in general in D_2O. Smaller volumes can be used for concentrated solutions or with the highest-intensity beams.

3.1.4 Buffer solutions

A typical SAS measurement consists of scattering from the macromolecular solution(s) followed by the scattering from a solution of exactly the same composition as that of the buffer at dialysis equilibrium.

Standard buffers. The most straightforward manner of preparation is to use the buffer obtained after final dialysis or the output of an HPLC. It is essential that the buffer composition be identical in all respects to the macromolecular solution (pH, salt content, small molecule content, temperature). Buffers with high temperature coefficients (such as Tris) should not be used. Care should be taken to avoid high salt concentrations where possible, as both for SAXS and SANS they can increase significantly the buffer scattering and may hence reduce the contrast. Note that for SANS, even though the change in scattering length density of the solvent may be significant, the difference between the scattering length density of the solvent and that of the dissolved macromolecule is usually proportionately larger, and hence the contrast is less influenced by salts and small molecules (see Section 1.3). For SAXS, high salt concentrations also increase absorption, which is undesirable, as this may lead to higher radiation damage. Still, measurements at elevated NaCl concentrations (up to 2–4 M) are possible on SR sources. However, if salts of heavier materials are present in significant amounts, proper buffer signal subtraction may become difficult.

Buffers requiring special attention. Solutions of membrane proteins solubilised in detergent present special difficulties both for SAXS and for SANS. When a membrane protein is solubilised with a detergent the solvent always contains free detergent molecules. These exist as free monomers if the residual detergent concentration is below the critical micellar concentration (cmc) or, more importantly, also as micelles if the concentration is above the cmc. Depending on the contrast these micelles themselves will scatter more or less strongly, and their contribution is difficult to remove. In either case it is important that the buffer contain exactly the same amount of free micelles as the membrane protein solution—a situation which is often difficult to attain. If the detergent is one which dialyses easily and rapidly, such as LDAO (lauryldimethylamine-oxide), an appropriate buffer can be prepared by dialysis. The other alternative is to treat the solution as a mixture of micelles and protein detergent complexes. One method is to determine the structure of the micelles independently and to use this as an input for determining the structures of the mixture. These issues arise mainly for SANS, where the high contrast between, for example, protein, detergent and D_2O allows the detergent binding to be investigated. For SAXS such experiments are less often carried out due to the lack of contrast between water and detergent. It is, however, advised not to use detergent in any purification step for SAXS. For more details on membrane protein/detergent complexes, see Section 6.4.

For protein folding studies, addition of unfolding agents in molar amounts is often required. For X-rays there is a major difference between the two most frequently used agents: guanidine hydrochloride (GnHCl) and urea. Whereas a 6 M GnHCl solution absorbs about 80% of 8 keV photons at 1 mm path length, 6 M urea increases the absorption by a few percent only. It is thus preferable to employ urea rather than GnHCl for unfolding in X-ray studies. Some additives tend to form their own ultrastructure in solution if added in significant amounts. Thus, starting from about 5 mM, Adenosine triphosphate (ATP) may aggregate in clusters, which provide noticeable scattering and make buffer subtraction a non-trivial procedure (negative intensities). In any case, most additives worsen the signal-to-noise ratio and may potentially yield problems with the buffer subtraction, so that only really needed substances should be used for the buffers.

Buffer solutions for H/D contrast variation. For SANS contrast variation the H/D composition of the buffer solution is particularly critical, and use of the final dialysis buffer is by far the best way to ensure that sample and buffer have the same H_2O/D_2O composition. It is important to ensure that not only all H/D atoms in the buffer have reached exchange equilibrium, but also that labile protons in the macromolecule (that is, those attached to N, O or S) have also exchanged. Two changes of dialysis at least are normally required, and total dialysis time should be $\sim$24 hours to ensure that H/D exchange has reached a plateau. Experiments have shown that complete H/D exchange in globular proteins can take many months or even years (particularly for protons involved in deeply buried secondary structures), but that within $\sim$24 hours 100% of water protons and 75–80% of labile protons on the proteins have exchanged. The rest exchange slowly over a period of time, but with very little change during the course of the experiment. It is particularly important to verify that the deuterium content of sample and buffer solution are identical, and this may be done using separate neutron transmission measurements (see Section 3.4.2). The background in SANS experiments is dominated by the incoherent scattering of hydrogen, so that although the contrast for a hydrogenated protein in H_2O is not very different from that in D_2O, the signal-to-noise ratio in D_2O is an order of magnitude higher due to the much smaller incoherent scattering background. pH is a parameter of importance in all biochemical or structural studies, and care must be taken to ensure that it is equivalent for solutions both in H_2O and in D_2O. When measuring with a pH meter calibrated with H_2O it is advisable to prepare solutions of the same pD using the relationship (Glasoe and Long 1960): $pD = pH + 0.4X$, where pH refers to the reading on the pH meter when measuring a D_2O solution, and X refers to the mole fraction of D_2O in solution.

3.2 Experimental protocols and instrumental corrections

The fundamental aim of an SAS experiment is to measure the scattering due to isolated macromolecules. These are always dissolved in an aqueous solution, and it is therefore necessary to subtract from the measured scattering curve of the sample any contribution from the solvent or indeed from other extraneous sources.

3.2.1 Electronic noise

The electronic noise, or dark current, produced by the detection system can be measured by recording the signal on the detector in the absence of beam. In many cases the electronic noise is very weak and can be neglected.

3.2.2 Environmental X-ray/neutron noise

Environmental noise is produced by X-rays or neutrons that arrive on the detector when the source is switched on. It is produced by radiation transmitted through the instrument independently of the presence of a sample, and is, for example, produced by other equipment in the neighbourhood. If this happens, one should ensure that these influences do not change during the measurement.

  The environmental noise can be measured after blocking the beam directly after the sample with a suitable absorber (lead for X-rays and cadmium or a borated plate for neutrons).

3.2.3 Scattering from apertures and collimation

Radiation that hits the walls of the collimation system can be reflected or scattered and reach the detector. Most of this secondary radiation is blocked by apertures along the primary beam system. However, the sharp edges of these apertures can themselves become the source of scattering, reflections or fluorescence. These effects can be measured by registering a scattering pattern with only the sample aperture in place (empty-beam scattering).

3.2.4 Scattering from the sample container

Biological samples for SAS are usually dilute solutions that need to be kept in a container, glass capillaries or thin cells with mica walls for X-rays or quartz cells for neutrons. Structures or scratches in these containers can contribute a background.

3.2.5 Solvent scattering

A priori, the solvent is homogeneous and does not produce coherent scattering. However, neutrons (and to a much smaller extent X-rays) produce inelastic incoherent scattering. This can be a predominant scattering contribution, in particular towards larger angles. The incoherent scattering from hydrogen is a very important source of scattering in SANS, and will be described in more detail in Section 3.4.

3.2.6 Absorption/transmission

True absorption by the material of the sample container, the solvent (or matrix) and the particles under study and their scattering weaken both the scattered and the direct beam that arrives on the detector. Whilst the absorption of X-rays by heavier elements can be very significant, true absorption for neutrons is negligible for the sample thicknesses employed. However, the strong incoherent scattering of neutrons by hydrogen effectively attenuates the beam in a way similar to true absorption. In both cases this attenuation must be taken into account when subtracting background contributions from the scattering curves. The transmission of a sample is defined as the ratio:

$$T_s = I_{0,s}/I_0 \qquad (3.1)$$

where $I_{0,s}$ is the intensity of the direct beam after the sample, and I_0 that of the corresponding incident beam. For practical reasons it is sometimes useful to replace the transmission of the sample T_s by its ratio to that of the empty cell:

$$T_{s/ec} = T_s/T_{ec} \qquad (3.1a)$$

When one counts with a constant incident flux (fixed monitor count) as is often done for SANS, $T_{s/ec} = I_{0,s}/I_{0,ec}$, whereas if one counts for a fixed time, as usually done with SR-SAXS, the intensities of the incident beams must also be taken into account.

For absolute scaling, it may be necessary to multiply with the transmission of the empty beam with respect to the empty cell; this depends on the method for absolute scaling that is used. While electronic and environmental noise do not depend on the transmission, the other background sources described here do. However, for neutrons, while the scattering from apertures, collimation and the sample container depend on it proportionally, the incoherent inelastic solvent scattering scales with $1 - T$, since the *non*-transmitted beam is (in a first approximation) scattered isotropically, that is, uniformly over a solid angle of 4π. Therefore, only a very small part ($<0.1\%$) of this scattering goes into the forward direction and so contributes a negligible amount to the transmitted beam.

The intensity scattered by the sample is proportional to the product $D\exp(-\mu D)$, where D is the thickness of the sample and μ the linear absorption coefficient of the material. This function has a maximum at $D = 1/\mu$ (Fig. 3.1), which defines the optimal thickness of the sample. For X-rays, rather conveniently, the optimum thickness of a water sample is very close to 1 mm at the photon energy of 8 keV, and most X-ray cells are fabricated to have about 1 mm path length. For neutrons, the optimum thickness is also typically in the mm range, but also depends on the fraction of D_2O in the solvent (see Section 3.4.2).

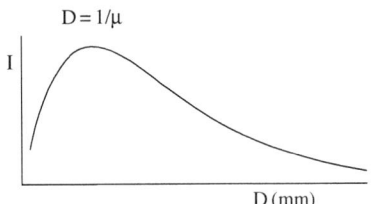

Fig. 3.1 Variation of scattered intensity as a function of sample thickness.

3.2.7 Experimental protocol and data reduction

As molecules may take all possible orientations in solution, the scattering is isotropic and there is no azimuthal dependence of the scattering. It should, however, be checked that fibres or filaments injected into capillaries do not orient under flow. The only dependence of the scattered intensity is on the radial position corresponding to the scattering angle (2θ) between the main beam and the scattered beam. The neutron or photon counts recorded on the detector may therefore be circularly averaged about the direct beam, reducing a two-dimensional array to one dimension. In earlier instruments (and, of course, for linear detectors in SAXS) all data reduction was carried out on circularly averaged data, but with the advent of virtually unlimited data storage facilities modern SAS data reduction programs usually treat all the measured data as two-dimensional arrays which are only reduced to a one-dimensional plot of $I(q)$ versus q after background subtraction and instrumental correction. Averaging the data at an early stage has the advantage, however, that errors can be more easily assessed and phenomena such as parasitic scattering more easily spotted. The experimental protocol remains the same, independently of the method of data reduction.

The basic protocol for data reduction in all SAS experiments from macromolecular solutions involves the subtraction of solvent, empty cell and extraneous background (electronic, parasitic, and so on) from that of the macromolecular solution. The scattering of the solvent buffer is usually the most important contribution to be subtracted but cannot always be reproduced exactly by a sample of buffer, due to, for example, incomplete dialysis. It is therefore often advisable to treat both the macromolecular solution and the solvent as separate samples from which the empty cell and noise scattering are subtracted. The corrected scattering due to the macromolecular solution ($I_s(q)$) is therefore given by

$$I_s(q) = \tilde{I}_s(q) - \tilde{I}_{noise}(q) - T_{s/ec}/\left(\tilde{I}_{ec}(q) - \tilde{I}_{noise}(q)\right) = \tilde{I}_s(q) - T_{s/ec}\tilde{I}_{ec}(q) - \left(1 - T_{s/ec}\right)\tilde{I}_{noise}(q)$$

$$(3.2)$$

and the solvent scattering $I_{solvent}(q)$ is similarly given by

$$I_{solvent}(q) = \tilde{I}_{solvent}(q) - T_{solvent/ec}\tilde{I}_{ec}(q) - (1 - T_{solvent/ec})\tilde{I}_{noise}(q) \quad (3.2a)$$

where the tildes ($\sim$) indicate measured quantities, s denotes 'solution', and ec denotes 'empty cell'.

If the buffer composition of solution and solvent are exactly matched, then the scattering due to the dissolved macromolecule is given by

$$I_{solute}(q) = I_s(q) - I_{solvent}(q) \qquad (3.3)$$

The final intensities must of course be normalised to unit concentration and sample transmission to be put on an absolute scale.

For SANS, pure water, H_2O, is a special solvent, as its scattering can be used for absolute scaling (see Section 3.7).

More detailed protocols for SAXS and SANS are given in Sections 3.3 and 3.4 respectively.

3.3 Special features of SAXS: radiation damage

The X-ray measurements on solutions of biological macromolecules, both on laboratory instruments and on SR sources, must include separate experiments on the solutions and matching solvents. The intensity of the incoming or the transmitted beam and the transmission of the sample should be monitored to normalise the experimental data and to properly subtract the background. The intensity of the beam in front of the sample is measured with an ionisation chamber or by fluorescence from a thin foil (e.g. made from Kapton). The transmitted direct beam can be monitored using a semi-transparent beamstop or by a photo-diode mounted in the beamstop. Alternatively, the transmission of the sample is obtained separately by measuring the scattering pattern of a strong scatterer (such as glassy carbon) placed downstream from the sample. In principle, it is always advisable to have the monitor values measured simultaneously with the actual data collection, and not as a separate experiment.

The generic scheme of a solution SAXS experiment series is presented in Fig. 3.2. Here, the first row represents a measurement of a calibration standard to define the angular axis of the experimental data—that is, the relation between the binning channel of a detector and the momentum transfer q. This conversion factor can, of course, be determined from the geometry of the instrument given the X-ray wavelength. As the parallax effects (and in some cases also depth of the detector) are not always easy to take into account, in practice one usually measures partially ordered or polycrystalline samples with sufficiently large repeat spacings such as dry collagen ($d = 65$ nm) or organic powders (such as Ag behenate, $d = 5.838$ nm, or tripalmitin, $d = 4.06$ nm). These samples provide a series of distinct peaks at q-values multiple of $2\pi/d$, such that the q-axis can be established easily.

The measurements of the empty cell containing neither sample nor buffer, and those of the scattering from pure water (rows 2–3 in Fig. 3.2) are always useful for assessing the background level of the camera and also to obtain the net water scattering. The latter can be useful for absolute calibration of the scattering data and determination of the molecular mass of the solute (see Section 3.7). Alternatively, scattering from well-characterised proteins can be employed to estimate the molecular mass (rows 4–6 in Fig. 3.2) (solutions of cytochrome C, lysozyme, bovine serum albumin (BSA) or glucose isomerase are used at different facilities). One must keep in mind that the scattering from buffers containing different concentrations of salts or other additives may differ significantly from each other. Measurements and subtraction of the matching buffer (ideally, the last dialysis buffer) is a must for any SAXS experiment (rows 7–9 in Fig. 3.2). It is also highly preferable to measure the sample and buffer in the same cell, as the background scattering of different cell windows (or capillaries) may vary.

There are two major differences between laboratory and SR-SAXS measurements. First, the background of the laboratory camera is usually relatively stable, and only variations in the primary intensity of the X-ray source are observed, which can be corrected by the appropriate normalisation over the monitor values. At SR sources, with much more complicated systems of X-ray production and transport, factors such as variations of the trajectory of particles in the storage ring, thermal load on the optical elements and so on, may lead not only to intensity changes but also to drifts of the background scattering with time. Therefore, it is useful to measure the relevant buffers close in time to the sample measurements—ideally, one buffer before the sample and one just after—which are averaged during the data processing. The scheme in Fig. 3.2 is related to the SR measurements. For laboratory sources these requirements are less critical, and the buffer measurement may be separated in time (also, one buffer may be used as reference for multiple samples).

Another important difference between SANS and SAXS lies in the X-ray radiation-induced damage. As discussed in Chapter 1, the X-rays interact with matter through absorption and scattering. For 12 keV photons interacting with light atoms in biological samples, only about 10% are scattered and 90% are absorbed, leading to a photoelectric effect followed by Auger emission, shake-up excitations and secondary electron cascades. These phenomena damage the sample, but with the relatively low intensities of laboratory sources they can usually be neglected. However, at SR sources, especially starting from third-generation sources yielding monochromatic X-ray beams with up to 10^{13} photons s^{-1} in a sub-mm^2 spot size, radiation damage becomes a serious problem. For MX the damage increases the mosaicity of the crystal, temperature B factors, disrupts the intramolecular disulphide bridges (Murray and Garman 2002) and photoreduces metal centres. These effects can be significantly reduced by cryo-cooling the samples, and most SR studies are performed on frozen crystals (Juers and Matthews 2004) even if this significantly affects the structure (Fraser *et al.* 2011). The major advantage of biological SAXS is that the samples are solutions, which also means that they cannot be frozen (studies on frozen solutions are in principle possible, but technically extremely complicated). Most SAXS measurements are performed close to room temperature, and means other than freezing must be employed to minimise radiation damage. The major effect of radiation damage in SAXS is not the deterioration of the structure itself (this is hardly seen at the low resolution of the method) but rather the radiation-induced aggregation. The interactions of the incident X-rays with water molecules in the sample create hydroxyl or hydroperoxyl radicals, and the radical-activated proteins tend to be cross-linked by covalent and/or non-covalent bonds (Garrison 1987). The aggregates are readily seen in the low-q region of the scattering patterns (large particles) but also in the wide-angle X-ray scattering curves (Fischetti *et al.* 2003). Interpretation of the scattering data affected by radiation effects can be impossible or misleading, and everything must therefore be done to minimise these effects.

One way to reduce radiation damage is to use a flow cell where the solute is constantly flowing through the irradiated volume. A disadvantage of this approach is the increased amount of material needed, depending on the flow rate. Addition of about 2 mM of reducing agents such as dithiothreitol (DTT) or tris(2-carboxyethyl)phosphine (TCEP) acting as scavengers capturing the free radicals reduces radiation damage. The use of DTT or TCEP is, however, not always possible, for example, with proteins containing disulphide bridges, as the latter may be reduced, thus compromising the integrity of the structure. Cryoprotectant molecules, such as glycerol, ethylene glycol and sucrose, are also able to reduce radiation damage, largely by slowing down the diffusion of free radicals. The use of such additives (typically, about 10% (w/w) is required) has a generic disadvantage of worsening the signal-to-noise ratio by reducing the contrast and increasing the absorption of the sample. Based on systematic measurements on lysozyme solutions, a noticeable protective effect starting already at 1% (w/w) has been observed (Kuwamoto *et al.* 2004) and attributed to the fact that small amounts of cryoprotectant enhance the repulsive terms of the protein–protein interactions. Cryoprotectants such as ethylene glycol may also have disadvantages when studying reactions in solution (West *et al.* 2008).

As evident from the above paragraph, there are different means of limiting radiation damage, all having advantages and shortcomings. A universal approach, employed at nearly all modern SR stations, is to slice the data collection time into individual successive time frames (for example, ten or twenty, depending on the possibilities of the experimental setup). The recorded patterns are processed separately and compared to each other (essentially to the first frame), and the frames showing systematic changes are not included in the subsequent averaging and further data processing. This analysis can be done automatically using standard statistical criteria, and the users are warned when radiation damage is observed. The measurements are usually done in frames for all samples, even for the buffers (which might also help, for example, when there is a beam loss during the measurements—at least some frames can be rescued). Of course, dividing into multiple frames requires a fast readout detector and storage space sufficient to store the experimental data from individual frames.

Radiation damage represents a problem on a completely different scale in experiments on the fourth-generation sources (XFELs). The huge incoming flux provokes a Coulomb explosion and complete disintegration of macromolecules within a few tens of femtoseconds, but it may be still possible to collect sufficient single particle diffraction data within this time frame. Simulations suggest that the flux of 10^{13} photons per 100 nm^2 area could be sufficient to produce a diffraction pattern from a single lysozyme molecule at up to 0.3 nm resolution

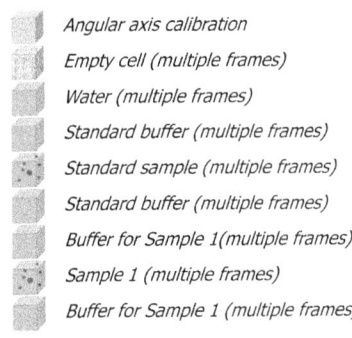

Angular axis calibration

Empty cell (multiple frames)

Water (multiple frames)

Standard buffer (multiple frames)

Standard sample (multiple frames)

Standard buffer (multiple frames)

Buffer for Sample 1(multiple frames)

Sample 1 (multiple frames)

Buffer for Sample 1 (multiple frames) ...

Fig. 3.2 Typical sequence of SAXS measurements on an SR source.

(Neutze *et al.* 2000). With XFEL sources presently becoming reality, the theoretical predictions will soon be verified.

3.4 Special features of SANS: the use of H_2O/D_2O mixtures

The outstanding advantage of SANS measurements lies mainly in the possibility of exploiting contrast variation as described in Chapter 1 by preparing a series of samples containing different proportions of D_2O/H_2O (see Section 3.1). In this book we shall systematically refer to the mole fraction of D_2O being the ratio of the number of moles of heavy water to the total number of moles of water $[D_2O]/([D_2O]+[H_2O])$. This is often also quoted as '%D_2O'. In order to extract the maximum information from a contrast variation series it is necessary to measure the scattering from samples dissolved in solutions of at least four different D_2O mole fractions.

3.4.1 Choosing contrasts

The choice of contrasts to be measured depends on the scattering length densities of the macromolecular solute. Usually, data should be measured from samples dissolved in H_2O buffer, pure D_2O buffer and in at least two other H_2O/D_2O mixtures. As SANS experiments are of most use with complexes of molecules with different scattering length densities, the buffers should be chosen to emphasise the contrasts of the individual components. The calculation of contrasts and match points has been described in Section 1.3. Note that one should not measure too close to the overall match point of the complex, as the signal is usually weak and the background scattering from hydrogen rather high. So, if investigating the structure of an RNA-containing virus, for example, one would usually measure the scattering in H_2O and D_2O buffers (high positive and negative contrasts) and in a buffer of ~0.4 mole fraction D_2O (highlighting the RNA) and ~0.68 mole fraction D_2O (highlighting the protein component).

3.4.2 Checking the D_2O content of solutions

Knowledge of the exact deuterium content of the sample and buffer is particularly important for SANS where different H_2O/D_2O ratios are being used. It must be verified that the deuterium content of sample and buffer are identical, which may be achieved by using separate neutron transmission measurements. The transmission of a sample is defined as the ratio of the flux through the sample and the incident flux. The transmission can be obtained from the intensities of the direct beam transmitted through a cell filled with sample or buffer and an empty cell. Most atomic nuclei have no or very little true absorption for neutrons, and the beam is attenuated primarily due to incoherent scattering by hydrogen. The transmission is therefore a quantitative measure of the hydrogen content of the sample or buffer, and is given by

$$T = \exp(-N_{H_2O}\sigma_{H_2O} - N_{D_2O}\sigma_{D_2O}).D \qquad (3.4)$$

where N_{H_2O} and N_{D_2O} are the numbers of H_2O molecules and D_2O molecules per cm^3 of solution, respectively, and σ_{H_2O} and σ_{D_2O} are the neutron scattering cross-sections of H_2O and D_2O, respectively. D is the path length of the neutron beam through the sample in cm.

For a sample of water containing a mole fraction X of D_2O, this can be written as

$$T = \exp[-(1-X)(N_{H_2O}+N_{D_2O})\sigma_{H_2O} - X(N_{H_2O}+N_{D_2O})\sigma_{D_2O}]D \quad (3.5)$$

which yields:

$$T = \exp[-(N_{H_2O}+N_{D_2O})\sigma_{H_2O} - X(N_{H_2O}+N_{D_2O})(\sigma_{H_2O}-\sigma_{D_2O})]D$$
$$(3.6)$$

$$-\frac{1}{D}\ln T = (N_{H_2O}+N_{D_2O})\sigma_{H_2O} - X(N_{H_2O}+N_{D_2O})(\sigma_{H_2O}-\sigma_{D_2O})$$
$$(3.7)$$

Hence, a plot of $-\frac{1}{D}\ln T$ versus X will be linear, as shown in Fig. 3.3.

It is therefore sufficient to measure the transmission of a buffer made up in H_2O and one made up in pure D_2O (or any other precisely known mole fraction) to determine the mole fraction of D_2O in a solution of any other transmission. Neutron transmission varies with wavelength and also with temperature (due to changes in water density and changes in inelastic scattering). It is therefore important to measure all transmissions used to estimate deuterium content at the same temperature and neutron wavelength. Typical values for the transmission of a 1 mm thick sample of H_2O at 25°C for different wavelengths are plotted in Fig. 3.4. Values at any wavelength can easily be interpolated from the fitted polynomial. These values are, however, only indicative and should always be determined experimentally, as they depend on temperature and on buffer composition of the solution.

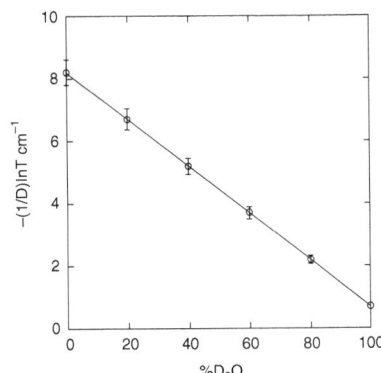

Fig. 3.3 Plot showing the linearity of log of transmission (normalised to sample thickness) with D_2O content of the buffer.

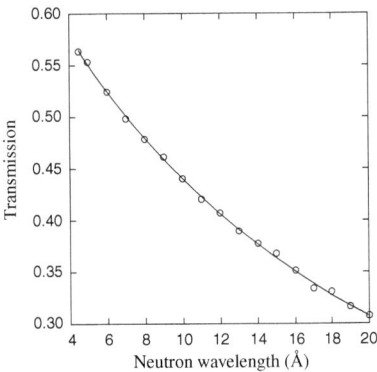

Fig. 3.4 Transmission of a 1 mm thick sample of water at 25°C as a function of neutron wavelength. Data replotted, with permission, from (Lindner 1998, 2000).

3.4.3 Choice of q-range: instrument configuration

The q-range over which data need to be measured will determine the configuration of the SANS instrument. There are a number of criteria which determine the q-range to be measured:

1. If the approximate maximum intraparticle dimension (D_{max}) of the particle is known then:

$$q_{min} < 1/D_{max}$$

q_{max} will be determined by the resolution required but should normally be at least 2 nm^{-1}

2. If measurements are to be made only in the Guinier zone (Section 3.6.1) then the required q-range will be much smaller such that:

$$q_{max} \leq 1.3/R_g$$
$$q_{min} \leq 0.5/R_g$$

where R_g is the estimated radius of gyration which is usually not known. A generous margin should therefore be left in the estimation of the q-range.

Most SANS instruments have a limited q-range in any particular configuration, so that several configurations are usually necessary to cover the full range. The parameters defining the obtainable q-range are principally the neutron wavelength and the sample-detector distance. Instruments on reactor sources usually exploit a monochromatic beam, whilst on neutron spallation sources they usually access a wide range of neutron wavelengths simultaneously and can therefore sample a wider q-range than instruments on a reactor source. However, the overall flux and more complicated corrections required make the two sources more or less equivalent. For special purposes instruments have been built on reactor sources exploiting the white beam (see Section 2.5.1).

The q-range accessible for a given instrument configuration can be derived from the definition of $q = (4\pi/\lambda)\sin\theta$ (see Chapter 1). If the points on the detector closest to and furthest from the main beam are r_{min} and r_{max} and the sample-detector distance is L, then

$$q_{min} = \frac{4\pi}{\lambda} \sin\left(0.5 \tan^{-1}\left(\frac{r_{min}}{L}\right)\right) \quad and \quad q_{max} = \frac{4\pi}{\lambda} \sin\left(0.5 \tan^{-1}\left(\frac{r_{max}}{L}\right)\right)$$
$$(3.8)$$

In most reactor-based SANS instruments the wavelengths available with optimum intensity lie between 0.5 and 1.0 nm, whereas for a spallation source the beam may contain all wavelengths from 0.2 to 2.0 nm. For example, on a 1 m^2 detector (see D22 at ILL), where the direct beam impinges on the centre of the detector, the scattered beam can be recorded at distances of 5–70 cm from the direct beam. Typical q-ranges available for different wavelengths and sample-detector distances are therefore as shown in Table 3.1.

Note that essentially the same q-range can be attained using a sample-to-detector distance L and wavelength λ or sample-to-detector distance $2L$ and wavelength $\lambda/2$. The final choice is usually made such that if several different distances are to be used to cover a wide q-range the same wavelength is maintained throughout all the series of measurements. The relative neutron flux at the different wavelengths and the different transmissions should also be considered. The q-range available at a given setting can be extended by a factor of ~50% if the detector can be offset such that the direct beam falls on the edge of the detector. The

Table 3.1 q-ranges available for a $1\,m^2$ detector centred on the direct beam as a function of sample–detector distance (L) and neutron wavelength (λ).

Sample–detector distance (m)	Wavelength (nm)	q_{min} (nm^{-1})	q_{max} (nm^{-1})
1.0	0.5	.628	7.556
	1.0	.314	3.778
2.0	0.5	.314	4.211
	1.0	.157	2.106
5.0	0.5	.126	1.747
	1.0	.063	.873
10.0	0.5	.063	.878
	1.0	.031	.439
20.0	0.5	.031	.440
	1.0	.016	.220

resultant scattering curve will, however, have less statistical precision as fewer detector cells contribute to each q-value.

Other important instrumental parameters to be determined are the beam divergence and the detector-offset angle with respect to the direct beam (see Fig. 1.3). The divergence of the beam is usually determined by the distance of the effective neutron source from the sample. In most instruments this is determined by a set of moveable neutron guides whose exit can be at a variable distance from the sample. The same effect can be obtained with an aperture of variable size at a fixed distance from the sample. The beam divergence is usually chosen such as to maximise beam intensity while maintaining a spot size that is smaller than the beam-stop dimension and gives the resolution required.

3.4.4 SANS experiments: the experimental protocol

Once the required instrument configuration has been chosen as a function of the q-range to be investigated, the precise series of measurements to be carried out must be planned. For SAXS and SANS the basic principles are the same—we wish to obtain the scattering from isolated particles in solution by subtracting the scattering of the buffer from that of the sample to obtain the excess scattering of the dissolved macromolecules. As explained above, this requires several different measurements to account for other sources of background, absorption effects, and so on. Background subtraction is not so critical when the signal-to-background ratio is very high, for example in high contrast or low angles, but is much more so for model-fitting at higher q-values. The basic experimental protocol for SANS, as summarised in Fig. 3.5, is as follows:

1. Transmissions of samples, buffers, empty sample cell and empty diaphragm are measured. These measurements can also serve to determine the point at which the neutron beam impinges on the detector so long as they are carried out in the same instrument

configuration as the scattering measurements. These measurements are carried out by recording an image of the attenuated direct beam on the detector with the beam-stop removed.

2. The scattering of each sample, corresponding buffer, an H_2O sample, an empty cell, the empty sample diaphragm and a piece of cadmium or boron carbide are recorded. The last sample allows the measurement of neutrons falling on the detector originating from any other source other than the sample; for example, neutrons from other instruments. Indeed, in a neutron facility there is usually a very dilute 'neutron gas' in the experimental hall which will be seen by the detector if not adequately shielded. The scattering from water is predominantly isotropic and therefore independent of q, and can therefore be used to correct for inhomogeneities in the detector response. All scattering is therefore normalised to that of water, which is indeed also one way of putting the scattering on an absolute scale (see Sections 3.7.2 and 3.7.3).

The most delicate part of data reduction is the treatment of background; in particular, incoherent scattering arising from hydrogen. As mentioned in Section 3.2.7 it is very often difficult to measure this background accurately, and small adjustments are often required. There are several ways in which the background can be estimated if no reliable direct measurement is available:

1. Assume that all scattering at high q ($>5\,\mathrm{nm}^{-1}$) is due to buffer and subtract this (flat) background level.
2. Remove a flat background such that the higher-angle scattering falls as q^{-4} (Porod asymptotic; see Section 4.1.6).
3. Adjust the buffer scattering by addition of H_2O or D_2O until buffer and sample transmissions are identical.
4. Carry out a series of scattering experiments at D_2O contents around those aimed for, measure their transmissions and plot the buffer scattering level (I_b) as a function of transmission (this should be approximately proportional to $1 - T_b$), and interpolate the buffer scattering from the transmission of the sample.

Transmission measurements

Empty cell (T_c)
Empty beam (T_e)
1mm pure water (T_w)
Samples (T_s)
Buffers (T_b)

Scattering measurements

Empty cell (I_c)
Empty beam (I_e)
Blocked beam (I_B)
1 mm pure water (I_w)

Scattering measurements for each D_2O content

Samples (I_s)
Buffers (I_b)

Fig. 3.5 Sequence of measurements for a typical SANS experiment with contrast variation.

3.5 Random and systematic errors

As any results of physical measurements, SAS patterns are experimental data and as such they must contain the associated errors. The basic assumption allowing one to define and then propagate the errors in the experimental data comes from the fact that the scattering intensities are measured as numbers of photons or neutrons registered by a detector. The number of counts per pixel K obeys a so-called Poisson distribution, which converges to a Gaussian distribution at sufficiently large counts ($K > 30$) (Bevington 1969). For most practical purposes one can assume that having measured K photons in a pixel, the associated uncertainty (standard deviation) of this value is $K^{1/2}$. Given the standard deviations of all measured experimental data (for example, of the pixel counts of an area detector), these can be propagated through all the manipulations with the data following the standard equations. For a sum or difference of two values a and b with standard deviations $\sigma(a)$ and $\sigma(b)$, respectively, the standard deviation will be $\sigma(a \pm b) = (\sigma(a)^2 + \sigma(b)^2)^{1/2}$, for a product $\sigma(ab) = (a^2 \sigma(a)^2 + b^2 \sigma(b)^2)^{1/2}$, for a ratio $\sigma(a/b) = (a/b)(\sigma(a)^2/a^2 + \sigma(b)^2/b^2)^{1/2}$ (Bevington 1969). Accurate error propagation throughout the data manipulations, starting with the radial averaging and going through the normalisations and subtractions, finally yields the scattering pattern as the set of data points $I(q_i)$ with the associated standard deviations $\sigma(q_i)$. Correctly propagated standard deviations are a necessary prerequisite for the subsequent data analysis and interpretation. The vast majority of X-ray and neutron data processing packages do provide the users with these estimates.

The standard deviations are required in the further analysis and interpretation to compute the most widespread goodness-of-fit criterion, discrepancy between the experimental and calculated data $I_{calc}(q)$.

$$\chi^2 = \frac{1}{N_k - 1} \sum_{j=1}^{N_k} \left[\frac{I(q_j) - cI_{calc}(q_j)}{\sigma(q_j)} \right]^2 \tag{3.9}$$

where N_k is the number of the experimental points, and c is a scaling factor. A lower χ means a better fit, and χ is a target to be minimised in most of the data fitting and model building procedures. Statistically, χ follows the so-called χ^2 probability distribution (Bevington 1969), and $\chi \approx 1$ indicates that there are no systematic deviations between $I(q)$ and $I_{calc}(q)$; that is, that the computed pattern agrees well with the experimental data.

Looking at the error propagation formulae it is easy to identify the processing step that significantly degrades the signal-to-noise ratio: namely, buffer subtraction. This step often involves subtraction of values which are close to each other, whereby the difference value decreases significantly compared to both of the initial values, whereas the associated error increases. Slight errors in the normalisation factors may lead to unphysical difference data. These considerations further underline the necessity of accurate measurements of the buffer scattering.

It must be noted that the error propagation scheme presented above is rather simplified and the correct estimate of the standard deviations may not be as straightforward. First of all, for some detectors, the number of counts could be just proportional to the number of scattered photons/neutrons, but not precisely represent the latter. This is true especially for integrating X-ray detectors such as CCD cameras or image plates, where the incoming X-rays first result in charge or excited state accumulation which is then read out by conversion into another type of signal (for example, visible light in the case of imaging plates). Moreover, for such detectors a (small) offset may be added to compensate for the electronic noise and, for example, to avoid negative counts. Integrating detectors are usually calibrated in such a way that their readout is close to the estimated numbers of counts. Deviations are of course possible at, for example, elevated count rates, but $\sigma = K^{1/2}$ should in most cases be considered a good practical estimate of the standard deviation.

Another cause of difficulties for error propagation is the presence of systematic errors. These include, for example, non-uniform response and dark current of the detector, and variations in the primary intensity values and in the transmission. The detector response is often calibrated using a uniform scattering source to correct for the efficiency of the individual channels, but the efficiency may change with time, while accurate response measurements may be rather time-consuming to be repeated often. The errors in the monitor values used for normalisation are also not easy to propagate in the data processing, and the monitors are often simply taken as measured. All these factors may lead to additional errors in the processed intensities, and the standard deviations estimated purely from statistical error propagation may become underestimates of the real uncertainty on the data. These underestimates may be very undesirable, leading to incorrect statistical assessment of the data; in particular, the discrepancy χ could exceed unity significantly even if the data are well fitted. In case of doubt, standard deviations in subtracted solution scattering data may be validated *a posteriori* by using the fact that the ideal $I(q)$ function is expected to be a smooth curve. Using any reasonable procedure to smooth the data, for example, splines (Schelten and Hossfeld 1971), polynomials (Rolbin *et al.* 1980), frequency filtering (Muller and Damaschun 1979), or the result of an indirect Fourier transform (IFT, see Chapter 4) one can compute the discrepancy (eq. 3.9) between the original and smoothed data. If this value is significantly different from unity, there must have been a problem in the computed standard deviations. As a possible remedy, these values may, for example, be rescaled (multiplied by a factor) to provide $\chi = 1$ against the smoothed data.

Finally, one should always keep in mind that statistical errors can be estimated and taken into account, whereas systematic errors can often not be corrected.

3.6 Basic structural information and data quality

SAS from compact, non-interacting particles obeys some rules that are helpful in judging the quality of the scattering curves themselves and obtaining some basic information about the sample. The corresponding parameters are partially model-independent and should be checked before trying a detailed analysis and interpretation of the data.

3.6.1 Initial slope and radius of gyration

As will be explained in detail in Chapter 4, the initial slope of the background-corrected normalised scattering curve from particles can be approximated by a Gaussian (Guinier 1939):

$$I(q) \approx I(0)\, e^{\frac{-q^2 R_g^2}{3}} \tag{3.10}$$

—('Guinier's law'), where R_g is the radius of gyration. Therefore, one can determine the zero-angle intensity or forward scattering and R_g from the zero intercept ($\ln I(0)$) and initial slope in a plot of $\ln I(q)$ versus q^2. This was particularly valuable before the routine use of computers. Equation (3.10) holds in a range of about $0 < q < 1/R_g$, commonly known as the 'Guinier zone' or 'Guinier range'. Depending on the shape of the particles, the higher limit can sometimes be a little bit larger; this can only be verified *a posteriori*, that is, one should check the range after having determined R_g and iterate if necessary.

In the soft-matter field, the so-called Zimm plot is often used, where $1/I(q)$ or $c/I(q)$ is plotted versus q^2 for different concentrations c (Zimm 1948). This yields very similar R_g values to those of Guinier's law, but not quite identical due to the different higher-order contributions in the series expansion.

Whenever the Guinier plot at low q is not linear, the sample has either considerable aggregation (upswing) or intermolecular repulsion (downswing), at least in the case of homogeneous particles. Detailed analysis of such scattering data should not be attempted, except when the interaction behaviour can be modelled exactly.

Molecules composed of two or more moieties with different scattering length densities also can display a strange R_g behaviour close to the match point of the particles.

One should be aware of the fact that a small proportion of aggregates, for example, 5% dimers in an otherwise monomeric solution, is difficult to detect in the Guinier plot, that is, a straight line does not guarantee monodispersity.

3.6.2 Zero-angle scattering

Equation (3.10) also yields the forward scattering of the particles, $I(0)$. As we shall see at the beginning of Section 3.7, this is proportional to the number N of particles times the square of the product of the particle volume and its excess scattering length density. Since N is inversely proportional to the molecular mass of the particles for a given particle weight concentration, $I(0)$ is proportional to the molecular mass, which can be calculated as shown in Section 4.1.5 so long as $I(0)$ is on an absolute scale. The forward scattering is therefore also a parameter that allows one to check the consistency of the data. If 5% of the particle concentration exist as dimers and 95% as monomers, the zero-angle scattering increases by 5%, since for the same concentration $N_{dimer} = N_{monomer}/2$, and the square of the scattering mass increases by 4. Due to the systematic errors involved in the determination of the concentration and of the absolute scattering intensity, there is also a limit on the accuracy of the estimate of the level of aggregation.

In SANS there is a particular relationship between $I(0)$ and the neutron contrast (that is, the D content of the buffer) which will be described in detail in Chapter 4. $I(0)$ (normalised to concentration, transmission and sample thickness) is a linear function of the (contrast)^2 and a plot of $I(0)^{1/2}$ versus ($[D_2O]/([H_2O] + [D_2O])$) should be linear, with $I(0)^{1/2}$ going to zero at the match point of the particle. Alternatively a plot of $I(0)$ versus $[D_2O]/([H_2O] + [D_2O])$ should be a parabola touching the abscissa at the match point. If, however, the macromolecule under investigation consists of two or more moieties with different contrast behaviour (for example, nucleic acid and protein) then the particle will only become contrast matched at a single solvent composition if the particle stoichiometry (such as nucleic acid/protein ratio) is the same for all particles. This is therefore a good measure of sample quality for macromolecular complexes. This is well illustrated in Fig. 3.6, the case of ferritin, an iron storage protein where different particles contain various amounts of iron (Stuhrmann 1974). A plot of $\sqrt{I(0)}$ versus D_2O mole fraction displays systematic deviations from linearity and a plot of $I(0)$ versus D_2O mole fraction never goes to zero, showing that at all D_2O contents there exist some unmatched particles.

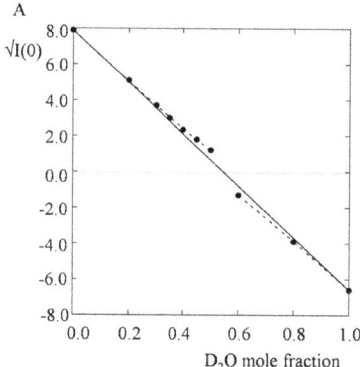

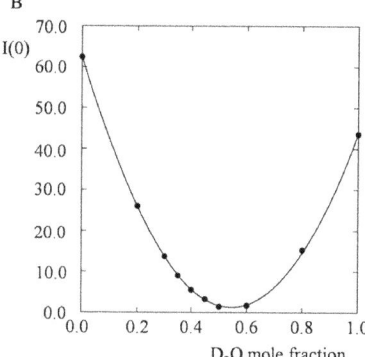

Fig. 3.6 Variation of $I(0)$ as a function of contrast for the iron storage protein ferritin. A. $I(0)^{1/2}$ *versus* D_2O mole fraction, B. $I(0)$ *versus* D_2O mole fraction. Data replotted, with permission, from Stuhrmann (1974).

3.6.3 Pair-distance distribution function $p(r)$

The scattering intensity of non-interacting particles in dilute solution can be described by an integral that is limited to the maximal dimension (maximal chord length, D_{max}) in the particles:

$$I(q) = 4\pi \int_0^{D_{max}} p(r) \frac{\sin(qr)}{qr} dr \qquad (3.11)$$

$p(r)$ is the so-called pair-distance distribution function (PDDF), or distance distribution, of the particles. The inverse Fourier transform that defines $p(r)$ as a sine Fourier integral of the intensity cannot be calculated because the integration range is infinite, but the intensity can only be measured in a limited q-range, and with errors.

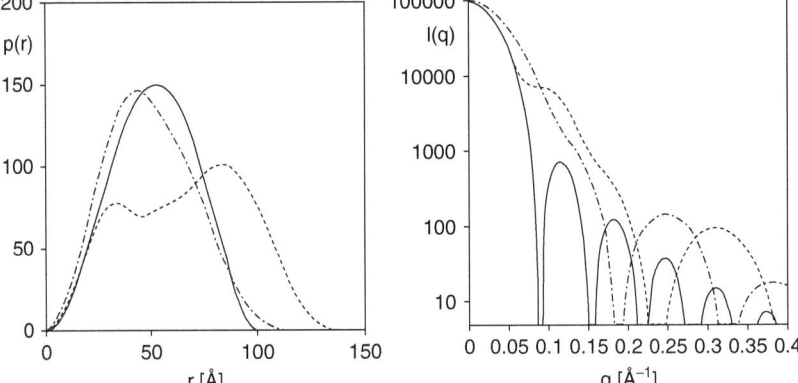

Fig. 3.7 Model pair-distance distribution functions (left) and intensity curves (right) for three bodies of equal volume, a sphere of radius 5 nm (full line), and two tori, a horn torus, where the ring touches the central axis (dotted–dashed), and a ring torus (dashed).

The $p(r)$ can be obtained from eq. (3.11) via an IFT procedure (Glatter 1977) as described in detail in Chapter 4.

Since the distance distribution is a function in real space, it is often easier to recognise features of the particles in the $p(r)$ function than in the scattering curve. This is illustrated in Fig. 3.7, where theoretical PDDFs and intensity curves for spheres and two different tori of equal volume are compared. The intensity curves display deep minima that normally would be damped by instrumental smearing, but also because real molecules are never perfectly symmetrical bodies. In particular, the ring torus can be identified by its $p(r)$ function.

Figure 3.8 shows scattering curves and PDDFs for a series of geometrical bodies each having the same D_{max} but different shapes and therefore different volumes.

By definition, PDDFs start smoothly with a value of zero at $p(0)$, and they should terminate smoothly at a maximal dimension D_{max}. A deviation of $p(0)$ from zero is a sign of an error or uncertainty in the

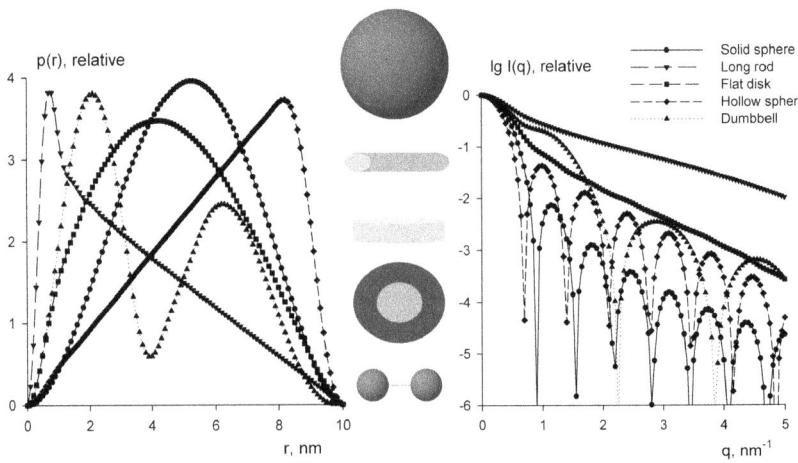

Fig. 3.8 Distance distribution functions (left) and scattering intensities (right) of geometrical bodies with the same maximum size ($D_{max} = 10$ nm) and different shapes. Adapted, with permission, from Svergun and Koch (2003), © IOP Publishing Ltd, 2003.

background subtraction which can be used to estimate the background or, alternatively, $p(0)$ can be forced to zero. A long tail or a shoulder at the high-r end of the PDDF should induce caution and may be a sign of aggregation. Aggregation is also visible in the Guinier range (see Section 3.6.1), but calculating the $p(r)$ function does not require a q_{min} as small as a Guinier fit.

A downswing below zero at the high-r end of the PDDF can be due to opposite contrast of two ends of a molecule (usually only observed in SANS), but in the case of homogeneous particles it is a sign of repulsive interactions between the particles, rendering it difficult to calculate correct *ab initio* models as in the case of aggregation. A tendency to dimerisation combined with volume exclusion effects at higher concentrations or electrostatic repulsion may have counterbalancing effects in the PDDF.

3.7 Calibration to absolute scale and molecular mass

The aim of a scattering experiment is to obtain structural information about the sample. In the case of biological macromolecules this includes the knowledge of the molecular mass. X-rays and neutrons provide intrinsic features that allow one to determine the absolute differential scattering cross-section of the sample, and by using additional information about the sample (chemical composition, concentration and so on) the apparent molecular mass M of the scattering particles.

3.7.1 Differential scattering cross-section

As explained in Chapter 1 (Fig. 1.4), the differential scattering cross-section $d\sigma/d\Omega(q)$ is a property of the sample and is defined as the ratio (as a function of q) of the number of scattered photons or neutrons per unit solid angle and time to the number of photons or neutrons in the incident beam per unit area and time.

Although this definition suggests that the number of photons or neutrons counted by the detector actually depends on the illuminated area it may be less obvious that the thickness of the sample also matters. Therefore, one defines a differential scattering cross-section per unit volume, $d\sigma_V/d\Omega(q)$, sometimes also called $d\Sigma/d\Omega(q)$ or $\tilde{\sigma}(q)$:

$$\frac{d\sigma_V}{d\Omega}(q) = \frac{1}{V}\frac{d\sigma}{d\Omega}(q) = \frac{1}{SD}\frac{d\sigma}{d\Omega}(q) \left(\text{sterad}^{-1}\text{cm}^{-1}\right) \qquad (3.12)$$

where S is the illuminated area and D the sample thickness. The number of photons or neutrons observed in a detector pixel per second $I_s(q)$ depends on the instrument geometry and the incoming beam flux and is given by:

$$I_S(q) = I_0 \eta S D \Delta \Omega T_s \frac{d\sigma_V}{d\Omega}(q)\,(\text{s}^{-1}) \tag{3.13}$$

where I_0 is the incident beam intensity (photons or neutrons $\text{cm}^{-2}\text{s}^{-1}$) at the sample position, η the detector sensitivity. I_0 and η depend on the wavelength, and I_0 also depends on the primary beam geometry.

$\Delta \Omega$ is the solid angle defined by the area of the detector element (pixel size, A) and the square of the sample-detector distance (L_2, see Fig. 1.04), $\Delta \Omega = A/L_2^2$.

One can determine the *effective* beam intensity ($I_0\eta S$) *in the direct-beam spot on the detector* by measuring the neutrons counted per second, using attenuators of known attenuation factor values. With this, and defining $\frac{d\sigma_V}{d\Omega}(q) = I_{\text{total}}(q)$ as in Chapter 1, one can therefore obtain the differential scattering cross-section from the observed scattered intensity:

$$I_{\text{total}}(q) = \frac{I_s}{I_0 \eta S \Delta \Omega D T_s}\,\left(\text{sterad}^{-1}\,\text{cm}^{-1}\right). \tag{3.13a}$$

3.7.2 Absolute calibration with H_2O and neutrons

According to (Jacrot and Zaccai 1981), the scattering from an incoherent scatterer into a solid angle $\Delta \Omega$ can be estimated as

$$I_{\text{inc}}(q) = \frac{(1 - T_{\text{inc}})I_0 S \Delta \Omega}{4\pi}(\text{s}^{-1}) \tag{3.14}$$

Equation (3.14) assumes that the neutrons that are not transmitted by the sample are scattered isotropically. In practice, one uses water as an incoherent scatterer. Since the scattering from water is inelastic, (Jacrot and Zaccai 1981) introduced a factor f that takes into account the non-isotropic nature of this scattering. However, the fact that the wavelength of inelastically scattered cold neutrons is on average shorter than that of the incoming beam, due to 'upscattering', may lead to a relative decrease of their detection probability as compared to elastically scattered neutrons; this also enters into f, which therefore also becomes instrument dependent. Entering the efficiency as in eq. (3.12):

$$I_{H_2O}(q) = \frac{(1 - T_{H_2O})I_0 \eta S \Delta \Omega}{4\pi f}(\text{s}^{-1}) \tag{3.15}$$

Note that the thickness of the water sample appears only indirectly in this equation, namely, through its transmission. If the water scattering and the sample are measured at identical conditions, I_0, S and $\Delta \Omega$ are the same for the sample and water. Therefore, with neutrons one can use the scattering from light water as a means of absolute scaling.

$$I_0 \eta \Delta \Omega S = I_{H_2O}(0)\frac{4\pi f}{(1 - T_{H_2O})}(\text{s}^{-1}) \tag{3.15a}$$

With eq. (3.15a) we obtain

$$I(0) = \frac{I_S(0)}{I_{H_2O}(0)} \frac{(1 - T_{H_2O})}{4\pi f} \frac{T_{ap}}{DT_S} (\text{sterad}^{-1}\,\text{cm}^{-1}), \qquad (3.16)$$

where T_s and T_{ap} are transmissions of the sample and aperture with respect to that of the empty cell.

Knowing f for a given instrument and wavelength one does not need the direct-beam intensity for calculating the differential cross-section if one divides by the water scattering, which is, in fact, the usual way to correct for the detector response (see 3.4.4).

The disadvantage of this method is that f depends on the wavelength and on the instrument/detector used. Although tables for this are available at many instruments it is preferable to determine $I_0\eta S$ in eq. (3.15a) using attenuators of known attenuation factors.

3.7.3 Absolute calibration for X-rays

For neutrons, the strong incoherent water scattering, although degrading the signal-to-noise ratio, provides an ideal way for the absolute calibration of the scattering data. For X-rays such a clearly preferable tool does not exist, but there are several approaches to obtain the absolute intensity, all of them having specific advantages and shortcomings. Absolute calibration is usually more difficult with laboratory sources where the signal is much weaker than on SR sources (Dreiss *et al.* 2006). The incoming flux of X-rays can be estimated by different attenuation techniques—for example, Stabinger and Kratky (1978)—and various strongly scattering samples can be calibrated as secondary standards. One of the best known is Lupolen (low-density polyethylene) (Kratky *et al.* 1966), which provides a strong peak corresponding to a spacing of 15 nm. Lupolen is stable to the X-rays on laboratory sources and has been rather popular in the past.

The X-ray scattering from water can also be employed for absolute scale calibration, as it is nearly a constant at small angles, though due to a completely different physical phenomenon compared to neutrons. For X-rays, incoherent scattering effects are negligible, but the coherent scattering which is measured is due to the scattering density fluctuations in the irradiated volume. A hypothetical completely uniform sample (and also an ideal single crystal) would produce a non-observable zero-angle peak (δ function) corresponding to the entire irradiated volume and no scattering otherwise (except Bragg reflections in the case of a single crystal). However, for fluids, mobility of the molecules leads to a constantly fluctuating density pattern, and these statistical fluctuations yield a forward intensity proportional to the isothermal compressibility χ_T (Guinier and Fournet 1955). The differential cross-section is expressed as $d\Sigma/d\Omega(0) = \rho^2 k_B T \chi_T$, where ρ is the scattering length density, k_B is the Boltzmann constant and T is the temperature (for pure water at room temperature, $I(0) = 1.65 \times 10^{-2}\,\text{cm}^{-1}$). The use of water scattering (after subtraction of the scattering of the empty cell) is perhaps the most straightforward way to absolute calibration.

Similarly to the inelastic water scattering for neutrons, the water measurements should be done in the same conditions as for the samples, by which all the geometrical factors cancel out and the ratio of the intensities is equal to the ratio of the differential cross-sections. Alternative methods, such as the measurements of silica particles (Dreiss *et al.* 2006) or glassy carbon (Fan *et al.* 2010), are available. One can determine the apparent molecular mass MM of non-interacting particles in dilute solution from the scattering cross-section if one knows the chemical composition and the concentration c of the particles (see Section 4.1.5).

References

Bevington, P. B. (1969). *Data Reduction and Error Analysis for the Physical Sciences*. New York: McGraw-Hill.

Dreiss, C. A., Jack, K. S. and Parker, A. P. (2006). 'On the absolute calibration of bench-top small-angle X-ray scattering instruments: A comparison of different standard methods', *J. Appl. Crystallogr.* 39: 32–38.

Fan, L., Degen, M., Bendle, S., Grupido, N. and Ilavsky, J. (2010). 'The absolute calibration of a small-angle scattering instrument with a laboratory X-ray source', *J. Phys. Conf. Ser.* 247: 012005.

Fischetti, R. F., Rodi, D. J., Mirza, A., Irving, T. C., Kondrashkina, E. and Makowski, L. (2003). 'High-resolution wide-angle X-ray scattering of protein solutions: Effect of beam dose on protein integrity', *J. Synchrotron Radiat.* 10: 398–404.

Fraser, J., S, Van Den Bedem, H., Samelson, A. J. *et al.* (2011). 'Accessing protein conformational ensembles using room-temperature X-ray crystallography', *P. Natl. Acad. Sci. USA* 108: 16247–52.

Garrison, W. M. (1987). 'Reaction mechanisms in the radiolysis of peptides, polypeptides, and proteins', *Chem. Rev.* 87: 381–98.

Glasoe, P. K., and Long, F. A. (1960). 'Use of glass electrodes to measure acidities in deuterium oxide', *J. Phys. Chem.* 64: 188–90.

Glatter, O. (1977). 'Data evaluation in small-angle scattering: calculation of radial electron-density distribution by means of indirect Fourier transformation', *Acta Physica Austriaca* 47: 83–102.

Guinier, A. (1939). 'La diffraction des rayons X aux très faibles angles: Applications à l'étude des phénomènes ultra-microscopiques', *Ann. Phys. (Paris)*: 12.

Guinier, A., and Fournet, G. (1955). *Small Angle Scattering of X-Rays*. New York: Wiley.

Jacrot, B., and Zaccai, G. (1981). 'Determination of molecular-weight by neutron-scattering', *Biopolymers* 20: 2413–26.

Juers, D. H., and Matthews, B. W. (2004). 'Cryo-cooling in macromolecular crystallography: Advantages, disadvantages and optimisation', *Q. Rev. Biophys.* 37: 105–19.

Kratky, O., Pilz, I. and Schmitz, P. J. (1966). 'Absolute intensity measurement of small angle x-ray scattering by means of a standard sample', *J. Colloid Interf. Sci.* 21: 24–34.

Kuwamoto, S., Akiyama, S. and Fujisawa, T. (2004). 'Radiation damage to a protein solution, detected by synchrotron X-ray small-angle scattering: Dose-related considerations and suppression by cryoprotectants', *J. Synchrotron Radiat.* 11: 462–8.

Lindner, P. (1998) Standard water calibration at D11. *ILL Technical Report*. Institut Laue-Langevin.

Lindner, P. (2000). 'Water calibration at D11 verified with polymer samples', *J. Appl. Crystallogr.* 33: 807–11.

Muller, J. J., and Damaschun, G. (1979). 'Comparison and optimisation of smoothing procedures for small-angle X-ray scattering curves. Polynomial fitting and modified frequency filtering', *J. Appl. Crystallogr.* 12: 267–74.

Murray, J., and Garman, E. (2002). 'Investigation of possible free-radical scavengers and metrics for radiation damage in protein cryocrystallography', *J. Synchrotron Radiat.* 9: 347–54.

Neutze, R., Wouts, R., Van Der Spoel, D., Weckert, E. and Hajdu, J. (2000). 'Potential for biomolecular imaging with femtosecond X-ray pulses', *Nature* 406: 752–7.

Rolbin, Y. A., Svergun, D. I. and Schedrin, B. M. (1980). 'Smoothing experimental curves of small-angle scattering', *Kristallografiya* 25: 231–9.

Schelten, J., and Hossfeld, F. (1971). 'Application of spline functions to the correction of resolution errors in small-angle scattering', *J. Appl. Crystallogr.* 4: 210–23.

Stabinger, H., and Kratky, O. (1978). 'A new technique for the measurement of the absolute intensity of x-ray small angle scattering: The moving slit method', *Makromolekul. Chem.* 179: 1655–9.

Stuhrmann, H. B. (1974). 'Neutron small-angle scattering of biological macromolecules in solution', *J. Appl. Crystallogr.* 7: 173–8.

Svergun, D. I., and Koch, M. H. J. (2003). 'Small angle scattering studies of biological macromolecules in solution', *Rep. Progr. Phys.* 66: 1735–82.

West, J. M., Xia, J., Tsuruta, H., Guo, W., O'Day, E. M. and Kantrowitz, E. R. (2008). 'Time evolution of the quaternary structure of Escherichia coli aspartate transcarbamoylase upon reaction with the natural substrates and a slow, tight-binding inhibitor', *J. Mol. Biol.* 384: 206–18.

Zimm, B. H. (1948). 'The scattering of light and the radial distribution function of high polymer solutions', *J. Chem. Phys.* 16: 1093–9.

DATA ANALYSIS METHODS

Part
II

In this part of the book, the methods to interpret SAXS and SANS experimental data in terms of structural models of biological macromolecules will be presented. Chapter 4 is devoted to the approaches to analyse the SAS data from dilute monodisperse solutions containing identical particles. For these systems, the measured intensity corresponds to the spherically averaged scattering from a single particle, allowing one to determine not only the overall parameters but also a low-resolution three-dimensional structure. Modern approaches to construct structural models from SAXS and SANS data using *ab initio* methods and hybrid modelling will be considered. Chapter 5 deals with cases where the monodisperse approximation is not applicable. These include mixtures of particles of different size, shape or conformation, as well as systems of interacting particles, where the interparticle interference effects contribute significantly to the measured intensity. The approaches presented in Chapter 5 do not aim at the determination of structure, but rather at the quantitative characterisation of the composition of mixtures and the analysis of particle interactions in solution.

Monodisperse systems

4

4.1 Overall parameters of particles   93

4.2 Multipole representation of SAS intensity   108

4.3 Shannon sampling   110

4.4 *Ab initio* shape analysis   112

4.5 Computation of scattering patterns from atomic models   125

4.6 Hybrid methods   128

4.7 Labelling and triangulation   134

References   145

Monodisperse solutions contain particles of identical size, shape and mass (see also Chapter 3). Such systems are the easiest to work with inasmuch as the experimental scattering is related to the structure of the individual particle. Biological systems often fulfil the requirement of monodispersity naturally by the way they are produced in the cell. The methods of extracting structural information from monodisperse macromolecular will be presented in this chapter.

4.1 Overall parameters of particles

For some simple geometrical shapes, the scattering intensity can easily be calculated using analytical equations; Table 4.1 presents the most important cases where such an analytical calculation is possible. The availability of these equations to rapidly compute scattering from simple homogeneous particles allowed modelling procedures in terms of three-parametric bodies, which were popular in the 1960s. More sophisticated approaches appeared later, and these will be presented in Sections 4.2–4.6.

SAS curves also directly yield several overall model-independent parameters (such as the R_g and V_p in Table 4.1), which require no or minimal assumptions about the sample. These parameters often provide a check of the consistency of the data and/or simple geometrical characteristics of the sample. Therefore, they should always be used to limit the choice of models and also to validate them. The derivation of these parameters is simplified by using the concept of the pair-distance-distribution function (PDDF), often also called the distance distribution (function).

4.1.1 The distance distribution function

As seen in Chapter 3, it is often easier to work with real-space information. Obtaining such information is, however, not straightforward, because the solution of the *inverse scattering problem* involves an infinite (Fourier) integration of the scattering curve that is neither known from zero to infinity nor without errors.

Table 4.1 Radii of gyration (R_g), volumes (V) and analytical equations for the scattering intensity of geometrical bodies.

Body	R_g^2	V	Scattering intensity
Sphere (radius R_0)	$(3/5)R_0^2$	$(4\pi/3)R_0^3$	$V^2\left[3\dfrac{\sin t - t\cos t}{t^3}\right]^2 = V^2\Phi^2(t),\ t = qR_0$
Hollow sphere (radii R_1 and R_2)	$(3/5)\left(R_2^5 - R_1^5\right)/\left(R_2^3 - R_1^3\right)$	$(4\pi/3)\left(R_2^3 - R_1^3\right)$	$[V_1\Phi_1(t) - V_2\Phi_2(t)]^2$
Ellipsoid (semi-axes a, b, c)	$(a^2 + b^2 + c^2)/5$	$(4\pi/3)(a\,b\,c)$	$V^2\displaystyle\int_0^1\int_0^1 \Phi^2\left[q\sqrt{\left[(a^2\cos^2(\pi x/2) + b^2\sin^2(\pi x/2))(1 - y^2) + c^2 y^2\right]}\right]dx\,dy$
Rectangular prism (edge lengths A, B, C)	$(A^2 + B^2 + C^2)/12$	$A\ B\ C$	$V^2\displaystyle\int_0^1 \Psi_P\left[q, B\sqrt{(1 - x^2)}, A\right] S^2(qBCx/2)\,dx,\ S(t) = (\sin t)/t,$ $\Psi_P[q, B, A] = \dfrac{2}{\pi}\displaystyle\int_0^{\pi/2} S^2(qA\sin y/2)\,S^2(qB\cos y/2)\,dy$
Elliptic cylinder (semi-axes a, b; height h)	$(a^2 + b^2)/4 + h^2/12 = R_c^2 + h^2/12$	$\pi\ a\ b\ c$	$V^2\displaystyle\int_0^1 \Psi_{EC}\left[q, a\sqrt{(1 - x^2)}\right] S^2(qhx/2)\,dx,\ \Lambda(t) = 2J_1(t)/t$ $\Psi_{EC}[q,a] = \dfrac{1}{\pi}\displaystyle\int_0^\pi \Lambda^2\left[qa\sqrt{\dfrac{1+v^2}{2} + \dfrac{1-v^2}{2}\cos y}\right]dy,\ v = b/a$
Hollow cylinder (radii R_1, R_2; height h)	$\left(R_1^2 + R_2^2\right)/2 + h^2/12$	$\pi\left(R_1^2 - R_2^2\right)h$	$V^2\displaystyle\int_0^1 \Psi_{HC}^2\left[q, R_1\sqrt{(1-x^2)}, R_2\sqrt{(1-x^2)}\right] S^2(qhx/2)\,dx,\ \Lambda(t) = 2J_1(t)/t$ $\Psi_{[HC]}[q, R_1, R_2] = \dfrac{1}{1-v^2}\left[\Lambda^2(qR_1) - v^2\Lambda^2(qR_2)\right],\ v = R_2/R_1$

R_c is the cross-sectional radius of gyration.
$J_1(t)$ is the Bessel function of the first kind.

From eqs. (1.1) and (1.4) it follows that

$$\frac{d\sigma_V}{d\Omega}(q) = \frac{\langle A(\mathbf{q}) A^*(\mathbf{q}) \rangle}{V} = \frac{1}{V} \left\langle \int_V d\mathbf{r} \int_V d\mathbf{r}' \Delta\rho(r)\Delta\rho(r') \exp(-i\mathbf{q}(\mathbf{r} - \mathbf{r}')) \right\rangle$$

(4.1)

Here, the integral is over the whole illuminated sample volume V, and $<\ldots>$ denotes the spatial and temporal average. For the sake of simplicity, we shall use $I_{\text{total}}(q)$ for $d\sigma_V/d\Omega$ (that is, fraction of neutrons or photons scattered from the incident beam/(unit solid angle and unit time)) in the following.

In the case of particles in solution, one can separate this expression—at least for centrosymmetric particles—into a part that describes scattering from N particles and one that describes the interaction between the particles:

$$I_{\text{total}}(q) = \frac{N}{V} \left\langle \int_{V_p} d\mathbf{r} \int_{V_p} d\mathbf{r}' \Delta\rho(\mathbf{r})\Delta\rho(\mathbf{r}') \exp(-i\mathbf{q}(\mathbf{r} - \mathbf{r}')) \right\rangle$$

$$\left\langle \frac{1}{N} \sum_{i=1}^{N} \sum_{j=1}^{N} \exp\left(-i\mathbf{q}\left(\mathbf{r}_i - \mathbf{r}_j\right)\right) \right\rangle$$

(4.2)

The first factor in brackets in eq. (4.2) is the form factor $I(q)$, and the second one is the structure factor $S(q)$, which will be discussed in Chapter 5:

$$S(q) = 1 + \frac{1}{N} \left\langle \sum_{i=1}^{N} \sum_{j \neq i}^{N} \exp\left(-i\mathbf{q}\left(\mathbf{r}_i - \mathbf{r}_j\right)\right) \right\rangle$$

$$= 1 + \frac{N-1}{V} \int g(\mathbf{r}) \exp\left(-\mathbf{q}\mathbf{r}\right) d\mathbf{r}$$

(4.3)

where $g(\mathbf{r})$ is the pair-correlation function of the centres of (scattering) mass of the particles (see Appendix 3). The form factor $I(q)$ is sometimes also written $F(q)$ or $V_p^2 P(q)$. Assuming $S(q) \approx 1$ (that is, absence of interactions) $I(q)$ can be assumed to be proportional to $I_{\text{total}}(q)$. This will be often the case in biological applications, where one usually deals with dilute samples. Note, however, that more concentrated protein solutions or even dilute solutions of charged species such as nucleic acids may exhibit significant interactions.

If one defines the correlation function of the particle $\gamma(\mathbf{r})$ (Debye and Bueche 1949) as

$$\gamma_p(\mathbf{r}) = \left\langle \int_{V_p} \Delta\rho(r')\Delta\rho(r' - r) dr' \right\rangle$$

(4.4)

it follows that the form factor $I(q)$ of the particles is

$$I(q) = \left\langle \int_{V_p} \gamma_p(\mathbf{r}) \exp\left(-i\mathbf{q}\mathbf{r}\right) d\mathbf{r} \right\rangle = \int_{V_p} \left(\frac{1}{4\pi} \int_{\Omega} \gamma_p(\mathbf{r}) d\Omega \right) \exp\left(-i\mathbf{q}\mathbf{r}\right) d\mathbf{r} \tag{4.5}$$

Solving the integral in round brackets, $I(q)$ becomes

$$I(q) = \int_0^{D_{max}} 4\pi r^2 \gamma_p(r) \frac{\sin(qr)}{qr} dr \tag{4.6}$$

or

$$I(q) = 4\pi \int_0^{D_{max}} p(r) \frac{\sin(qr)}{qr} dr \tag{4.6a}$$

with $p(r) = r^2\gamma(\mathbf{r})$, the PDDF or distance distribution representing the number of distances between two points in the particle within the interval between r and $r + dr$ weighted by the product of the excess scattering-length densities at these two points. This function obviously extends over the interval $[0, D_{max}]$, where D_{max} is the largest distance between any two points inside the particle, weighted by the product of the excess scattering-length densities at these two points. Hence, $p(0) = 0$ and $p(D_{max}) = 0$. In an homogeneous particle, $p(r)$ simply counts the number of distances within the interval r and $r + dr$.

Interestingly, and in contrast to the intensity calculations, analytical equations are rarely available for the $p(r)$ functions, even for simple homogeneous geometrical bodies. One of the very few cases is a classical PDDF of a solid sphere of radius R: $p(r) = Vr^2(1 - 3t/4 + t^3/16)$, where $V_p = 4\pi R^3/3$ is the volume and $t = r/R$.

Equation (4.6) can be written in a discrete way for particles that are modelled using N_{sph} spheres. $I(q)$ is then given by Debye's formula (Debye 1915):

$$I(q) = \sum_{i=1}^{N_{sph}} F_i^2(q) + 2 \sum_{i=1}^{N_{sph}-1} \sum_{j=i+1}^{N_{sph}} F_i(q) F_j(q) \frac{\sin(qr_{ij})}{qr_{ij}} \tag{4.6b}$$

where $F_i(q)$ are the form factors of the spheres (including their scattering power) and r_{ij} the distance between the centres of sphere i and j. The calculation in eq. (4.6b) can be simplified and accelerated if the spheres are of the same size and scattering power, and also by binning the distances r_{ij} into a limited number (such as 10,000) of channels (Glatter 1972, 1980). Further applications of the Debye formula are described in Section 4.4.

Since eqs. (4.6) and (4.6a) are sine Fourier transforms, one can write the inverse Fourier transform of (4.6a) as

$$p(r) = \frac{1}{2\pi^2} \int_0^\infty I(q)\, qr \sin(qr)\, dq \tag{4.7}$$

However, $p(r)$ cannot be determined directly using this equation, because $I(q)$ is not known over the full interval $0 \leq q \leq \infty$, and because it can only be obtained with statistical and systematic errors.

The 'indirect Fourier transform' (IFT) method was therefore introduced by Glatter (1977): The distance distribution is approximated by a limited series of functions in real space, applying coefficients that are to be determined by fitting the theoretical scattering curve obtained via eq. (4.6a) to the measured scattering intensities. Since the ideal scattering curves are 'smeared' due to the instrumental conditions (source, sample and detector-pixel dimensions and wavelength distribution and, in the case of neutrons, even gravity) it is important to be able to apply easily the corresponding convolutions to the theoretical (approximated) scattering curves.

In mathematical terms, the approximated distance distribution $p_a(r)$ can be written as

$$p_a(r) = \sum_{i=1}^{n_s} c_i \varphi_i(r) \text{ for } 0 \leq r \leq D_{\max} \tag{4.8}$$

where n_s is the number of functions φ_i Glatter (1977) uses cubic B splines (Greville 1969, Schelten and Hossfeld 1971). These functions consisting of cubic segments offer a good combination of flexibility and computational simplicity to parameterise smooth functions on a finite support.

Equation (4.6a) allows one to calculate a scattering function $I_a(q)$ for any given set of coefficients. $I_a(q)$ is itself a sum of partial scattering functions ψ_i with the *same* coefficients:

$$I_a(q) = 4\pi \sum_{i=1}^{n_s} c_i \int_0^{D_{\max}} \varphi_i(r) \frac{\sin(qr)}{qr}\, dr = 4\pi \sum_{i=1}^{n_s} c_i \psi_i(q) \tag{4.9}$$

Similarly, every other linear transformation, in particular the convolutions due to geometrical and wavelength smearing, lead to sums of functions $\zeta_i(q)$ that always keep the same coefficients $c_i, i = 1, \ldots n_s$. The coefficients of a sum of these individual convoluted scattering curves are determined by fitting it to the observed scattering curves using a least-squares approximation; applying these coefficients in eq. (4.8) one obtains the approximated distance distribution. Figure 4.1 presents a schematic view of the procedure.

The number of splines n_s is not fixed, but it should neither be too small nor too large. If n_s is too small, eq. (4.8) cannot provide the necessary resolution; if it is too large, the number of spline coefficients

Fig. 4.1 Procedure of IFT with the use of cubic splines. Bottom right: A set of equidistant cubic splines (φ_i, $i = 1 \ldots n_s$) with coefficients c_i (to be determined) are summed up to yield a distance distribution $p(r)$, above. Bottom, centre: For each spline, an individual scattering function ψ_i is calculated by Fourier transformation. Bottom left: The ψ_i are convoluted using the instrument functions to yield 'smeared' individual intensities ζ_i. A sum of these smeared intensities is fitted to the experimental curve $I(q)$ (top left), yielding the coefficients c_i. If one applies the same coefficients to the set of scattering functions ψ_i one obtains the 'desmeared' scattering curve $I_{app}(q)$ (top, centre), and in the same way the fitted distance distribution $p(r)$ by applying the now known coefficients (top right). Note that the fit requires a regularisation procedure described in the text. Adapted, with permission, from Glatter (1979).

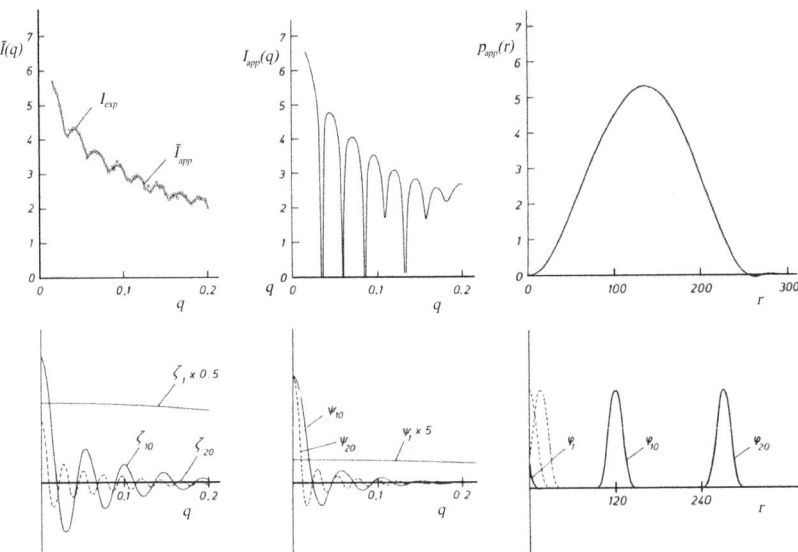

may exceed the number of independent parameters that can be deduced from the scattering curve. For most practical applications, 20–30 splines are sufficient to parameterise the $p(r)$ function.

Evidently, D_{max} needs to be estimated for defining the real-space region that is spanned by the sum of spline functions (and for the upper integration limit). Due to the sampling theorem of Fourier transformation (Shannon 1949, Bracewell 1986), D_{max} cannot be chosen larger than π/q_{min}, where q_{min} is the smallest measured q-value (see also Section 4.3). Inversely, this also means that q_{min} must be smaller than π/D_{max} if a D_{max} is known by other methods; for example, EM.

The weighted least-squares fit consists in calculating and minimising the discrepancy χ (eq. 3.9):

$$\chi^2 = \frac{1}{N_k - 1} \sum_{j=1}^{N_{obs}} \left[\frac{I(q_j) - I_{calc}(q_j)}{\sigma_j} \right]^2 = \sum_{j=1}^{N_{obs}} \left[\frac{I(q_j) - 4\pi \sum_{i=1}^{n_s} c_i \zeta_i (q_j)}{\sigma_j} \right]^2$$

(4.10)

Here $I(q_i)$ are the N_{obs} observed scattering intensities with errors σ_i at the q-values q_i. It is important to note that the fit does not require data points in the very-low and very-high q-regions.

Solving eq. (4.10) is a so-called *ill-posed problem*. If one were to use the equation as is, there would be a high risk of finding a solution where the coefficients c_i lead to a strongly oscillating $p(r)$ function. The aim, though, is to find the smoothest solution fitting the observed data. This is obtained by introducing an additional constraint

$$\chi + \alpha N_{c'} = \text{Min}$$

(4.11)

where $N_{c'}$ is the minimised norm of the derivatives of the solution coefficients:

$$N_{c'} = \sum_{i=1}^{n_s-1} \left(c_{i+1} - c_i\right)^2 \tag{4.11a}$$

and α is a so-called Lagrange multiplier. Equation (4.11) is a variant of the regularisation formalism (Tychonoff 1943). For $\alpha = 0$, the minimum condition (4.11a) does not come into play at all, whereas its contribution increases with α. One calculates solutions for different values of α and looks for a solution where χ increases only little while the oscillations in the coefficients diminish.

Practically, the Lagrange multiplier reduces the effective number of parameters (coefficients) used in the approximation, since it links them to each other. The method has been used in the program ITP. For further practical details of the method we refer to the original paper (Glatter 1977).

Svergun has provided an alternative with the program GNOM that is now frequently used and employs automatic criteria for the Lagrange parameter, D_{max} and so on, thus helping the non-experienced user (Svergun *et al.* 1988, Svergun 1992). The GNOM solution is also an intermediary step in the calculation of *ab initio* structure models described in Section 4.3.

Other approaches to IFT have been reported by several authors (Moore 1980, Taupin and Luzzati 1982, Provencher 1982a, Provencher 1982b, Mangani *et al.* 1988, Hansen and Pedersen 1991). These algorithms are less widely spread, and not all of them are applicable to biological systems. Although Moore's approach allows one to approximate the data well and understand error propagation, it suffers from termination effects and oscillations in the $p(r)$ function (Hansen and Pedersen 1991). Promising new approaches to IFT consist in the use of Bayesian and maximum-entropy methods for estimating the parameters for indirect transformation; see, for example, Hansen (2000) and Vestergaard and Hansen (2006).

Brunner-Popela and Glatter (1997) have extended the indirect Fourier transform method to allow the treatment of the scattering of interacting particles in the generalised indirect Fourier transform method (GIFT), which allows one to simultaneously analyse the form factor and the structure factor of such a solution (Weyerich *et al.* 1999). The speed and reliability of GIFT, which is also referred to in Sections 5.5 and 8.1, was further improved by replacing the non-linear least-squares approach for finding the parameters of the structure factor by the 'Boltzmann simplex simulated annealing (BSSA) algorithm' (Bergmann *et al.* 2000).

Note that some features in real space, like hollow spheres (such as virus capsids), rod-like or lamellar structures, appear more clearly in a plot of $p(r)/r$ *versus r*.

4.1.2 Cross-sectional distance distribution

The scattering from elongated particles with constant cross-section can be expressed as (Porod 1948, Kratky and Porod 1948):

$$I(q) = \frac{l\pi}{q} I_c(q) \tag{4.12}$$

$I_c(q)$ depends only on the cross-section and its structure. l is the length of the particle. By analogy to eqs. (4.1)–(4.4) one can define a cross-sectional correlation function $\gamma_c(r)$ and a cross-sectional distribution $p_c(r)$:

$$\gamma_c(r) = 2\pi \int_{A_c} \Delta\rho_c(r')\Delta\rho_c(r' + r)dr = \frac{p_c}{r} \tag{4.13}$$

where the integration is over the cross-sectional area A_c of the elongated particles. With the two-dimensional average in the plane perpendicular to the long axis,

$$\langle \exp(-i\mathbf{q}\mathbf{r}_c) \rangle = J_0(qr_c) \tag{4.14}$$

where $J_0(x)$ is the Bessel function of the first kind of order 0 (Watson 1995), $I_c(q)$ can be written as a Hankel transformation (Glatter 1982):

$$I_c(q) = 2\pi \int_0^\infty p_c(r) J_0(qr) \, dr \tag{4.15}$$

4.1.3 Distance distribution of platelets (lamellae)

By analogy to eq. (4.12) the scattering from flat particles with constant thickness and an extension much larger than the thickness can be expressed as (Glatter 1982)

$$I(q) = \frac{2\pi A_l}{q^2} I_t(q) \tag{4.16}$$

where A_l is the area of the lamellar particles and $I_t(q)$ is the thickness factor that is related to thickness pair-distribution by

$$I_t(q) = 2 \int_0^\infty p_t(r) \cos(qr) \, dr \tag{4.17}$$

(Glatter 1982), where the thickness distribution $p_t(r)$ is equal to the thickness correlation function $\gamma_t(r)$

$$p_t(r) = \gamma_t(r) = 2\pi \int_0^\infty \Delta\rho_t(r')\Delta\rho_t(r' + r)dr \tag{4.18}$$

IFT procedures allow one to obtain $p_c(r)$ and $p_t(r)$ from $I_c(q)$ and $I_t(q)$ (Glatter 1982).

4.1.4 Forward-scattering intensity and radius of gyration

As shown in eq. (4.6a), the scattering from a dilute, interaction-free solution of homogeneous particles can be written as

$$I(q) = 4\pi \int_0^{D_{max}} p(r) \frac{\sin(qr)}{qr} dr \qquad (4.19)$$

The function $(\sin x)/x$ can be approximated by:

$$\frac{\sin x}{x} \approx 1 - \frac{x^2}{3!} + \frac{x^4}{5!} - \ldots \approx 1 - \frac{x^2}{6} + \frac{x^4}{120} - \ldots \qquad (4.20)$$

With this, eq. (4.19) becomes

$$I(q) = 4\pi \left(\int_0^{D_{max}} p(r)dr - \frac{q^2}{6} \int_0^{D_{max}} r^2 p(r)dr + \frac{q^4}{120} \int_0^{D_{max}} r^4 p(r)dr - \ldots \right) \qquad (4.21)$$

or

$$I(q) = 4\pi \int_0^{D_{max}} p(r)dr \left(1 - \frac{q^2}{3}\frac{1}{2}\frac{\int_0^{D_{max}} r^2 p(r)dr}{\int_0^{D_{max}} p(r)dr} + \frac{q^4}{120}\frac{\int_0^{D_{max}} r^4 p(r)dr}{\int_0^{D_{max}} p(r)dr} - \ldots \right) \qquad (4.21a)$$

This is

$$I(q) = I(0) \left(1 - \frac{q^2 R_g^2}{3} + \ldots \right) \qquad (4.22)$$

since the third term in eq. (4.21) contributes only very little to the sum, and R_g, the second moment of $p(r)$, is the radius of gyration.

Guinier (1939) recognised that eq. (4.22) resembles the expansion of an exponential

$$e^{-x} = 1 - x + \frac{x^2}{2!} - \frac{x^3}{3!} = \ldots \qquad (4.23)$$

that is,

$$I(q) \approx I(0) \exp \left(-\frac{q^2 R_g^2}{3} \right) \qquad (4.24)$$

('Guinier's law') so that one can determine the zero-angle intensity and R_g from the zero intercept (ln $I(0)$) and initial slope in a plot of ln $I(q)$ *versus* q^2. This was particularly valuable in times where one used graphical solutions before computers became generally available. Equations (4.22) and (4.24) hold in a range of about $0 < q < 1/R_g$.

In the soft matter field, one often uses the so-called Zimm plot, where $1/I(q)$ or $c/I(q)$ is plotted *versus* q^2 for different concentrations c (Zimm

1948b, Zimm 1948a). In fact, since $1/(1 - x)$ can be expanded as $1 + x + x^2 + \ldots$, one finds from eq. (4.22) that $1/(I(q)) \approx I(0)(1 + (qR_g)^2/3 + \ldots)$. The Zimm approximation yields very similar but not quite identical R_g values due to the different second-order contribution.

If the R_g is not determined from the second moment of the $p(r)$ function (where aggregates appear as shoulders or long tails at high r-values), but from a logarithmic plot, care is to be taken when the low-q intensities deviate, even slightly, from a straight line; this is a sign of aggregation (deviation towards higher intensity) or repulsive interactions (deviation towards lower intensity).

The R_g obtained from a scattering experiment is analogous to the mechanical radius of gyration where mass is replaced by scattering length or electron density. The mechanical R_g is given by

$$R_g^2 = \sum_j m_j r_j^2 / \sum_j m_j \tag{4.25}$$

where m_j is a mass located at distance r_j from the centre of mass. $\sum_j m_j r_j^2$ is known as the moment of inertia, and $m_p = \sum_j m_j$ is the total mass of the particle.

Replacing m_j by the excess scattering density the expression for the R_g of the scattering particle becomes

$$R_g^2 = \int_{V_p} (\rho(\mathbf{r}) - \rho_s) r^2 d\mathbf{r} / \int_{V_p} (\rho(\mathbf{r}) - \rho_s) d\mathbf{r} \tag{4.26}$$

where $\rho(\mathbf{r})$ is the scattering length density or electron density at position $\mathbf{r}$ (measured from the centre of gravity in terms of scattering) of the particle, and V_p the particle volume including also atoms of the hydration layer which may have a density different from the bulk solvent. ρ_S is the scattering length density of the solvent. It should be noted that for neutron scattering where the contrast can be positive or negative, R_g^2 can also be negative. The radii of gyration of a number of simple geometrical bodies are given in Table 4.1.

The R_g obtained from a scattering experiment is a first indication of the compactness of a macromolecule. For a given volume the smallest R_g is that of the solid sphere; the more anisometric the body, the larger its R_g. For X-rays, most single-component macromolecules (proteins, nucleic acids) can be considered as having close to uniform electron density, and the R_g is related to the geometrical shape of the particle.

For neutrons, the contrast between solvent and macromolecule can vary enormously, as described in Chapter 1. The R_g of a macromolecular complex comprising two or more components such as protein, nucleic acid, lipid or deuterated versions of these moieties depends very strongly on the relative contrasts of the components. As a consequence of the characteristic functions (see Section 1.3) the R_g varies with the contrast $\bar{\rho}$ (Stuhrmann and Kirste 1965):

$$R_g^2 = R_v^2 + \frac{\alpha}{\bar{\rho}} + \frac{\beta}{\bar{\rho}^2} \tag{4.27}$$

Equation (4.27) was first expressed completely (including the third term) by Ibel and Stuhrmann (1975) and Cotton and Benoit (1975). In other words, the square of the R_g is a quadratic function of the inverse of the contrast. For a complex having two components with strongly different scattering densities the constant β is an indication of the distance d between the centres of mass of the two components.

$$d = \left(\frac{\beta}{\bar{\rho}_1^2} \right)^{1/2} + \left(\frac{\beta}{\bar{\rho}_2^2} \right)^{1/2} \tag{4.28}$$

where $\bar{\rho}_1$ and $\bar{\rho}_2$ are the match points of components 1 and 2 (Koch and Stuhrmann 1979). If β is zero the two components have the same centre of mass. In this case, a plot of R_g^2 *versus* $\frac{1}{\bar{\rho}}$ ('Stuhrmann plot') will be linear. If its slope α is positive then the component of higher scattering density is farther from the centre of mass, and the component of lower scattering density is closer and *vice versa* if α is negative. Examples are discussed in Section 6.2.6.

Note, however, that if the exchange of hydrogen by deuterium is not homogeneous throughout the particle the geometrical parameters of the particle can no longer be simply extracted from a contrast variation study (Witz 1983).

Another analogy between R_g in scattering and in mechanics lies in the use of the parallel axis theorem. Briefly, if the R_g of a complex of two particles is known as well as the radii of gyration of the two components, the distance between the two components can be calculated without recourse to any other structural information (see Section 4.7.3).

The forward scattering $I(0)$ is directly proportional to the square of the contrast, as shown previously (Section 1.3). The point at which it vanishes is therefore the point at which the solvent and solute have the same scattering length density ,and is known as the contrast match or isopicnic point. For a complex composed of two components each of known scattering length density, the contrast match point can be used to determine the stoichiometry of the complex.

4.1.5 Obtaining the molecular mass

If one divides the differential scattering cross-section per volume by the particle concentration c_S (in g cm^{-3}) one obtains the differential scattering cross-section per unit mass:

$$\frac{d\sigma_m}{d\Omega}(q) = \frac{1}{c_S} I(q) (cm^2 g^{-1} \text{ sterad}^{-1}) \tag{4.29}$$

The forward scattering from N non-interacting particles in solution is

$$I_s(0) = I_0 \eta \frac{T_s}{T_{ap}} \Delta \Omega N \left(\int_{V_p} \rho(\mathbf{r}) d\mathbf{r} - \rho_s V_p \right)^2 \quad \text{(neutrons or photons s}^{-1}\text{)}$$

(4.30)

Comparison with eq. (3.12) yields the differential scattering cross-section per unit volume from N non-interacting particles in solution in the forward direction as

$$I(0) = \frac{N}{V} \left(\int_{V_p} \rho(\mathbf{r}) d\mathbf{r} - \rho_s V_p \right)^2 \quad \text{(sterad}^{-1} \text{cm}^{-1}\text{)}$$

(4.31)

For neutrons, $\int_{V_p} \rho(\mathbf{r}) \, d\mathbf{r} = \sum_P b_i$, the sum over the scattering lengths of all atoms in one particle (including bound water).

With $\rho_p = \int_{V_p} \rho(\mathbf{r}) d\mathbf{r} / V_p$, the average scattering-length or electron density of the particle, one obtains

$$\int_{V_p} \rho(\mathbf{r}) d\mathbf{r} - \rho_s V_p = (\rho_p - \rho_s) V_p = \Delta \rho_p V_p = \Delta \rho_p \bar{v}_p m_p = \Delta \rho_p \bar{v}_p \frac{M}{N_A}$$

(4.32)

with M, the molecular mass (MM) and N_A, Avogadro's number, m_p, the mass of one particle, $\Delta \rho_p = (\rho_p - \rho_s)$ its average excess scattering density, and $\bar{v}_p$ its partial specific volume. The number of particles in the beam N is:

$$N = \frac{V c_s}{m_p} = \frac{V c_s N_A}{M}$$

(4.33)

Equation (4.27) results in

$$I(0) = \frac{c_s N_A}{M} \left(\Delta \rho_p \bar{v}_p \frac{M}{N_A} \right)^2 = \frac{c_s M}{N_A} \left(\Delta \rho_p \bar{v}_p \right)^2 \text{(sterad}^{-1} \text{cm}^{-1}\text{)}.$$

(4.34)

or

$$M = I(0) \frac{N_A}{c_s \left(\Delta \rho_p \bar{v}_p \right)^2} = \frac{d\sigma_m}{d\Omega}(0) \frac{N_A}{\left(\Delta \rho_p \bar{v}_p \right)^2} \text{(g Mol}^{-1}\text{)}$$

(4.35)

Note that because of hydrogen exchange, b_i and therefore ρ_p and $_{\Delta\rho}$ depend on the contrast for neutrons.

The molecular mass so obtained is an *effective* molecular mass. Molecules composed of two moieties (protein, lipids, sugar, nucleic acids) measured at the matching point of one of them may only reveal the molecular mass of the visible part, using the respective partial specific volume. This MM can in principle be recalculated into the MM of the entire molecule if its stoichiometric composition is known.

In practical SAXS applications, standard proteins with known MMs (such as cytochrome C, lysozyme, bovine serum albumin or glucose isomerase) are often employed to determine the experimental MM of the given protein. Assuming that the $\bar{v}_p$ values of the standard and actually measured protein are similar, a ratio $M_p = M_{st} \times \frac{I(0)_p/c_p}{I(0)_{st}/c_{st}}$ is used where $I(0)_p, I(0)_{st}$ are the scattering intensities at zero angle of the studied and the standard protein, respectively, M_p, M_{st} are the corresponding molecular masses and c_p, c_{st} are the concentrations.

Alternatively, scattering from water can be used to obtain the scattering from the solute on the absolute scale (Feigin and Svergun 1987, Orthaber *et al.* 2000) and then to calculate the MM. The water sample is measured in the same cell as the protein samples, and the scattering from the empty holder is subtracted. As indicated in Section 3.7.3, scattering from water in the small angle range is a constant defined by the isothermal compressibility, and forward scattering from water $I(0)$ in absolute scale is known to be $1.632 \times 10^{-2}\,\mathrm{cm}^{-1}$ at 293 K (Orthaber *et al.* 2000). By dividing the relative $I(0)$s of the solute with the experimental $I(0)$ of water and then multiplying by the absolute scattering of water one obtains the $I(0)$s of the proteins in absolute scale. Following (4.35), the MM of a protein in kDa is expressed as $M = (I(0) \times N_A)/(10^3 \times \Delta\rho_M^2)$, where $I(0)$ is the absolute forward scattering (after normalising against concentration), $\Delta\rho_M = [\rho_{M,prot} - (\rho_{solv} \times \bar{v}_p)] \times r_o$ is the scattering contrast per mass, $N_A = 6.023 \times 10^{23}\,\mathrm{mol}^{-1}$ is the Avogadro number, $\rho_{M,prot} = 3.22 \times 10^{23}\,\mathrm{e\,g}^{-1}$ is the electron density per mass of proteins, $\rho_{solv} = 3.34 \times 10^{23}\,\mathrm{cm}^{-3}$ is the electron density per volume of the aqueous solvent, and $r_o = 2.8179 \times 10^{-13}\,\mathrm{cm}$ is the scattering length of an electron.

The accuracy of the calculation of the protein MM from SAXS or SANS data is about 10% for the two methods, largely limited by the accuracy of the determination of the solute concentrations. The partial specific volume of a protein $\bar{v}_p$ can in most cases be fixed to a consensus value $0.7425\,\mathrm{cm}^3/\mathrm{g}$, yielding good results for protein solutions (Mylonas and Svergun 2007). For nucleic acids and other macromolecular systems, the appropriate molecular parameters (ρ_M and $\bar{v}_p$) must be taken into account. Roughly, an RNA or DNA molecule scatters X-rays about twice as much compared to a protein with the same MM in an aqueous solution.

With neutrons, one can directly apply eq. (4.35), or one can make use of the formula by Jacrot and Zaccai (1981):

$$\frac{I(0)}{c_s \times I_{H_2O}(0)} = f \frac{4\pi T_S}{\left(1 - \frac{T_{H_2O}}{T_{ec}}\right)^2} M N_A D \times 10^{-3} \left(\frac{\sum b - \rho_S V_p}{M}\right)^2 \quad (4.35a)$$

where $I_{H_2O}(0)$ is the forward scattering of water using the same aperture and instrument settings as for the sample, T are the transmissions,

f is a tabulated correction factor depending on the wavelength and the instrument used, c_s is the sample concentration in $g\,L^{-1}$, D is the sample thickness, and Σ_b is the sum over all scattering lengths in the excluded particle volume V_p. The expression in brackets on the right-hand side is nearly a constant for different proteins in H_2O. From eq. (4.35a) follows

$$M = \frac{I(0)}{c_s \times I_{H_2O}(0)} \frac{\left(1 - \frac{T_{H_2O}}{T_{ec}}\right)^2}{4\pi T_S f\,N_A D} \times 10^3 \left(\frac{M}{\sum b - \rho_S V_p}\right)^2 \qquad (4.35b)$$

Given that water scattering belongs to standard measurement protocols for SANS, reference proteins are rarely used for the MM determination with neutrons.

4.1.6 Porod invariant

Following the Parseval theorem (see Appendix 1, eq. A1.39) the total scattered intensity is equal to the total fluctuation of the scattering length density. The integral

$$Q = \int_0^\infty I(q) q^2 dq = 2\pi^2 \int_V \Delta\rho(\mathbf{r}) d\mathbf{r} \qquad (4.36)$$

yields a constant, Q, called Porod's invariant (Porod 1951), irrespective of the nature of the scattering sample. For a binary particulate system— that is, in the absence of the variation of the scattering density in the solvent nor in the solute—Q is related directly to the excluded particle volume, and, using that $I(0) = (\Delta\rho)^2 V_p^2$, one obtains

$$Q = 2\pi^2(\Delta\rho)^2 V_p \quad \text{and} \quad V_p = 2\pi^2 I(0)/Q \qquad (4.36a)$$

Equation (4.36a) yields the particle volume and provides a consistency test of the data, but since eq. (4.36) is an infinite integral, the measured intensity must be extrapolated towards zero angle and infinity. The zero-angle extrapolation can be done using the Guinier approximation (see eq. 4.24), whereas employing the Porod asymptotic (see next section 4.1.7) one can approximate the behaviour at large q.

4.1.7 Porod asymptotic and Kratky plot

Disregarding scattering contributions from the internal structure of the particle that would be visible at large $q(>10\text{nm}^{-1})$, and assuming the presence of a rather sharp interface between the particles and the solvent, one should expect the scattering intensity to decay proportionally to q^{-4}. This has been derived independently by Porod (1951) and Debye *et al.* (1957).

$$\lim_{q\to\infty} I(q) = \frac{2\pi(\Delta\rho)^2}{q^4} \frac{S_{int}}{V} \qquad (4.37)$$

where $\Delta\rho$ is the excess scattering length density, S_{int} is the sum of internal scattering surfaces, and V is the (illuminated) volume of the sample.

Porod stated in his original paper that this relation would hardly be of practical use. It turns out, however, that it is useful in many cases—in studies of two phase-systems, for example—as it does not depend on the presence of particles, and in studies of granular systems where the contribution of the grain surface scattering can be separated from the internal structure of the grains.

In the case of biological systems it is often difficult to measure the exact contribution from the solvent, in particular with neutrons, since the composition of a reference buffer may differ slightly from that of the solution. Even a small difference can lead to a different incoherent-scattering level. In a plot of $I(q) \times q^4$ *versus* q^4 (Porod plot) a residual background appears as a slope and $2\pi(\Delta\rho)^2 S/V$ as the zero intercept. This allows one to estimate a flat background.

Rough surfaces, fractal systems and the influence of curvature in some symmetric bodies lead to a behaviour that is different from eq. (4.37). Since this is of less importance for biological systems, it is not further treated here; details can be found in, for example, Schmidt (1995). However, of much practical importance is the fact that the asymptotic behaviour of the scattering intensity is sensitive to the folding state of the macromolecule.

To qualitatively distinguish between globular particles and disordered states a Kratky plot $q^2 \times I(q)$ *versus* q is used (Kratky 1982). The intensity of a globular protein decays approximately as q^{-4}, yielding a bell-shaped Kratky plot with a well-defined maximum (see the example in Fig. 4.2). Unfolded proteins show a much slower intensity decay: for

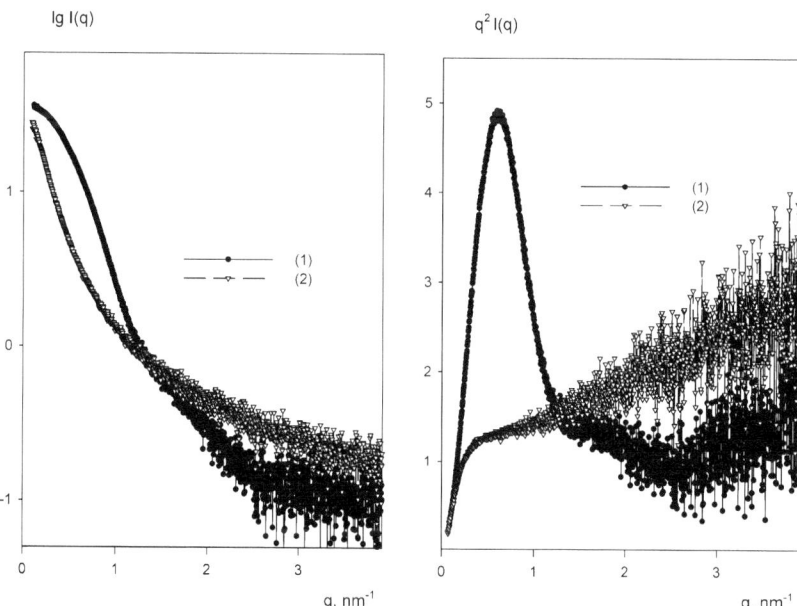

Fig. 4.2 Conventional semi-logarithmic plot (left panel) and Kratky plot (right panel) of a folded protein (BSA, curve 1) and from an intrinsically unfolded protein (τ, curve 2). Note that the scattering from the folded protein (curve 1) decays slower than q^{-4} at higher angles because of the contribution from the internal structure.

example, an ideal random (Gaussian) chain (a random chain) decays as q^{-2} (Debye 1947). The Kratky plot of an unfolded protein therefore presents a plateau instead of the peak. At yet higher angles (higher resolution), SAS 'sees' the scattering from the individual chain with the asymptotic of the rod (q^{-1}). The Kratky plot from unfolded proteins therefore has a plateau over a specific range of q, which is followed by a monotonic increase. Partial unfolding or flexibility of the macromolecule lead to an increase of the scattering at higher angles and display an intermediate behaviour between that of the folded protein and of the random chain. This graphical representation enhances the relevant features at higher angles and is a very good tool to qualitatively assess compactness, for example, in unfolding experiments. The Kratky plot is routinely used in SAS data analysis and provides the first estimate of the folded state of the macromolecule. More advanced approaches to quantitatively characterise flexible and (intrinsically) unfolded macromolecules are described in Section 5.4.

4.2 Multipole representation of SAS intensity

Scattering and the transition between real and reciprocal space are described by the Fourier transformation introduced in Chapter 1, and this transformation forms the mathematical basis for data treatment. The scattering from monodisperse systems, which results from the spherical averaging over all orientations of a single particle, is also conveniently described in spherical coordinates (q, Ω) in reciprocal space as

$$I(q) = \langle A(\mathbf{q}) \cdot A^*(\mathbf{q}) \rangle_\Omega \tag{4.38}$$

In these coordinates the natural representation of trigonometric functions relies on spherical harmonics (Harrison 1969, Stuhrmann 1970a, Stuhrmann 1970b), which form a complete set of orthogonal angular functions $Y_{lm}(\Omega)$ defined on the surface of the unit sphere. They are combinations of trigonometric functions of orders l and m, where the lower-order harmonics define the gross structural features of the particle and the higher harmonics describe finer details (see Appendix 2).

The scattering density of a particle can be expressed as a sum of spherical harmonics (also called multipole expansion) as:

$$\rho(\mathbf{r}) \approx \rho_L(\mathbf{r}) = \sum_{l=0}^{L} \sum_{m=-l}^{l} \rho_{lm}(r) \, Y_{lm}(\omega) = \sum_{l=0}^{L} \rho_l(\mathbf{r}) \tag{4.39}$$

where $(r, \omega) = (r, \theta, \varphi)$ are spherical coordinates and

$$\rho_{lm}(r) = \int_\omega \rho(\mathbf{r}) \, Y_{lm}^*(\omega) d\omega \tag{4.40}$$

are the radial functions. The truncation value L defines the accuracy of the expansion (due to completeness of the spherical harmonics,

$\rho_L(\mathbf{r}) \rightarrow \rho(\mathbf{r})$ when $L \rightarrow \infty$). The scattering amplitude can similarly be represented in reciprocal space:

$$A(\mathbf{q}) = \sum_{l=0}^{L} \sum_{m=-l}^{l} A_{lm}(q) Y_{lm}(\Omega) \qquad (4.41)$$

The partial amplitudes $A_{lm}(q)$ are related to the radial functions by the Hankel transformation

$$A_{lm}(q) = i^l \sqrt{\frac{2}{\pi}} \int_{0}^{\infty} j_l(qr)\rho_{lm}(r) r^2 dr \qquad (4.42)$$

where the $j_l(qr)$ are spherical Bessel functions (Stuhrmann 1970b). Substituting eq. (4.41) into eq. (4.38), all cross-terms in the average vanish due to the orthogonality of the spherical harmonics yielding a simple expression for the intensity:

$$I(q) = \sum_{l=0}^{L} I_l(q) = 2\pi^2 \sum_{l=0}^{L} \sum_{m=-l}^{l} \left| A_{lm}(q) \right|^2 \qquad (4.43)$$

The scattering intensity of a particle is thus a sum of independent contributions from substructures corresponding to different spherical harmonics $Y_{lm}(\omega)$. The absence of cross-terms is a unique important property of the multipole expansion (note that cross-terms are always present when decomposing the scattering density into other additive components, for example, in contrast variation; see Chapter 1, eq. (1.6)).

The use of eqs. (4.39)–(4.43) allows one not only to rapidly compute scattering patterns from a given structure but also to meaningfully approach the inverse problem (that is, reconstructing the structure from a given solution scattering pattern). The ambiguity of the inverse problem has been clearly demonstrated using spherical harmonics (Stuhrmann 1970b). Indeed, the partial densities $\rho_l(r)$ in the multipole expansion in eq. (4.39) can be rotated arbitrarily without affecting the partial intensities $I_l(q)$ (Stuhrmann 1970b), which means that an infinite number of distributions $\rho(r)$ exists, providing one and the same intensity $I(q)$. It is therefore clear that without assumptions or additional information, $\rho(r)$ cannot be reconstructed uniquely from solution scattering data. An exception here is the case of a spherically symmetric particle where all $I_l(q)$ but the monopole term ($l = 0$) are equal to zero; then only the signs of the function $[I_0(q)]^{1/2}$ need to be determined in order to reconstruct the radial density distribution $\rho_0(r)$.

Let us consider for illustration one of the most vivid ways of representing a structure by spherical harmonics using the particle envelope (Fig. 4.3). The envelope function $r = F(\omega)$ corresponds for each direction ω to the distance r from the centre to the outer surface of the particle (most distant point in the particle) along this direction. Let us assume that the surface is convex and hence the function is single-valued; that is, the border is not crossed more than once (which is true for many

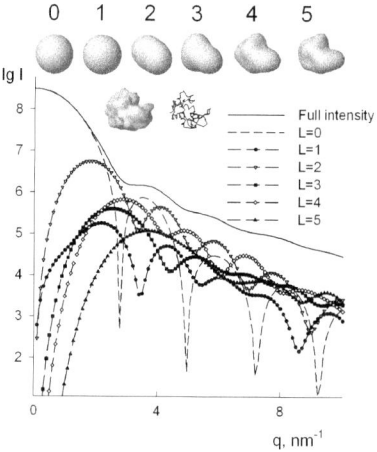

Fig. 4.3 Accuracy of shape representation using spherical harmonics. Top row: surface representations of truncated envelope functions of lysozyme. Second row: high-resolution envelope function and C_α-trace of the protein. The shape scattering intensity from lysozyme is plotted together with the individual contributions from different multipole components $I_L(q)$, $L = 0 \ldots 5$. Note that only the monopole component ($L = 0$) has a non-zero forward scattering value. Adapted from Koch, Vachette, *et al.* (2003), with permission from Cambridge University Press, © 2003.

shapes at low resolution). Then, for homogeneous particles, the density $\rho(r)$ can be represented as

$$\rho(r) = \begin{cases} 1, \ 0 \leq r < F(\omega) \\ 0, \ r \geq F(\omega) \end{cases} \tag{4.44}$$

and the envelope function is

$$F(\omega) = \sum_{l=0}^{L} \sum_{m=-l}^{l} f_{lm} Y_{lm}(\omega) \tag{4.45}$$

Here, combining eqs. (4.40) and (4.44), the multipole coefficients f_{lm} are the complex numbers

$$f_{lm} = \int_{\omega} F(\omega) \, Y^*_{lm}(\omega) d\omega \tag{4.46}$$

This remarkable representation, first introduced in SAS by Stuhrmann (1970a), allows one rather unexpectedly to 'save' one dimension in the description of a particle structure. Indeed, for convex homogeneous particles with a single-valued outer surface function (both assumptions being not too restrictive at low resolution) the structure can be described by a two-dimensional function $F(\omega)$ instead of a three-dimensional distribution $\rho(r)$. Moreover, the envelope is represented just by a set of (complex) numbers f_{lm} such that the particle shape at low resolution can be described by a few parameters only.

The truncation value L defines the accuracy of the structure representation (the higher L, the more detail can be obtained). An example of the evolution of the envelope function of lysozyme with increasing L is displayed in Fig. 4.3 along with the partial intensities $I_l(q)$. In the monopole approximation ($L = 0$), the particle is described just by the radius of an equivalent sphere R_0, and $I_0(q)$ is the scattering intensity from this sphere. All partial intensities with $l > 0$ are equal to zero at $q = 0$ and grow as q^{2l}, so that the contribution of higher harmonics increases with the scattering angle (that is, with the resolution they provide, which can be estimated as approximately $2\pi R_0/(l + 1)$). As is evident in Fig. 4.3, a quadrupole approximation ($L = 2$) largely describes the particle anisometry, whereas at $L = 4$ the main low-resolution features of the shape are already reasonably well represented. Further increase of L provides an increasingly detailed description of the envelope at the cost of a quadratically growing number of parameters (equal to $(L + 1)^2$).

4.3 Shannon sampling

Before proceeding to the methods allowing one to build three-dimensional models, let us consider the question of the information content of SAS patterns from monodisperse systems. The amount of information can be quantitatively assessed from Shannon's sampling theorem (Shannon and Weaver 1949), making use of the fact that the scattering curve $I(q)$ is the Fourier transform of the distance distribution function. Indeed, eq. (4.19) can be rewritten as

$$qI(q) = 4\pi \int_0^{D_{max}} [p(r)/r] \sin qr \, dr \qquad (4.47)$$

Given that the $p(r)$ function is equal to zero beyond D_{max}, after double integration by parts eq. (4.47) yields

$$qI(q) = \sum_{k=1}^{\infty} q_k I(q_k) \left[\frac{\sin D_{max}(q - q_k)}{D_{max}(q - q_k)} - \frac{\sin D_{max}(q + q_k)}{D_{max}(q + q_k)} \right] \qquad (4.48)$$

In this representation a continuous function $I(q)$ is described fully by its values in a discrete set of points q_k (called Shannon channels), where $q_k = k\pi/D_{max}$. $I(q)$ is thus a so-called analytical function (Frieden 1971), which is a direct consequence of the boundedness of the $p(r)$ function to the interval $[0, D_{max}]$. The intensity values $I(q)$ can therefore not be independent from each other, and the Shannon theorem allows one to assess the number of *independent* parameters (or *degrees of freedom*) associated with a data set. Indeed, an analytical function can be reasonably well represented on an interval $[q_{min}, q_{max}]$ by eq. (4.48) using $N_s = D_{max}(q_{max} - q_{min})/\pi$ values of $I(q_k)$ at the Shannon channels in this interval. In practice, solution scattering curves decay rapidly with q, and they are normally recorded only at resolutions below 1 nm, so that the typical number of the Shannon channels is around a dozen. As an example, an SAS curve from an hypothetical protein with $D_{max} = 10$ nm, measured in the range of momentum transfer from $q = 0.1$–5 nm^{-1}, contains $4.9 \times 10/\pi \approx 15$ Shannon channels.

It is obvious that larger values of N_s correspond to more information in the scattering data, but could one go as far as stating that N_s represents the number of independent parameters that can be extracted from an SAS pattern? It is tempting to have such a clear-cut estimate of the information content, and the issue has been a subject of extensive debate in the past (Damaschun *et al.* 1968, Moore 1980, Taupin and Luzzati 1982). In fact, it can be shown readily that the series in eq. (4.48) must contain an infinite number of terms for the scattering intensity to have physical sense.

Figure 4.4 displays Shannon representations of the scattering intensity from an ellipsoid model, truncated at $N_s = 10$ and 20. Inside the Shannon interval the approximation in eq. (4.48) provides a reasonable approximation of $I(q)$. However, systematic deviations are observed when approaching q_{max}, and for $q > q_{max}$ the truncated Shannon representation oscillates around zero, which is unphysical, as the scattering intensity cannot be negative.

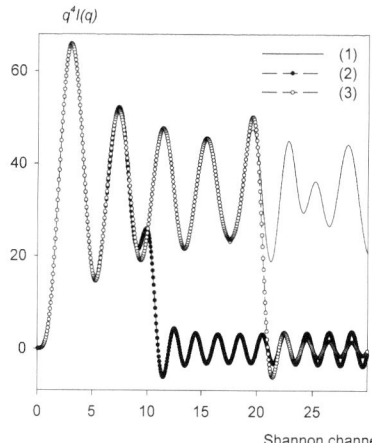

Fig. 4.4 Shannon approximation of the scattering intensity from an ellipsoidal particle (1) using ten (2) and twenty (3) Shannon channels. Adapted from Koch, Vachette, *et al.* (2003), with permission from Cambridge University Press, © 2003.

The experimental SAS patterns (especially SR-SAXS data) are usually vastly oversampled; that is, the angular increment in the data sets is much smaller than the Shannon increment $\Delta q = \pi/D_{max}$. As known from optical image reconstruction (Frieden 1971), this oversampling allows in principle to extend the data beyond the measured range (so-called 'super-resolution'), which effectively increases the formal number of Shannon channels. As indicated long ago (Sayre 1952), oversampling also allows one to reconstruct the phases of the continuous intensity data. Relatively simple iterative hybrid input–output methods (Fienup 1982) can be employed for the phase retrieval from the data collected from X-ray lasers (Miao *et al.* 1999).

Shannon's sampling theorem does provide very useful guidance for performing SAS measurements. In particular, the value of q_{min} should not exceed that of the *first* Shannon channel—for example, q_{min} must be smaller than π/D_{max}—in order to not lose information. This condition should be observed also when using the indirect transformation methods (see Section 4.1.1). However, the value of N_s alone cannot predict the number of parameters which can be extracted from a given data set. As we shall see below, extraction of the structural parameters from SAS is a highly non-linear problem, and the level of detail of the models depends on many factors. These include not only the experimental angular range available but also the accuracy of the data, the *a priori* information available, the possibility to simultaneously fit multiple data sets and so on.

4.4 *Ab initio* shape analysis

As indicated in the previous chapter, reconstruction of three-dimensional structures from SAS data is inherently ambiguous, and strong simplifying assumptions have to be made if one wishes to retrieve $\rho(\mathbf{r})$ from a given scattering pattern $I(q)$. At low resolution (say, about 2–3 nm), the search can usually be limited to homogeneous models; that is, one assumes that the intensity comes from a uniform body. Recalling the basic scattering functions in Chapter 1, one assumes that $I(q) = I_C(q)$ and discards the second and third terms in eq. (1.6). In principle, contrast variation can be performed to obtain $I_C(q)$ experimentally; in practice, for example for sufficiently large proteins, the measured data to 2–3 nm resolution (that is, q about 2–3 nm^{-1}) can provide a good approximation to $I_C(q)$. Often, a constant term is subtracted from $I(q)$ to enforce a q^{-4} decay of the intensity in conformity with Porod asymptotic (eq. 4.37).

Shape modelling was performed in SAS starting from 1960s, first on a trial-and-error basis by computing scattering patterns from different shapes and comparing them with the experimental data. Originally, this modelling was limited to the simplest three-parameter bodies, such as prisms, ellipsoids, or elliptical or hollow circular cylinders. The experimental scattering curves were screened against charts containing log $I(q)$ *versus* log (qR_g) plots of the scattering patterns from such bodies (see, for example, Kratky and Pilz 1978). Later, with improved computational resources, it became possible to reasonably rapidly compute the scattering intensity from more complex models represented by regularly packed spheres using Debye's formula, eq. (4.6) (Glatter 1972, Rolbin *et al.* 1973) The trial-and-error modelling was usually constrained by additional information (for example, from EM or hydrodynamic data).

4.4.1 Envelope reconstruction

A very elegant idea of an automated procedure to reconstruct low-resolution molecular envelopes *ab initio* was proposed by Stuhrmann (1970a). Using the definition of the envelope function introduced in eqs. (4.40)–(4.42), $\rho(\mathbf{r}) = 1$ in the range of integration of eq. (4.36) (that is, from 0 to $r = F(\omega)$). Inserting eq. (4.36) into eq. (4.24) and expanding $j_l(qr)$ as a Taylor power series, the partial scattering amplitudes of the particle can be expressed as

$$A_{lm}(q) = (i \cdot q)^l \cdot \sqrt{2/\pi} \cdot \sum_{p=0}^{\infty} \left((-1)^p \cdot f_{lm}^{(l+2p+3)} \cdot \left\{ 2^p \cdot p! \cdot (l+2p+3) \cdot [2(l+p)+1]!! \right\}^{-1} \cdot q^{2p} \right) \tag{4.49}$$

Here the coefficients of the kth power of the shape function are defined as

$$f_{lm}^{(k)} = \int_{\omega} [F(\omega)]^k \cdot Y_{lm}^*(\omega) \, d\omega \tag{4.50}$$

For an envelope function defined by a set of coefficients f_{lm}, it is possible to analytically compute the values $f_{lm}^{(k)}$ using a recurrence formula (Svergun and Stuhrmann 1991) based on Wigner coefficients (see Appendix 2). Equations (4.49) and (4.42) therefore allow fast computation of the partial amplitudes and hence of the scattering intensity from the f_{lm} coefficients; that is, from the given envelope function.

The ability to rapidly compute scattering patterns from the envelopes parameterised by shape coefficients opens a way for an *ab initio* envelope determination from the scattering data. The search algorithm (Svergun *et al.* 1996, Svergun *et al.* 1997) starts from a spherical shape for which all coefficients but f_{00} are equal to zero. Subsequently, the f_{lm} coefficients are obtained, which minimise the discrepancy between the experimental and calculated curves.

Only a few parameters are required to describe envelope functions at low resolution. The shape can be restored at an arbitrary position and orientation, and due to rotational and translational degrees of freedom the number of independent f_{lm} values in eq. (4.45) is $N_p = (L+1)^2 - 6$. For $L = 3$ and $L = 4$, this leads to ten and nineteen parameters respectively—values comparable to the typical numbers of

Shannon channels in the experimental data. One may therefore expect that unique envelope determination from the scattering data could be possible.

When speaking about uniqueness of the SAS-based modelling, one should note first that for any distribution $\rho(\mathbf{r})$ and its mirror image (enantiomorph) $\rho(-\mathbf{r})$ yields exactly the same scattering intensity $I(q)$. The shapes of the enantiomorph are described by a set with the coefficients

$$f_{lm} = (-1)^m f_{l,-m} \tag{4.51}$$

Ab initio SAS reconstructions must always be considered together with their mirror images, and this should always be kept in mind when considering these models. Taking into account this ambiguity, computer simulations on model bodies by Svergun *et al.* (1996) demonstrated that unique shape reconstruction is possible for the class of bodies described by envelopes up to $L = 4$.

Information on particle symmetry, if available, allows one to impose restrictions on the multipole coefficients f_{lm} in eq. (4.45) and to improve the reliability of the *ab initio* shape restoration. Consider, for example, an homodimeric particle with a two-fold symmetry axis along z. In this case, all f_{lm} coefficients with odd m are zero, and the particle shape at $L = 4$ is described by twelve independent parameters instead of nineteen for the general case. The higher the symmetry, the more multipole coefficients would vanish (an extensive list of symmetry selection rules for spherical harmonics is available: Spinozzi *et al.* 1998). Moreover, the quaternary structure of symmetric particles can be restored in terms of the envelope function of their asymmetric unit. Thus, scattering from a symmetric homodimer is readily expressed *via* the shape of a monomer and a single additional parameter, the distance Δd between the monomers (Kozin *et al.* 1997).

The envelope determination approach was implemented in the first publicly available shape-determination program SASHA (Svergun 1997), which was an important milestone allowing not only method developers but also regular users to construct low-resolution models directly from the scattering data. Indeed, several groups successfully employed the envelope function method to generate low-resolution models of proteins (Grossmann *et al.* 2000, Bernocco *et al.* 2001, Krueger *et al.* 2000, Aparicio *et al.* 2002).

The reliability of low-resolution envelope determination has been demonstrated in numerous model calculations, but even better proof is provided by the cases where high-resolution crystallographic models became available *a posteriori*. One such example, presented in Fig. 4.5, displays a low-resolution model of the dimeric macrophage infectivity potentiator (MIP) from *Legionella pneumophila* (Schmidt *et al.* 1995) and the crystal structure of the enzyme, determined six years later (Riboldi-Tunnicliffe *et al.* 2001).

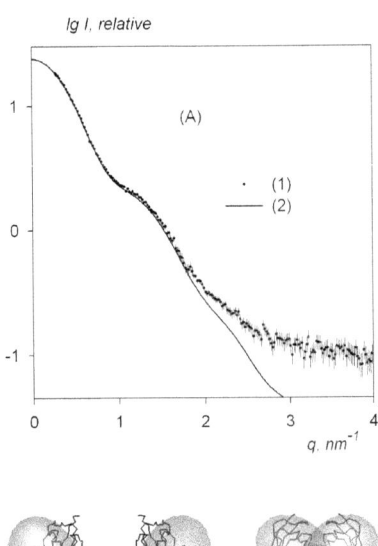

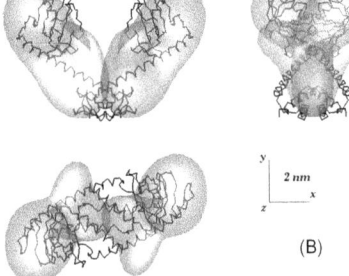

Fig. 4.5 A: X-ray scattering from macrophage infectivity potentiator (MIP) (1) and the computed scattering from the *ab initio* envelope model (2) obtained by the program SASHA, assuming two monomers related by a two-fold symmetry axis. B: Atomic model of homodimeric MIP (C_α chain, Protein Data Bank (PDB) entry 1FD9) superimposed on the *ab initio* envelope model (the two monomers are displayed as a semi-transparent envelope). The right and bottom views are rotated counterclockwise by 90° around Y and X, respectively. Adapted from Koch, Vachette, *et al.* (2003), with permission from Cambridge University Press, © 2003.

4.4.2 Shape determination using finite elements

Shape analysis in terms of envelopes has a limited ability to describe complicated shapes, where the particle border can be crossed several times when moving along a given angular direction (for example, crescent-shaped particles or those with internal cavities). A more flexible representation can be achieved by describing the shape by a collection of finite elements (such as small spheres, similar to the trial-and-error Debye modelling). The concept of the trial-and-error finite elements (bead) modelling was turned into an *ab initio* shape determination method by Chacon *et al.* (1998). The basic idea was to perform an intelligent automated shape search in a confined volume by delegating the choice of the model to a computer. A search volume can be taken which should definitely enclose the particle shape—for example, a sphere with the diameter equal to the maximum dimension of the particle D_{max}, which is readily obtained from its scattering pattern. The search volume is filled with $N \gg 1$ densely packed spheres of radius $r_0 \ll D_{max}$, each of which may belong either to the particle (index = 1) or to the solvent (index = 0), and the particle shape is thus completely described by a string, **X**, of N bits (Fig. 4.6). In this search volume one should now find a shape—that is, the binary string—such that the computed scattering from this configuration fits the experimental data. It is clear that a brute-force screening of all possible configurations would not be possible even with the modern computers: for a search volume containing a thousand beads, one would needs to test $2^{1000} \approx 10^{301}$ configurations. Therefore, heuristic Monte Carlo-like algorithms are employed, where, starting from a random distribution of 1s and 0s, the model is randomly modified to find a binary string that fits the experimental data. In the original method (Chacon *et al.* 1998), implemented in the program DALAI_GA, a genetic algorithm (GA) (Jones 1998) was employed, and it was demonstrated that the search does allow one to reconstruct low-resolution shapes from the scattering data.

Here we shall describe a more general 'dummy atoms' procedure (Svergun 1999), which is similar to the original approach of Chacon *et al.* (1998) but also allows one to construct models of inhomogeneous particles based on contrast variation data; *ab initio* shape determination is a particular case of this procedure. Assume a particle consisting of K components with distinctly different scattering-length densities. In a nucleoprotein complex, for example, these components are the protein and nucleic acid moieties ($K = 2$). A search volume sufficiently large to enclose the particle (for example, a sphere of radius $R = D_{max}/2$) is filled with N 'dummy atoms' or spherical beads, closely packed with an hexagonal packing radius $r_0 \ll R$. Each dummy atom is assigned an index X_j indicating the phase to which it belongs (X_j ranges from 0 (solvent) to K). Given the fixed bead positions, which are defined by the packing grid, the structure of this dummy atom model (DAM) is fully described by a phase assignment (configuration) vector **X** with $N \approx (R/r_0)^3$ components.

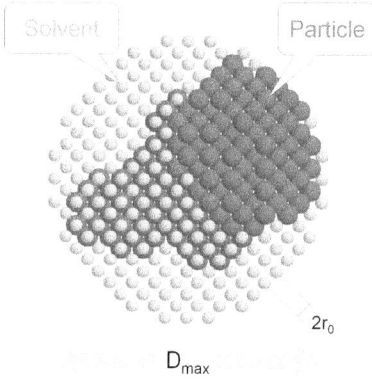

Fig. 4.6 Search volume and particle shape representation in the *ab initio* analysis by bead modelling.

Denoting the contrast of the beads in the kth phase as $\Delta\rho_k$, the scattering intensity from the entire DAM is

$$I(q) = \left\langle \sum_{k=1}^{K} \Delta\rho_k A_k^2(\mathbf{q}) \right\rangle_\Omega \tag{4.52}$$

where $A_k(\mathbf{q})$ is the scattering amplitude from the volume occupied by the kth phase. Representing the amplitudes with spherical harmonics as in eq. (4.42), one obtains

$$I(q) = 2\pi^2 \sum_{l=0}^{\infty} \sum_{m=-l}^{l} \left\{ \sum_{k=1}^{K} \left[\Delta\rho_k A_{lm}^{(k)}(q) \right]^2 + 2 \sum_{n>k} \Delta\rho_k A_{lm}^{(k)}(q) \Delta\rho_n \left[A_{lm}^{(n)}(q) \right]^* \right\} \tag{4.53}$$

The partial amplitudes from the volume occupied by the kth phase in a DAM are simply evaluated from the bead positions following eq. (4.45),

$$A_{lm}^{(k)}(q) = i^l \sqrt{2/\pi}\, v_a \sum_j j_l(qr_j) Y_{lm}^*(\omega_j) \tag{4.54}$$

where the sum runs over all atoms of that phase, $(r_j\omega_j) = \mathbf{r}_j$ are their polar coordinates and $v_a = (4\pi r_0^3/3)/0.74$ is the displaced volume per dummy atom. Equations (4.52)–(4.54) therefore allow one to rapidly compute the scattering curves from a multiphase DAM for an arbitrary configuration X and arbitrary contrasts $\Delta\rho$.

This formalism can be employed to design an algorithm to reconstruct not only the shape but also the internal organisation of multicomponent particles by simultaneously fitting multiple data sets (for example, obtained by contrast variation). Given a set of $M \geq 1$ scattering patterns $I_{\exp}^{(m)}(q), m = 1, \ldots M$, one searches for a configuration X minimising the overall discrepancy between the experimental and calculated data

$$\chi^2 = \frac{1}{M-1} \sum_{m=1}^{M} \frac{1}{N_m - 1} \sum_{j=1}^{N} \left[\frac{I_{\exp}^{(m)}(c_j) - c_k I_{\text{calc}}^{(m)}(q_j)}{\sigma^{(m)}(q_j)} \right]^2 \tag{4.55}$$

where the index m runs over the scattering curves, N_m are the numbers of experimental points, c_m are scaling factors and $I_{\text{calc}}(q_j)$ and $\sigma(q_j)$ are the intensities calculated from the subsets of the beads belonging to the appropriate phases and the experimental errors at the momentum transfer q_j, respectively.

For an adequate description of a structure the number of dummy atoms must, however, be large ($N \approx 10^3$), and thus the number of parameters describing such a model significantly exceeds the number of Shannon channels. It is clear that constraints must be applied to meaningfully discuss the uniqueness of reconstructions with such an approach. The most important and also most natural constraints are compactness and interconnectivity of the models, and these character-istics can be computed rapidly (Svergun 1999).

By its very nature, SAS provides low-resolution information only, and any meaningful *ab initio* model should provide an overall shape instead of trying to represent fine details of the object. The require-ment of low resolution for a model built from equal beads of a size r_0 can be imposed from simple geometrical considerations. First, a list of neighbours (atoms at a distance $2r_0$) is generated for each dummy atom. The looseness or degree of isolation of each non-solvent atom is calculated as $P(N_e) = \exp(-0.5N_e) - \exp(-0.5N_c)$, where N_e is the number of neighbours having the same index, and $N_c = 12$ is the coordination number for hexagonal packing. The looseness of the con-figuration $\mathbf{X}$ is characterised by the average value $P(\mathbf{X}) = <P(N_e)>$ over all non-solvent atoms. Another requirement, interconnectivity, presents a possibility of connecting two arbitrarily selected atoms belonging to a phase through neighbouring atoms belonging to the same phase. For this, one can build graphs (interconnected fragments) inside each phase, using the lists of neighbours computed to calculate the loose-ness. A quantitative measure of connectivity of the kth phase can be conveniently evaluated as $G_k(\mathbf{X}) = ln(N_k/M_k) \geq 0$, where N_k and M_k are the numbers of dummy atoms in the entire phase and in the largest connected fragment, respectively.

The two restraints may be added as additional terms (penalties) to be minimised such that the task of retrieving a low-resolution model from the scattering data can be formulated as follows: given a DAM, find a configuration $\mathbf{X}$ minimising the goal function,

$$f(\mathbf{X}) = \chi^2(\mathbf{X}) + \alpha \sum_K \|P_k(\mathbf{X}) + G_k(\mathbf{X})\| \tag{4.56}$$

where $\alpha > 0$ is the weight of the looseness and interconnectivity penalty. This is usually selected in such a way that the second term yields a significant (about 10–50%) contribution to the function value $f(\mathbf{X})$ at the end of the minimisation.

The problem of finding a configuration $\mathbf{X}$ minimising $f(\mathbf{X})$ is rather complex, due to the large number of variables (phase assignments of the $N \approx 10^3$ beads). The most straightforward technique to minimise $f(\mathbf{X})$ is a plain Monte Carlo search: one starts from a random config-uration and modifies it (for example, changing the assignment of a single bead). If the change improves $f(\mathbf{X})$, it is accepted, otherwise it is rejected. Then, another random change is made and the procedure is repeated until no better configuration is found. The pure Monte Carlo search is very simple, but it tends to be trapped by local minima in $f(\mathbf{X})$; as in the search of a global minimum of such a multimodal func-tion only heuristic global minimisation algorithms can be employed. As indicated above, the original method (Chacon *et al.* 1998) used GA (Jones 1998); another appropriate technique, simulated annealing (SA, Kirkpatrick *et al.* 1983) has been employed by Svergun (1999). The idea of SA is to perform random modifications of the system (that is, of the vector **X**) while always moving to configurations that decrease f(**X**) but sometimes also to those increasing f(**X**). The probability of accepting this 'worsening' move decreases in the course of the minimisation (the system is 'cooled'). Initially, the 'temperature' (a parameter regulating the probability of accepting the 'worsening move') is high and the changes almost random, whereas at the end (low temperature) one nearly resorts to a pure Monte Carlo search. It was shown that SA is a powerful approach in the search of global minima for multivariate functions described by many parameters (Press *et al.* 1992, Ingber 1993). The full SA protocol for shape determination is:

1. Start from a random configuration $\mathbf{X}_0$ at a 'high' temperature T_0 (for example, $T_0 = f(\mathbf{X}_0)$).
2. Select a bead at random, randomly change its index (that is, the phase to which it belongs) to obtain configuration $\mathbf{X}'$ and compute $\Delta = f(\mathbf{X}') - f(\mathbf{X})$.
3. If $\Delta < 0$, move to $\mathbf{X}'$; if $\Delta > 0$, do this with probability $\exp(-\Delta/T)$. Repeat step 2 from $\mathbf{X}'$ (if accepted) or from $\mathbf{X}$.
4. Hold T constant for $100\,N$ reconfigurations or $10\,N$ successful reconfigurations—whichever comes first. Then, cool the system ($T' = 0.9T$), and continue cooling until no improvement in $f(X)$ is observed.

The use of the spherical harmonics in expansion eq. (4.53) allows one to rapidly compute the intensity from the bead model. Indeed, the Debye formula in eq. (4.6b) has a computational 'cost' proportional to the squared number of beads N^2, whereas the computing time for eq. (4.53) is proportional to N times the number of the harmonics used. For large N, spherical harmonics provide a significant gain in computational speed. Moreover, during SA, only a single bead is changed at each move and hence only a single summand in eq. (4.54) must be updated to calculate the partial amplitudes. As this calculation is the most time-consuming operation in the evaluation of $f(\mathbf{X})$, the entire procedure is accelerated about N times. It is this acceleration, which makes it possible to use robust SA, which would otherwise be prohibitively slow, as millions of function evaluations are required for a typical minimisation run (Ingber 1993).

The above SA-based method (Svergun 1999) is rather general, having a capacity to analyse the data from multi-component particles, and some examples of its full-scale application will be presented in Chapter 6. In the particular case of a single-component particle ($K = 1$), the 'dummy atoms' approach reduces to the *ab initio* shape determination procedure implemented in the program DAMMIN (Svergun 1999). This program became rather popular, and starting from the year 2000 was actively employed by numerous groups to restore low-resolution shapes of macromolecules from solution scattering data (Funari *et al.* 2000, Egea *et al.* 2001, Sokolova *et al.* 2001, Fujisawa *et al.* 2001, Aparicio *et al.* 2002, Svergun *et al.* 2000, Scott *et al.* 2002).

DAMMIN—similar to the envelope reconstruction algorithm SASHA—can take into account the point symmetry of the particle. The symmetry is implemented as a hard restraint by coupling the phase assignments of the symmetry-related beads in the DAM. By appropriate positioning of the symmetry elements, selection rules are applied to the partial amplitudes similar to those in SASHA (for example, for a rotational axis of nth order around z, all harmonics vanish except for those with $l = kn, k = 0, 1, \ldots$). Therefore, the symmetry allows one not only to reduce the number of beads, but also to further speed up the calculations.

A number of symmetries, including rather high ones (such as cubic and icosahedral), can be taken into account in DAMMIN. Symmetry is a very powerful tool in the *ab initio* shape reconstruction, but it must be employed with caution. Indeed, correlations between the symmetry-related beads make the search space anisotropic, and this may lead to artefacts in the reconstruction, which are well illustrated by the shape reconstruction of the hydrophilic (V_1) portion of *Manduca sexta* ATPase. The X-ray scattering pattern of this large protein complex with a molecular mass of about 550 kDa is displayed in Fig. 4.7A. The *ab initio* low-resolution model restored by DAMMIN without symmetry restrictions (Fig. 4.7B, left) displays an elongated mushroom-like shape, which agrees well with the results of cryo-EM (Grüber *et al.* 2000). As the major portion of the enzyme was expected to have quasi-threefold symmetry, this restriction was imposed during shape reconstruction. Surprisingly, the shape restored assuming P3 symmetry was a flat particle (Fig. 4.7B, middle panel), with both shapes yielding practically the same fit to the experimental data. This example indicates that anisotropy of the search space due a symmetry axis may force the method to converge to a shape with an incorrect anisometry.

To cope with such an artefact, anisometry selection is also introduced in DAMMIN as a penalty function. For each DAM, the three principal moments of inertia (eigenvalues of the inertia tensor) can be calculated and sorted in descending order ($I_1 \geq I_2 \geq I_3$). For prolate particles, I_1 and I_2 (two largest moments of inertia) are perpendicular to the longest axes, and are both much larger than I_3. For oblate particles the largest moment of inertia I_1 significantly exceeds I_2 and I_3. From these moments it is therefore possible to rapidly determine whether the particle is prolate, oblate or globular (all three moments are close to each other). The requirement of anisometry is added as an extra penalty to the function $f(\mathbf{X})$, and can be used when *a priori* information about the anisometry is available. This information may be especially important when using symmetry in *ab initio* reconstructions. In the above case of V_1 ATPase, a P3 reconstruction with the prolateness requirement yields a model (Fig. 4.7B, right) which is compatible with the P1 reconstruction but displays better-defined structural features due to the symmetry imposed. The P3 analysis further detected a major structural transition due to redox modulation in the headpiece and at the bottom of the stalk (Grüber *et al.* 2000). One should therefore always, even for symmetric

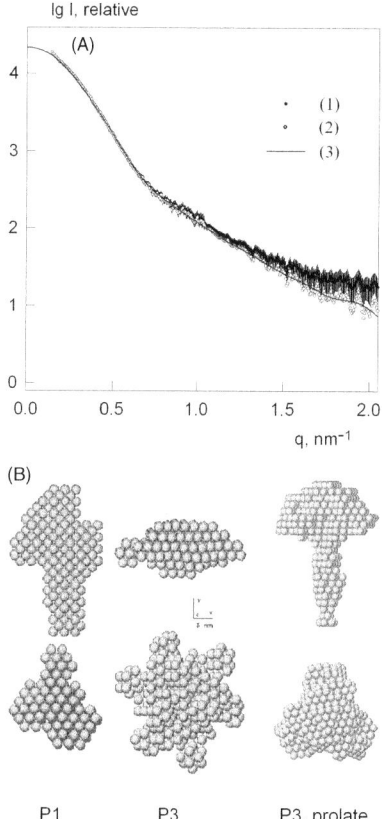

lg I, relative

(A)

(1)
(2)
(3)

q, nm⁻¹

(B)

P1   P3   P3, prolate

Fig. 4.7 A: X-ray scattering patterns from V1 ATPase: experimental scattering before (1) and after (2) subtracting a constant and scattering from the *ab-initio* models (3). B: The models obtained by DAMMIN with different symmetry and anisometry restrictions (see text for explanations). The bottom view is rotated counter-clockwise by 90° around X. Adapted from Koch, Vachette, *et al.* (2003), with permission from Cambridge University Press, © 2003.

particles, start with a P1 *ab initio* analysis, such that proper anisometry of the symmetric models can be selected.

Clearly, 'dummy atom' or 'bead' models are described by a number of parameters N significantly exceeding the number of Shannon channels from the sampling theorem N_s (Section 4.3), and it is natural to ask how such models can be reconstructed *ab initio*. Indeed, as noted above, oversampling of the experimental data (that is, measurements with the angular increment much smaller than π/D_{max}) in general increases the information content. The increase is not easy to decipher in numbers for real experimental data, especially if instrumental smearing is present (the experimental points cannot be considered independent observations in this case). It was demonstrated that the effective number of degrees of freedom may range from zero for a signal-to-noise ratio of 1 to $15N_s$ for a signal-to-noise ratio of 10^3 (Frieden 1971). This observation does, of course, not justify the use of models described by $15N_s$ parameters, but rather indicates that the number of degrees of freedom strongly depends on data accuracy. Further, the number of independent parameters in a DAM is much lower than the number of beads N because of the required compactness of the models. In fact, at low temperatures SA does not try to improve the fit to the data but rather to make the model as compact as possible, whereas the fit acts as a constraint, the major reductions in $f(\mathbf{X})$ being due to reducing the penalty and not the discrepancy. The higher the information content of the data, the more stringent the data constraint should be, and the more detail the DAM should keep. The compactness condition is therefore crucial for a reliable shape reconstruction using bead modelling. In DALAI_GA (Chacon *et al.* 1998), the first bead modelling program, the model was implicitly constrained by a gradual decrease of the packing radius r_0 during minimisation, whereas the later version of this program also had explicit constraints, similar to those in DAMMIN or MONSA (Chacon *et al.* 2000).

There are different flavours of *ab initio* modelling using finite volume elements, including those not restraining the search space. The first attempt was a 'give-and-take' procedure (Bada *et al.* 2000) implemented in the program SAXS3D, which places beads on an unlimited hexagonal lattice, whereby at each Monte Carlo step a new bead is added, removed or relocated to improve the agreement with the data (connectivity is not explicitly imposed, which may result in unconnected models). The SASMODEL program (Vigil *et al.* 2001) represents the model by a superposition of interconnected ellipsoids and employs a Monte Carlo search of their positions and dimensions to fit the experimental data. A later implementation of DAMMIN, called DAMMIF (Franke and Svergun 2009), also employs an unrestricted volume that can grow in size as needed during the SA procedure. Upon each random modification, disconnected models are immediately rejected, without calculation of scattering amplitudes, which is performed only for connected models. Moreover, the necessary scattering amplitudes in terms of spherical harmonics are intelligently precomputed for each bead, contributing to the total scattering at least once. These measures accelerate the DAMMIF modelling by 25–40 times compared to the original DAMMIN.

It may be argued whether it is better to use a pre-defined or an un-limited search space. On one side, improper restriction of the search space may lead to boundary effects and artefacts in the models. On the other side, the value of D_{max} is usually reliably determined from the experimental data and provides a valuable constraint reducing the uncertainty of the shape reconstruction. The uncertainty is, of course, an extremely critical issue, which should be discussed in more detail. Even though the *ab initio* methods using finite elements were proven to work well in numerous test and practical examples, running any of the above shape determination programs several times with different seeds for random number generators produces a manifold of models corres-ponding to nearly identical scattering curves. Such a diversity of the *ab initio* models is illustrated in Fig. 4.8 for the shape determination of HIV reverse transcriptase. The question naturally arises about how differ-ent these models are, whether they define a common shape, or whether variations between them can serve as an indicator of the stability of the reconstruction.

First of all, let us recall that the computed scattering patterns (and thus the bead models) are invariant to handedness such that for any model its enantiomorph model must always be considered. The mod-els obtained in independent runs (and also their mirror images) can be superimposed and averaged to obtain a most probable model, and this procedure is automated in the program DAMAVER (Volkov and Sver-gun 2003). The superposition protocol utilises the program SUPCOMB (Kozin and Svergun 2001), which is able to align two arbitrary low- or high-resolution models represented by ensembles of points (such as beads) without information about correspondence between these points. The superposition is done by minimising a dissimilarity meas-ure between two models called a normalised spatial discrepancy (NSD). Here, for every point in the first model, the minimum value among the distances between this point and all points in the second model is found, and the same is done for the points in the second model. These distances are added and normalised against the average distances between the neighbouring points for the two models. Generally, an NSD value close to 1 indicates that the two models are similar (like the dis-crepancy value in the comparison of the scattering curves, see eq. (3.9)). Given a set of models provided by a shape determination algorithm, they are all pairwise superposed, a reference model, with lowest aver-age NSD, is selected and possible outliers with high NSD against the rest of the set are marked. After superposition of all models except outliers, the entire assembly of beads can be remapped onto a grid to build an occupancy map, and an average model can be constructed by filtering out the low-occupancy grid points. Further, grid points with non-zero occupancy can be used as a new search volume to construct a refined model. The results of the averaging procedure for the shape reconstruction of HIV reverse transcriptase are illustrated in Fig. 4.8.

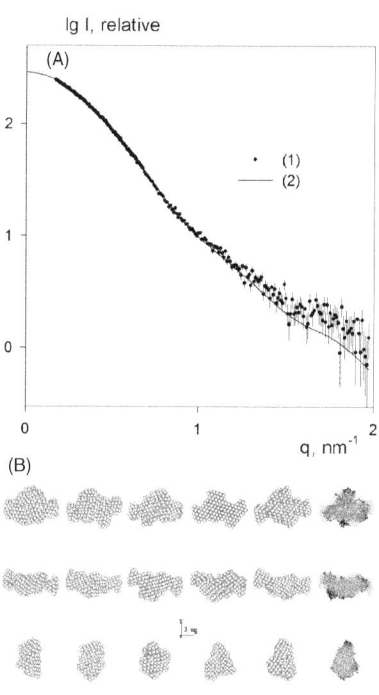

Fig. 4.8 Shape determination of HIV re-verse transcriptase. A: Synchrotron X-ray scattering patterns: experimental data (1) and the scattering from the models re-stored by DAMMIN (2). B: five models restored by DAMMIN and (the right-most column) the average model superimposed with the atomic model of the enzyme (PDB entry 3HVT) (Wang *et al.* 1994). The middle and bottom views are ro-tated counterclockwise by 90° around X and Y, respectively. Adapted from Koch, Vachette, *et al.* (2003), with permission from Cambridge University Press, © 2003.

Using DAMAVER, a systematic study of the uniqueness of the shape reconstruction was performed on numerous model bodies (Volkov and Svergun 2003). It was found that there is one class of shapes—namely, flat particles (starting with the lamella size to thickness ratio exceeding 1:5), for which *ab initio* shape reconstruction without symmetry restrictions is impossible (DAMMIN was used in this case, but the same holds for other methods). However, such unreliable shape analysis can be identified by the instability of the reconstruction, yielding an average NSD exceeding unity for multiple runs of the *ab initio* program. Performing and averaging multiple shape reconstructions therefore provides important practical guidance for assessing the uniqueness of the shape analysis.

4.4.3 Dummy-residues approach

The above shape determination methods represent the model as an homogeneous particle and, strictly speaking, should be used to fit the shape scattering curve (see basic-function decomposition in eq. (1.6)). The latter could in principle be obtained by contrast variation experiments, as demonstrated by Ibel and Stuhrmann (1975). Such experiments are, however, not performed in practice, not only because of the cumbersome experimental procedures but also due to the low accuracy of this decomposition at higher angles. In most practical cases the scattering data measured at sufficiently high contrasts can be used 'as is'; for example, for the X-ray scattering on proteins with MM exceeding ~30 kDa, a reasonable approximation to the 'shape scattering' curve can be obtained by subtracting an appropriate constant enforcing the Porod-like q^{-4}-decay of the intensity (see eq. 4.37). However, as the scattering data at higher resolution are dominated by the contribution from the internal structure, only a restricted part of the scattering pattern can be used (typically, up to $q = 2 - 3 \, \text{nm}^{-1}$; that is, a resolution of about 2–3 nm. This, of course, limits the information content of the data and thus not only the resolution but also the reliability of the models.

In an alternative *ab initio* approach (Svergun *et al.* 2001) a concept of dummy residues (DR) was introduced to construct models of proteins from the SAXS patterns accounting for higher-resolution data. The major building blocks of proteins are folded polypeptide chains where the C_α atoms of adjacent amino acid residues in the primary sequence are separated by about 0.38 nm. At moderate resolution (~0.5 nm), a protein can be represented as an assembly of DRs centred at the C_α positions. The scattering from a protein with K residues at positions $\mathbf{r}_i$ can be computed approximately by the Debye formula,

$$I_{DR}(q) = \sum_{i=1}^{K} \sum_{j=1}^{K} g_i(q) g_j(q) \frac{\sin qr_{ij}}{qr_{ij}} \tag{4.57}$$

where $g_j(q)$ is the spherically averaged form factor of the jth residue, and $r_{ij} = |\mathbf{r}_i - \mathbf{r}_j|$ is the distance between the ith and jth residues. To further simplify the calculations one may also ignore differences in the scattering between individual residues and assume that they all have an averaged form factor $g(q)$ (Fig. 4.9A). The latter can be computed from the spherically averaged scattering amplitudes of the amino acid residues after solvent subtraction weighted according to their abundance. Svergun *et al.* (2001) computed $I_{DR}(q)$ from twenty-five different proteins and deduced average correction factors to best represent the all-atom computed curves also at higher angles (Fig. 4.9A, curve 2). To account for the bound solvent, the model is surrounded by dummy solvent atoms representing the first hydration shell (see Section 4.6 for a more detailed discussion of the computation of the scattering patterns from atomic models and of bound solvent effects). Simulations with known protein structures indicated that such rather simple DR model adequately represents the scattering patterns up to 0.5 nm resolution.

The concept of DRs (which is similar to a 'globbic' approach in protein crystallography; Guo *et al.* 2000) became very useful in various algorithms to represent proteins or their fragments in SAXS data modelling. For the *ab initio* analysis a three-dimensional protein model can be constructed by finding a spatial arrangement of the DRs that fits the experimental solution scattering pattern (Svergun *et al.* 2001). The reconstruction procedure is similar to the 'dummy atoms' modelling in the previous section inasmuch as SA is employed to randomly modify the search model inside a sphere of diameter D_{max}. However, for the DAM reconstructions the number of particle beads is not known *a priori* and their positions are fixed to the hexagonal grid. For the DR modelling the number of residues K is considered fixed (known from the primary sequence), and they are not bound to a fixed grid. The set of parameters to be found are the $3K$ coordinates of the DRs, $\mathbf{X} = [\mathbf{r}_j]$. One starts from a random configuration, and each move consists in relocating a DR taken at random to an arbitrary point at a distance of 0.38 nm from another randomly selected DR within the search volume. Another important difference is that instead of being compact, the distribution of the DRs is required to be locally similar to that in a protein. Indeed, the spatial arrangement of C_α atoms in the vicinity of a given atom in a real protein is rather specific, due to the separation along the chain, excluded volume effects and local interactions. A histogram of the average number of neighbours in a 0.1-nm-thick spherical shell surrounding a given C_α atom as a function of the shell radius $<N(R_k)>$ for $0 < R_k < 1$ nm is presented in Fig. 4.9B. In order to obtain a locally protein-like DR model the histogram $N_{DR}(R_k)$ should be similar to $<N(R_k)>$.

The function to be minimised is similar to that in eq. (4.50):

$$f(\mathbf{X}) = \chi^2(\mathbf{X}) + \alpha \left[P(\mathbf{X}) + G(\mathbf{X}) + R_{g0}^2 \right] \qquad (4.58)$$

The first term in the penalty part of eq. (4.58) imposes a protein-like nearest-neighbour distribution, while the second term, $G(\mathbf{X})$, ensures that the model is connected; that is, each DR has at least one neighbour

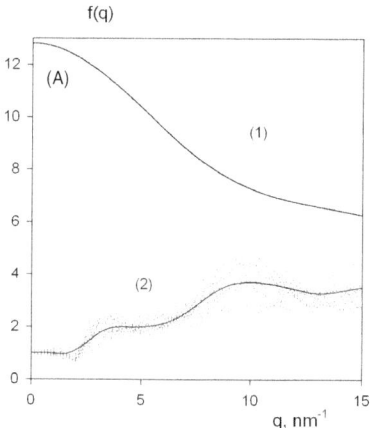

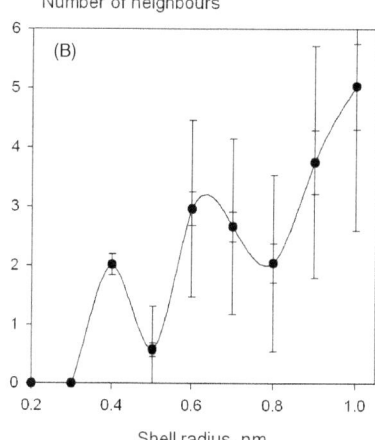

Fig. 4.9 A: Averaged X-ray form factor of a residue (1) and the average correction factor (2). Dotted curves represent individual correction functions computed for twenty-five different proteins. B: Histogram of an average number of C_α atoms in 0.1-nm thick spherical shells around a given C_α atom in a protein. Smaller error bars: variation of the averaged values over all proteins; larger error bars: averaged variation within one protein. Reproduced from Svergun *et al.* (2001), with permission from Elsevier, © 2001.

at a distance of 0.38 nm. The third term, proportional to the R_g of the model, keeps the centre of mass of the DR model close to the origin, and its weight is gradually decreased during the SA procedure. The algorithm, implemented in the program GASBOR, can also take particle symmetry into account by generating symmetry mates for the DRs in the asymmetric unit (the same point groups are supported as for the shape determination by DAMMIN).

As usual with SA, millions of function evaluations are required for minimisation, and it would take prohibitively long to fully recompute $f(\mathbf{X})$ each time, especially using the Debye formula. To speed up the computations the table of off-diagonal distances $[r_{ij}, i > j]$ is computed only once and updated when moving one DR at a time. Still, the performance of the method depends significantly on the size of the protein. For smaller proteins (<50 kDa) the GASBOR algorithm may be even faster than the multipole expansion in DAMMIN (though still slower than DAMMIF), whereas for larger proteins (>100 kDa), the Debye formula is much slower. In a real space version of GASBOR (Petoukhov and Svergun 2003) a gain in speed by a factor of 5 is obtained by fitting the distance distribution function $p(r)$ instead of the intensity $I(q)$.

The major advantage of the DRs compared to bead-shape determination methods is the possibility of taking into account higher angle scattering data (up to $q = \sim 10\,\mathrm{nm}^{-1}$), so usually GASBOR employs fewer free parameters while accounting for more experimental information than bead modelling. As a result, higher-resolution features depicted by the DR models are more reliable than those from bead modelling. The difference in the low-resolution reconstruction between the three approaches (envelope, bead and DR models) is presented in Fig. 4.10 on a classical example: hen egg-white lysozyme. The methods provide similar overall shapes, but the less detailed shape models only fit the scattering at very low angles whereas the more detailed DR model neatly fits the entire experimental scattering pattern.

Similar to the bead modelling methods discussed above, the DR method produces a manifold of spatial distributions of DRs, rather than a single solution. Calculations on simulated and experimental scattering patterns indicate that the differences between the DR models are substantially smaller than those observed in low-resolution shape determination. The variations between the DR models preserve the domain structure of the protein, and the average (most probable) model can be generated by averaging the results of independent SA runs as described in the previous section.

The concept of DR modelling is useful not only for *ab initio* shape reconstruction but also for other applications, which will be considered in subsequent sections. At present, the available algorithms for *ab initio* DR modelling are restricted to the analysis of the X-ray scattering from proteins, but this principle can be extended to represent nucleic acids as assemblies of nucleotides and also to the neutron scattering-based computations (deuteration has to be adequately taken into account in the latter case).

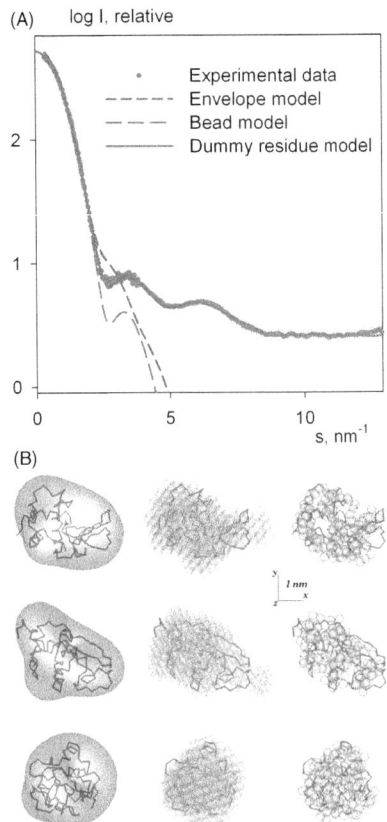

Fig. 4.10 A: Synchrotron X-ray scattering from lysozyme (1) and scattering from the *ab-initio* models: (2) envelope model (SASHA); (3) bead model (DAMMIN); (4) dummy residue model (GASBOR). B: Atomic model of lysozyme (C_α chain, PDB entry 6LYZ (Diamond 1974)) superimposed to the *ab initio* shapes obtained by SASHA (left column, semi-transparent envelope), DAMMIN (middle column, semi-transparent beads) and GASBOR (right column, semi-transparent dummy residues). The models were aligned by SUPCOMB (Kozin and Svergun 2001); the middle and bottom rows are rotated counterclockwise by 90° around X and Y, respectively.

4.5 Computation of scattering patterns from atomic models

Calculation of the scattering from atomic models and their comparison with experimental patterns was long used to determine the degree of similarity between the crystal and solution structures and also for validation of the predicted homology models (Langridge *et al.* 1960, Ninio *et al.* 1972, Müller 1983, Pavlov *et al.* 1986, Pavlov 1985). Moreover, given the high-resolution models of individual domains or subunits, quaternary structures can be constructed by rigid-body modelling (Boehm *et al.* 1999, Chamberlain *et al.* 1998, Krueger *et al.* 1999, Svergun *et al.* 1998a, Svergun *et al.* 1998b), which has become one of the most powerful hybrid methods in the study of macromolecular complexes (see Section 4.6).

The accurate computation of scattering patterns from atomic structures is by no means a trivial procedure, as the solvent scattering must be adequately taken into account. Generally, both for X-rays and neutrons, the scattering intensity from a particle in solution can be expressed as

$$I(q) = \left\langle \left| A_a(q) - \rho_b A_{ex}(q) + \delta \rho_h A_h(q) \right|^2 \right\rangle_\Omega \tag{4.59}$$

where $A_a(q)$, $A_{ex}(q)$ and $A_h(q)$ are, respectively, the scattering amplitudes from the particle *in vacuo*, from the excluded volume and from the hydration shell. The latter term may appear due to the fact that the scattering density of the bulk solvent, ρ_b, may differ from that of the bound solvent (denoted as ρ_h) resulting in a non-zero contrast for the shell $\delta \rho_h = \rho_h - \rho_b$ (Svergun *et al.* 1995).

In the early studies the term due to the bound solvent was usually neglected, and most attention was paid to the calculation of the scattering due to the excluded volume (the volume inaccessible to the solvent). Given the fact that the experimental scattering is observed as an average over a large number of independent macromolecules, the excluded volume for these calculations should be represented as a continuum filled by uniform solvent density ρ_b. In the effective atomic scattering method, the excluded volume is built from dummy solvent atoms located at the positions of the atoms in the macromolecule (Langridge *et al.* 1960, Fraser *et al.* 1978). This approach is perhaps the simplest and fastest way of computing $A_{ex}(q)$, and is implemented in several methods (Schmidt *et al.* 1995, Lattman 1989, Svergun *et al.* 1995), and, more recently, by Poitevin *et al.* (2011) and Schneidman-Duhovny *et al.* (2010). The effective atomic scattering method non-uniformly fills the excluded volume and thus may not represent adequately the scattering at wider angles (resolutions better than 1 nm). In the cube method the particle surface accessible to the solvent is defined by rolling a sphere simulating a water molecule on the van der Waals surface of the particle following (Lee and Richards 1971). The excluded volume can then be represented by cubes with a small edge (down to 0.05 nm) to ensure its precise and uniform filling. Approaches using the cube method can be more accurate for computations of WAXS patterns (Ninio *et al.* 1972, Müller 1983, Pavlov and Fedorov 1983, Virtanen *et al.* 2011). There are also attempts to compute the excluded volume scattering by averaging the water ensembles from molecular dynamic (MD) simulations (Grishaev *et al.* 2010), but it is questionable whether these calculations are physically more justified than the simple cube method.

The importance of the accounting for the scattering by the bound solvent was understood at the end of 1980s, when it was observed that the scattering patterns computed from the crystallographic models of proteins yield poor fits to the experimental SAXS data. The crystal structures often appeared 'too small', and better fits could be obtained by adding some water molecules to the protein surface (Hubbard *et al.* 1988, Grossmann *et al.* 1993, Fujisawa *et al.* 1994). A general program, CRYSOL (Svergun *et al.* 1995), was developed simulating the first solvation shell by surrounding the particle by a continuous outer envelope. For this, the angular envelope function of the particle $F(\omega)$ is computed as in eq. (4.44) and surrounded by a 0.3-nm-thick homogeneous layer with the scattering density ρ_b. Given the atomic coordinates, $A_{ex}(q)$ is computed using the effective atomic factors and $A_b(q)$ from the envelope layer, using the multipole expansion with expressions similar to eqs. (4.54) and (4.49). The program predicts the scattering profile or fits the experimental data by adjusting the excluded volume of the particle and the density of the hydration layer. The use of multipole expansion speeds up the calculations and, additionally, the partial amplitudes computed by CRYSOL can be used further for rapid computation of scattering from complexes (see Section 4.6). The scattering contributions from the different terms of eq. (4.59) and the influence of the hydration shell contribution for hen egg-white lysozyme are presented in Fig. 4.11.

Analysis of the SAXS data from numerous known protein crystal structures using CRYSOL indicated that inclusion of the hydration shell significantly improved the agreement between the experimental and calculated X-ray scattering curves. Typically, the density of the border layer of 1.05–1.25 times that of the bulk was obtained, suggesting that the hydration shell around proteins is denser than the bulk solvent. However, SAXS studies alone are not able to provide unequivocal proof of the physical origin of this higher density. Indeed, a similar effect on the scattering curves could result from higher mobility or disorder of the surface side chains in solution compared to their average structure in the crystal, which would also increase the apparent particle size. This ambiguity was resolved with the development of the program CRYSON (Svergun *et al.* 1998b)—an analogue of CRYSOL for computing neutron scattering in H_2O/D_2O mixtures (taking H/D exchange effects into account). As the protein has a negative contrast in D_2O, a denser hydration shell should lead to the opposite effect compared to X-rays; for example, the crystal structure appearing larger than the SANS data predict. The SAXS/SANS results were fully compatible with a denser

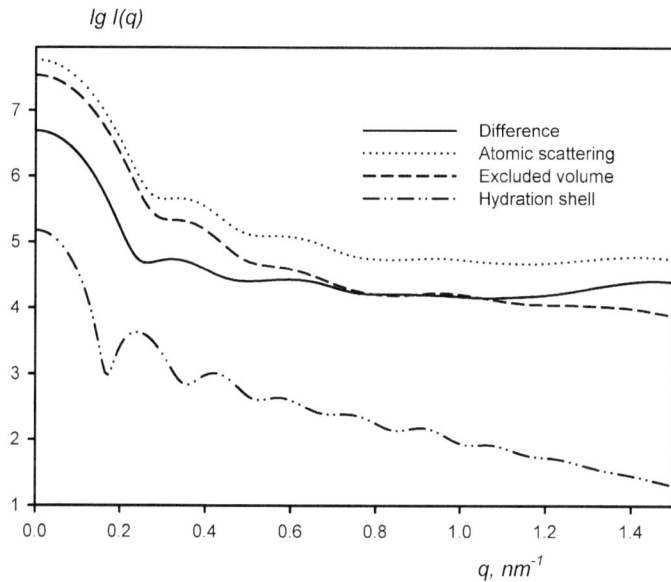

Fig. 4.11 The scattering components computed from the atomic model of lysozyme in solution by CRYSOL. The curve marked "Difference" is the net scattering accounting for the solvent contribution.

hydration shell rather than with a higher mobility of the side chains on the protein surface (Svergun *et al.* 1998b).

The experimental SAXS/SANS data confirm the higher density in the hydration shell predicted from extensive MD simulations (Levitt and Sharon 1988, Merzel and Smith 2002), and accounting for this shell, which was considered controversial in the past, became a must in all calculations. Several groups contributed to the further development of programs to compute SAXS/WAXS patterns from macromolecular structures, with various ways of representing the contribution of the hydration shell. These include positioning of dummy waters at exposed surfaces in program FOXS (Schneidman-Duhovny *et al.* 2010), explicit MD simulations in program AXES (Grishaev *et al.* 2010), MD simulations-based empirical hydration predictions (Virtanen *et al.* 2010, Virtanen *et al.* 2011), and the electrostatic-based solvent distribution calculations utilising a Poisson–Boltzmann–Langevin formalism (Poitevin *et al.* 2011). A number of programs are available (as downloads or on Web servers, see Appendix 4), allowing one to compute the SAXS/WAXS patterns, and all yield comparable results in most cases. Less development has taken place in SANS, where CRYSON (Svergun *et al.* 1998b) remains effectively the only choice.

One should also mention other tools available for dealing with bead models. The HYDRO program suite (Garcia De La Torre *et al.* 2000) can be employed to calculate hydrodynamic parameters (diffusion coefficients, intrinsic viscosity, relaxation time and so on), providing a direct link to these complementary biophysical methods and thus an additional consistency check. The program EM2DAM (Petoukhov *et al.* 2012) converts cryo-EM maps into bead models and computes their scattering patterns. For a list of programs to compute the scattering and the structural parameters from either atomic or bead models, see Appendix 4.

4.6 Hybrid methods

During the last decades, high-resolution methods—especially X-ray crystallography but also NMR—have generated tremendous numbers of structures of individual proteins or their domains in the large-scale structural genomics initiatives (Levitt 2007). Most cellular functions are, however, accomplished by macromolecular complexes, and the focus of modern structural biology has shifted largely towards their study (Aloy and Russell 2006). They are often too large for structural NMR and may possess inherent structural flexibility, making them difficult to crystallise. As solution scattering is sensitive to changes in the quaternary structure of macromolecules it should be particularly useful for the analysis of such complexes. Numerous hybrid approaches have been developed for using the high-resolution models in the SAS analysis, including modelling solely with subunits of known structure (rigid-body analysis), and also approaches accounting for the missing portions, such as linkers, loops or entire domains. The two cases will be considered in the ensuing sections.

4.6.1 Rigid-body modelling

The idea of rigid-body refinement is best illustrated by considering a binary complex consisting of subunits A and B where the high-resolution structures of the two subunits are available but the structure of the complex is unknown. The scattering amplitudes from the subunits centred at the origin in some reference orientations, denoted $A(q)$ and $B(q)$, respectively, can be precomputed by the methods described in the previous section. An arbitrary complex can be constructed by fixing the first subunit and rotating and translating the second one (Fig. 4.12).

The rotation is described by the Euler angles α, β and γ (Edmonds 1957) and the shift by a vector $u = (u_x, u_y, u_z)$, so that the complex is defined by six parameters. Denoting $C(\mathbf{q})$ the scattering amplitude from the displaced second subunit, the scattering from the complex can be expressed as (Svergun 1991)

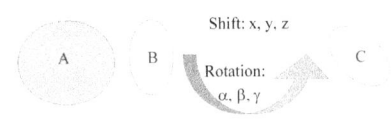

Fig. 4.12 The principle of rigid-body modelling illustrated for a two-body complex A and B described by three translational and three rotational parameters.

$$I(q) = I_A(q) + I_B(q) + 2\langle A(\mathbf{q}) C^*(\mathbf{q})\rangle_\Omega \qquad (4.60)$$

As in the previous sections, the spherical average is best handled by using the spherical harmonics $Y_{lm}(\Omega)$. The multipole representation of the amplitudes $A(q)$ and $C(q)$ as in eq. (4.43) yields the closed expression

$$I(q) = 2\pi^2 \sum_{l=0}^{\infty} \sum_{m=-l}^{l} \left(|A_{lm}(q)|^2 + |B_{lm}(q)|^2 + 2Re\left[A_{lm}(q)C_{lm}^*(q)\right] \right) \quad (4.61)$$

Importantly, the partial functions $C_{lm}(q)$ of the rotated and translated second subunit can be expressed analytically via the $B_{lm}(q)$ functions, the elements of the finite rotation matrix and the Wigner $3j$ coefficients (Edmonds 1957, Svergun 1994, Svergun 1991); see Appendix 2. Rigid-body modelling of a binary complex against solution scattering data therefore requires determination of the six rotational and positional parameters of the second subunit. This idea can be generalised easily for a complex consisting of K rigid bodies, which, in the general case, can be described by $6(K-1)$ parameters.

If a rapid method for computation of $I(q)$ is available one can try to perform a search of these parameters by fitting the experimental scattering from the complex. It is clear, however, that to have sensible models the search space must be restrained—at the very least by the conditions of non-overlapping and interconnected subunits. The computational methods not employing spherical harmonics are relatively slow but should often simplify the search models. Historically, one of the first approaches to speed up the computations on atomic models, which was, however, limited to binary complexes, was published by Pavlov (1985). In an 'automated constrained fit' procedure (Boehm *et al.* 1999), atomic models are substituted with bead models and thousands of configurations generated in an exhaustive search for the best fit, often combined with hydrodynamic data (Perkins *et al.* 2011). In another approach (Krueger *et al.* 1998, Tung *et al.* 2000, Tung *et al.* 2002), a representation of the domains in terms of triaxial ellipsoids was used to find their approximate arrangement in the complex. The atomic models of the domains are subsequently positioned within the ellipsoids utilising information from other methods including NMR, homology modelling and energy minimisation (Wall *et al.* 2000). A multi-method modelling approach also using hydrodynamic data has been described by (Nollmann *et al.* 2005). Methods utilising NMR residual dipolar coupling (RDC) information, which restricts the possible mutual orientation of protein domains, have also been developed (Mattinen *et al.* 2002, Grishaev *et al.* 2005).

Computations using spherical harmonics are sufficiently rapid to allow for different modelling strategies. First, programs were designed allowing for interactive rigid-body modelling—an approach which is in a way similar to interactive graphic interpretation of electron density maps in protein crystallography; for example, Potterton *et al.* (2004). These programs (ASSA for Unix (Kozin *et al.* 1997, Kozin and Svergun 2000) and MASSHA for Windows (Konarev *et al.* 2001)) allowed three-dimensional display and manipulation of high- and low-resolution models, being coupled to computational modules to rapidly compute the scattering from the complex and the fit to the experimental data. In the purely interactive mode, the user may shift and rotate the subunits while observing corresponding changes in the fit to the experimental data. A refinement mode is also available, where the program performs an automated search for the best fit in the vicinity of the current configuration.

Interactive programs were a milestone in rigid-body modelling, and they still play an important role, especially in tutorials, helping one to understand which kinds of motions/rotations influence the scattering pattern. However, as in protein crystallography, where the interactive chain building was superseded by automated refinement programs (for example, Langer *et al.* 2008, Adams *et al.* 2011, Murshudov *et al.* 2011), global search algorithms were also developed and became most used for SAXS/SANS. Perhaps the most comprehensive program is SAS-REF (Petoukhov and Svergun 2005, Petoukhov and Svergun 2006, Petoukhov *et al.* 2012), which, starting from an arbitrary positioning of subunits, conducts random rigid-body movements and rotations, using SA to search for the best fit of the computed complex scattering to the experimental data. In the simplest case the program fits a single scattering pattern from a complex of two and more subunits by constructing an interconnected non-overlapping model (the absence of steric clashes and interconnectivity are always ensured by adding appropriate penalties similar to eq. 4.56). However, the algorithm is also able to account for a variety of additional information, including

- simultaneous fitting of multiple scattering curves, when multiple constructs such as deletion mutants or sub-complexes have been measured, and assuming that the organisation of substructures is unchanged compared to the full complex;
- simultaneous fitting accounting for multiple data sets from contrast variation by SANS and even simultaneous fitting of SAXS and SANS data;
- the use of symmetry, whereby only the asymmetric part is modelled and the rest of the complex is obtained by symmetry operations;
- orientational constraints, if probable relative orientations of some of the subunits are known (for example, from RDCs measured by NMR);
- inter-residue contacts (for example, from mutagenesis, cross-linking experiments or chemical shifts in NMR);
- distances between specific residues or domains (for example, from Fourier transform infrared spectroscopy (FTIR) and FRET analysis).

There are also other publicly available programs for rigid-body modelling, including, for example, exhaustive search methods DIMFOM and GLOBSYMM (Petoukhov and Svergun 2005), and the program FOXSDOCK screening possible docking models against SAXS data (Schneidman-Duhovny *et al.* 2011). Rigid-body analysis has become one of the major tools in the SAXS/SANS studies of macromolecular complexes, and several applications will be presented in Chapter 6. Hybrid modelling in SAS differs from model construction in EM studies, where the high-resolution structures are usually 'docked' into the low-resolution envelopes (Wriggers *et al.* 2011). In principle, docking into the *ab initio* SAS shapes is also possible but less straightforward, as the SAS shapes are usually of lower quality than the EM envelopes. Further, because of the ambiguities of the reconstruction the SAS shapes may be lacking reliable higher-resolution features to confidently position the available subunits. However, in contrast to EM, rapid calculation of theoretical scattering from possible models and its comparison with experiments is possible in SAS. This 'data-driven' modelling is physically perhaps more meaningful than docking into shapes derived from the data.

Still, a word of caution must be given here against overinterpretation of the rigid-body models constructed using SAS. They apparently retain the resolution of the initial (high-resolution) structures of the subunits or domains, but the rigid-body models are based on low-resolution SAS information. Therefore, the latter models can by no means be called 'high-resolution', and drawing far-fetched conclusions (at atomic or residue level of detail) is only possible if the models are appropriately validated. Similar to the *ab initio* modelling, running rigid-body programs such as SASREF several times, starting from different random approximations, may yield different models, providing similar fits to the data. These multiple models may give an indication on the possible ambiguity of the modelling. There are convenient tools available for this analysis which distribute the rigid-body models into clusters with distinctly different quaternary structures (such as the program DAMCLUST, developed as an extension of DAMAVER; Petoukhov *et al.* 2012). Further analysis and ranking of the different clusters requires screening against the criteria not taken into account in the SAS modelling (such as surface and charge complementarity and energetic considerations) and/or additional experiments to further validate the models (see examples in Chapters 6 and 9).

4.6.2 Addition of missing fragments

'Pure' rigid-body modelling is possible only when the full high-resolution structures (or reliable homology models) of all the subunits are available. However, inherent flexibility and conformational heterogeneity of macromolecules or complexes can often result in the absence of loops and even entire domains in structures determined by MX or NMR. Further, in the studies of multi-domain or multi-subunit macromolecules consisting of globular domains/subunits linked by flexible loops, high-resolution structures or homology models may be available for the individual domains/subunits but usually not for the linkers. Potentially disordered regions (such as termini) may also not be seen at high resolution, due to flexibility. Still, the regions missing in the high-resolution models do contribute to the scattering of the macromolecule in solution, and their approximate configurations can be reconstructed. During such modelling, precise conformations of the missing terminal loop or interdomain linker are not required, but approximate 'placeholders' should be added to the model to adequately compute the scattering intensity from the entire molecule. In this very important practical case a combination of rigid-body and *ab initio* modelling can be employed to determine the overall structure of the entire assembly.

The idea behind this approach is that the unknown portions of the structure are represented as beads or chains of DRs, as introduced in Section 4.4.3. The domains or subunits are then moved as rigid bodies to find their optimal positions and orientations together with the probable conformations of the flexible linkers or missing domains. Using spherical harmonics, the scattering from such a model contains two contributions:

$$I(q) = 2\pi^2 \sum_{l=0}^{\infty} \sum_{m=-l}^{l} \left| \sum_k A_{lm}^{(k)}(q) + \sum_j D_{lm}^{(j)}(q) \right|^2 \qquad (4.62)$$

where $A_{lm}^{(k)}(q)$ and $D_{lm}^{(j)}(q)$ are the partial amplitudes of the domains with known structure and from the missing portions, respectively. The former contribution is changing with the rigid-body movements and rotations of the domains; the latter contribution varies when the moieties representing the missing portions are modified.

In the first implementation (Petoukhov *et al.* 2002), missing loops or domains were added by fixing a known structure and building the unknown fragments to fit the scattering data from the entire particle (program suite CREDO). Different scenarios were used depending on the availability of information about the missing portions. If the interface between the known and unknown parts was not available, the missing domain was represented by a gas of DRs. If the interface was known, loops or domains were represented as interconnected chains (or ensembles of residues with spring forces between the C_α atoms), attached to known position(s) in the available structure. Residue-specific constraints were employed to ensure native-like folds of the missing fragments. The fit to the experimental data was obtained by SA to minimise a scoring function similar to that in eq. (4.58) containing the discrepancy and the relevant penalty terms.

In a later approach (Petoukhov and Svergun 2005), missing linkers or protein domains with unknown structure are represented by a flexible chain of interconnected DRs (program BUNCH). In the SA-driven global minimisation, a single step modifying the model is performed by two types of move (Fig. 4.13).

First, a DR is selected at random dividing the entire chain in two parts, and the smaller part of the chain is rotated by a random angle about a random axis drawn through this DR. Alternatively, a random rotation of the part of the structure is performed between two randomly selected DRs about the axis connecting them. In the generated fragments, consecutive DRs are always centred at 0.38 nm from their neighbours, and appropriate penalties are added to ensure native-like folds. In particular, absence of steric clashes is required such that the distances between all non-neighbouring DRs do not exceed 0.4 nm. The allowed combinations of bond and dihedral angles are selected to comply with the predictions of the quasi-Ramachandran C_α–only plot (Kleywegt 1997). If the scattering data from partial constructs or

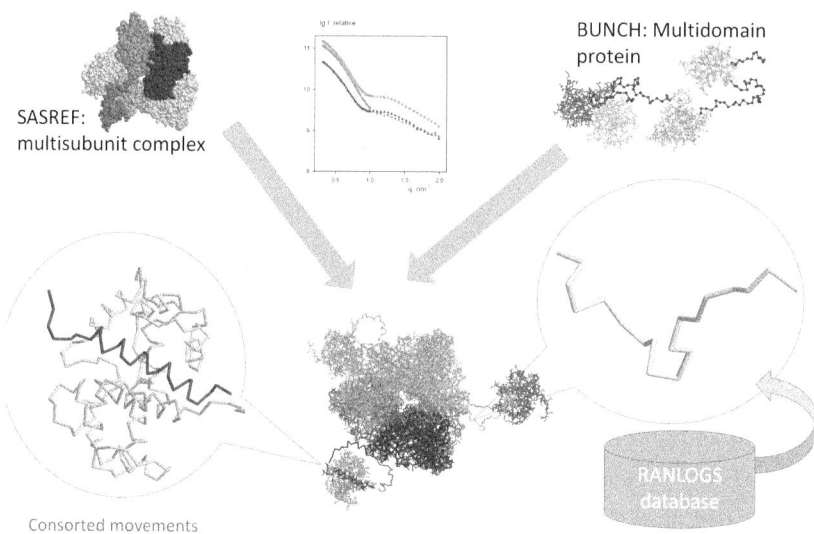

SASREF:
multisubunit complex

BUNCH: Multidomain
protein

Consorted movements

RANLOGS
database

Fig. 4.13 Schematic representation of different methods for rigid-body modelling and addition of missing fragments (see text for explanations). Adapted, with permission, from Petoukhov, Franke, *et al.* (2012) © International Union of Crystallography.

deletion mutants are available, BUNCH can simultaneously fit multiple scattering data sets by that computed from the relevant portions of the model. As in SASREF, the symmetry of the particle can be taken into account such that only the symmetry-independent part is modified and the rest generated automatically.

Whereas BUNCH does allow one to amend rigid-body models with missing fragments, the range of application is limited inasmuch as the algorithm is primarily oriented towards single-chain proteins or symmetric assemblies containing one polypeptide chain per asymmetric part. There are important cases in practice—for example, complexes consisting of multiple subunits where one or more components have missing fragments of noticeable length—where neither SASREF nor BUNCH would be suitable. A more general approach combining the functionality of the two techniques has recently been developed (program CORAL, Petoukhov *et al.* 2012). Here, the atomic models of individual domains belonging to multiple components of a complex are translated and rotated similarly to SASREF. These rearrangements are, however, not fully random inasmuch as the distances between the N- and C-terminal portions of the subsequent domains belonging to one chain are constrained by the maximum length of the missing portion. The algorithm employs a pregenerated library of self-avoiding random loops (called RANLOGS). This library provides DR chains with linker lengths from five to a hundred amino acids sampling twenty structures for every possible end-to-end distance for the given length with a sampling step of 0.2 nm. When a domain is translated/rotated in CORAL, the corresponding random loop(s) connecting this domain with the preceding/following one is inserted as a placeholder for the missing linker, and its contribution is added to the computed scattering intensity of the system (Fig. 4.13). The steps yielding configurations where no linker can be found in the library are immediately rejected. The C- and N- terminal portions of the subunits, if missing, can also be randomly selected from the library, but they do not constrain the domain movements. In CORAL, a possibility is introduced of consorted motion of selected domains, which would keep their mutual arrangement while changing their position and orientation with respect to the rest. This feature is useful, for example, when there are several known oligomerisation interfaces in one system. The modelling is done with SA using the usual constraints available in SASREF (for example, overlaps, contact restraints and so on) as additions to the target function to be minimised. It should be noted here that the techniques utilising DR-based modelling (BUNCH and CORAL) are, similarly to the *ab initio* GASBOR method (Section 4.4.3), limited to X-ray scattering from proteins. The principle is readily extendable for nucleic acids and for neutrons, but, perhaps due to the lack of demand, more general algorithms are not yet available.

In this section we described the basic approaches used for hybrid modelling of the SAS data. As can be seen, the hybrid methods cover a broad range of applications, from multidomain proteins to complexes. Additional information from different sources, if available, can readily be incorporated into the SAS-based modelling to constrain possible solutions, but these hybrid models must still be carefully validated against independent data—most desirably, the data not used in the construction of the model. Some of the validation approaches will be considered in Chapter 9.

4.7 Labelling and triangulation

Nearly all X-ray studies and most of the neutron SAS studies of biological macromolecules are performed on samples obtained by conventional biochemical preparations. It turns out that much information can be gained by labelling parts of the molecules or specific sites. In the case of X-rays, one can use heavy atoms or heavy-atom complexes, whereas in the case of neutrons one usually employs deuterium (^{2}H, D) to specifically label parts of a larger molecule composed of several subunits. This allows one to specifically highlight parts of the molecule and render others 'invisible' at particular contrasts; that is, in buffers of specific ^{1}H/^{2}H composition (see Chapter 1). Deuterium labelling is often used to create a contrast between subunits of identical chemical composition (such as proteins), but it can also be helpful to adjust the contrasts of naturally different components (such as protein and nucleic acids or sugars).

4.7.1 Deuteration by biochemical procedures

Whilst there are brute-force methods to replace natural-abundance hydrogen (>99.98% ^{1}H) by deuterium in organic compounds, it is preferable (and much easier) to achieve this replacement in biological macromolecules by growing microorganisms in culture solutions containing D_2O and deuterated substrates (Daboll *et al.* 1962, Lederer *et al.* 1986, Moore 1977, Shcherbakova and Serdyuk 2000). Moreover, microorganisms can be forced to overproduce the desired molecules using recombinant DNA, thus needing much fewer resources than a classical culture.

The growing need of deuterated molecules for neutron scattering, but also for NMR, has led to the establishment of deuteration laboratories at neutron centres—first at the ILL, Grenoble, France, and then also at other places: ANSTO, Australia; CSBM, Oak Ridge, USA; Los Alamos, USA. Deuteration laboratories provide sophisticated computer-controlled devices for optimal and cost-effective fermentation, but also help with adapting microorganisms to growth in D_2O.

4.7.2 Highlighting parts of molecules

When studying large macromolecules or macromolecular complexes, it is difficult to obtain information about the shape and location of their substructures by SAS from the whole molecule. However, if the (protein) particle can be separated into its subunits and be reconstituted by biochemical or biophysical methods, SANS allows one to study the shapes and positions of the subunits *in situ*. This requires growing the microorganism in H_2O and in D_2O, isolating the macromolecules from both conditions, separating the subunits and reconstituting chimeric molecules containing one (or several) subunit(s) in deuterated form and the others ('matrix') with natural-abundance hydrogen. The scattering of this molecule at the matching condition of the H-subunits corresponds basically to the contribution of the deuterated ones.

In fact, one can also use partial deuteration of the matrix by growing the macromolecules in a medium with a D_2O content slightly below 100% to avoid a scattering length density (SLD) of the molecules higher than that of pure D_2O. One can then use H subunits as labels and measure in a buffer matching the SLD of the partially deuterated matrix. This has the advantage of a much lower incoherent background at the expense of a lower contrast between the labelled subunits and the matrix.

This last strategy is also useful with complexes of protein and nucleic acids, where the protein contrast can be doubled when using (partially) deuterated RNA or DNA in an SLD-matching buffer solution close to 100% D_2O. The contrast of H-nucleic acids with respect to H-protein or partially deuterated protein is similar, but the incoherent background is much less at the matching conditions for the latter.

4.7.3 Label triangulation

Large biological macromolecular complexes are often difficult to crystallise, and even if one can do so it is not easy to obtain their structure by X-ray crystallography. If, however, the distances between a sufficient number of components can be determined, their three-dimensional arrangement—that is, the quaternary structure of the complex—can be reconstructed by a technique termed 'label triangulation' (Hoppe 1972). The method was described in detail by May (1991), and the content of this reference (copyright Elsevier, 1991) is used in this section.

Ribosomes are the protein-synthesising machinery in all living cells (Rodnina *et al.* 2011). They can be separated into their constituents, proteins and nucleic acids, and reconstituted into biochemically active entities. Label triangulation was used to determine the three-dimensional arrangement of the individual proteins in the ribosome long before the X-ray crystallographic structures of the ribosome were solved.

Origin of the technique: heavy-atom labels. Kratky and Worthmann (1947) labelled organic compounds with heavy-metal atoms (iodine) and measured the scattering curves of these molecules with X-rays. These were dominated by the heavy-atom contributions. The distances between the labels could therefore be calculated from the periodicity of the differences between the compounds containing two and one heavy atoms. Since these depend on the conformations of the molecules, the latter can be deduced from such scattering experiments. Later, Vainshtein *et al.* (1970) determined heavy-atom distances in the antibiotic polypeptide gramicidine C, and even larger proteins (haemoglobin, histidine decarboxylase) were studied by Vainshtein *et al.* (1980) with an organomercury compound containing four Hg atoms.

Quite recently, Mathew-Fenn *et al.* (2008) measured the interference between two gold nanocrystal labels attached to the ends of a series of DNA molecules of different lengths by SAXS. Their study revealed a difference in stiffness of at least an order of magnitude in the absence of applied tension compared to single-molecule stretching experiments. Grishaev *et al.* (2012) matched the protein SLD for X-rays by a 65% sucrose solution and determined the distance of Pb labels in a lead-substituted calmodulin–peptide complex.

Hoppe (1972) proposed to reconstruct the three-dimensional arrangement of subunits in macromolecules from distances determined by SAS. Engelman and Moore (1972) were the first to propose deuterium rather than heavy-metal complexes for labelling the components, and to use neutron scattering for determining the distances. Later, Hoppe (1973) showed that measuring the scattering curve from an equimolar mixture of unlabelled particles (complexes) with particles containing two labelled components, and subtracting the scattering from a mixture of the two particles with either of the two selected components labelled, results in a difference curve containing the inter-component interference only.

Distance determination by neutron scattering. The distance distribution $p(r)$ (see Section 4.1.1) of a dumbbell composed of two spheres with a rigid centre-of-gravity distance d, as shown in Fig. 4.14, displays two

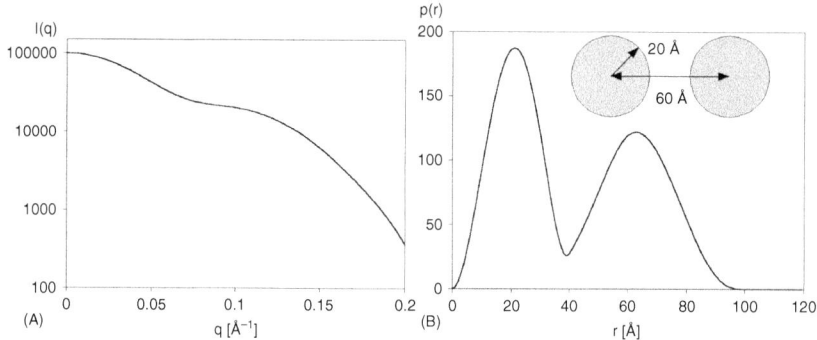

Fig. 4.14 A) Scattering curve and B) pair-distance distribution function for a dumb-bell of two spheres of radius 20 Å and a centre-of-gravity distance d of 60 Å.

peaks if the distance between the spheres is sufficiently large, and these are completely separated if the distance is more than twice the larger radius. The peak at larger r is due to all the distance vectors connecting volume elements of one sphere with volume elements of the other sphere, and the maximum of this peak corresponds closely to the distance separating the centres of gravity of the two spheres. The low-r peak corresponds to the superposition of the two intra-sphere pair-distance distributions. For two points instead of two spheres at a distance d, one would observe a δ function at $r = d$.

In real systems the situation is more complicated: The components are in reality neither point-like nor spherical, and the centre-of-gravity distance between the two components can be shorter than the maximal chord length within both components; this can lead to an overlap of the intra- and inter-subunit distance distributions as in Fig. 4.14. Furthermore, the two components are part of a larger system that is in general also visible to neutrons. This difficult problem was finally solved by different approaches.

The interference term, which contains all the distance information, can be obtained from a linear combination of the scattering curves from four different samples at equal concentration (Hoppe 1973); see Fig. 4.15. Let us assume that the particles under study are monodisperse and in dilute solution. The scattering intensity $I_a(q)$ from the unlabelled complex (sample a in Fig. 4.13) can then be written as

$$I_a(q) = \frac{N}{V} \sum_{i=1}^{n} \sum_{j=1}^{n} b_i b_j \sin \frac{q r_{ij}}{q r_{ij}}, r_{ij} = |\mathbf{r}_i - \mathbf{r}_j| \qquad (4.63)$$

b_i are the scattering lengths of the n atoms i, $i = 1 \ldots n$, in the entire complex, assuming the particles to be in vacuum. In reality, we rather would have to use scattering length densities ρ, and to consider only the excess scattering-length density with respect to the solvent.

Distinguishing two components within the complex, with n_1 and n_2 atoms, respectively, one can rewrite eq. (4.59) to obtain

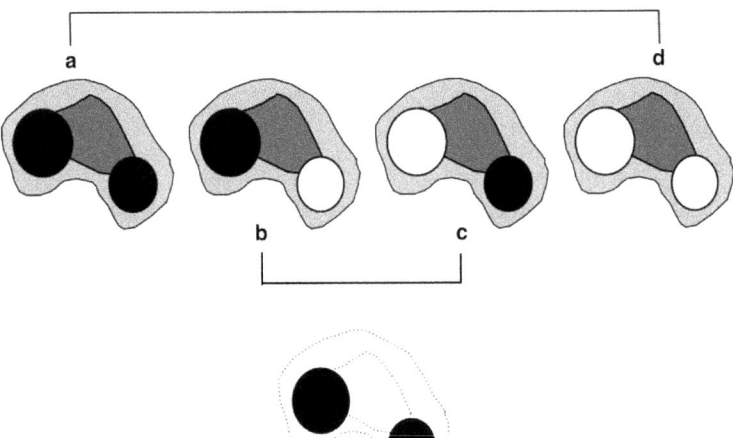

Fig. 4.15 Isomorphous replacement. Sample *a* contains complexes with two natural-abundance (black) subunits, sample *b* and *c* complexes with one deuterated (white) and one natural-abundance subunit, sample *d* complexes with two deuterated subunits.

$$I_a(q) = \frac{N}{V}\left(\sum_{i=1}^{n-n_1-n_2}\sum_{j=1}^{n-n_1-n_2} b_ib_j\frac{\sin qr_{ij}}{qr_{ij}} + \sum_{k=1}^{n_1}\sum_{l=1}^{n_1} b_kb_l\frac{\sin qr_{kl}}{qr_{kl}} + \sum_{m=1}^{n_2}\sum_{n=1}^{n_2} b_mb_n\frac{\sin qr_{mn}}{qr_{mn}} \right)$$

$$+ 2\frac{N}{V}\left(\sum_{i=1}^{n-n_1-n_2}\sum_{k=1}^{n_1} b_ib_k\frac{\sin qr_{ik}}{qr_{ik}} + \sum_{i=1}^{n-n_1-n_2}\sum_{m=1}^{n_2} b_ib_m\frac{\sin qr_{im}}{qr_{im}} + \sum_{k=1}^{n_1}\sum_{m=1}^{n_2} b_kb_m\frac{\sin qr_{km}}{qr_{km}} \right) \tag{4.64}$$

The first term corresponds to the scattering of the complex without the two selected components ('stripped' complex), and the second and third terms to that of the two components. The fourth and fifth terms are due to the interference between the stripped complex and the two components, and the sixth term to that of the interference between the two selected components.

Only the last term contains the information about the distance between the two components. If one exchanges some of the atoms of the components by isotopes with a different scattering length, one can distinguish it from the other scattering contributions. There are not many isotopes suitable for labelling, because these atoms have to be present in a sufficient number, and the difference in scattering length between the isotopes must be important. In practice, only the light and heavy isotopes of hydrogen (^{1}H and ^{2}H, deuterium) are useful for label triangulation with neutrons. In the case of X-rays, heavy-atom compounds can be attached to selected components of the complex. This, however, yields the centre-to-centre distances of these compounds rather than those of the components themselves.

By exchanging natural-abundance hydrogen by deuterium in one component one obtains for the scattering from such a complex (sample *b*):

$$I_b(q) = \frac{N}{V}\left(\sum_{i=1}^{n-n_1-n_2}\sum_{j=1}^{n-n_1-n_2}b_ib_j\frac{\sin qr_{ij}}{qr_{ij}} + \sum_{k=1}^{n_1}\sum_{l=1}^{n_1}b_k'b_l'\frac{\sin qr_{kl}}{qr_{kl}} + \sum_{m=1}^{n_2}\sum_{n=1}^{n_2}b_mb_n\frac{\sin qr_{mn}}{qr_{mn}}\right)$$

$$+2\frac{N}{V}\left(\sum_{i=1}^{n-n_1-n_2}\sum_{k=1}^{n_1}b_ib_k'\frac{\sin qr_{ik}}{qr_{ik}} + \sum_{i=1}^{n-n_1-n_2}\sum_{m=1}^{n_2}b_ib_m\frac{\sin qr_{im}}{qr_{im}} + \sum_{k=1}^{n_1}\sum_{m=1}^{n_2}b_k'b_m\frac{\sin qr_{km}}{qr_{km}}\right) \tag{4.65}$$

Here, b_i' replaces b_i in component 1, but most atoms (all non-H atoms, and all H atoms that exchange with the solvent) are supposed to remain unchanged. A nearly identical equation holds for component 2 (sample c). Labelling both components (sample d) yields

$$I_d(q) = \frac{N}{V}\left(\sum_{i=1}^{n-n_1-n_2}\sum_{j=1}^{n-n_1-n_2}b_ib_j\frac{\sin qr_{ij}}{qr_{ij}} + \sum_{k=1}^{n_1}\sum_{l=1}^{n_1}b_k'b_l'\frac{\sin qr_{kl}}{qr_{kl}} + \sum_{m=1}^{n_2}\sum_{n=1}^{n_2}b_m'b_n'\frac{\sin qr_{mn}}{qr_{mn}}\right)$$

$$+2\frac{N}{V}\left(\sum_{i=1}^{n-n_1-n_2}\sum_{k=1}^{n_1}b_ib_k'\frac{\sin qr_{ik}}{qr_{ik}} + \sum_{i=1}^{n-n_1-n_2}\sum_{m=1}^{n_2}b_ib_m'\frac{\sin qr_{im}}{qr_{im}} + \sum_{k=1}^{n_1}\sum_{m=1}^{n_2}b_k'b_m'\frac{\sin qr_{km}}{qr_{km}}\right) \tag{4.66}$$

Adding the scattering from a complex with two labelled components (sample d) to that of sample a ('reference particle') and subtracting the sum of the scattering of the complexes with either of the components labelled (b and c) results in a scattering-intensity difference $\Delta I(q)$:

$$\Delta I(q) = 2\frac{N}{V}\left(\sum_{k=1}^{n_1}\sum_{m=1}^{n_2}b_kb_m\frac{\sin qr_{km}}{qr_{km}} + \sum_{k=1}^{n_1}\sum_{m=1}^{n_2}b_k'b_m'\frac{\sin qr_{km}}{qr_{km}} - \sum_{k=1}^{n_1}\sum_{m=1}^{n_2}b_k'b_m\frac{\sin qr_{km}}{qr_{km}} - \sum_{k=1}^{n_1}\sum_{m=1}^{n_2}b_kb_m'\frac{\sin qr_{km}}{qr_{km}}\right)$$

$$\tag{4.67}$$

or

$$\Delta I(q) = 2\frac{N}{V}\left(\sum_{k=1}^{n_1}\sum_{m=1}^{n_2}(b_k - b_k')(b_k - b_k')\frac{\sin qr_{km}}{qr_{km}}\right) \tag{4.68}$$

Only non-exchangeable hydrogen atoms in both components are left; this assumes that deuteration does not change the atomic coordinates of the atoms in the components. At least at the level of low resolution this should be true, and eq. (4.68) reduces to

$$\Delta I(q) = 2\frac{N}{V}(b_D - b_H)^2\left(\sum_{k=1}^{n_1'}\sum_{m=1}^{n_2'}\frac{\sin qr_{km}}{qr_{km}}\right) \tag{4.69}$$

The summation is only over the n_1' and n_2' non-exchangeable hydrogen atoms in components 1 and 2. The DNA-length determination by Mathew-Fenn *et al.* (2008) also followed this scheme, but used heavy atom clusters.

Equation (4.69) was obtained by calculating the difference of two sums. As shown by Hoppe (1973), mixing equimolar amounts of samples a and d, as well as b and c, and subtracting the scattering of the second mixture from the first, also yields eq. (4.69). The difference is that using this 'mixed isomorphous replacement principle', as Hoppe termed it, also inter-particle interferences vanish. One can therefore measure the scattering curves of concentrated samples, in order to compensate for the relatively low effective label concentration. This assumes, however, that deuteration does not change the interaction behaviour, which is presumably valid at least at low resolution.

A model of the large (50S) *subunit of Escherichia coli*. The feasibility and the limitations of the distance determination by SAS can be demonstrated with an approximate model of the large (50S) ribosomal subunit from the bacterium *E. coli* shown in Fig. 4.16. The model volume has been obtained by calculating the dry volume from the RNA and protein masses, which are all known from their sequences (Voss 2006), and by multiplying them with estimated partial specific volumes of 0.525 and $0.7\,\mathrm{cm^3 g^{-1}}$ for the RNA (assumed inside) and all proteins, respectively. The model was then blown up so that the radius of gyration R_G of the homogeneous molecule (all weights equal to 1) gave a value of 7.2 nm found for fully deuterated 50S, which is nearly homogeneous in 100% H_2O. The linear increase due to this is about 20%. Although this model does not to exactly reproduce the 50S scattering, it serves to illustrate the arguments put forward here realistically.

356 spheres were used to build the model and to calculate the scattering curve with a modified version of the program MULTIBODY (Glatter 1980). Ribosomal RNA (rRNA) makes up two thirds of the mass, which, due to the lower partial specific volume, occupies only 58% of the volume (207 spheres). The residual mass of the 50S subunit consists of thirty-four different proteins—one of them existing in four copies, and all others in one copy per ribosome (Arnold and Reilly 1999). One single protein occupies, on average, only about one hundredth of the mass of the subunit. In our example, two proteins (dark in Fig. 4.16) are assumed to consist of two and three unit spheres, respectively; that is, both are smaller than the average for one protein.

Contrast optimisation. The method is most useful for large complexes, for which $n \gg n_1, n_2$ holds in eq. (4.64). In this case the component interference term is usually dominated by the scattering from the stripped complex (the first term in eq. 4.64), but the other interference terms can also become important.

Minimising the scattering contributions from parts of the molecule other than the labelled component(s) is important to reduce the influence of statistical errors by avoiding differences between large numbers. This means that one must measure the scattering at the match point of the reference particle. In the case of heterogeneous complexes, the scattering does not even vanish at the lowest contrast (consider ribosomes consisting of nucleic acids and proteins which differ in their scattering length density). This is demonstrated in Fig. 4.17, where the different

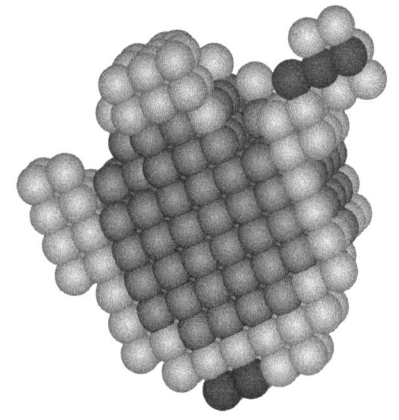

Fig. 4.16 An approximate model of the large (50S) ribosomal subunit from *E. coli*. It consists of 356 spheres with a volume of about 6.9 nm^3 each, 220 for RNA (grey) and 136 for protein (light grey). Two labelled proteins, consisting of two and three spheres respectively, are shown in dark grey.

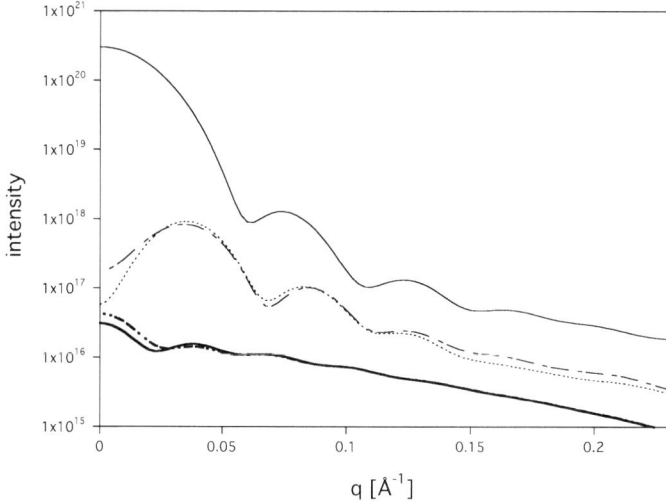

Fig. 4.17 Scattering curves (in arbitrary units) for the 50S model presented in Fig. 4.16 at different contrasts and with different labelling conditions. Thin continuous line: native 50S subunit in H_2O; thin dotted line: native 50S subunit near the match point (about 60% D_2O); thin dotted–dashed line: the same with two deuterated proteins; thick continuous line: 'isolated proteins', two protonated proteins within a completely invisible ribosomal 50S subunit. Thick dotted–dashed line: 'Glassy' 50S containing two protonated proteins; that is, all ribosomal proteins except the labels and the whole of rRNA are deuterated to a scattering-length density corresponding to ~95% D_2O, with a realistic difference between the protein and RNA match points of ~0.5% on the D_2O scale.

scattering curves for the model shown in Fig. 4.16 are plotted on a logarithmic scale *versus* momentum transfer.

Using a natural abundance ribosome and deuterated labels, the matrix scattering and the label-to-matrix interferences largely dominate the scattering from the labelled proteins even at the reference-particle match point in eq. (4.66). These are the conditions chosen by Capel *et al.* (1987) for their study of the small (30S) ribosomal subunit from *E. coli*. Since this subunit is only half the mass of the 50S, the situation is more favourable but still far from optimal.

The 'glassy ribosome'. Improving this awkward situation requires producing particles where the scattering powers of the ribosomal bulk protein and of the RNA are matched, and suppressing the high incoherent (and inelastic) isotropic scattering from protons in the solvent by moving the match-point conditions towards a buffer containing a high fraction of D_2O.

Producing large amounts of bacteria for extracting deuterated bulk proteins and RNA rather than only those needed for the label proteins makes this strategy more expensive. Nierhaus *et al.* (1983) determined the necessary growth conditions and reconstituted 'glassy ribosomes' from cells produced at different levels of D_2O in the growth medium for ribosomal proteins and RNA.

As can be seen in Figs. 4.17 and 4.18, the signal-to-noise ratio is substantially improved by using glassy ribosomes, and the intra- plus inter-component terms dominate the scattering intensity even at a realistic level of protein/RNA mismatch.

The signal/noise gain was partially used up by reducing the sample concentration (but measuring larger sample volumes). This had two effects: 1) Hoppe's mixing method is no longer indispensable, reducing the effort of sample preparation, and 2) approximate radii of gyration of single ribosomal proteins within the large subunit could be determined (Nierhaus *et al.* 1983).

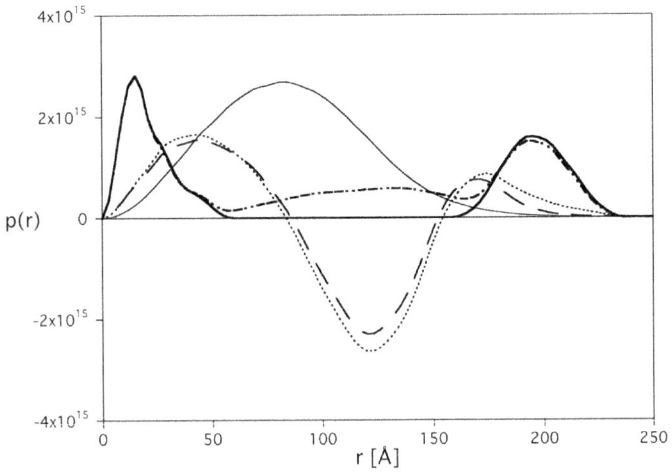

Fig. 4.18 Distance distributions for the 50S model presented in Fig. 4.16 at different contrasts and with different labelling conditions; legend as in Fig. 4.17. The distance distribution for native 50S is divided by 100, and those for the isolated label pair and the labels in a slightly mismatched glassy 50S are multiplied by 50.

Since it is difficult to exactly adjust the D_2O content in the samples before the measurement, the contrast of all samples was varied slightly around the theoretical match point of the reference particle. The first measurement was recorded slightly above the match point, and small quantities of H_2O buffer (about 5% volume) were added twice to the sample. The measurements near the expected match point were recorded with the longest preset. The difference calculation is then performed at the *a posteriori* calculated match point.

Extracting distance information. Stöckel *et al.* (1979) obtained component distances within DNA-dependent RNA polymerase from the radii of gyration of the scattering curves of the four different complexes described in Fig. 4.15. According to the parallel axis theorem of mechanics, applied to SAS, the distance d can be found from the formula

$$R_{G1,2}^2 = f_1 R_{G1}^2 + f_2 R_{G2}^2 + f_1 f_2 d^2 \tag{4.70}$$

where $f_i = \Delta\rho_i V_i / (\Delta\rho_1 V_1 + \Delta\rho_2 V_2)$, $i = 1,2$ are the relative scattering 'masses' of components 1 and 2; $R_{G1,2}$, R_{G1} and R_{G2} are the radii of gyration of the dumbbell and of the two single components, respectively.

Another expression holds for the radii of gyration (May 1991) of the scattering-intensity difference curve $\Delta I_{ij}(q)$ (eq. 4.69) for the component pair ij (May 1978, Moore *et al.* 1978):

$$2\Delta R_{Gij}^2 = R_{Gi}^2 + R_{Gj}^2 + d_{ij}^2 \tag{4.71}$$

which is independent of the coefficients f_i. The component radii of gyration, R_{Gi}, can be regarded as unknowns in the set of equations described by eq. (4.71). The R_{Gi} can be found if a sufficient number of ΔR_{Gij} are determined. The number of distances d_{ij} separating N components—that is, the maximal number of measurements—is $N(N-1)$. Since only $(4N-10)$ values of ΔR_{Gij} are needed for positioning $N > 3$ components (see *Model-building* below), the R_{Gi} can be (in principle) calculated if the number o of observations (measurements) is greater than $5N - 10$.

The range where the Guinier approximations behind equation system (4.71) (and 4.70), hold is rather narrow and is mainly determined by the values of d_{ij}. Ramakrishnan and Moore (1981) therefore proposed to obtain the ΔR_{ij} as the second moments of the pair-distance distribution function $p_{ij}(r)$ corresponding to $\Delta I_{ij}(q)$.

An alternative approach was developed by May and Nowotny (1989), who showed that a B-spline approximates the shape of the 'distance peak' sufficiently well and thus directly yields a value for the inter-component distance. One or two B-splines represent the intra-component distances, and residual scattering from imperfect matching can be taken into account for by an additional, usually flat peak (see Fig. 4.18). Fitting the Fourier transform of pair-distance distributions composed of three or four B-splines of variable location, height and width to the scattering curves of the complex containing two labels allows one to extract the distance information (see Fig. 4.19).

Model-building. The values in a table of distances obtained in either way represent a partially filled triangular matrix, as the distance d_{ij} is identical to d_{ji}. The schematic drawing in Fig. 4.20 illustrates the

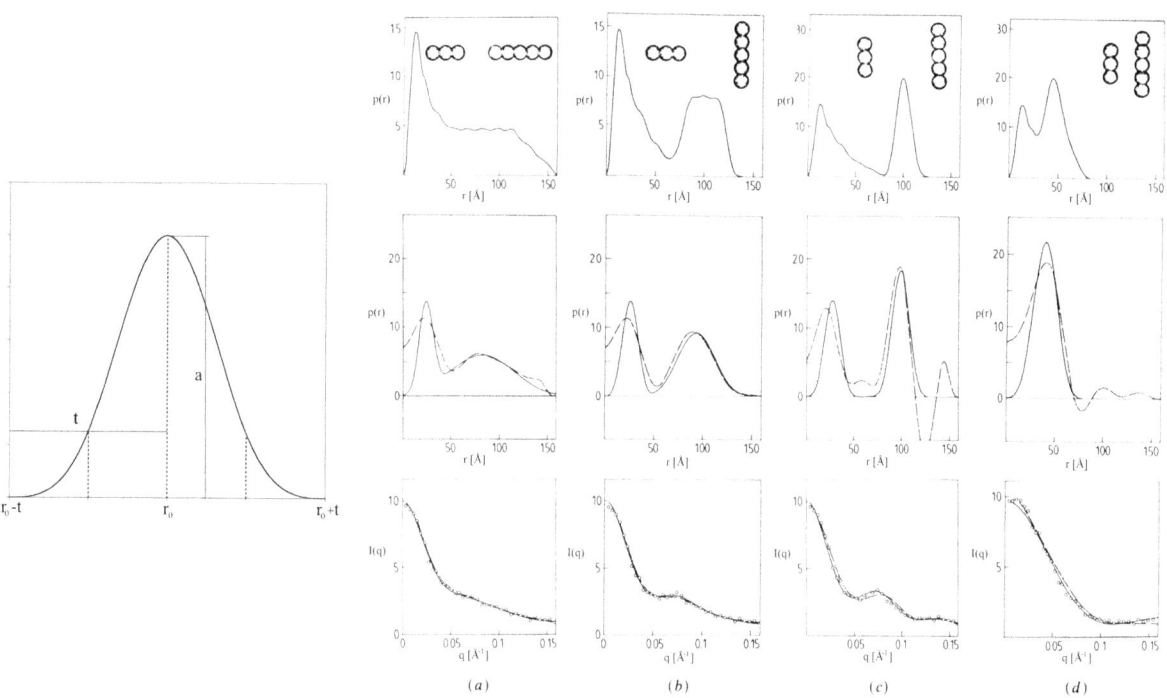

Fig. 4.19 Left: Sketch of a B-spline in real space. B-splines are assembled from four parts, each being a third-order polynomial. r_0 is the location of the B-spline, a its height, and t its half-width. These parameters are varied for the fit of the Fourier transform of a sum of three to four peaks to the scattering curves. Right: Four examples of model fits obtained with the 'moving spline' method. Reproduced from May and Nowotny (1989), with permission of the International Union of Crystallography, © 1989.

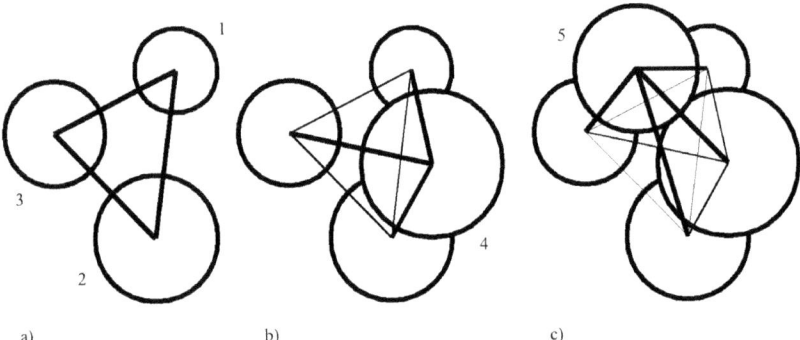

Fig. 4.20 Triangulation process. a) Three distances result in a basic triangle, from which a basic tetrahedron is built with three more distances to a fourth component in the complex (b). c) The fifth and every further component are fixed by (at least) three more distance values plus a fourth link that determines the side with respect to the binding plane.

triangulation process, which starts with the construction of a triangle from three distances. Three more distances fix the position of a fourth component and thus define a basic tetrahedron. The handedness of the tetrahedron fixes that of the whole model, but it cannot be determined on the basis of the distances obtained from neutron scattering; this requires other methods, such as three-dimensional EM, where the phases are known.

Since a tetrahedron has twelve coordinate values in space, but one disposes of only six distances, one component is arbitrarily chosen as the origin of the whole model. Its coordinates are (0, 0, 0). The second component points in x direction: $(x_2, 0, 0)$. The third one defines the $z = 0$ plane: $(x_3, y_3, 0)$, and the fourth the z coordinate (x_4, y_4, z_4) (Fig. 4.20b). For obvious reasons, the basic tetrahedron should be constructed with components that are not too close to each other. The fifth (Fig. 4.20c) and every further component require three distances to a triangle of components that is already fixed, plus the knowledge on which side of the plane determined by these component it lies. A fourth distance (dashed in Fig. 4.20c) resolves the uncertainty. Therefore, the spatial arrangement of N components is safely determined from $4N - 10$ distances.

Mathematically, the triangulation algorithm can be written as a system of equations

$$\sqrt{\left(x_i - x_j\right)^2 + \left(y_i - y_j\right)^2 + \left(z_i - z_j\right)^2} - d_{ij} = \sigma_{ij} \qquad (4.72)$$

with $x_1 = 0, y_1 = 0, z_1 = 0, y_2 = 0, z_2 = 0$, and $z_3 = 0$. The σ_{ij} represent the uncertainty of the distance determinations. Equation (4.72) can be minimised with a least-squares procedure:

$$\sum_{k=1}^{o} \frac{\left(\sqrt{\left(x_i - x_j\right)^2 + \left(y_i - y_j\right)^2 + \left(z_i - z_j\right)^2} - d_{ij}\right)^2}{\sigma_{ij}^2} = \text{Min.} \qquad (4.73)$$

o is the number of observations (measurements). It is puzzling that only $o > 3N - 6$ distances are required to fulfil eqs. (4.67) and (4.73), whereas $4N - 10$ are required from the above considerations. In fact, since only the orientation of the new component with respect to the triangle formed by the three others is needed, the fourth distance value contributes to the redundancy and therefore the quality of the data set used for the model construction. If only three distances are known for a given component it can still be attached, but its position with respect to the triangle to which it is bound remains arbitrary; that is, it must be known from other considerations.

In the approach of Ramakrishnan and Moore (1981), d_{ij}^2 can be replaced by $\Delta R_{ij}^2 - R_{Gi}^2 - R_{Gj}^2$, where the radii of gyration R_{Gi}^2 are additional unknowns. As eq. (4.73) can be written in a quadratic form, it becomes

$$\sum_{k=1}^{o} \frac{\left((x_i - x_j)^2 + (y_i - y_j)^2 + (z_i - z_j)^2 + R_i^2 + R_j^2 - 2\Delta R_{ij}^2 \right)}{\sigma_{\Delta R_{ij}}^2} = \text{Min}$$

(4.74)

where $\sigma_{\Delta Rij}$ are the errors of the determination of the moments $M_{ij} = 2\Delta R_{ij}$. Since now the R_{Gi} have to be determined as well, $o > 4N - 6$ measurements of moments are necessary.

According to Serdyuk and Pavlov (1988) their 'triple isomorphous replacement' method (Pavlov and Serdyuk 1987) can also improve the results of distance measurements. However, three different samples must be prepared for every distance pair and every single protein to be measured. Whilst a noticeable improvement may be expected for the determination of shape parameters *in situ*, the method requires substantially more effort for triangulation than the mixing method of Hoppe (1973).

Label triangulation is a rather complicated technique, both in terms of sample preparation and experimental measurements. The approach is, however, capable of yielding exciting results, as will be illustrated in practical examples presented in Section 6.2.

References

Adams, P. D., Afonine, P. V., Bunkóczi, G. *et al.* (2011). 'The Phenix software for automated determination of macromolecular structures', *Methods* 55: 94–106.

Aloy, P. and Russell, R. B. (2006). 'Structural systems biology: Modelling protein interactions', *Nat. Rev. Mol. Cell Bio.* 7: 188–97.

Aparicio, R., Fischer, H., Scott, D. J. *et al.* (2002). 'Structural insights into the beta-mannosidase from *T. reesei* obtained by synchrotron small-angle X-ray solution scattering enhanced by X-ray crystallography', *Biochemistry US* 41: 9370–5.

Arnold, R. J. and Reilly, J. P. (1999). 'Observation of Escherichia coli ribosomal proteins and their posttranslational modifications by mass spectrometry', *Anal. Biochem.* 269: 105–12.

Bada, M., Walther, D., Arcangioli, B., Doniach, S. and Delarue, M. (2000). 'Solution structural studies and low-resolution model of the *Schizosaccharomyces pombe* sap1 protein', *J. Mol. Biol.* 300: 563–74.

Bergmann, A., Fritz, G. and Glatter, O. (2000). 'Solving the generalized indirect Fourier transformation (GIFT) by Boltzmann simplex simulated annealing (BSSA)', *J. Appl. Crystallogr.* 33: 1212–6.

Bernocco, S., Finet, S., Ebel, C. *et al.* (2001). 'Biophysical characterization of the C-propeptide trimer from human procollagen III reveals a tri-lobed structure', *J. Biol. Chem.* 276: 48930–6.

Boehm, M. K., Woof, J. M., Kerr, M. A. and Perkins, S. J. (1999). 'The Fab and Fc fragments of IgA1 exhibit a different arrangement from that in IgG: A study by X-ray and neutron solution scattering and homology modelling', *J. Biol. Chem.* 286: 1421–47.

Bracewell, R. N. (1986). *Fourier Transform and Its Applications*. New York: McGraw-Hill.

Brunner-Popela, J. and Glatter, O. (1997). 'Small-angle scattering of interacting particles. 1. Basic principles of a global evaluation technique', *J. Appl. Crystallogr.* 30: 431–42.

Capel, M. S., Engelman, D. M., Freeborn, B. R. *et al.* (1987). 'A complete mapping of the proteins in the small ribosomal subunit of *Escherichia coli*', *Science* 238: 1403–6.

Chacon, P., Diaz, J. F., Moran, F. and Andreu, J. M. (2000). 'Reconstruction of protein form with X-ray solution scattering and a genetic algorithm', *J. Mol. Biol.* 299: 1289–302.

Chacon, P., Moran, F., Diaz, J. F., Pantos, E. and Andreu, J. M. (1998). 'Low-resolution structures of proteins in solution retrieved from X-ray scattering with a genetic algorithm', *Biophys. J.* 74: 2760–75.

Chamberlain, D., Ullman, C. G. and Perkins, S. J. (1998). 'Possible arrangement of the five domains in human complement factor I as determined by a combination of X-ray and neutron scattering and homology modeling', *Biochemistry US* 37: 13918–29.

Cotton, J. P. and Benoit, H. (1975). 'Étude du contraste et de son influence sur les déterminations de l'intensité diffusée et du rayon de giration dans les systèmes de macromolécules hétérogènes', *J. Phys.-Paris* 36: 905–10.

Daboll, H. F., Crespi, H. L. and Katz, J. J. (1962). 'Mass cultivation of algae in pure heavy water', *Biotechnol. Bioeng.* 4: 281–97.

Damaschun, G., Müller, J. J. and Pürschel, H. V. (1968). 'Über die Mess-Strategie bei der Untersuchung der Röntgen-Kleinwinkelstreuung von verdünnten monodispersen Lösungen von Makromolekülen', *Monatsh. Chem.* 99: 2343–8.

Debye, P. (1915). 'Zerstreuung von Röntgenstrahlen', *Annalen der Physik* 351: 809–23.

Debye, P. (1947). 'Molecular-weight determination by light scattering', *J. Phys. Colloid. Chem.* 51: 18–32.

Debye, P., Anderson, H. R. and Brumberger, H. (1957). 'Scattering by an inhomogeneous solid. 2. The correlation function and its application', *J. Appl. Phys.* 28: 679–83.

Debye, P. and Bueche, A. M. (1949). 'Scattering by an inhomogeneous solid', *J. Appl. Phys.* 20: 518–25.

Diamond, R. (1974). 'Real-space refinement of the structure of hen egg-white lysozyme', *J. Mol. Biol.* 82: 371–91.

Edmonds, A. R. (1957). *Angular Momentum in Quantum Mechanics*. Princeton, NJ: Princeton University Press.

Egea, P. F., Rochel, N., Birck, C., Vachette, P., Timmins, P. A. and Moras, D. (2001). 'Effects of ligand binding on the association properties and conformation in solution of retinoic acid receptors RXR and RAR', *J. Mol. Biol.* 307: 557–76.

Engelman, D. M. and Moore, P. B. (1972). 'A new method for the determination of biological quarternary structure by neutron scattering', *P. Natl. Acad. Sci. USA* 69: 1997–9.

Feigin, L. A. and Svergun, D. I. (1987). *Structure Analysis by Small-Angle X-Ray and Neutron Scattering*. New York: Plenum Press.

Fienup, J. R. (1982). 'Phase retrieval algorithms: A comparison', *Appl. Optics* 21: 2758–9.

Franke, D. and Svergun, D. I. (2009). 'DAMMIF, a program for rapid *ab-initio* shape determination in small-angle scattering', *J. Appl. Crystallogr.* 42: 342–6.

Fraser, R. D. B., Macrae, T. P. and Suzuki, E. (1978). 'An improved method for calculating the contribution of solvent to the X-ray diffraction pattern of biological molecules', *J. Appl. Crystallogr.* 11: 693–4.

Frieden, B. R. (1971). 'Evaluation, design and extrapolation methods for optical signals, based on the use of the prolate functions', in Wolf, E. (ed.), *Progress in Optics*. Amsterdam: North Holland.

Fujisawa, T., Kostyukova, A. and Maeda, Y. (2001). 'The shapes and sizes of two domains of tropomodulin, the P-end-capping protein of actin-tropomyosin', *FEBS Lett.* 498: 67–71.

Fujisawa, T., Uruga, T., Yamaizumi, Z., Inoko, Y., Nishimura, S. and Ueki, T. (1994). 'The hydration of Ras p21 in solution during GTP hydrolysis based on solution X-ray scattering profile', *J. Biochem.-Tokyo* 115: 875–80.

Funari, S. S., Rapp, G., Perbandt, M. *et al.* (2000). 'Structure of free *Thermus flavus* 5S rRNA at 1.3 nm resolution from synchrotron X-ray solution scattering', *J. Biol. Chem.* 275: 31283–8.

Garcia De La Torre, J., Huertas, M. L. and Carrasco, B. (2000). 'Calculation of hydrodynamic properties of globular proteins from their atomic-level structure', *Biophys. J.*, 78: 719–730.

Glatter, O. (1972). 'X-ray small angle scattering of molecules composed of subunits', *Acta Phys. Austriaca*, 36: 307–15.

Glatter, O. (1977). 'A new method for the evaluation of small-angle scattering data', *J. Appl. Crystallogr.* 10: 415–21.

Glatter, O. (1979). 'Auswertung physikalischer Experimente mit besonderer Berücksichtigung der Streuung an räumlich beschränkten Streumedien', Habilitation thesis, Universität Graz.

Glatter, O. (1980). 'Computation of distance distribution functions and scattering functions of models for small angle scattering experiments', *Acta Phys. Austriaca* 52: 243–56.

Glatter, O. (1982). 'Data treatment', in Glatter, O. and Kratky, O. (eds.), *Small-Angle X-Ray Scattering*. London: Academic Press.

Greville, T. N. E. (1969). *Theory and Application of Spline Functions*. New York: Academic Press.

Grishaev, A., Anthis, N. J. and Clore, G. M. (2012). 'Contrast-matched small-angle X-ray scattering from a heavy-atom-labeled protein in structure determination: Application to a lead-substituted calmodulin-peptide complex', *J. Am. Chem. Soc.* 134: 14686–9.

Grishaev, A., Guo, L. A., Irving, T. and Bax, A. (2010). 'Improved fitting of solution X-ray scattering data to macromolecular structures and structural ensembles by explicit water modeling', *J. Am. Chem. Soc.* 132: 15484–6.

Grishaev, A., Wu, J., Trewhella, J. and Bax, A. (2005). 'Refinement of multidomain protein structures by combination of solution small-angle X-ray scattering and NMR data', *J. Am. Chem. Soc.* 127: 16621–8.

Grossmann, J. G., Abraham, Z. H., Adman, E. T. *et al.* (1993). 'X-ray scattering using synchrotron radiation shows nitrite reductase from *Achromobacter xylosoxidans* to be a trimer in solution', *Biochemistry US* 32: 7360–6.

Grossmann, J. G., Ali, S. A., Abbasi, A. *et al.* (2000). 'Low-resolution molecular structures of isolated functional units from arthropodan and molluscan hemocyanin', *Biophys. J.* 78: 977–81.

Grüber, G., Svergun, D. I., Godovac-Zimmermann, J., Harvey, W. R., Wieczorek, H. and Koch, M. H. (2000). 'Evidence for major structural changes in the *Manduca sexta* midgut V1 ATPase due to redox modulation. A small angle X-ray scattering study', *J. Biol. Chem.* 275: 30082–7.

Guinier, A. (1939). 'La diffraction des rayons x aux très faibles angles: Applications à l'etude des phénomènes ultra-microscopiques', *Ann. Phys.-Paris* 12: 161–236.

Guo, D. Y., Blessing, R. H. and Langs, D. A. (2000). 'Globbic approximation in low-resolution direct-methods phasing', *Acta Crystallogr. D* 56: 1148–55.

Hansen, S. (2000). 'Bayesian estimation of hyperparameters for indirect Fourier transformation in small-angle scattering', *J. Appl. Crystallogr.* 33: 1415–21.

Hansen, S. and Pedersen, J. S. (1991). 'A comparison of three different methods for analysing small-angle scattering data', *J. Appl. Crystallogr.* 24: 541–8.

Harrison, S. C. (1969). 'Structure of tomato bushy stunt virus. I. The spherically averaged electron density', *J. Mol. Biol.* 42: 457–83.

Hoppe, W. (1972). 'A new X-ray method for the determination of the quaternary structure of protein complexes', *Israel J. Chem.* 10: 321–33.

Hoppe, W. (1973). 'The label triangulation method and the mixed isomorphous replacement principle', *J. Mol. Biol.* 78: 581–5.

Hubbard, S. R., Hodgson, K. O. and Doniach, S. (1988). 'Small-angle x-ray scattering investigation of the solution structure of troponin C', *J. Biochem.* 263: 4151–8.

Ibel, K. and Stuhrmann, H. B. (1975). 'Comparison of neutron and X-ray scattering of dilute myoglobin solutions', *J. Mol. Biol.* 93: 255–65.

Ingber, L. (1993). 'Simulated annealing: Practice versus theory', *Math. Comput. Model.* 18: 29–57.

Jacrot, B. and Zaccai, G. (1981). 'Determination of molecular weight by neutron scattering', *Biopolymers* 20: 2413–26.

Jones, G. (1998). *Genetic and Evolutionary Algorithms. Encyclopedia of Computational Chemistry*. Chichester, UK: Wiley.

Kirkpatrick, S., Gelatt, C. D., Jr. and Vecci, M. P. (1983). 'Optimization by simulated annealing', *Science* 220: 671–80.

Kleywegt, G. J. (1997). 'Validation of protein models from C-alpha coordinates alone', *J. Mol. Biol.* 273: 371–6.

Koch, M. H., Vachette, P. and Svergun, D. I. (2003). 'Small-angle scattering: A view on the properties, structures and structural changes of biological macromolecules in solution', *Q. Rev. Biophys.* 36: 147–227.

Koch, M. H. J. and Stuhrmann, H. B. (1979). 'Neutron-scattering studies of ribosomes', *Method. Enzymol.* 59: 670–706.

Konarev, P. V., Petoukhov, M. V. and Svergun, D. I. (2001). 'MASSHA: a graphic system for rigid body modelling of macromolecular complexes against solution scattering data', *J. Appl. Crystallogr.* 34: 527–32.

Kozin, M. B. and Svergun, D. I. (2000). 'A software system for automated and interactive rigid body modeling of solution scattering data', *J. Appl. Crystallogr.* 33: 775–7.

Kozin, M. B. and Svergun, D. I. (2001). 'Automated matching of high- and low-resolution structural models', *J. Appl. Crystallogr.* 34: 33–41.

Kozin, M. B., Volkov, V. V. and Svergun, D. I. (1997). 'ASSA: a program for three-dimensional rendering in solution scattering from biopolymers', *J. Appl. Crystallogr.* 30: 811–15.

Kratky, O. (1982). 'Natural high polymers', in Glatter, O. and Kratky, O. (eds.), *Small-Angle X-ray Scattering.* London: Academic Press.

Kratky, O. and Pilz, I. (1978). 'A comparison of X-ray small-angle scattering results to crystal structure analysis and other physical techniques in the field of biological macromolecules', *Q. Rev. Biophys.* 11: 39–70.

Kratky, O. and Porod, G. (1948). 'Die Abhängigkeit der Röntgen-Kleinwinkelstreuung von Form und Größe der kolloidalen Teilchen in verdünnten Systemen. III', *Acta Phys. Austriaca* 2: 113–47.

Kratky, O. and Worthmann, W. (1947). 'Über die Bestimmbarkeit der Konfiguration gelöster organischer Moleküle durch interferometrische Vermessung mit Röntgenstrahlen', *Monatsh. Chem.* 76: 263–81.

Krueger, J. K., Gallagher, S. C., Wang, C. A. and Trewhella, J. (2000). 'Calmodulin remains extended upon binding to smooth muscle caldesmon: A combined small-angle scattering and Fourier transform infrared spectroscopy study', *Biochemistry US* 39: 3979–87.

Krueger, J. K., Mccrary, B. S., Wang, A. H., Shriver, J. W., Trewhella, J. and Edmondson, S. P. (1999). 'The solution structure of the Sac7d/DNA complex: A small-angle X-ray scattering study', *Biochemistry US* 38: 10247–55.

Krueger, J. K., Zhi, G., Stull, J. T. and Trewhella, J. (1998). 'Neutron-scattering studies reveal further details of the Ca2+/calmodulin-dependent activation mechanism of myosin light chain kinase', *Biochemistry US* 37: 13997–14004.

Langer, D. L., Van Der Kwast, T. H., Evans, A. J. *et al.* (2008). 'Intermixed normal tissue within prostate cancer: Effect on MR imaging measurements of apparent diffusion coefficient and T2-sparse versus dense cancers', *Radiology* 249: 900–08.

Langridge, R., Marvin, D. A., Seeds, W. E. *et al.* (1960). 'The molecular configuration of deoxyribonucleic acid: Molecular models and their Fourier transforms', *J. Mol. Biol.* 2: 38–62.

Lattman, E. E. (1989). 'Rapid calculation of the solution scattering profile from a macromolecule of known structure', *Proteins* 5: 149–55.

Lederer, H., May, R. P., Kjems, J. K., Schaefer, W., Crespi, H. L. and Heumann, H. (1986). 'Deuterium incorporation into *Escherichia coli* proteins. A neutron-scattering study of DNA-dependent RNA polymerase', *Eur. J. Biochem.* 156: 655–9.

Lee, B. and Richards, F. M. (1971). 'The interpretation of protein structures: Estimation of static accessibility', *J. Mol. Biol.* 55: 379–400.

Levitt, M. (2007). 'Growth of novel protein structural data', *P. Natl. Acad. Sci. USA* 104: 3183–8.

Levitt, M. and Sharon, R. (1988). 'Accurate simulation of protein dynamics in solution', *P. Natl. Acad. Sci. USA* 85: 7557–61.

Mangani, M., Puliti, P. and Stefanon, M. (1988). 'Numerical solution of the inverse problem in the analysis of neutron small angle scattering experiments.', *Nucl. Instrum. Meth. A* 271: 611–16.

Mathew-Fenn, R. S., Das, R. and Harbury, P. a. B. (2008). 'Remeasuring the double helix', *Science* 322: 446–9.

Mattinen, M. L., Paakkonen, K., Ikonen, T. *et al.* (2002). 'Quaternary structure built from subunits combining NMR and small-angle X-ray scattering data', *Biophys. J.* 83: 1177–83.

May, R. P. (1978). 'Die Bestimmung der Abstände von Proteinen der großen ribosomalen Untereinheit durch Neutronenkleinwinkelstreuung mit Hilfe der Markentriangulations-Methode', Dr. rer. nat. thesis, Technische Universität, München.

May, R. P. (1991). 'Label triangulationn, in Lindner, P. and Zemb, T. (eds.), *Neutron, X-Ray and Light Scattering: Introduction to an Investigative Tool for Colloidal and Polymeric Systems.* Amsterdam: Elsevier Science Publishers.

May, R. P. and Nowotny, V. (1989). 'Distance information derived from neutron low-q scattering', *J. Appl. Crystallogr.* 22: 231–7.

Merzel, F. and Smith, J. C. (2002). 'Is the first hydration shell of lysozyme of higher density than bulk water?', *P. Natl. Acad. Sci. USA* 99: 5378–083.

Miao, J., Charalambous, P., Kirz, J. and Sayre, D. (1999). 'Extending the methodology of X-ray crystallography to allow imaging of micrometre-sized non-crystalline specimens', *Nature* 400: 342–4.

Moore, P. B. (1977). 'A simple technique for estimating deuterium incorporation levels in macromolecules', *Anal. Biochem.* 82: 101–8.

Moore, P. B. (1980). 'Small-angle scattering: Information content and error analysis', *J. Appl. Crystallogr.* 13: 168–75.

Moore, P. B., Langer, J. A. and Engelman, D. M. (1978). 'The measurement of the locations and radii of gyration of proteins in the 30S ribosomal subunit of *E. coli* by neutron scattering', *J. Appl. Crystallogr.* 11: 479–82.

Müller, J. J. (1983). 'Calculation of scattering curves for macromolecules in solution and comparison with results of methods using effective atomic scattering factors', *J. Appl. Crystallogr.* 16: 74–82.

Murshudov, G. N., Skubak, P., Lebedev, A. A. *et al.* (2011). 'REFMAC5 for the refinement of macromolecular crystal structures', *Acta Crystallogr. D* 67: 355–67.

Mylonas, E. and Svergun, D. I. (2007). 'Accuracy of molecular mass determination of proteins in solution by small-angle X-ray scattering', *J. Appl. Crystallogr.* 40: s245–9.

Nierhaus, K. H., Lietzke, R., May, R. P. *et al.* (1983). 'Shape determinations of ribosomal proteins in situ', *P. Natl. Acad. Sci. USA* 80: 2889–93.

Ninio, J., Luzzati, V., and Yaniv, M. (1972). 'Comparative small-angle X-ray scattering studies on unacylated, acylated and cross-linked *Escherichia coli* transfer RNA I Val', *J. Mol. Biol.* 71: 217–29.

Nollmann, M., Stark, W. M. and Byron, O. (2005). 'A global multi-technique approach to study low-resolution solution structures', *J. Appl. Crystallogr.* 38: 874–7.

Orthaber, D., Bergmann, A. and Glatter, O. (2000). 'SAXS experiments on absolute scale with Kratky systems using water as a secondary standard', *J. Appl. Cryst.* 33: 218–25.

Pavlov, M. Y. (1985). 'Determination of the relative position of the domains in 2-domain proteins based on diffuse X-ray scattering data', *Dokl. Akad. Nauk SSSR* 281: 458–62.

Pavlov, M. Y. and Fedorov, B. A. (1983). 'Improved technique for calculating X-ray scattering intensity of biopolymers in solution: Evaluation of the form, volume, and surface of a particle', *Biopolymers* 22: 1507–22.

Pavlov, M. Y. and Serdyuk, I. N. (1987). 'Three-isotopic-substitutions method in small-angle neutron scattering', *J. Appl. Crystallogr.* 20: 105–10.

Pavlov, M. Y., Sinev, M. A., Timchenko, A. A. and Ptitsyn, O. B. (1986). 'A study of apo- and holo-forms of horse liver alcohol dehydrogenase in solution by diffuse X-ray scattering', *Biopolymers* 25: 1385–97.

Perkins, S. J., Nan, R., Li, K., Khan, S. and Abe, Y. (2011). 'Analytical ultracentrifugation combined with X-ray and neutron scattering: Experiment and modelling', *Methods* 54: 181–99.

Petoukhov, M. V., Eady, N. A., Brown, K. A. and Svergun, D. I. (2002). 'Addition of missing loops and domains to protein models by X-ray solution scattering', *Biophys. J.* 83: 3113–25.

Petoukhov, M. V., Franke, D., Shkumatov, A. V. *et al.* (2012). 'New developments in the ATSAS program package for small-angle scattering data analysis', *J. Appl. Crystallogr.* 45: 342–50.

Petoukhov, M. V. and Svergun, D. I. (2003). 'New methods for domain structure determination of proteins from solution scattering data', *J. Appl. Crystallogr.* 36: 540–4.

Petoukhov, M. V. and Svergun, D. I. (2005). 'Global rigid body modeling of macromolecular complexes against small-angle scattering data', *Biophys. J.* 89: 1237–50.

Petoukhov, M. V. and Svergun, D. I. (2006). 'Joint use of small-angle X-ray and neutron scattering to study biological macromolecules in solution', *Eur. Biophys. J.* 35: 567–76.

Poitevin, F., Orland, H., Doniach, S., Koehl, P. and Delarue, M. (2011). 'AquaSAXS: A web server for computation and fitting of SAXS profiles with non-uniformly hydrated atomic models', *Nucleic Acids Res.* 39: W184–9.

Porod, G. (1948). 'Die Abhängigkeit der Röntgen-Kleinwinkelstreuung von Form und Größe der kolloidalen Teilchen in verdünnten Systemen. IV.', *Acta Phys. Austriaca* 2: 255–92.

Porod, G. (1951). 'Die Röntgenkleinwinkelstreung von dichtgepackten kolloidalen Systemen, 1. Teil', *Kolloid Z.* 124: 83–114.

Potterton, L., McNicholas, S., Krissinel, E. *et al.* (2004). 'Developments in the CCP4 molecular-graphics project', *Acta Crystallogr. D* 60: 2288–94.

Press, W. H., Teukolsky, S. A., Wetterling, W. T. and Flannery, B. P. (1992). *Numerical Recipes*. Cambridge: University Press.

Provencher, S. W. (1982a). 'A constrained regularization method for inverting data represented by linear algebraic or integral-equations', *Comput. Phys. Commun.* 27: 213–27.

Provencher, S. W. (1982b). 'CONTIN: a general-purpose constrained regularization program for inverting noisy linear algebraic and integral-equations', *Comput. Phys. Commun.* 27: 229–42.

Ramakrishnan, V. R. and Moore, P. B. (1981). 'Analysis of neutron distance data', *J. Mol. Biol.* 153: 719–38.

Riboldi-Tunnicliffe, A., Konig, B., Jessen, S. *et al.* (2001). 'Crystal structure of Mip, a prolylisomerase from *Legionella pneumophila*', *Nat. Struct. Mol. Biol.* 8: 779–83.

Rodnina, M., Wintermeyer, W. and Green, R. (Eds.) (2011). *Ribosomes Structure, Function, and Dynamics*. Wien: Springer.

Rolbin, Y. A., Kayushina, R. L., Feigin, L. A. and Schedrin, B. M. (1973). 'Computer calculations of the X-ray small-angle scattering by macromolecule models', *Kristall.* 18: 701–5.

Sayre, D. (1952). 'Some implications of a theorem due to Shannon', *Acta Crystallogr.* 5: 843.

Schelten, J. and Hossfeld, F. (1971). 'Application of spline functions to correction of resolution errors in small-angle scattering', *J. Appl. Crystallogr.* 4: 210–23.

Schmidt, B., König, S., Svergun, D. I., Volkov, V. V., Fischer, G. and Koch, M. H. J. (1995). 'Small-angle X-ray solution scattering study on the dimerization of the FKBP25mem from *Legionella pneumophila*', *FEBS Lett.* 372: 169–72.

Schmidt, P. W. (1995). 'Some fundamental concepts and techniques useful in small-angle scattering studies of disordered solids', in Brumberger, H. (ed.), *Modern Aspects of Small-Angle Scattering*. Dordrecht: Kluwer Academic Publishers.

Schneidman-Duhovny, D., Hammel, M. and Sali, A. (2010). 'FoXS: A web server for rapid computation and fitting of SAXS profiles', *Nucleic Acids Res.* 38: W540–4.

Schneidman-Duhovny, D., Hammel, M. and Sali, A. (2011). 'Macromolecular docking restrained by a small angle X-ray scattering profile', *J. Struct. Biol.* 173: 461–71.

Scott, D. J., Grossmann, J. G., Tame, J. R., Byron, O., Wilson, K. S. and Otto, B. R. (2002). 'Low resolution solution structure of the apo form of *Escherichia coli* haemoglobin protease Hbp', *J. Mol. Biol.* 315: 1179–87.

Serdyuk, I. N. and Pavlov, M. Y. (1988). 'A new approach in small-angle neutron scattering: A method of triple isotopic substitutions', *Makromol. Chem.-M. Symp.* 15: 167–84.

Shannon, C. E. (1949). 'Communication in the presence of noise', *P. IRE* 37: 10–21.

Shannon, C. E. and Weaver, W. (1949). *The Mathematical Theory of Communication*. Urbana, IL: University of Illinois Press.

Shcherbakova, I. V. and Serdyuk, I. N. (2000). 'Biosynthetic deuteration of *Thermus thermophilus* ribosomes for SANS', *J. Appl. Crystallogr.* 33: 552–5.

Sokolova, A., Malfois, M., Caldentey, J. *et al.* (2001). 'Solution structure of bacteriophage PRD1 vertex complex', *J. Biol. Chem.* 276: 46187–95.

Spinozzi, F., Carsughi, F. and Mariani, P. (1998). 'Particle shape reconstruction by small-angle scattering: Integration of group theory and maximum entropy to multipole expansion method', *J. Chem. Phys.* 109: 10148–58.

Stöckel, P., May, R. P., Strell, I. *et al.* (1979). 'Determination of intersubunit distances and subunit shape parameters in DNA-dependent RNA polymerase by neutron small-angle scattering', *J. Appl. Crystallogr.* 12: 176–85.

Stuhrmann, H. B. (1970a). 'Ein neues Verfahren zur Bestimmung der Oberflächenform und der inneren Struktur von gelösten globulären Proteinen aus Röntgenkleinwinkelmessungen.', *Z. Phys. Chem. Neue Fol.* 72: 177–98.

Stuhrmann, H. B. (1970b). 'Interpretation of small-angle scattering of dilute solutions and gases. A representation of the structures related to a one-particle scattering functions', *Acta Crystallogr. A* 26: 297–306.

Stuhrmann, H. B. and Kirste, R. G. (1965). 'Elimination der intrapartikulären Untergrundstreuung bei der Röntgenkleinwinkelstreuung an kompakten Teilchen (Proteinen)', *Z. Phys. Chem. Neue Fol.* 46: 247–50.

Svergun, D. I. (1991). 'Mathematical methods in small-angle scattering data analysis', *J. Appl. Crystallogr.* 24: 485–92.

Svergun, D. I. (1992). 'Determination of the regularization parameter in indirect-transform methods using perceptual criteria', *J. Appl. Crystallogr.* 25: 495–503.

Svergun, D. I. (1994). 'Solution scattering from biopolymers: Advanced contrast variation data analysis', *Acta Crystallogr. A* 50: 391–402.

Svergun, D. I. (1997). 'Restoring three-dimensional structure of biopolymers from solution scattering', *J. Appl. Crystallogr.* 30: 792–7.

Svergun, D. I. (1999). 'Restoring low resolution structure of biological macromolecules from solution scattering using simulated annealing', *Biophys. J.* 76: 2879–86.

Svergun, D. I., Aldag, I., Sieck, T., *et al.* (1998a). 'A model of the quaternary structure of the *Escherichia coli* F1 ATPase from X-ray solution scattering and evidence for structural changes in the delta subunit during ATP hydrolysis', *Biophys. J.* 75: 2212–19.

Svergun, D. I., Barberato, C. and Koch, M. H. J. (1995). 'CRYSOL: a program to evaluate X-ray solution scattering of biological macromolecules from atomic coordinates', *J. Appl. Crystallogr.* 28: 768–73.

Svergun, D. I., Becirevic, A., Schrempf, H., Koch, M. H. J. and Grüber, G. (2000). 'Solution structure and conformational changes of the *Streptomyces* chitin-binding protein (CHB1)', *Biochemistry US* 39: 10677–83.

Svergun, D. I., Petoukhov, M. V. and Koch, M. H. J. (2001). 'Determination of domain structure of proteins from X-ray solution scattering', *Biophys. J.* 80: 2946–53.

Svergun, D. I., Richard, S., Koch, M. H. J., Sayers, Z., Kuprin, S. and Zaccai, G. (1998b). 'Protein hydration in solution: Experimental observation by X-ray and neutron scattering', *P. Natl. Acad. Sci. USA* 95: 2267–72.

Svergun, D. I., Semenyuk, A. V. and Feigin, L. A. (1988). 'Small-angle-scattering-data treatment by the regularization method', *Acta Crystallogr. A* 44: 244–50.

Svergun, D. I. and Stuhrmann, H. B. (1991). 'New developments in direct shape determination from small-angle scattering 1. Theory and model calculations.', *Acta Crystallogr. A* 47: 736–44.

Svergun, D. I., Volkov, V. V., Kozin, M. B. and Stuhrmann, H. B. (1996). 'New developments in direct shape determination from small-angle scattering 2. Uniqueness.', *Acta Crystallogr. A* 52: 419–26.

Svergun, D. I., Volkov, V. V., Kozin, M. B., Stuhrmann, H. B., Barberato, C. and Koch, M. H. J. (1997). 'Shape determination from solution scattering of biopolymers', *J. Appl. Crystallogr.* 30: 798–802.

Taupin, D. and Luzzati, V. (1982). 'Information content and retrieval in solution scattering studies. 1. Degrees of freedom and data reduction', *J. Appl. Crystallogr.* 15: 289–300.

Tung, C. S., Wall, M. E., Gallagher, S. C. and Trewhella, J. (2000). 'A model of troponin-I in complex with troponin-C using hybrid experimental data: The inhibitory region is a beta-hairpin', *Protein Sci.* 9: 1312–26.

Tung, C. S., Walsh, D. A. and Trewhella, J. (2002). 'A structural model of the catalytic subunit-regulatory subunit dimeric complex of the cAMP-dependent protein kinase', *J. Biol. Chem.* 277: 12423–31.

Tychonoff, A. N. (1943). 'On the stability of inverse problems' (in Russian), *Dokl. Akad. Nauk SSSR* 39: 195–98.

Vainshtein, B. K., Feigin, L. A., Lvov, Y. M., Gvozdev, R. I., Marakushev, S. A. and Likhtenshtein, G. I. (1980). 'Determination of the distance between heavy-atom markers in haemoglobin and histidine decarboxylase in solution by small-angle X-ray scattering', *FEBS Lett.* 116: 107–10.

Vainshtein, B. K., Sosfenov, N. I. and Feigin, L. A. (1970). 'X-ray method for determination of the spacing between heavy atoms in macromolecules in solution, and its application for the investigation of gramicidin C derivatives (in Russian)', *Dokl. Akad. Nauk SSSR* 190: 574–7.

Vestergaard, B. and Hansen, S. (2006). 'Application of bayesian analysis to indirect Fourier transformation in small-angle scattering', *J. Appl. Crystallogr.* 39: 797–804.

Vigil, D., Gallagher, S. C., Trewhella, J. and Garcia, A. E. (2001). 'Functional dynamics of the hydrophobic cleft in the N-domain of calmodulin', *Biophys. J.* 80: 2082–92.

Virtanen, J. J., Makowski, L., Sosnick, T. R. and Freed, K. F. (2010). 'Modeling the hydration layer around proteins: HyPred', *Biophys. J.* 99: 1611–19.

Virtanen, J. J., Makowski, L., Sosnick, T. R. and Freed, K. F. (2011). 'Modeling the hydration layer around proteins: Applications to small- and wide-angle X-ray scattering', *Biophys. J.* 101: 2061–9.

Volkov, V. V. and Svergun, D. I. (2003). 'Uniqueness of *ab initio* shape determination in small angle scattering', *J. Appl. Crystallogr.* 36: 860–4.

Voss, N. R. (2006). 'Geometric studies of RNA and ribosomes, and ribosome crystallization', PhD thesis, Yale University, New Haven.

Wall, M. E., Gallagher, S. C. and Trewhella, J. (2000). 'Large-scale shape changes in proteins and macromolecular complexes', *Annu. Rev. Phys. Chem.* 51: 355–80.

Wang, J., Smerdon, S. J., Jager, J. *et al.* (1994). 'Structural basis of asymmetry in the human immunodeficiency virus type 1 reverse transcriptase heterodimer', *P. Natl. Acad. Sci. USA* 91: 7242–6.

Watson, G. N. (1995). *A Treatise on the Theory of Bessel Functions.* Cambridge: Cambridge University Press.

Weyerich, B., Brunner-Popela, J. and Glatter, O. (1999). 'Small-angle scattering of interacting particles. II. Generalized indirect Fourier transformation under consideration of the effective structure factor for polydisperse systems', *J. Appl. Crystallogr.* 32: 197–209.

Witz, J. (1983). 'Contrast variation of the small-angle neutron-scattering of globular particles: the influence of hydrogen-exchange', *Acta Crystallogr. A* 39: 706–11.

Wriggers, W., Alamo, L. and Padron, R. (2011). 'Matching structural densities from different biophysical origins with gain and bias', *J. Struct. Biol.* 173: 445–50.

Zimm, B. H. (1948a). 'Apparatus and methods for measurement and interpretation of the angular variation of light scattering; preliminary results on polystyrene solutions', *J. Chem. Phys.* 16: 1099–116.

Zimm, B. H. (1948b). 'The scattering of light and the radial distribution function of high polymer solutions', *J. Chem. Phys.* 16: 1093–99.

Polydisperse and interacting systems

5.1 Size, shape and conformation
    polydispersity   152

5.2 Size distribution functions   153

5.3 Shape polydispersity
    and oligomeric mixtures   155

5.4 Conformational polydispersity
    and flexible systems   159

5.5 Interacting systems
    and structure factor   163

References   166

In Chapter 4, methods were considered to interpret the scattering from ideal monodisperse systems for which the measured intensity is directly related to the single-particle scattering. The data analysis from monodisperse systems usually aims at obtaining information about the low-resolution structure of the particle. A dilute solution of purified biological macromolecules is a unique type of object, where the requirement of monodispersity is often fulfilled (which makes SAXS/SANS so attractive for biological solution scattering analysis). Still, in many practical cases one has to deal with non-ideal situations when the monodisperse approximation is not applicable. The most important cases are (i) when the particles differ in size, shape or conformation (such as equilibrium oligomeric mixtures or solutions of flexible macromolecules) and (ii) when interparticle interactions cannot be neglected (such as concentrated solutions of charged particles). It is extremely difficult to try to reconstruct the shape or model the structure of a single particle based on the scattering from a mixture or a strongly interacting system. To characterise such systems, different types of questions are addressed (for example, composition of the mixture or type or interaction potential) and, accordingly, different data analysis methods are required.

5.1 Size, shape and conformation polydispersity

Let us first consider a general system consisting of different types of non-interacting particles with arbitrary structures. Here, the interference effects are neglected, all particles scatter independently, and the scattering pattern from such a mixture can be written as a linear combination

$$I(q) = \sum_{k=1}^{K} n_k I_k(q) \tag{5.1}$$

where $n_k > 0$ and $I_k(q)$ are the number fraction and the scattering intensity from the kth type of particle (component), respectively, and K is the number of components. For such polydisperse systems, three important cases can be distinguished:

(i) Size polydispersity, whereby all particles in the system have similar shapes but differ in size. In this case the mixture is characterised by a size distribution function telling how many particles of the given size are present in the system. This type of polydispersity is more common for non-biological objects, such as metal nanoparticles or colloids, where the particle shape is typically spherical and the task is to find the size distribution function.

(ii) Shape polydispersity, when the system contains particles with distinctly different shapes and sizes. Typical systems of this type are oligomeric mixtures (for example, monomer–dimer equilibrium) or weakly bound complexes dissociating into individual components.

(iii) Conformational polydispersity, when the system contains particles of identical molecular mass and sequence though these particles may adopt different conformations in solution. The number of conformations may be small (such as co-existence of open and closed form of an enzyme) or huge (for multidomain proteins with flexible linkers or IDPs).

In all these cases, schematically depicted in Fig. 5.1, the SAS data are usually interpreted assuming that the intensities of the components $I_k(q)$ are known, and the aim is to determine the fractions of the components in the mixture. The different approaches employed to this end are presented in the following sections.

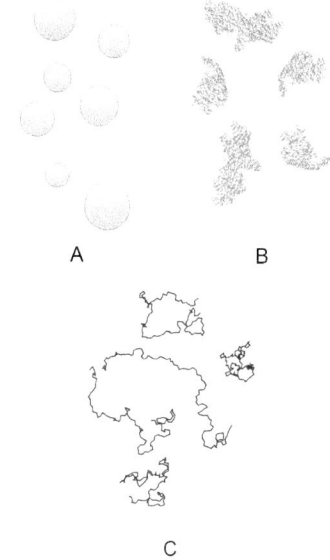

Fig. 5.1 Types of polydisperse systems containing mixtures of particles. A: Size polydispersity (here, spherical objects of different radii). B: Shape polydispersity (here, a mixture of monomers and dimers). C: Conformational polydispersity (here, various conformers of a disordered protein).

5.2 Size distribution functions

For the systems with size polydispersity, the particles are assumed to have similar shapes and differ only in size. The scattering from a particle of size R can be written as $(\Delta\rho)^2 V^2(R) i_0(qR)$, where $\Delta\rho$ is the particle contrast, $V(R)$ is its volume, and $i_0(qR)$ is the normalised scattering intensity of the particle ($i_0(0) = 1$). In the general case, the sizes are not just adopting discrete values but may vary continuously. They are conveniently described by a volume distribution function $D(R) = N(R)V(R)$, where $N(R)$ is the number of particles with characteristic size R. The scattering intensity is given by the integral

$$I(q) = (\Delta\rho)^2 \int_{R_{\min}}^{R_{\max}} D(R)\, V(R)\, i_0(qR)\, dR \qquad (5.2)$$

where R_{min} and R_{max} are the minimum and maximum particle sizes, respectively. This type of size polydispersity is often observed for colloidal systems such as micelles, microemulsions, block copolymers or metal nanoparticles. In most practical cases one assumes that the particle shape is known (very often, the particles are simply considered to be spherical). If the function $i_0(qR)$ is known (or assumed to be known) the integral equation (5.2) can be solved with respect to the volume distribution function $D(R)$ to fit the observed experimental data. There are simple ways of computation based on representation of $D(R)$ by two- or three-parameter monomodal distributions like Gaussian, Schultz and so on. A more advanced indirect transformation method was proposed as an extension of the technique for reconstruction of the distance distribution functions from the SAS data considered in Section 4.1.1 (Glatter 1980). Indeed, the function $D(R)$ can be expanded into orthogonal functions on the interval $[R_{min}, R_{max}]$, as in eq. (4.8). Then one can follow the same formalism as described in Section 4.1.1 to reconstruct a smooth $D(R)$ distribution fitting the given experimental data. Major indirect transformation programs (such as the original IFT method (Glatter 1980) or the program GNOM (Svergun 1992)) have the options to compute size distribution functions. It should be noted that the reconstruction of $D(R)$, similar to the computation of $p(r)$, is an ill-posed problem and, depending on the form factor $i_0(qR)$, solving eq. (5.2) may be even more complicated than solving eq. (4.6a).

The most common application of this size polydispersity concept is the characterisation of systems for which one can safely assume that the particles are spherical but have different size. For the case of a solid sphere, R is the particle radius, $V(R) = 4\pi R^3/3$, and the form factor $i_0(x) = 3[\sin(x) - x\cos(x)]/x^3$, where $x = qR$. Size polydispersity may lead to drastic changes in the scattering patterns compared to those from monodisperse systems, as illustrated in a simulated example in Fig. 5.2. This example also demonstrates that despite the apparently featureless scattering from a polydisperse system, the size distribution function of spherical particles can be reconstructed reliably from the scattering data.

The structural parameters determined from the experimental data correspond to the averaging over the distribution function. Thus, for a polydisperse system of solid spheres the R_g value is a co-called z-average, $R_g = (3 < R^2 >_z /5)^{1/2}$, where the average sphere radius is expressed as

$$< R^2 >_z = \int_{R_{min}}^{R_{max}} R^5 D(R)dR \left[\int_{R_{min}}^{R_{max}} R^3 D(R)dR \right]^{-1} \qquad (5.3)$$

Given that solutions of biological macromolecules such as proteins or nucleic acids rarely display such size polydispersity, we shall not consider this case further. The readers are referred to textbooks and

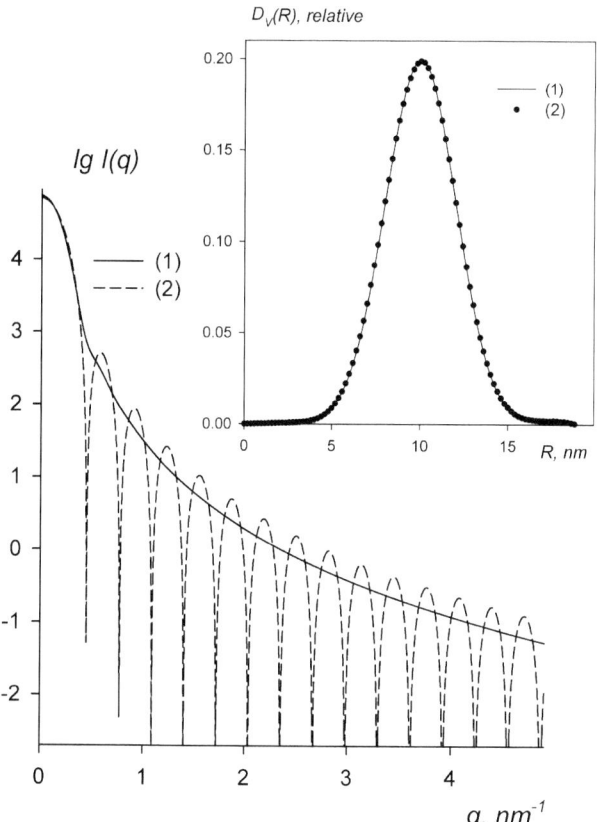

Fig. 5.2 Scattering from a polydisperse mixture of spherical particles. (1), simulated scattering from a polydisperse system with the Gaussian volume distribution with the average radius $R_0 = 10$ nm and FWHM of 50%; (2) scattering from a monodisperse system of spheres with $R = R_0$. The volume distribution is displayed as a solid line in the insert; dotted distribution was obtained by inverting the scattering by a polydisperse system using an indirect transformation (program GNOM).

reviews considering SAS applications to polymers and colloidal systems for more detail (for example, Glatter and Kratky 1982, Feigin and Svergun 1987, Pedersen 2002).

5.3 Shape polydispersity and oligomeric mixtures

Mixtures exhibiting shape polydispersity are one of the most important cases for biological solutions. This concept applies to equilibrium systems such as oligomeric mixtures or partially dissociating complexes, and also to non-equilibrium processes such as assembly/disassembly of large macromolecular objects. In all these cases the solutions contain (usually small) numbers of particles with distinctly different shapes and sizes (for example, for a monomer–dimer mixture there are two distinct particle types). Most often, the aim of a SAS study is to characterise the system in terms of the fractions of the components in the mixture.

5.3.1 Scattering from components and volume fractions

In principle, some information about the distribution of species can be obtained from the overall parameters like forward scattering and radius of gyration. It follows from eq. (5.1) that

$$I(0) = \sum_{k=1}^{K} n_k I_k(0) \tag{5.4}$$

$$I(0)R_g^2 = \left\{ \sum_{k=1}^{K} n_k I_k(0) R_{gk}^2 \right\} \tag{5.5}$$

Theoretically, for binary and even ternary mixtures consisting of two or three components with known MM and R_g one should be able to deduce the number fractions using eqs. (5.4) and (5.5). However, in practice the uncertainties in the absolute scaling and in the computation of R_g often make such analysis insufficiently accurate. In the previous chapter we saw that for monodisperse systems, going beyond the analysis of $I(0)$ and R_g allows one to obtain very useful information. The same holds for mixtures, where invoking full scattering patterns yields more reliable information than the analysis of the overall parameters.

In most practical applications one is interested not in number concentrations n_k but rather in the volume fractions of species $v_k = n_k V_k$, where V_k is the particle volume of the kth species. In the discrete form, eq. (5.1) can be rewritten as

$$I(q_i) = \sum_{k=1}^{K} v_k i_k(q_i) \tag{5.6}$$

where the index $i = 1, \ldots N$, runs over N experimental points, and $i_k(q) = I_k(q)/V_k$ are appropriately renormalised scattering intensities from the species. If the latter intensities are known, the volume fractions v_k can be determined by fitting the experimental scattering from the mixture $I_{\exp}(q_i)$. This leads to a system of linear equations, which is usually overdetermined, as the number of points N in most cases significantly exceeds the number of components K. Standard non-negative linear least squares (Lawson and Hanson 1974) can be used to minimise the discrepancy between the calculated and the experimental data, as in eq. (3.9). Given that the fitting is often performed on a relative scale, one can without loss of generality rescale the volume fractions such that their sum is equal to unity. This formalism is useful for characterising well-defined systems such as oligomeric equilibrium mixtures of proteins (for example, program OLIGOMER; Konarev *et al.* 2006).

Figure 5.3 displays an example of the scattering pattern from an equimolar mixture of monomers and dimers (computed from the crystallographic structure of kinesin, taken from the PDB entry 3KIN; Kozielski *et al.* 1997). The scattering from the mixture represents a 'blended' scattering from monomers and dimers, and, accordingly, the

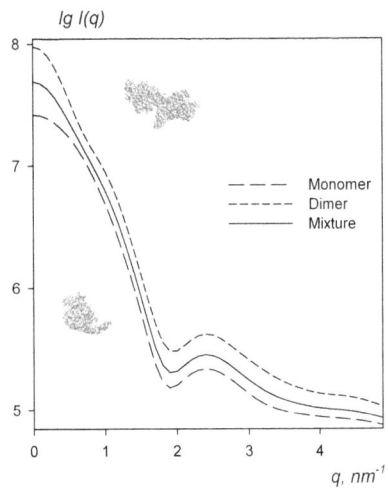

lg I(q)

Monomer
Dimer
Mixture

q, nm⁻¹

Fig. 5.3 Scattering curves computed from a monomeric and dimeric protein and their equimolar mixture. The scattering data are computed from the crystallographic structures of kinesin (PDB entry 3KIN, (Kozielski *et al.* 1997) displayed as inserts) using CRYSOL (Svergun *et al.* 1995).

overall structural parameters computed from the mixture would lie between the individual components. This allows one to rapidly distinguish between a mixture and a complex. The quality of fitting using eq. (5.6) provides another simple means to check whether a complex is formed. Indeed, let us assume that two proteins, A and B, with known structures, are mixed and the scattering from such a mixture is measured. To rapidly assess the complex formation one may try to fit the experimental data just as a linear combination of the scattering intensities computed from A and from B. If this fitting is successful, then one can be sure that no complex is formed.

5.3.2 Singular value decomposition

The formalism in the previous section assumes that the number of components and also their structures (or at least their scattering patterns) are known. This may often not be the case, for example, for the SAS applications to study processes (such as assembly or folding). The limiting (initial and final) states are often known (such as completely folded and completely unfolded protein) and the primary task is to find whether there are some intermediates (that is, the value of K is not known *a priori*). In these studies, a series of experiments is typically performed to monitor the change of the system (with time, temperature, pH and so on) and a number of scattering patterns $I^m(q_i)$ $(m = 1, \ldots M)$ is recorded, each of them obeying eq. (5.5) with different weights v_k. A mathematical approach called singular value decomposition (SVD) (Golub and Reinsh 1970) permits one to assess the number of components based on the experimental data only, without further assumptions. The matrix $\mathbf{A} = [A_{mi}] = [I^{(m)}(q_i)]$, $(i = 1, \ldots N, m = 1, \ldots M$, where N is the number of experimental points) is represented as a matrix product $\mathbf{A} = \mathbf{U}*\mathbf{S}*\mathbf{V}^T$. Here, the matrix $\mathbf{S}$ is diagonal, and the columns of the orthogonal matrices $\mathbf{U}$ and $\mathbf{V}$ are the eigenvectors of the matrices $\mathbf{A}*\mathbf{A}^T$ and $\mathbf{A}^T*\mathbf{A}$, respectively. The matrix $\mathbf{U}$ yields a set of so-called left singular vectors—orthonormal base curves $U^{(k)}(q_i)$—that spans the range of matrix $\mathbf{A}$, whereas the diagonal of $\mathbf{S}$ contains their associated singular values in descending order. The larger the singular value, the more significant the singular vector, and the number of significant singular vectors yields the minimum number of independent curves required to represent the entire data set by their linear combinations. In addition to the sufficiently high singular value, significant components must be non-random curves, and these can be identified by non-parametric tests like the one due to Wald and Wolfowitz (Larson 1975). An example of an SVD analysis of experimental data in Fig. 5.4 was obtained on fifteen scattering patterns from lumazine synthase in different buffers with varying pH. This protein progressively assembles into capsids upon pH change, and the five major components detected by SVD represent the intermediates of the assembly (Zhang *et al.* 2006).

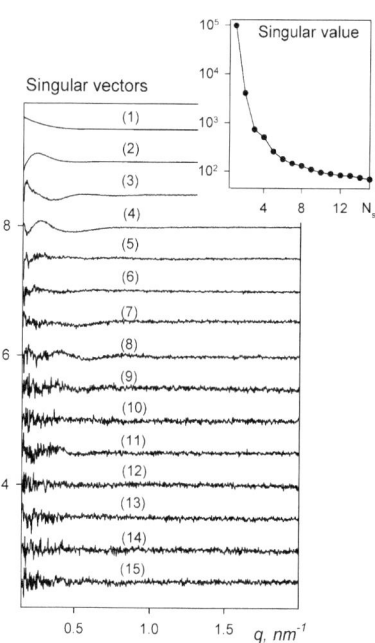

Fig. 5.4 SVD analysis of multiple components upon assembly of lumazine synthase capsid (Zhang *et al.* 2006). Fifteen scattering patterns collected at different pH and in different buffers were subject to SVD. The successive singular vectors displayed in logarithmic scale are displaced by one order of magnitude for better visualisation. The vectors from 1 to 5 contain systematic contributions (major components); those from 6 to 9 still have some systematic (presumably, minor components); the vectors starting from ten are pure noise.

The number of significant singular vectors M_S assessed by SVD provides an estimate of the minimum number of independent components in equilibrium or non-equilibrium mixtures. Clearly, M_S must be smaller than the number of measured curves M; the case $M_S = M$ means that all the scattering curves are independent and eq. (5.6) does not hold. SVD was initially employed in SAXS in the early 1980s (Fowler *et al.* 1983), being especially popular in the analysis of titration and time-resolved experiments (Chen *et al.* 1996, Bilgin *et al.* 1998, Perez *et al.* 2001). Programs for SVD analysis of SAS data are publicly available (for example, Konarev *et al.* 2003). Of course, SVD analysis provides only a lower limit, and the actual number of components (such as the number of intermediates in (un)folding or assembly of proteins) may also be larger than M_S.

Another caveat to be made is that SVD gives the number of components but does not tell anything about the scattering patterns of these components. The singular vectors, being nominally also scattering curves, do not correspond to any particles and have no physical meaning in this sense. To extract information about the individual profiles, time-resolved or titration experiments are usually carried out driving the equilibrium as far as possible towards one of the states (some practical examples are considered in Chapter 6).

5.3.3 Low-resolution shape modelling for mixtures

In some cases it is possible to build low-resolution three-dimensional models from the scattering by oligomeric mixtures even if the system cannot be driven to a monodisperse state. Some *ab initio* and rigid-body modelling programs presented in Chapter 4 are able to take polydispersity into account (Petoukhov *et al.* 2012). Thus, a DR modelling program GASBORMX—an extension of GASBOR (Section 4.3.3)—is capable of building *ab initio* DR models of multimers taking into account the presence of free monomers in solution. In the search procedure the scattering of the multimer and of the monomer are calculated and the scattering from the mixture is computed as their linear combination following eq. (5.6). The volume fractions of the components are either determined from the least squares fitting or they are fixed to specified values, if these are known *a priori*. A similar idea is used in SASREFMX—an extension of SASREF (Section 4.6.1)—but the approach is not limited to a multimer–monomer mixture. In SASREFMX, a subset of subunits (dissociation products) can be selected as an additional component (for example, a scenario is possible when a ternary complex partially dissociates into a binary part and a free third component). During the SA-driven modification of the mutual subunit arrangement, the experimental scattering pattern is computed as a linear combination of the scattering of the complex and individual components. The volume fractions of the bound and the dissociated states in the mixture are determined by linear least-square fitting using eq. (5.6), similar to the above *ab initio* case.

An alternative approach applicable to transient complexes has been proposed recently in (Blobel *et al.* 2009), and employs a so-called multivariate curve resolution with alternating least squares (MCR-ALS) algorithm. MCR-ALS is known from chemometrics to analyse sets of multivariate (multichannel, multiwavelength and so on) signals arising from weighted linear combinations of pure components. It was demonstrated that the approach allows one to extract the scattering patterns from individual species for a system with monomer–dimer equilibrium with the volume fraction of dimers not exceeding 15% in the mixture (Blobel *et al.* 2009).

The above approaches extend the possibilities of modelling of oligomeric mixtures, but one should not forget that accounting for the possible polydispersity adds further to the uncertainty of the SAS data interpretation. Care must therefore be taken when using these techniques, especially in cases when the volume fraction(s) of some of the components are small (below 10%).

5.4 Conformational polydispersity and flexible systems

Equilibrium between a few well-defined macromolecular states (such as co-existence of open and closed states of an enzyme) is perhaps the simplest case of conformational polydispersity of biological macromolecules in solution. If the structures of the distinct conformational states are available (for example, from MX studies), their scattering patterns can be evaluated and eq. (5.6) can be solved to determine the occupancies. These systems can therefore be treated in the same way as oligomeric mixtures considered in the previous section. If the individual structures are not known, one usually tries to change conditions in such a way that the equilibrium is shifted (if possible, completely) towards one of the states.

5.4.1 Challenges in the analysis of flexible macromolecules

A very different and much more challenging situation arises in studies of systems exhibiting significant flexibility. SAS is often applied to multidomain proteins with flexible linkers or to IDPs, as the flexibility precludes the use of high-resolution techniques such as MX. We have already seen in Section 4.1 that the appearance of the Kratky plot allows one to distinguish between folded and unfolded macromolecules, and also to identify qualitatively the presence of flexible parts of the structure. Clearly, elevated R_g-values and monotonous scattering curves lacking any features also suggest possible flexibility (though this behaviour can also be observed for rigid but highly anisometric structures). The major difficulty in the structural characterisation of flexible systems lies in the fact that the scattering data can hardly be analysed in terms of a single model. In Section 4.6, hybrid methods were considered, providing information about structural fragments (such as loops or termini) not 'seen' by high-resolution methods, but still sensed by SAS. It is, however, clear that these fragments must exhibit limited flexibility, otherwise the approximate conformation would not be meaningful. Also, *ab initio* methods, if applied to the scattering from flexible systems, would provide 'an average' shape, but what is the physical sense of a shape averaged over an astronomical number of conformations experienced, for example, by an IDP in solution? It is clear that any method to interpret the SAS data from flexible systems cannot look for a single model but instead must take the conformational average into account.

Averaging over randomly generated ensembles has been considered in several publications; for example, a program called Flexible-Meccano was developed to generate randomised models by consecutively assembling peptide units, considered as rigid entities (Bernado *et al.* 2005). The algorithm employs a force field accounting for the residue-specific Ramachandran plot and a coarse-grained description of the side-chains to avoid steric clashes. This algorithm was employed to generate ensembles of IDPs and demonstrate that the averages over the random ensembles correlate well with the experimental SAXS and NMR (RDC) data.

5.4.2 Ensemble-based methods

A general methodology accounting for the coexistence of multiple conformations to quantitatively describe flexible systems was proposed in (Bernado *et al.* 2007). The approach, called the Ensemble Optimisation Method (EOM), is schematically depicted in Fig. 5.5. The aim of the method is to represent the solute by an ensemble consisting of K conformations of the same molecule. The representative number of conformers K is not known *a priori*, and can be either fixed to a reasonable value—for example, about 20–50, or refined during the optimisation procedure. First, a random pool is generated containing $M \gg K$ conformers, which are supposed to cover the conformational space explored by the flexible macromolecule in solution. The scattering curves $I_k(q)$ from all the structures in the pool are precomputed by an appropriate program; for example, CRYSOL (Svergun *et al.* 1995) or CRYSON (Svergun *et al.* 1998), for X-rays and neutrons, respectively. The scattering from any given subset can be computed by the sum of the selected $I_k(q)$ following eq. (5.6) and assuming that the conformers are uniformly populated. GA (Jones 1998) is then employed to select an 'optimal' ensemble; that is, a subset of configurations that, after averaging their individual scattering profiles, fit the experimental data. This algorithm involves 'genetic evolution' of the system, which usually requires millions of modifications to improve the so-called 'fitness function' (the fit to the experimental data in our

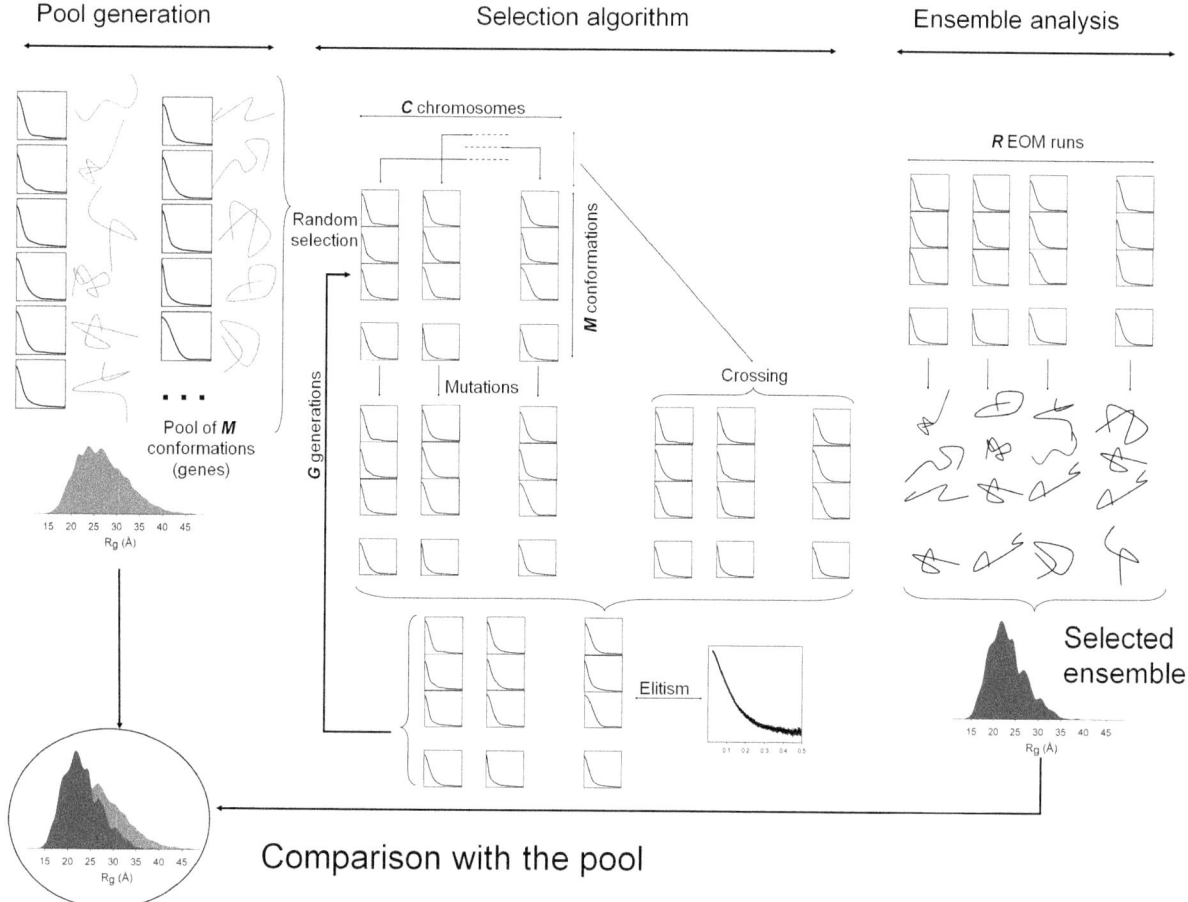

Fig. 5.5 An overall scheme of the ensemble optimisation method. A random pool covering the conformational space of the system is generated (left panel) and sub-ensembles are selected using GA (middle panel). The results of repetitive runs of the algorithm are averaged and the obtained R_g distributions are compared with those in the initial pool. Adapted from Bernado and Svergun (2012), with permission of the Royal Society of Chemistry.

case). Importantly, the intensities $I_k(q)$ become the representatives of the conformers for all GA selection operators, i.e. the latter are performed without the need to recompute the scattering patterns, and this speeds up the calculations significantly.

The GA procedure works as follows. Each subset (called chromosome, in GA terminology) contains K scattering profiles (genes) from the conformers. In the first generation a fixed number C (usually, $C = 50$) chromosomes are created by randomly picking up conformers from the pool. In each generation G these C chromosomes are submitted to the standard GA operations, mutation and crossing. In the former operation the genes of each chromosome are randomly exchanged for others either from the pool or from the chromosomes of the same generation. In the latter operation, genes of two randomly selected chromosomes are exchanged. These operations increase the pool to 3C chromosomes, and, for each chromosome, the sum of the individual SAS profiles computed with eq. (5.6) is compared with the experimental scattering, yielding the discrepancy χ^2 (eq. 3.9) playing the role of the 'fitness function'. The best C chromosomes yielding the lowest χ^2 are selected for further evolution ('elitism', in GA terms). This creation of the elite population is repeated typically for up to $G = 1,000$ to 5,000 generations to yield the chromosome (that is, the ensemble of structures) with the best agreement to the experimental data. The entire GA selection is then repeated from the very beginning with different random seeds (typically about $R \approx 200$ times). The distributions of the low-resolution parameters like the R_g or D_{max} in the selected ensembles provide information about the flexibility and abundance of conformations in the solute.

It is important to stress here that EOM is not supposed to provide information about the individual conformations in solutions; given the low resolution of SAXS, this would be a severe overinterpretation of the data. The selected conformers are neither the most populated conformations, nor even are they supposed to necessarily exist in solution. Instead, the EOM is a tool to describe the size and shape distributions sampled by the unfolded molecule. The ensembles selected in the repeated EOM runs normally contain different conformations, but they all provide similar R_g and D_{max} distributions.

It is instructive to compare the EOM-derived distributions with those of the random ensembles (such as initial pools). The latter are usually broad single mode functions. Obtaining an EOM distribution, which is significantly narrower than the initial pool, indicates that the macromolecule is rather rigid in solution. The position of the main peak in the EOM ensemble (with respect to the peak position in the pool) allows one to judge whether the macromolecule is compact or extended. For IDPs, for example, local transient structures display broader distributions of R_g and normally shifted towards larger values, whereas long-range contacts and residual tertiary structure provide more compact distributions than the random coils. Bimodal EOM distributions may indicate a macromolecule flipping between a closed and an open configuration.

The major improvement of EOM over the earlier approaches is that the method describes the global properties of the macromolecules in terms of size and shape distributions rather than condensing all structural characteristics into single averaged parameters (such as R_g value). The technique is applicable to various types of systems, most notably to multidomain proteins and complexes with flexible linkers, and the IDPs. The method consists of two independent modules. The program RANCH generates the pool of random configurations utilising high-resolution models for regions of known structure when available, and DRs for the flexible segments. Originally, this program generated single-chain C_α-only models accounting for the C_α–C_α Ramachandran plot (Kleywegt 1997) and avoiding steric clashes. Later, RANCH was extended to allow the generation of random pools from oligomeric multi-chain conformations, with options for the incorporation of symmetry and specification of inter-domain contacts (Petoukhov *et al.* 2012). The selection of the optimal ensemble is performed by the program GAJOE implementing a GA procedure, which was recently further optimised to have a possibility of automatically adjustable size of the chromosome (Petoukhov *et al.* 2012). The division into the two modules allows one to employ other structures than those generated by RANCH (for example, models from molecular dynamics simulation) to select the optimum ensemble.

The concept of ensemble fitting of the SAS data became very useful in solution-scattering studies of flexible systems. Several other approaches utilising the EOM-like strategy have also become available (Pelikan *et al.* 2009, Yang *et al.* 2010, Rozycki *et al.* 2011). Often, the SAS-based ensemble analysis is coupled with experimental NMR data on chemical shifts, RDCs and other information (Blackledge 2010, Marsh and Forman-Kay 2011, Bertini *et al.* 2010). The ensemble-based approaches significantly broadened SAS applications to flexible objects, as will be illustrated in examples presented in Chapters 6 and 9.

5.5 Interacting systems and structure factor

Interactions between macromolecules in solution may be specific or non-specific (Leckband and Israelachvili 2001), and they involve the macromolecular solute and co-solutes (salts, small molecules, polymers), the solvent and, where applicable, co-solvents. The formation of complexes as a result of specific interactions involving cooperative interactions between complementary surfaces is described in the previous section dealing with mixtures and equilibria. In contrast, non-specific interactions are more difficult to describe, and drastic simplifications are usually required.

In general, the spherically averaged scattering from a volume of a solution of anisotropic objects like macromolecules that is coherently illuminated is given by:

$$I(q,t) = \left\langle \sum_{i=1}^{N} \sum_{j=1}^{N} A_i(\mathbf{q}, \mathbf{u_i}, \mathbf{v_i}, \mathbf{w_i}) \cdot A_j(\mathbf{q}, \mathbf{u_j}, \mathbf{v_j}, \mathbf{w_j}) \exp(i\mathbf{q}\mathbf{r_{ij}}(t)) \right\rangle_\Omega \quad (5.7)$$

where A is the scattering amplitude of the individual particles computed as in eq. (1.4), and $\mathbf{u,v,w}$ are unit vectors giving their orientation relative to the reference coordinate system in which the momentum transfer vector $\mathbf{q}$ is defined. Equation (5.7) is very useful for the description of complexes and rigid-body modelling, as explained in Section 4.6. For particles in solution, which have a distribution of orientations and distances, this would lead to unwieldy calculations.

Interparticle interactions result in a modulation of the scattering pattern of isolated particles by the structure factor which reflects their distribution and to a much lesser extent their relative orientation in solution. One therefore assumes that the particles are spherical and that the interactions between pairs of particles can be described by a central potential depending only on their distance, as discussed in more detail in Chapter 8. This potential takes into account the mutual impenetrability of the macromolecules, the screened electrostatic interactions between charges at the surfaces of the macromolecules and the shorter-ranged van der Waals interactions. Non-specific interactions essentially determine the behaviour at larger distances, whereas in the case of attractive interactions leading to, for example, crystallisation, specific interactions dominate at short range.

For identical spherical particles eq. (5.7) simplifies to the product of the square of the form factor of the isolated particles and of the structure factor of the solution which reflects their spatial distribution. This is valid for globular proteins and weak or moderate interactions in a limited q-range (Tardieu 1994, Veretout *et al.* 1989). The structure factor can then be obtained readily from the ratio of the experimental intensity at a concentration c to that obtained by extrapolation to infinite dilution or measured at a sufficiently low concentration c_0 where all correlations between particles have vanished:

$$S(q,c) = \frac{c_0 I_{\exp}(q,c)}{c I(q,c_0)} \tag{5.8}$$

A detailed treatment of the structure factor for spherical particles and its relationship to the pair distribution function and the Ornstein-Zernike equation is given in Appendix 3.

It is extremely difficult to obtain analytical expressions for the structure factor even for spherical particles. This becomes possible only in very rare cases; for example, when the individual molecules are treated as hard spheres. Based on the statistical–mechanical treatment of hard-sphere fluids (Percus and Yevick 1958), $S(q)$ was expressed as an analytical function of two parameters, the sphere radius R and the volume fraction of the spheres ϕ, $S(q,R,v) = (1 - C(q,R,\phi))^{-1}$ (Ashcroft and Leckner 1966). Here, $C(q,R,\phi)$ is the Fourier transformation of the interparticle correlation function given by the equation

$$
\begin{aligned}
C(q,R,\phi) = -\frac{24\phi}{x^6} & \Big\{ \alpha x^3 \left[\sin x - x\cos x\right] \\
& + \beta x^2 \left[2x\sin x - (x^2 - 2)\cos x - 2\right] \\
& + \gamma \left[(4x^3 - 24x)\sin x - (x^4 - 12x^2 + 24)\cos x + 24\right]\Big\}
\end{aligned} \tag{5.9}
$$

where $x = 2qR$, $\alpha = (1 + 2\phi)^2/(1 - \phi)^4$, $\beta = -6\phi(1 + \phi/2)^2/(1 - \phi)^4$ and $\gamma = \phi(1 + 2\phi)^2/[2(1 - \phi)^4]$.

Figure 5.6A presents the theoretical structure factors computed for the systems of hard spheres using eq. (5.9) for different volume fractions. Although the hard spheres do not interact with each other, there

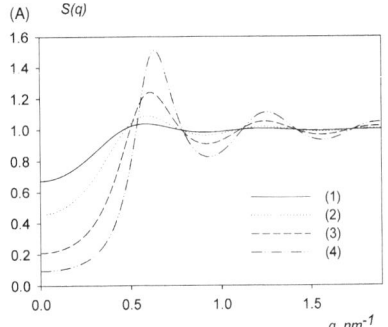

(A) $S(q)$

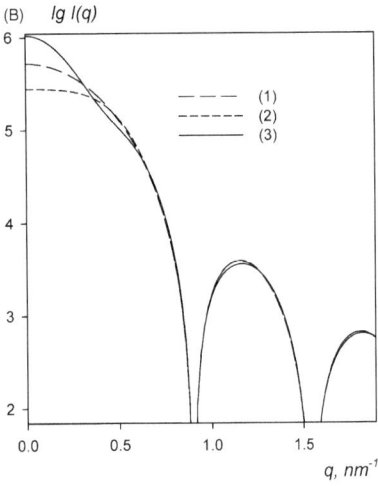

(B) $\lg I(q)$

Fig. 5.6 Interparticle interference effects for a system of solid hard spheres. A: Structure factors computed for systems of hard spheres with radius $R = 5$ nm (curves 1–4 correspond to $\phi = 0.05, 0.1, 0.2, 0.3$, respectively). B: Computed scattering from a solid sphere with $R = 5$ nm (1), from a solution of hard spheres with $\phi = 0.2$ (2) and from a dumbbell (3); the latter curve is divided by two to normalise to the two other curves (1) at higher angles.

is an apparent repulsive effect due to the excluded volume, diminishing $I(0)$. At higher angles, modulations appear due to non-random distribution of the spheres, and the amplitude of the modulations increases with concentration. In contrast, attractive interactions, where particles tend to form clusters or aggregates, lead to an increase in $I(0)$ due to the larger coherently scattering volume. In case of specific aggregates, oscillations around the ideal scattering curve from an infinitely diluted solution of particles are observed reflecting the interference term between the individual particles. The difference between attractive and repulsive interactions is exemplified in Fig. 5.6B for the hard-sphere case. This figure also reveals that the interparticle interactions are barely noticeable beyond the very low angle portion of the scattering data. Indeed, although the interactions are extremely pronounced (a rather dense system *versus* scattering from pure dimers) the differences become rather subtle beyond $q = 0.5$ nm^{-1}. This fact further underlines the importance of accurate measurements and careful analysis of the very small angle portions of the SAS data.

If separation of the structure factor and the form factor using eq. (5.9) is straightforward in the case of monodisperse solutions and repulsive interactions, this is no longer the case when the interactions are attractive and the polydispersity of the solution depends on its concentration. For spherical particles the GIFT method has been proposed, which is a generalisation of the indirect transformation technique described in Section 4.1. The structure factor is also parameterised similarly to the characteristic function, and non-linear data fitting is employed to find both the distribution function and the structure factor. For non-spherical (for example, rod-like) particles the method yields an effective structure factor (Brunner-Popela and Glatter 1997, Weyerich *et al.* 1999).

The interaction of rod-like molecules has been studied in detail, and a pair potential of the form $V(r,\mathbf{u}_1,\mathbf{u}_2)$ can be used to describe the interactions between molecules where r is the distance between the centre of mass and $\mathbf{u}_1$ and $\mathbf{u}_2$ denote the orientation of their axis (Weyerich *et al.* 1990). Unfortunately, for filaments SAS usually only yields cross-section information and an effective structure factor must be used.

For thin rods such as DNA at low ionic strength, the length distribution has little influence on the effective structure factor (Koch *et al.* 1995). In the dilute regime the position of its first maximum, determined by the centre-to-centre separation between rigid segments, varies as the square root of the concentration. The length distribution has, however, a strong influence on the relaxation times observed in electric field scattering (Koch *et al.* 1988, Koch *et al.* 1995) and on the slow mode observed in dynamic light scattering (Skibinska *et al.* 1999).

For mixtures of different types of particles with possible polydispersity and interactions between particles of the same component, the scattering intensity from a component entering eq. (5.6) can be represented as

$$I_k(q) = S_k(q) \cdot \int_0^\infty D_k(R) \cdot V_k(R) \cdot [\Delta \rho_k(R)]^2 \cdot i_{0k}(q, R) \ dR \qquad (5.10)$$

where $\Delta\rho_k(R)$, $V_k(R)$ and $i_{0k}(q, R)$ denote the contrast, volume and form factor of the particle with size R (these functions are defined by the shape and internal structure of the particles, and $i_{0k}(0, R) = 1$), whereas $S_k(q)$ is the structure factor describing the interference effects for the kth component. It is clear that quantitative analysis of such systems is only possible if assumptions are made about form and structure factors and about the size distributions. A parametric approach was proposed (Svergun *et al.* 2000) to characterise mixtures of particles with simple geometrical shapes (spheres, cylinders, dumbbells). Each component is described by its volume fraction, form factor, contrast, polydispersity and, for spherical particles, potential for interparticle interactions. The functions $D_k(R)$ are represented by two-parametric monomodal distributions characterised by the average dimension R_{0k} and dispersion ΔR_k. The structure factor for spherical particles $S_k(q)$ is represented in the Percus–Yevick approximation using the sticky hard sphere potential (Baxter 1968) described by two parameters, hard sphere interaction radius R_k^{hs} and 'stickiness' τ_k. The approach was developed in the study of AOT water-in-oil microemulsions and applied for quantitative description of the droplet to cylinder transition (Svergun *et al.* 2000). A general program MIXTURE based on this method is publicly available (Konarev *et al.* 2003). It should be noted that the parametric analysis is more suitable for soft condensed matter applications such as microemulsion systems; biological macromolecules often require more specialised approaches, and some of them will be considered in Chapter 8.

References

Ashcroft, N. W. and Leckner, J. (1966). 'Structure and resistivity of liquid metals', *Phys. Rev.* 145: 83–90.

Baxter, R. J. (1968). 'Percus–Yevick equation for hard spheres with surface adhesion', *J. Chem. Phys.* 49: 2270–3.

Bernado, P., Blanchard, L., Timmins, P., Marion, D., Ruigrok, R. W. and Blackledge, M. (2005). 'A structural model for unfolded proteins from residual dipolar couplings and small-angle X-ray scattering', *P. Natl. Acad. Sci. USA* 102: 17002–7.

Bernado, P., Mylonas, E., Petoukhov, M. V., Blackledge, M. and Svergun, D. I. (2007). 'Structural characterization of flexible proteins using small-angle X-ray scattering', *J. Am. Chem. Soc.* 129: 5656–64.

Bernado, P. and Svergun, D. I. (2012). 'Structural analysis of intrinsically disordered proteins by small-angle X-ray scattering', *Mol. Biosyst.* 8: 151–67.

Bertini, I., Giachetti, A., Luchinat, C. *et al.* (2010). 'Conformational space of flexible biological macromolecules from average data', *J. Am. Chem. Soc.* 132: 13553–8.

Bilgin, N., Ehrenberg, M., Ebel, C. *et al.* (1998). 'Solution structure of the ternary complex between aminoacyl-tRNA, elongation factor Tu, and guanosine triphosphate', *Biochemistry* 37: 8163–72.

Blackledge, M. (2010). 'Mapping the conformational mobility of multidomain proteins', *Biophys. J.* 98: 2043–4.

Blobel, J., Bernado, P., Svergun, D. I., Tauler, R. and Pons, M. (2009). 'Low-resolution structures of transient protein–protein complexes using small-angle X-ray scattering', *J. Am. Chem. Soc.* 131: 4378–86.

Brunner-Popela, J. and Glatter, O. (1997). 'Small-angle scattering of interacting particles. I. Basic principles of a global evaluation technique', *J. Appl. Crystallogr.* 30: 431 – 42.

Chen, L., Hodgson, K. O. and Doniach, S. (1996). 'A lysozyme folding intermediate revealed by solution X-ray scattering', *J. Mol. Biol.* 261: 658–71.

Feigin, L. A., and Svergun, D. I. (1987). *Structure Analysis by Small-Angle X-Ray and Neutron Scattering*. New York: Plenum Press.

Fowler, A. G., Foote, A. M., Moody, M. F. *et al.* (1983). 'Stopped-flow solution scattering using synchrotron radiation: apparatus, data collection and data analysis', *J. Biochem. Biophys. Methods* 7: 317–29.

Glatter, O. (1980). 'Determination of particle-size distribution functions from small-angle scattering data by means of the indirect transformation method', *J. Appl. Crystallogr.* 13: 7–11.

Glatter, O. and Kratky, O. (1982). *Small Angle X-Ray Scattering*. London: Academic Press.

Golub, G. H. and Reinsh, C. (1970). 'Singular value decomposition and least squares solution', *Numer. Math.* 14: 403–20.

Jones, G. (1998). *Genetic and Evolutionary Algorithms. Encyclopedia of Computational Chemistry*. Chichester: Wiley.

Kleywegt, G. J. (1997). 'Validation of protein models from Calpha coordinates alone', *J. Mol. Biol.* 273: 371–6.

Koch, M. H. J., Dorrington, E., Klaering, R. *et al.* (1988). 'Electric field X-ray scattering measurements on tobacco mosaic virus', *Science* 240: 194–6.

Koch, M. H. J., Sayers, Z., Sicre, P. and Svergun, D. (1995). 'A synchrotron radiation electric field X-ray solution scattering study of DNA at very low ionic strength', *Macromolecules* 28: 4904–7.

Konarev, P. V., Petoukhov, M. V., Volkov, V. V. and Svergun, D. I. (2006). 'ATSAS 2.1, a program package for small-angle scattering data analysis', *J. Appl. Crystallogr.* 39: 277–86.

Konarev, P. V., Volkov, V. V., Sokolova, A. V., Koch, M. H. J. and Svergun, D. I. (2003). 'PRIMUS: a Windows-PC based system for small-angle scattering data analysis', *J. Appl. Crystallogr.* 36: 1277–82.

Kozielski, F., Sack, S., Marx, A. *et al.* (1997). 'The crystal structure of dimeric kinesin and implications for microtubule-dependent motility', *Cell* 91: 985–94.

Larson, H. J. (1975). *Statistics: An Introduction*. New York: John Wiley.

Lawson, C. L. and Hanson, R. J. (1974). *Solving Least Squares Problems*. Englewood Cliffs, NJ: Prentice-Hall.

Leckband, D. and Israelachvili, J. (2001). 'Intermolecular forces in biology', *Quart. Rev. Biophys.* 34: 105–267.

Marsh, J. A. and Forman-Kay, J. D. (2011). 'Ensemble modelling of protein disordered states: Experimental restraint contributions and validation', *Proteins* 80: 556–72.

Pedersen, J. S. (2002). 'Scattering methods applied to soft condensed matter', in Lindner, P. and Zemb, T. (Eds.) *Neutrons, X-Rays and Light*. Amsterdam: North Holland Delta Series.

Pelikan, M., Hura, G. L. and Hammel, M. (2009). 'Structure and flexibility within proteins as identified through small angle X-ray scattering', *Gen. Physiol. Biophys.* 28: 174–89.

Percus, J. K., and Yevick, G. J. (1958). 'Analysis of classical statistical mechanics by means of collective co-ordinates', *Phys. Rev.* 110: 1–13.

Perez, J., Vachette, P., Russo, D., Desmadril, M. and Durand, D. (2001). 'Heat-induced unfolding of neocarzinostatin, a small all-beta protein investigated by small-angle X-ray scattering', *J. Mol. Biol.* 308: 721–43.

Petoukhov, M. V., Franke, D., Shkumatov, A. V. *et al.* (2012). 'New developments in the ATSAS program package for small-angle scattering data analysis', *J. Appl. Crystallogr.* 45: 342–50.

Rozycki, B., Kim, Y. C. and Hummer, G. (2011). 'SAXS ensemble refinement of ESCRT-III CHMP3 conformational transitions', *Structure* 19: 109–16.

Skibinska, L., Gapinski, J., Liu, H., Patkowski, A., Fischer, E. W. and Pecora, R. (1999). 'Effect of electrostatic interactions on the structure and dynamics of a model polyelectrolyte. II Intermolecular correlations', *J. Chem. Phys.* 110: 1794–1800.

Svergun, D. I. (1992). 'Determination of the regularization parameter in indirect-transform methods using perceptual criteria', *J. Appl. Crystallogr.* 25: 495–503.

Svergun, D. I., Barberato, C. and Koch, M. H. J. (1995). 'CRYSOL: a program to evaluate X-ray solution scattering of biological macromolecules from atomic coordinates', *J. Appl. Crystallogr.* 28: 768–73.

Svergun, D. I., Konarev, P. V., Volkov, V. V. *et al.* (2000). 'A small angle X-ray scattering study of the droplet-cylinder transition in AOT microemulsions', *J. Chem. Phys.* 113: 1651–65.

Svergun, D. I., Richard, S., Koch, M. H. J., Sayers, Z., Kuprin, S. and Zaccai, G. (1998). 'Protein hydration in solution: experimental observation by X-ray and neutron scattering', *P. Natl. Acad. Sci. USA* 95: 2267–72.

Tardieu, A. (1994). 'Thermodynamics and structure—concentrated solutions—structured disorder in vision', *Neutron and Synchrotron Radiation for Condensed Matter Studies*. Les editions de Physique (France), Springer.

Veretout, F., Delaye, M. and Tardieu, A. (1989). 'Molecular basis of eye lens transparency. Osmotic pressure and X-ray analysis of alpha-crystallin solutions.', *J. Mol. Biol.* 205: 713–28.

Weyerich, B., Canessa, E. and Klein, R. (1990). 'Structure and dynamics of suspensions of charged rod-like particles', *Faraday Discussions* 90: 245–59.

Weyerich, B., Brunner-Popela, J. and Glatter, O. (1999). 'Small-angle scattering of interacting particles. II. Generalized indirect Fourier transformation under consideration of the effective structure factor for polydisperse systems.', *J. Appl. Crystallogr.* 32: 197–209.

Yang, S., Blachowicz, L., Makowski, L. and Roux, B. (2010). 'Multidomain assembled states of Hck tyrosine kinase in solution', *P. Natl. Acad. Sci. USA* 107: 15757–62.

Zhang, X., Konarev, P. V., Petoukhov, M. V. *et al.* (2006). 'Multiple assembly States of lumazine synthase: a model relating catalytic function and molecular assembly', *J. Mol. Biol.* 362: 753–70.

BIOLOGICAL APPLICATIONS OF SOLUTION SAS

Part III

In this part, practical applications of SAXS and SANS to study the structure of biological macromolecules in solution will be presented. These are grouped in sections covering different types of experiment and, accordingly, different methodological approaches. In Chapter 6, studies of the structural organisation of various systems are reviewed, aiming at elucidation of the static structure of individual macromolecules and complexes or at providing the equilibrium composition of mixtures. The characterisation of the kinetics of biological processes in time-resolved experiments is considered in Chapter 7. Interacting systems, where SAS is an indispensable tool to determine the factors influencing interparticle interactions and the corresponding potentials, are covered in Chapter 8. Finally, Chapter 9 is devoted to the joint use of SAXS and SANS together with other structural, biophysical and computational techniques to illustrate the advantages of multipronged approaches in modern structural biology, and to describe the recent trends in biological SAS applications.

Static structural studies

6

6.1 Applications of *ab initio* shape
    determination   171

6.2 Quaternary structure analysis
    of proteins and complexes   181

6.3 Equilibrium mixtures and
    oligomeric composition   192

6.4 Membrane proteins and
    lipoproteins   198

6.5 Flexible systems   205

References   212

There are numerous excellent applications of SAXS and SANS to the study of the static structure of biological macromolecules in solution. In this chapter we present some of the classical examples (like the ribosome studies) but concentrate largely on more recent applications to different types of systems, where modern analysis methods are employed. The aim is not to present a review of the field, as this would not have been possible within a single chapter. Rather, the examples illustrate the potential of solution scattering as applied not only to solutions of various types of macromolecules including proteins, nucleic acids and lipids and their complexes, but also to mixtures and flexible systems.

6.1 Applications of *ab initio* shape determination

In this section we consider mostly stand-alone SAS applications, where SAXS and SANS are employed largely on their own to generate *ab initio* models. As indicated in Chapter 4, *ab initio* methods are able to reconstruct low-resolution shapes without *a priori* knowledge and with minimal assumptions about the macromolecule. Before considering the practical applications, we first briefly recall the major features and assumptions of *ab initio* analysis.

6.1.1 Bead *versus* dummy residues modelling

The most general and presently most frequently used *ab initio* shape determination by bead modelling (Section 4.4.2) assumes a uniform particle density. This is a good approximation for X-ray scattering from sufficiently large macromolecules (such as proteins with MM exceeding $\sim$30 kDa) which also holds for SANS at higher contrasts (such as proteins or nucleic acids in H_2O or in D_2O). In all these cases one typically uses the scattering patterns in a range up to $q = 2.5$–$3\,nm^{-1}$ (that is, a resolution of $\sim$2.5–2.0 nm). In this range the scattering contribution from the internal structure is eliminated largely by subtracting an appropriate constant to enforce the q^{-4}-decay of the intensity, eq. (4.37). The subtraction is usually done automatically by the shape-determination programs (Svergun 1999).

The dummy residues technique (Section 4.4.3) accounts for the internal structure of proteins. This technique is at present restricted to X-ray studies on proteins, but can be employed without size limitations and also on a broader range of scattering data (up to $q = 10\,\text{nm}^{-1}$ or about 0.6 nm resolution).

For systems consisting of components with significantly different contrasts such as nucleoprotein complexes, multiphase modelling is possible (see Section 4.4.2). To highlight specific portions in an entire structure this analysis requires, however, the simultaneous fitting of multiple scattering patterns taken under conditions where the contrasts of the different parts of the structure are varied.

6.1.2 Oligomeric state and conformational changes

The information that can be obtained most directly from the *ab initio* shape is the oligomeric state of the macromolecule. Sometimes the shape also provides direct evidence for the organisation of oligomers as in the case of Pur-α, a protein acting as transcriptional regulator, host factor for viral replication and cofactor for mRNP localisation in dendrites (Graebsch *et al.* 2009). The crystal structure of residues 40 to 185 from *Drosophila melanogaster* Pur-α, which constitutes a major part of the core region responsible for nucleic acid binding contains two almost identical structural motifs (PUR repeats I and II), which interact to form a PUR domain. The shape of the Pur-α I–II construct in solution reconstructed *ab initio* (Fig. 6.1A, B) clearly indicated that the construct is a

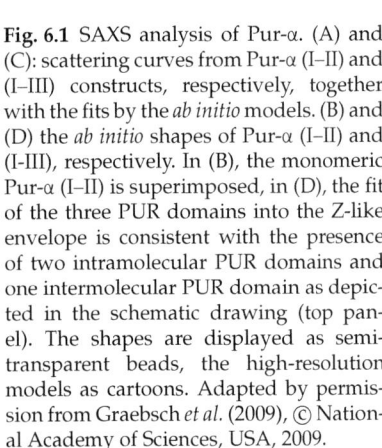

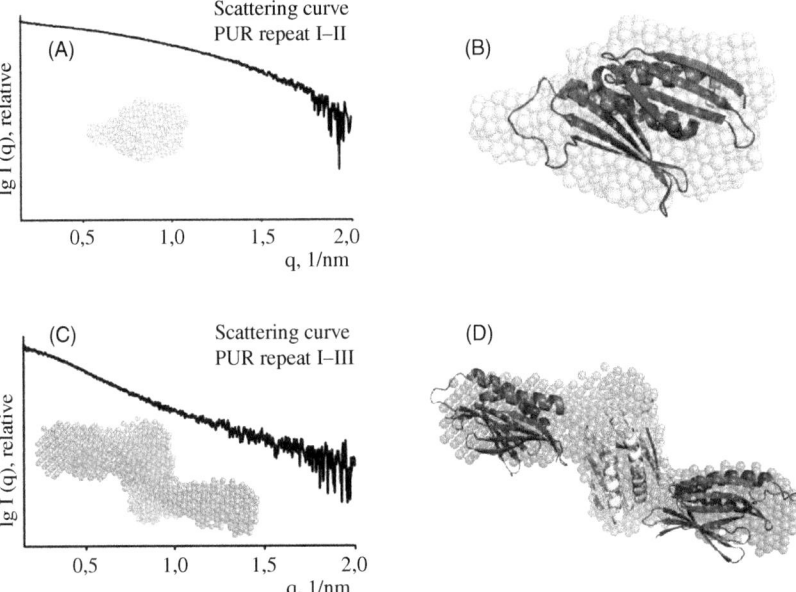

Fig. 6.1 SAXS analysis of Pur-α. (A) and (C): scattering curves from Pur-α (I–II) and (I–III) constructs, respectively, together with the fits by the *ab initio* models. (B) and (D) the *ab initio* shapes of Pur-α (I–II) and (I-III), respectively. In (B), the monomeric Pur-α (I–II) is superimposed, in (D), the fit of the three PUR domains into the Z-like envelope is consistent with the presence of two intramolecular PUR domains and one intermolecular PUR domain as depicted in the schematic drawing (top panel). The shapes are displayed as semi-transparent beads, the high-resolution models as cartoons. Adapted by permission from Graebsch *et al.* (2009), © National Academy of Sciences, USA, 2009.

monomer in solution with the same overall appearance as in the crystal. Multiple sequence alignment of Pur-α homologues from different species indicated that the *Drosophila* core region C-terminal to PUR repeat II contains a third PUR repeat sharing an almost identical predicted secondary structure and probably also a fold similar to repeats I and II. SAXS analysis of the Pur-α (I–III) construct revealed that the construct forms dimers in solution, suggesting that the dimerisation is probably mediated by repeat III. The three PUR domains can be well placed in the Z-like *ab initio* shape of the construct (Fig. 6.1C, D). For a good fit into the envelope, the PUR-domain in the middle must be oriented perpendicularly to the two flanking PUR domains. The Pur-α dimers are stable at concentrations down to 30 nM and may therefore exist *in vivo*. The elongated arrangement of PUR domains detected by SAXS appears well suited to the recognition of repetitive sequences as reported for Pur-α.

The *ab initio* shapes often provide sufficient resolution to identify conformational states of macromolecules, especially if information from homologous structures is available. This is illustrated by the structural study of the ATPase superfamily 2 (SF2) domain of RIG-I in complex with a nucleotide analogue (Civril *et al.* 2011). RIG-I is a receptor detecting cytosolic viral dsRNA and initiating an antiviral innate immune response. The crystal structure of the SF2 domain of mouse RIG-I determined to 0.22-nm resolution revealed an opened C-shaped configuration of three constituent domains. The envelope of the human RIG-I^{SF2} determined *ab initio* from SAXS data revealed the same overall shape as the mouse RIG-I^{SF2} in the crystal, suggesting that this conformation represents a 'signal off' state of RIG-I^{SF2} domain, found in the absence of an RNA ligand. The results, further corroborated by mutational analysis, indicate that the activation of RIG-I occurs through an RNA- and ATP-driven structural switch in the SF2 domain.

SAXS shapes can help visualising conformational changes. An example is provided by troponin C—an 18.4 kDa protein which senses Ca^{2+} and regulates contraction in striated muscle (Fig. 6.2). The crystallographic model of the Ca^{2+}-bound troponin C (Satyshur *et al.* 1994)

(a)

(b)

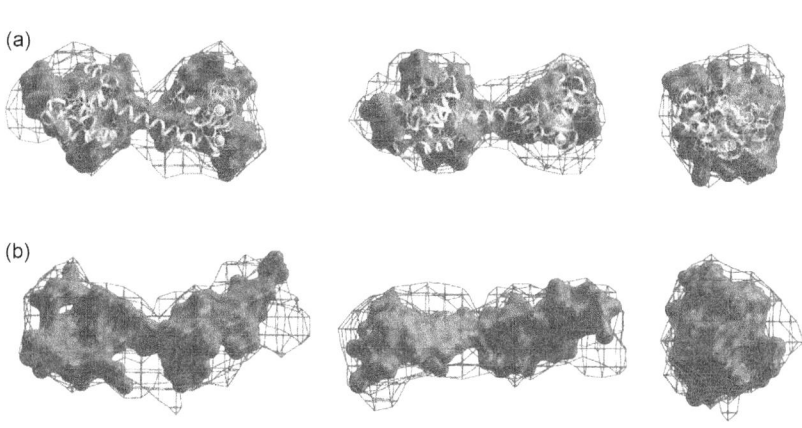

Fig. 6.2 *Ab initio* models of troponin C in solution with (a) and without (b) Ca^{2+} ions (Chacon *et al.* 2000). The single models are displayed as grey beads, three-model overlaps from repetitive runs as wire envelopes. In (a), the crystal structure of Ca^{2+}-bound troponin C is displayed as light grey ribbon Scale bar, 2 nm. Adapted from Chacon *et al.* (2000), with permission from Elsevier, © 2000.

has a dumbbell shape containing two globular domains connected by a helical linker. The shape obtained from the SAXS data by the first bead modelling program DALAI_GA (top row in Fig. 6.2) agrees well with the structural organisation and approximate dimensions of the crystal structure (Chacon *et al.* 2000). The models retrieved from the SAXS data on Ca^{2+}-free troponin C display a large expansion, suggesting an opening of one of the domains (bottom row in Fig. 6.2). This structural change was attributed to the dissociation of the two Ca^{2+} ions from their high affinity sites at the C-terminal domain of intact troponin C, in agreement with earlier modelling (Fujisawa *et al.* 1990). Interestingly, subsequent crystallographic models of Ca^{2+} binding to troponin complexes (Vinogradova *et al.* 2005) indicated that the loss of Ca^{2+} ions causes the rigid central helix of troponin C to collapse. This suggests that the apparent size increase of the Ca^{2+}-free troponin C in solution can be attributed to the increased flexibility of the molecule, and the *ab initio* model in Fig. 6.2 (bottom) may represent a mid-resolution conformational average (Chacon *et al.* 2000).

6.1.3 From individual proteins to complexes

In some cases, *ab initio* models help visualising rather subtle differences in shape, as illustrated by a deletion mutant of another muscle protein, titin. The latter has a MM of about 3–4 MDa, and is the largest gene product found in the human genome (Labeit and Kolmerer 1995, Bang *et al.* 2001) spanning one half of a muscle sarcomere and serving as a molecular ruler. SAXS was employed to model the solution structure of a tandem construct at the N-terminus of titin containing the two Z-disk Ig-like domains, about 10 kDa each (Z1Z2), and also complexes of Z1Z2 with telethonin, a 167 residue protein interacting specifically with Z1Z2 (Zou *et al.* 2003). Low-resolution models of Z1Z2 and His-Z1Z2 (the latter with a His-tag attached at the N-terminal end of Z1) were constructed *ab initio* from the scattering data using DAMMIN and GASBOR (Fig. 6.3A, B). A seven-residue-long poly-histidine tag could be localised at the tip of the Z1 domain by comparison of models of native Z1Z2 (left) and his-tagged Z1Z2 (right). Although this localisation was facilitated by the position at the end of an elongated molecule, this example is illustrative of the possibility of detecting binding even of a small peptide to a 200-residue-long protein. The shape analysis of the Z1Z2 complex with a truncated 90-residue telethonin construct TE(90) is also interesting (Fig. 6.3C). According to AUC and gel-filtration data, a 1:1 complex was expected, but the overall parameters from SAXS clearly indicated another stoichiometry. The shapes reconstructed by DAMMIN and GASBOR are very similar, and can accommodate two Z1Z2 molecules in an antiparallel association with telethonin acting as a central linker (Fig. 6.3C). These results suggest a cross-linking function for telethonin, connecting two titin molecules at their N-termini and leading to a telethonin-mediated auto-anchoring of

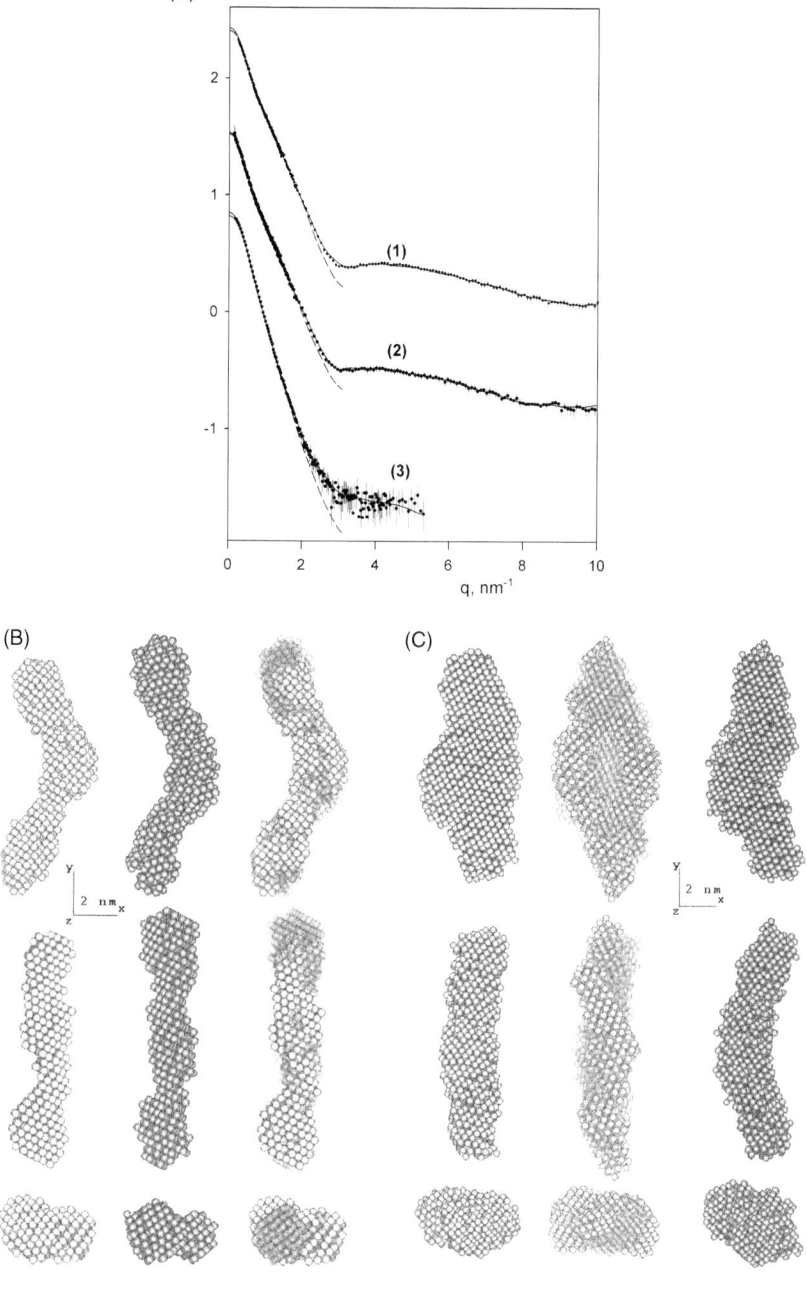

Fig. 6.3 *Ab initio* low-resolution models of Z1Z2 and its complexes with telethonin. (A), X-ray scattering patterns from Z1Z2 (1) His-Z1Z2 (2) and TE(90)-Z1Z2 complex (3). The experimental data are displayed as dots with error bars, DAMMIN fits as dashed lines, GASBOR fits as full lines. The scattering patterns are displaced by one logarithmic unit for better visualisation; (B) Averaged DR models of Z1Z2 (left) and His-Z1Z2 (middle), and their overlap (the extra seven residues due to the His-tag correspond to the extra volume at the top of the molecule in the upper and middle rows); (D) Low-resolution shape of TE(90)-Z1Z2 obtained by averaging twelve DAMMIN models (left panel) and the same model as semitransparent beads superimposed with two antiparallel DAMMIN models of Z1Z2 (solid beads, middle panel). The right panel displays the model of TE(90)-Z1Z2 obtained by averaging twelve GASBOR models. In all panels the middle and bottom rows are rotated counterclockwise by 90° around the Y- and X-axes, respectively. This research was originally published in Zou *et al.* (2003), © American Society for Biochemistry and Molecular Biology.

titin dimers in the Z-disk. This example also illustrates how functional questions can be addressed by a low-resolution technique such as SAXS. Interestingly, a later high-resolution crystal structure of the complex of TE(90) with Z1Z2 fully confirmed both the stoichiometry and cross-linking of Z1Z2 via telethonin (Zou *et al.* 2006).

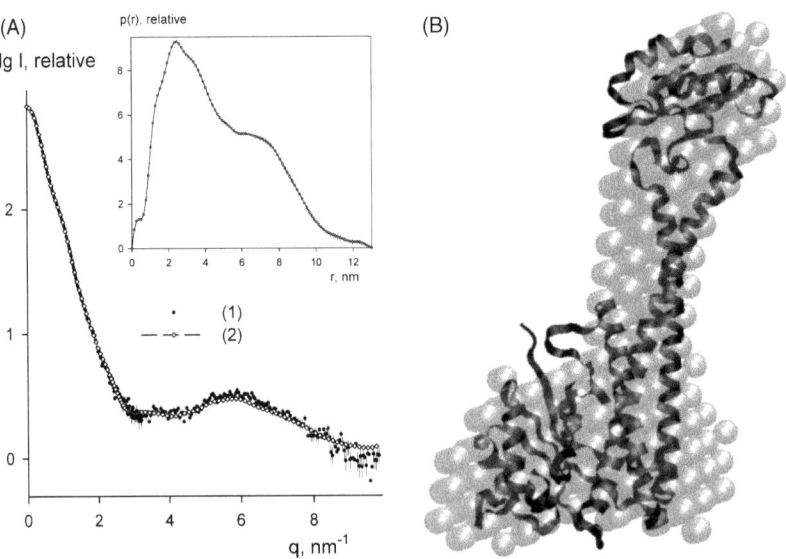

Fig. 6.4 Shape reconstruction of the C subunit C of the yeast V-ATPase (Armbruster *et al.* 2004). (A), SAXS pattern of the C subunit (1) and fit of the GAS-BOR model (2). The calculated p(r) function is displayed in the insert. Reprinted from Armbruster *et al.* (2004), with permission from Elsevier, © 2004. (B) *Ab initio* low-resolution model (average over twelve GASBOR runs, semitransparent beads) superimposed on the later crystal structure of the C subunit displayed as ribbons (PDB entry 1UL7; Drory *et al.* 2004).

An interesting example with *a posteriori* validation of an *ab initio* SAXS model by crystallography is given by a study of the stalk subunit Vma5p (subunit C) of the yeast V-ATPase (Armbruster *et al.* 2004), which plays a crucial role in the vacuolar system of eukaryotic cells. The cytoplasmic part V_1 of the V-ATPase is an assembly of eight subunits (A-H)—some of them homologous to those of the ATP synthase, but subunit C belongs to the non-homologous subunits, which determine the unique features of V-ATPase. The SAXS data collected from the 42 kDa subunit C are presented in Fig. 6.4A, and the p(r) function in the insert indicates that the protein is rather elongated and has two distinct domains, as revealed by a pronounced shoulder at 7 nm. The GASBOR reconstruction yielded the boot-shaped particle (Armbruster *et al.* 2004) in Fig. 6.4B superimposed on the crystal structure of subunit C published later (Drory *et al.* 2004). The comparison further underlines the ability of SAXS to reconstruct protein shapes *ab initio* (note that about fifty residues are missing in the crystal structure).

A further study of V-ATPase demonstrates the possibility of multi-component modelling with SAXS (Diepholz *et al.* 2008). In this work, the overall structure of the entire enzyme was analysed by EM. The conformation of the stator sub-complex containing subunits E, G and C was investigated by SAXS, recording separate patterns for the EGC and EG complexes (Fig. 6.5A). The compatibility of the models of EGC and EG reconstructed independently by DAMMIN suggests that there is no change in the EG part upon complexation with subunit C (Fig. 6.5B). This fact opens the possibility to conduct two-component modelling with MONSA by representing the EGC as two phases corresponding to the EG and C parts. The *ab initio* model is required to simultaneously fit the scattering data from the EG and EGC sub-complexes and also the previously recorded scattering from subunit C (Armbruster *et al.* 2004) (Fig. 6.5A). The MONSA model depicts subunit C in the EGC

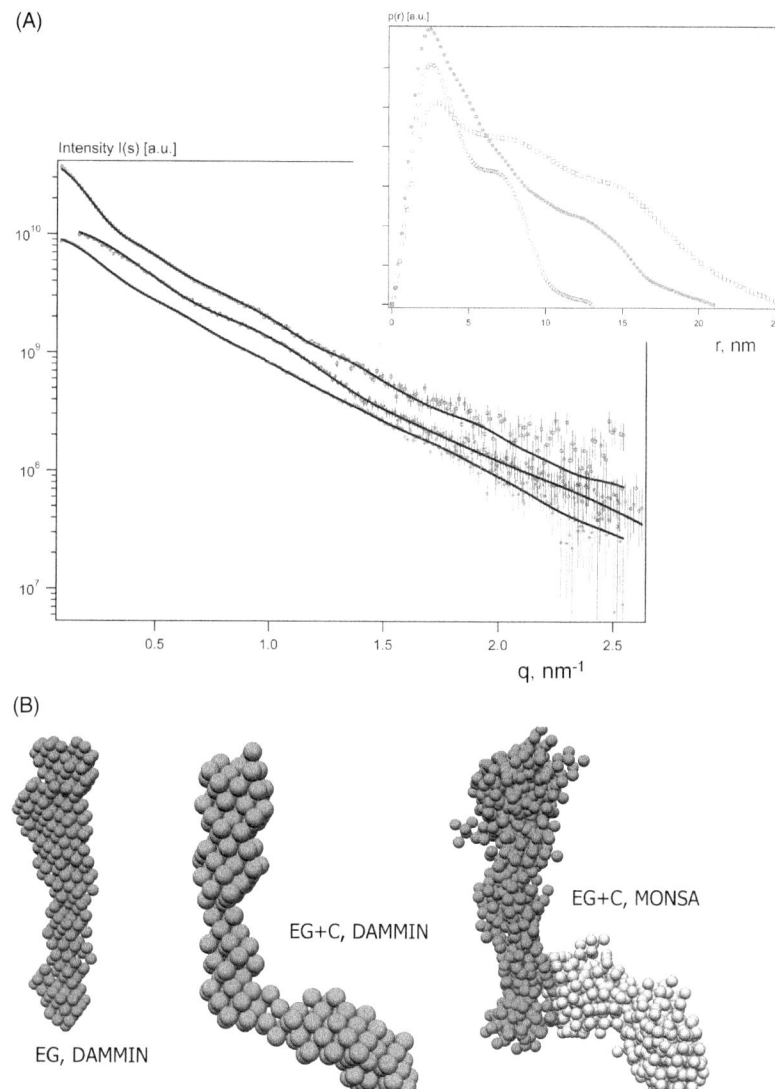

(A)

(B)

EG, DAMMIN

EG+C, DAMMIN

EG+C, MONSA

Fig. 6.5 SAXS analysis of the EGC sub-complexes. (A): SAXS patterns with error bars and fits computed from the *ab initio* models (solid lines). The curves from EGC (1), C (2) and EG (3) are displaced along the ordinate axis for better visualisation The p(r) functions computed from the experimental data are displayed in the insert. The scattering data for the C subunit are taken from (Armbruster *et al.* 2004). (B) *Ab initio* low-resolution models of EG and EGS sub-complexes computed by DAMMIN (left and middle panels, respectively). The MONSA model of the EGC sub-complex (depicting the C subunit in light grey beads), which simultaneously fits the three scattering patterns in (A) is displayed in the right panel (after Diepholz *et al.* 2008).

sub-complex revealing the EG part as a slightly kinked rod, which assembles with C into an L-shaped structure. This model is further supported by the EM data showing two out of three copies of EG linked by subunit-C. Interestingly, the relative arrangement of the EG and C subunits in solution is more open than that in the EM model of the holo-enzyme, suggesting a conformational change of EGC during regulatory assembly/disassembly (Diepholz *et al.* 2008).

6.1.4 Reconstruction of complex shapes

In contrast with some of the earlier shape determination versions, modern *ab initio* methods can also reliably reproduce very anisometric structures, as illustrated by human tropoelastin (Baldock *et al.* 2011).

Tropoelastin is a 60 kDa soluble precursor to elastin, the main elastic protein conveying extensional elasticity to mammalian tissues. Elastin is constructed by an assembly of many tropoelastin monomers that accumulate on a microfibrillar skeleton. Tropoelastin is often believed to be an unstructured protein, largely because models of elasticity invoke an element of disorder within the structure. However, X-ray and neutron scattering on the full length tropoelastin (containing the sequence of exons 2 to 36) yielded Kratky plots compatible with a non-globular but still folded protein, perhaps with some flexible linkers (Fig. 6.6A, B). The *ab initio* shapes reconstructed from SAXS and SANS data were highly reproducible (Fig. 6.6C), and also compatible with the experimental sedimentation data (hydrodynamic properties were computed from the SAS models using HYDROPRO (Garcia de la Torre *et al.* 2000)). Further SAXS analysis was performed on defined fragments that extended from a fixed N-terminus. The shapes obtained for the two deletion mutants (exons 2–18 and 2–24) were fully compatible with that of the full-length protein, and allowed the authors to define the C- and N-termini in the low-resolution models (Fig. 6.6D). Based on this shape, a head-to-tail model was proposed for a nascent elastic fibre assembly. From the methodological point of view it is interesting that GASBOR was employed here to analyse also SANS data, though it is designed for X-ray data only. This approach seemed to work well, largely because of the fact that the neutron data were available in a rather limited range of scattering vectors (Fig. 6.6A).

Another example of determination of a rather complicated and extended shape is given by DC-SIGN—a C-type lectin receptor of dendritic cells. This important component of the immune system has a cytoplasmic region, a transmembrane segment and an extracellular domain (ECD). This last one consists of a neck domain, acting as pH-sensor controlling the tetramerisation of the receptor, and a calcium-dependent carbohydrate recognition (CRD) domain. The solution structure of a construct of the entire ECD and its oligomerisation

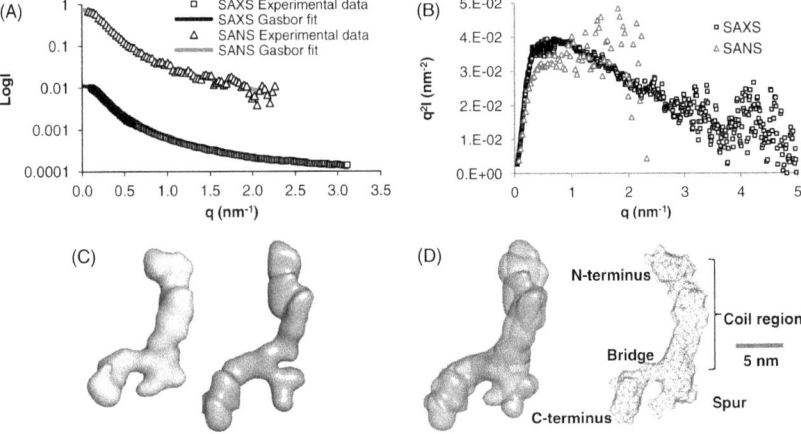

Fig. 6.6 SAXS and SANS analysis of human tropoelastin. (A): Scattering patterns (symbols) and fits computed from the *ab initio* models (solid lines); (B): SAXS and SANS Kratky plots; (C) averaged GASBOR shapes from the X-ray (left) and neutron (right) data; (D) superposition of semi-transparent SAXS and SANS shapes (left) and a labelled diagram of the model for full-length tropoelastin based on the shapes from the deletion mutants. The diagram shows the locations of the N-terminus, of the spur region containing exons 20–24, and of the C-terminus. Scale bar, 5 nm. Adapted with permission from Baldock *et al.* (2011), © 2011, National Academy of Sciences, USA.

were analysed by size exclusion chromatography, AUC and SAXS, while circular dichroism (CD) was used to monitor pH-dependent changes in secondary structure (Tabarani *et al.* 2009). The solution structure of the tetramer (MM ~155kDa) was calculated *ab initio* from the SAXS pattern and the results of several runs were averaged, yielding the model in Fig. 6.7A, consisting of a head in which four CRDs can be accommodated and of the elongated (coiled coil) neck region. This model gives an excellent fit to the experimental SAXS pattern, as illustrated in Fig. 6.7B, and its calculated sedimentation coefficient agrees well with the experimental value.

In many cases, complicated shapes reconstructed from SAS allow one to visualise the domain structure of proteins and using available high-resolution information, these models may allow one to make functional hypotheses. Examples are the analysis of the S-layer proteins from *Clostridium difficile* (Fagan *et al.* 2009) and of the toxin B from this bacterial pathogen (Albesa-Jove *et al.* 2010). Rigid-body modelling provides an even more powerful approach to the positioning of known domains, as illustrated by the numerous examples in Section 6.2.

6.1.5 Use of symmetry

Symmetry is a very important restraint and, though it should be used with caution, it can improve significantly the quality of the *ab initio* models. The dodecameric Ca/calmodulin-dependent kinase II (CaMKII) is a good example (Chao *et al.* 2011). This enzyme responds not only to the amplitude but also to the frequency of the activating signal, and plays a major role in neuronal signalling. In the crystal structure of an autoinhibited full-length human CaMKII holoenzyme the kinase domains are docked against a central hub. The *ab initio* shapes reconstructed with GASBOR using P62 symmetry demonstrate that this unexpected compact conformation is preserved in solution, and also remains unchanged upon addition of the tyrosine kinase inhibitor bosutinib used in the treatment of cancer (Fig. 6.8A–C). In contrast, mutation of residues at the kinase–hub interface leads to a conversion of the enzyme from a compact to an extended state (Fig. 6.8D). Further SAXS analysis of the CaMKII isoforms, which differ in the length of the linker

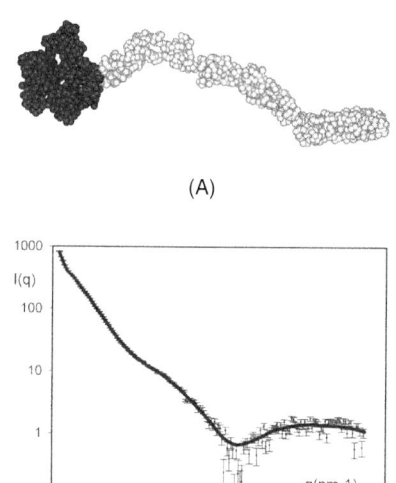

(A)

(B)

Fig. 6.7 Shape of the tetrameric extracellular domain of DC-SIGN. (A) The model calculated *ab initio* from the SAXS pattern in (B). The dark part is the head in which four CRDs can be accommodated and the elongated part is the neck domain (for details see text). To emphasise the agreement between the experimental pattern (dots) and the calculated one (full line) in (B) only half the number of experimental points has been plotted. Adapted from Tabarani *et al.* (2009), with permission from the American Society for Biochemistry and Molecular Biology, © 2009.

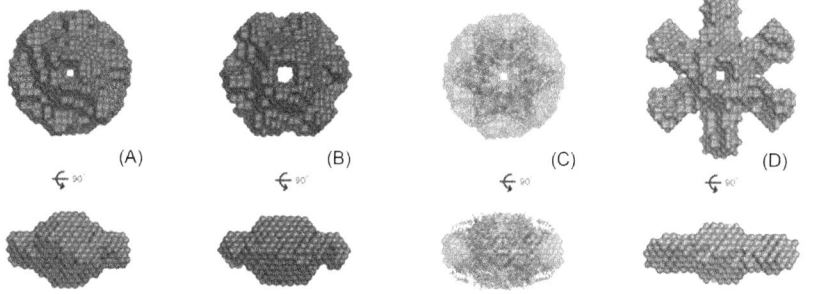

(A) (B) (C) (D)

Fig. 6.8 *Ab initio* shapes of the full-length CaMKII holoenzyme reconstructed under symmetry restriction. (A) Short linker human CaMKII; (B) the enzyme upon addition of the inhibitor; (C) the shape in (A) displayed in semi-transparent beads superimposed on the crystal structure of holoenzyme (ribbons); (D) the extended shape of CaMKII upon central hub mutation. Adapted from Chao *et al.* (2011), with permission from Elsevier, © 2011.

between the kinase domain and the hub, demonstrate that the constructs with longer linker have disc-like conformations similar to that depicted in Fig. 6.8D. It is proposed that the more extended form is able to bind calmodulins, and that the response of the enzyme to the frequency of activating calcium spikes is governed by the equilibrium between the compact and extended conformations.

Overall, symmetry averaging is a powerful tool for visualising macromolecular shapes, and symmetric reconstructions are employed very frequently for oligomeric macromolecules. Highly symmetric structures such as in CaMKII allow one to generate sophisticated models, but in practice lower-symmetry cases occur more often; for example, homodimeric proteins with P2 symmetry, as in study of the yeast Hsp40 proteins Ydj1 and Sis1 (Silva *et al.* 2011). As indicated in Chapter 4, it is always necessary to confront the symmetric models with those obtained in P1 to ensure proper anisometry of the reconstructions. The asymmetric models should display shapes similar to those of symmetric reconstructions even if only at a lower resolution. An example is provided by a study of the quaternary structure of the trifunctional proline utilisation A (PutA) flavoprotein (Singh *et al.* 2011), where both P1 and P2 models revealed an elongated V-shaped molecule.

6.1.6 Shapes of nucleic acids

All previous examples involved proteins, but *ab initio* shape determination is, of course, also applicable to nucleic acids (DNA and RNA). These may even be considered somewhat more favourable objects for shape determination with X-rays due to their higher contrast in aqueous solution compared to that of proteins. This leads to a better signal/noise ratio in the scattering, but also diminishes the relative contribution from the internal structure, which is undesirable for shape analysis (see above Section 6.1.1). Figure 6.9 presents the shape of a free *Thermus flavus* 5S ribosomal RNA in solution (Funari *et al.* 2000). This study was among the first ones where systematic averaging was employed to analyse the distribution of the reconstructed shapes (Fig. 6.9B), and based on its results an automated analysis and averaging procedure was developed (Volkov and Svergun 2003). The average shape of the 5S rRNA depicts a bent elongated molecule with a compact central region and two projecting arms, similar to those of tRNA, and tentative models of 5S rRNA A–D–E and B–C domains in the form of elongated helices can be well positioned within the shape (Fig. 6.9C). *A posteriori* comparison with the crystal structures of 5S rRNA inside the ribosome (see Section 6.2.6) indicated that the ribosomal RNA, not unexpectedly, becomes essentially more compact upon complexation with ribosomal proteins.

The *ab initio* shape determination by SAS became a very popular tool for characterising RNA molecules in solution, and numerous recent applications are reviewed in Rambo and Tainer (2010). Coarse-grained approaches tailored specifically to model the tertiary structure of RNAs based on SAXS data are being developed (Yang *et al.* 2010a).

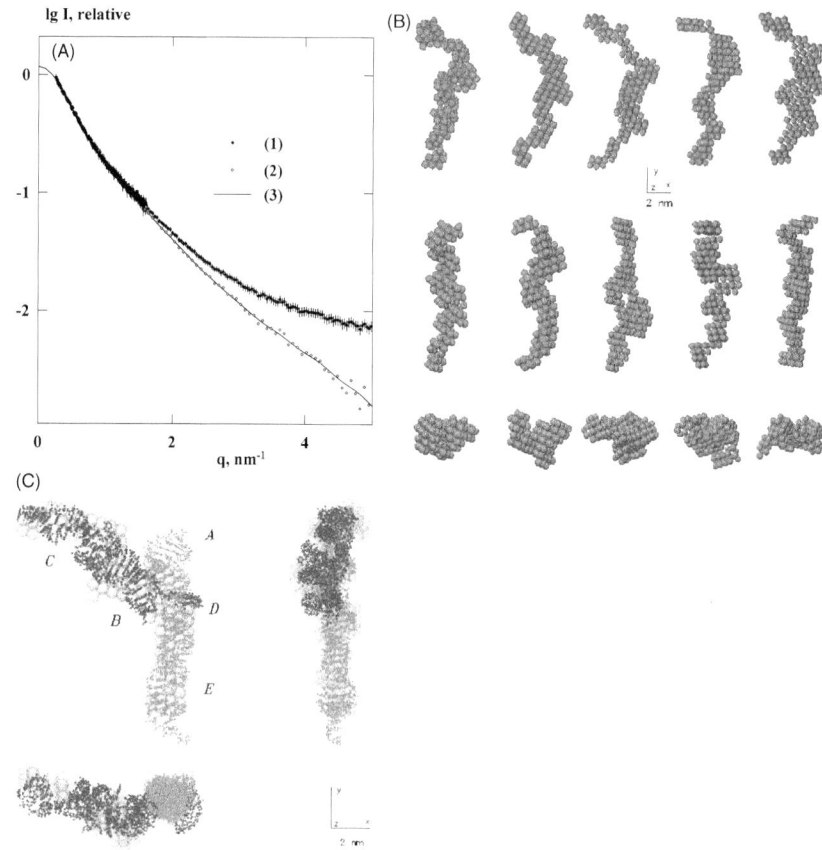

Fig. 6.9 Shape determination of a free *Thermus flavus* 5S ribosomal RNA in solution (Funari *et al.* 2000). (A) X-ray scattering data: (1) experimental curve; (2) shape scattering curve after subtraction of a constant to force the q^{-4} decay; (3) scattering from an *ab initio* bead model. (B) Low-resolution models of the 5S rRNA obtained in five independent shape determination runs (from left to right). The middle and bottom rows are rotated counterclockwise by 90° around the Y- and X-axes, respectively. (C) The final overall shape of the 5S rRNA (semitransparent beads) superimposed with the atomic models of the domains A–D–E (dark dots) and of a fragment containing domains B–C (grey dots). The upper left panel is displayed in the orientation as in Fig. 4A, rotated counterclockwise by 55° in the figure plane. The right and bottom panels are rotated counterclockwise by 90° around the Y- and X-axes with respect to that in the upper left panel. This research was originally published in Funari *et al.* (2000), ⓒ American Society for Biochemistry and Molecular Biology.

The studies presented in this section vividly demonstrate the possibilities and value of shape determination from solution scattering data. The main advantage of this approach is that no or little *a priori* information about the object is required, and very often the low-resolution models already yield sufficient information for drawing functional conclusions. In many cases, especially for complex particles, the use of information from other methods allows one to go further in the interpretation and to answer more complex biological questions, as illustrated in the following sections.

6.2 Quaternary structure analysis of proteins and complexes

In this section we consider SAXS and SANS applications to proteins and macromolecular complexes, where the interpretation was advanced far beyond *ab initio* shape analysis. In most of these cases, high-resolution structures of subunits or domains were utilised to construct hybrid models. For SANS analysis, studies relying on extensive use of contrast variation and selective deuteration are presented.

6.2.1 *Ab initio* and rigid-body modelling

The example of yeast V-ATPase in Section 6.1 illustrates that *ab initio* bead modelling of large systems is similar to that of single chains. For oligomers it is possible to make use of symmetry to simplify the calculations as illustrated in the case of the RuvBL1/RuvBL2 complex (Gorynia *et al.* 2011) or of the enzyme lumazine synthase, which catalyses the penultimate step in the synthesis of riboflavin (Zhang *et al.* 2006). Systems like protein–nucleic acids or protein–lipid complexes can be modelled using a two-phase approach. This is, of course, of greater use when the components present large differences in contrast, as in the case of ribosomes (see Section 6.2.6).

As illustrated in Section 6.1, the methods for modelling single chains or complete complexes are well established, and further developments aim mainly at higher computational speed and ease of use. The situation is different when it comes to integrating structural information about partial structures, isolated subunits, or protein domains which have become available through crystallography, NMR and SAS, into large scale models of multidomain proteins and multisubunit complexes that are often difficult to crystallise. This has led to a renewed emphasis of the advantages of integrating data from independent techniques (MX, NMR, SAS, AUC, EM, FRET and so on) into consistent models (Mattinen *et al.* 2002; Grishaev *et al.* 2005; Foerster *et al.* 2008; Mertens and Svergun 2010; Madl *et al.* 2011). Integration of SAS profiles with computational methods has been discussed in a recent review (Schneidman-Duhovny *et al.* 2012).

The availability of numerous atomic-resolution structures has not only changed the starting point of most modelling attempts but also considerably influenced the demands on modelling software. Initially, modelling of scattering patterns of biological macromolecules and large assemblies, such as the quaternary structure of tubulin assemblies (Bordas *et al.* 1983) or the higher-order structure of chromatin (for a review see Koch (1989)), for example, was largely based on input from EM. In contrast, in recent work one usually starts from models at atomic resolution.

The assumption underlying the construction of large models is that the partial structures behave as rigid bodies (that is, all the distances between pairs of points inside each part remain constant). This is often the case, but should not be taken for granted. Failure to obtain a reasonable fit by rigid body calculations may indicate that large-scale conformational changes have taken place. In this case, specific deuteration and *in situ* SANS observation of a subunit, masking other parts of the structure may provide a way to solve the problem.

6.2.2 Importance of data quality and proper reporting

The plausibility of the final models will, obviously, depend much on the reliability of the partial structures obtained from crystallographic,

NMR or homology modelling. In any case, it is useful to have an experimental SAS pattern of the different components in the assembly and in subassemblies. SAS data alone often do not suffice to interpret the results of rigid body modelling and additional information must be obtained from other methods. This is well illustrated by the study of the complex between ceruloplasmin and lactoferrin (Sabatucci *et al.* 2007). The scattering calculated from the crystallographic models of the individual proteins fitted well to their SAS curves. The *ab initio* envelope of the complex was in good agreement with the results of rigid-body modelling with the crystal structures. As the symmetry of the proteins precluded an unambiguous determination of their relative orientation, independent biochemical information was used to discriminate between possible models. Interestingly, docking programs available at that time failed to produce models compatible with the SAS pattern.

One should also be aware of the effect of restraints on the results. Not unexpectedly, methods that impose strong restraints making use of available structural information, tend to perform better than *ab initio* reconstructions when the quality of the SAS data is lower (see, for example, Nöllmann *et al.* (2005)). This should, however, not be taken as a sign of better performance of a method, but as a reminder of the importance of collecting the best possible experimental data.

To collect reliable data for increasingly demanding projects on complexes it is advisable to connect a size-exclusion chromatography column directly to the sample, making use of the short recording times (sub-second range) on modern instruments. The advantage of doing so is well illustrated in a study of different β-thymosin-actin complexes combining crystallography, SAXS and NMR (Didry *et al.* 2012), where small conformational changes detected by SAXS, which might otherwise have gone unnoticed, could be correlated with functional differences observed by biochemical methods.

This importance of the quality of experimental data and the need to properly report the exact procedures and external information used during modelling has recently been emphasised in the IUCr publication guidelines for SAS data from biomolecules in solution (Jacques *et al.* 2012). Provision of proper information is, of course, also crucial for comparing the results obtained with different programs.

6.2.3 Quaternary structure of proteins

The components of the immune system present some of the most challenging structural problems. These have been studied extensively by combining SAS with hydrodynamic methods and constrained (rigid-body) refinement. A good example is provided by secretory immunoglobulin A (SIgA1, MM ~420 kDa)—one of the most important components of the immune system (Bonner *et al.* 2009). High glycosylation, multiple domains and flexible hinge and tailpiece regions make it an unlikely candidate for crystallisation. Starting from known structures

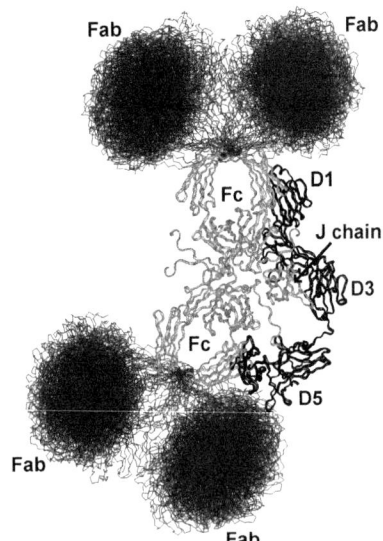

Fig. 6.10 Model of secretory immuno-globulin A (SIgA1) constructed by constrained refinement. The location of three of the five immunoglobulin variable type domains of the secretory component (D1, D3 and D5) domains and the J-chain are indicated. The Fab fragments represent the best fit of different trial conformations. Adapted from Bonner *et al.* (2009), with permission from Nature Publishing Group, © 2008.

of dimeric IgA (dIgA), the best location of the secretory domain (SC) in terms of radius of gyration and fit between the calculated and experimental SAXS and SANS data was first obtained. Subsequently, the best-fit structure was refined by randomisation of the four Fab hinges in the SIgA1 model. The final model in Fig. 6.10 represents a limited best-fit conformational set of structures that also yields a calculated sedimentation coefficient in excellent agreement with the experimental one. In this modelling approach (reviewed in (Perkins *et al.* 2011)) the parts of the structure for which crystallographic models are available are represented by beads of about 0.5 nm diameter to obtain an unhydrated model to fit neutron scattering curves. Hydration spheres are added to obtain fits to the X-ray patterns (Ashton *et al.* 1997).

It should be noted here that bead models used in SAS may not always be directly usable for hydrodynamic simulations (see Section 9.5.1). Different programs use various numbers of spheres with different sizes for a coarse-grained representation of the structure. Sometimes two spheres rather than a single DR are used for each residue—one for the main chain segment and another for the side chain. This leads to small systematic deviations in the calculated scattering patterns, requiring the introduction of correction factors, especially at higher q-values (see, for example, Grishaev *et al.* (2005)), which can be avoided by using explicit residue structure factors (Yang *et al.* 2009). It has been suggested that differences between results obtained with X-ray and NMR bead models in hydrodynamic modelling are related to the role of side chain flexibility on these properties, but that the effects are more important for small macromolecules (Rai *et al.* 2005).

The synergy between SAS and EM has evolved in recent years, helping one to analyse quaternary structure of proteins. SAS data can be used to scale the amplitudes of cryoelectron micrographs to improve the quality of the images (Schmid *et al.* 1999; Thuman-Commike *et al.* 1999). SAS models provide a simple way to verify the quaternary structures obtained by EM, as shown in the case of P-glycoprotein (ABCB1) (McDevitt *et al.* 2008) or of α-crustacyanin, where the results helped to resolve a controversy concerning the structure of the protein (Rhys *et al.* 2011). Here, the best agreement with the SAXS data was obtained by rigid-body modelling, independently fitting eight crystallographic models of β-crustacyanin dimers to the α-crustacyanin data using SASREF. The scattering patterns calculated from the EM model obtained by negative staining with single particle averaging or from a model obtained by extracting eight β-crustacyanin dimers from the crystal lattice yielded scattering patterns with systematic deviations. Other examples illustrating the advantages of combining SAS, EM and other methods for protein complexes can be found in a study suggesting a helical conformation in the Titin 11-domain super-repeat (Tskhovrebova *et al.* 2010), in extensive studies of the human pyruvate dehydrogenase complex core assembly (Vijayakrishnan *et al.* 2010) or of the post-translational circadian oscillator complex of *Synechococcus elongatus* (Pattanayek *et al.* 2011). Section 9.2 provides further examples of the synergistic use of SAS with other structural, biophysical and computational techniques.

The oxygen transporters (haemoglobin, haemocyanin) are probably the class of proteins whose quaternary structure has attracted most attention. Hitherto, modelling seems to have been largely limited to the use of ellipsoids to fit the scattering patterns of these sometimes very large ($\sim$2 MDa) proteins (Beltramini *et al.* 1996). Changes in the quaternary structures of the oxygenated and deoxygenated forms of the 4×6 haemocyanin of the tarantula *Eurypelma californicum* found that the oxy form is less compact than the deoxy form (Decker *et al.* 1996). Structures of both forms were constructed by fitting SAXS patterns on the basis of know crystallographic structures and EM results for related proteins (Hartmann and Decker 2002). The differences between the two structures revealed that the oxy–deoxy transition involves movements at all hierarchical levels of the quaternary structure. As the protein has recently been crystallised (Jaenicke *et al.* 2012) it should soon be possible to verify the findings and to detect any possible differences between the situation in solution and in the crystal.

Haemocyanins of different organisms have very different structures, and it is not clear how they are built up from their functional units. *Ab initio* and rigid-body modelling of fragments containing one to three functional units of a gastropod (*Rapana venosa*) haemocyanin (Sabatucci *et al.* 2005), using the crystal structure of a single functional unit, revealed that the end product of reconstitution depends on the mode of preparation of the functional units. This also illustrates that SAS is an indispensable quality-control tool in (re)assembly studies.

6.2.4 Nucleoprotein complexes and macromolecular machines

Protein–nucleic acid interaction studies have much benefited from SAS (for a recent review see, for example, Rambo and Tainer (2010)). Many proteins that interact with nucleic acids are multimers having disordered regions, at least when not bound to their targets. Examples of *ab initio* modelling of oligomers include the hexameric replicative helicase of some viruses (Whelan *et al.* 2012) and the hexameric *E. coli* RNA chaperone Hfq (Beich-Frandsen *et al.* 2011). In both cases the missing parts in crystal structures could be completed using BUNCH, leading to an improved agreement between the SAXS patterns and those calculated from the models.

AUC and hydrodynamic modelling are useful sources of additional information for the modelling of nucleoprotein complexes (see Section 9.5.1). This is clearly illustrated by a SAXS/SANS study of the Tn3 resolvase-crossover site synaptic complex (Nöllmann *et al.* 2004). The X-synapse, which is responsible for catalysis of strand exchange, consists of two DNA crossover sites (each about 30 base pairs) held together by a synaptic tetramer of resolvase subunits. Extensive rigid body simulations using crystallographic models

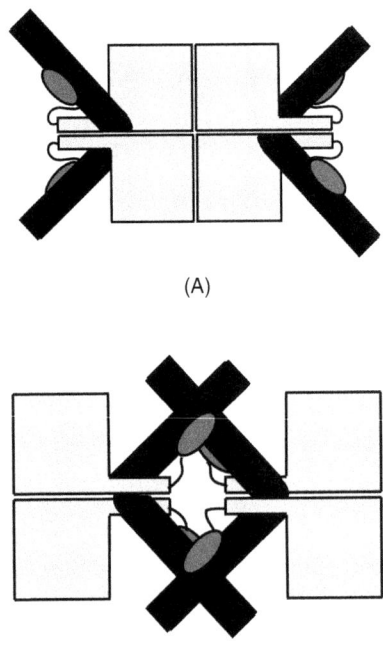

(A)

(B)

Fig. 6.11 Schematic models of the DNA-out (A) and DNA-in (B) models of the Tn3 resolvase-crossover site synaptic complex. The grey parts represent protein subunits in the tetramer and the black bent cylinders correspond to fifty base pair stretches of DNA. Adapted from Nöllmann *et al.* (2004), with permission from Elsevier, © 2004.

of B-DNA and of γδ-resolvase (Yang and Steitz 1995) revealed that the DNA-out model in Fig. 6.11A fitted the combined experimental data better than the alternative DNA-in model proposed earlier on the basis of crystal packing considerations in Fig. 6.11B (Rice and Steitz 1994). This was later confirmed by the crystal structure of a synaptic γδ-resolvase tetramer covalently linked to two cleaved DNAs (Li *et al.* 2005).

Shape determination of DNA-bound gyrase (Baker *et al.* 2010) illustrates a very different and somewhat more speculative approach where known crystallographic structures were manually placed in an envelope determined *ab initio* with DAMMIN. This process, which also integrated knowledge of symmetry and EM information, was complemented by parallel calculations with SASREF. As an additional verification hydrodynamic simulations with the final model yielded a good fit to the experimental values. Another example illustrating the advantage of combining SAXS, SANS and EM for the study of the structure and operation of nucleoprotein complexes is provided by the DNA-translocating type I DNA restriction enzymes (Kennaway *et al.* 2012).

Modelling of large nucleoprotein complexes is greatly facilitated by the information gained from contrast variation studies. These are possible with X-rays using, for example, sucrose, as done for the 50S ribosomal subunit from *Escherichia coli* (Svergun *et al.* 1994), but the full power of contrast variation is revealed in neutron scattering studies using H_2O/D_2O mixtures and/or specific perdeuteration. Here we shall consider the applications of neutrons (often after or in combination with X-rays) to study large macromolecular complexes made up of either proteins only or of proteins and nucleic acid or lipids. The SANS applications to membrane proteins and lipoproteins will be presented in Section 6.4. In the present section we shall deal with viruses, nucleosomes, ribosomes and chaperones.

Before considering them we shall mention SAS work on another important nucleoprotein complex, chromatin, which is composed of fundamental repeat units, the nucleosome core particles (Kornberg 1974). These contain about 146 base pairs of DNA, complexed with two copies each of four basic proteins, the histone octamer. Twenty years before the first near-atomic X-ray crystallographic structure of nucleosomes was published (Luger *et al.* 1997), neutron and X-ray solution studies showed the overall dimensions of the nucleosome, and were the first to demonstrate that the DNA was situated on the outside of a protein core (Hjelm *et al.* 1977, Pardon *et al.* 1977). Nucleosome core particles also were the first complexes to be studied successfully using contrast variation in single crystals (see Section 9.4).

6.2.5 Viruses

Viruses are made of nucleic acids (DNA or RNA), a protein coat and in some cases a lipid envelope, and have dimensions of 20–400 nm.

Although they had been studied with EM, X-ray crystallography and SAXS for a long time (see the example of the bacteriophage T7 in the Introduction), in particular the location of the nucleic acids (RNA or DNA) and the lipids (as far as they are present) remained unclear. Therefore, viruses were among the first biological objects to be studied with SANS and also neutron low-angle crystallography (see Section 9.3)—techniques ideally suited to study their composition and structures. Since it is easy to determine molecular masses by SANS (Jacrot and Zaccai 1981), one can, for example, obtain the number of copies of protein monomers that make up the protein capsid of viruses.

The first SANS studies on viruses dealt with processes in small plant viruses, such as a change of the localisation of RNA upon swelling of bromegrass mosaic virus following an increase in pH (Jacrot *et al.* 1976), or with comparative studies of small spherical RNA viruses (Jacrot *et al.* 1977). Such studies showed differences in the protein–nucleic acid distribution. In the Tomato Bushy Stunt Virus (TBSV) there is a bilobal radial distribution of protein (Chauvin *et al.* 1978), whereas the distribution is more homogeneous in the Southern Bean Mosaic Virus (SBMV) (Kruse *et al.* 1982).

Later, much larger (animal) viruses, such as the influenza virus, with a diameter of 120 nm, or adenovirus, were investigated with SANS. Such particles were originally less well characterised than the smaller plant viruses, and even parameters such as molecular mass were poorly known. Neutron scattering allowed the molecular mass of adenovirus to be determined, as well as the radial distribution of DNA and protein (Devaux *et al.* 1983). The influenza virus contains a lipid bilayer and has glycoprotein spikes. Therefore, a new method for deriving spherical shell models was developed, and the position of the lipid bilayer was precisely determined. The molecular mass was redetermined, and at the same time the chemical composition and the protein/lipid ratio (Mellema *et al.* 1981; Cusack *et al.* 1985).

Recent SANS studies describe the differences in the structures of Sindbis virus depending on whether it is grown in mammalian or insect cells (He *et al.* 2010), and the action of pH on the virus structure (He *et al.* 2012). The authors conclude that more cholesterol is incorporated in the mammalian-grown form, the outer protein is more extended, and there is a closer interaction between the RNA and the nucleocapsid protein. Lowering the pH from 7.2 to 6.4, the RNA undergoes a conformational change, and the protein is redistributed. Earlier, the Sindbis virus, which causes Sindbis fever and is transmitted by mosquitoes, had also been studied by SAXS (Stubbs *et al.* 1991).

6.2.6 Ribosomes

Ribosomes are the protein factories in all living cells (Rodnina *et al.* 2011). Due to their importance in the life cycle, the possibility of obtaining structural information by SANS was immediately exploited after the first SANS instruments, such as D11 at the ILL, Grenoble, France (Ibel 1976), became available.

Most neutron studies concern the ribosome of the bacterium *Escherichia coli*, which has two subunits—a larger one called 50S (after the Svedberg units measured by analytical centrifugation) and a smaller one, 30S. Together they form the 70S particle. Two thirds of the ribosome's mass consists of RNAs—two large chains (16S in the 30S and 23S in the 50S subunit), and the small 5S RNA in the 50S; the other third consists of about 50 individual proteins.

Separation of RNA and proteins. The early SANS studies dealt with the gross distribution of RNA and protein in the subunits and in the whole ribosome. A first publication concluded that there were protein and RNA-rich sides in the 50S subunit separated by about 6 nm (Moore *et al.* 1974). This was soon followed by a report indicating that the RNA is mainly in the centre of the particle, surrounded by a shell of protein, the centres of both moieties being separated by about 2 nm (Stuhrmann *et al.* 1976). In a joint study the two groups confirmed this last result using contrast variation on specifically deuterated *E. coli* ribosomal subunits and explained the cause of the initial discrepancy (Crichton *et al.* 1977).

Overall models of the ribosome. In one of the earliest applications of spherical harmonics to shape determination, the shape of the 50S subunit was modelled (Crichton *et al.* 1977). The scattering at infinite contrast, extrapolated to zero concentration, was extracted from a neutron contrast and concentration series, and the data were fitted by a Fourier transform of the shape function described by spherical harmonics; for the mathematics see the original paper or Section 4.4.1.

A model of the 30S subunit of the *E. coli* ribosome based on the protein positions and the subunit shape as obtained by EM, and on the scattering of the particles by X-rays and by neutrons (four contrasts), was published in (Spirin *et al.* 1979). A study of the 30S subunit from *Thermus thermophilus* employing spherical harmonics was presented in (Fan *et al.* 2000).

Protein positions within the ribosome. As the prokaryotic ribosome contains more than 50 individual proteins—only a few of them are multiple—it is possible to detect their positions in the two subunits or the whole 70S. Two or three decades before the crystallographic structures were solved (and one did not even imagine that this would ever be feasible) (Ban *et al.* 2000, Schluenzen *et al.* 2000, Wimberly *et al.* 2000), the only ways to determine the protein positions were immuno-EM and SANS.

Immuno-EM employs antibodies directed against specific ribosomal proteins, and detects their position by EM (Stöffler and Stöffler-Meilicke 1984). This method allowed one to obtain information about surface sites of the proteins at low resolution.

With the label triangulation method (see Section 4.7), however, it was possible to identify the centre-of-gravity locations of ribosomal proteins within the 30S and 50S subunits. A complete map of all twenty-one

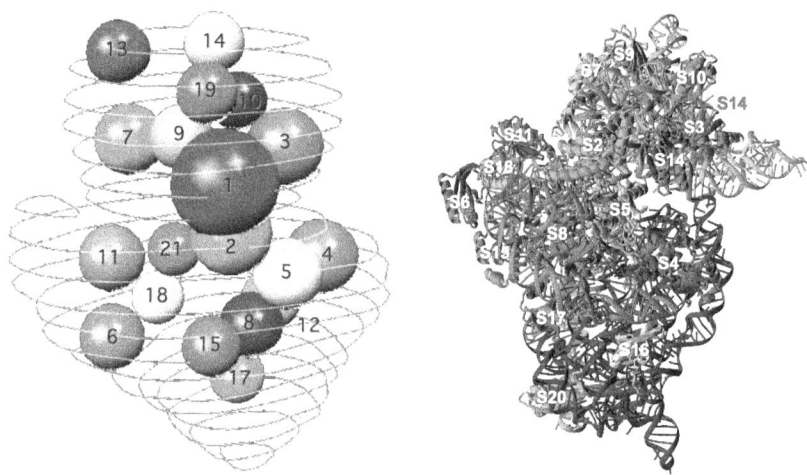

Fig. 6.12 Protein triangulation in the small ribosomal subunit. Left panel: positions of the twenty-one proteins in the ribosomal 30S subunit of *E. coli* as determined by SANS (Capel 1987). Picture courtesy M. Capel, modified. Proteins are drawn as spheres with radii corresponding to the molecular mass. S20 is hidden behind S1, S16 behind S2. Right panel: Crystal structure of the 30S subunit with the protein positions (Wimberly *et al.* 2000). Picture courtesy V. Ramakrishnan, labelled according to Brodersen *et al.* (2002). S12 is hidden, roughly behind S5; S13 is hidden behind S9 and S10.

proteins of the reconstituted 30S subunit of *E. coli* ribosomes was published by Capel *et al.* (1987). Only five of the protein positions determined by label triangulation deviated from the X-ray crystallographic structure (Wimberly *et al.* 2000), and only one of them (S20) was found at a completely different position (Fig. 6.12).

In the study of the 50S subunit using the 'glassy-ribosome' technique (Nierhaus *et al.* 1983), and the 'moving-splines' approach (May and Nowotny 1989, May *et al.* 1992) determined fifty protein distances that yielded a model with the positions of seven proteins within the 50S subunit. The suppression of the contrast between proteins and RNA also allowed the authors to determine (approximate) radii of gyration of H-labelled 50S proteins *in situ* (Nierhaus *et al.* 1983, Nowotny *et al.* 1994).

The map was expanded to fourteen protein positions in the 50S subunit (Willumeit *et al.* 2001), and the protein radii of gyration were determined *in situ*, using spin-contrast variation (Knop *et al.* 1992). This technique requires freezing the samples to very low temperatures (∼1 K) so that their spins can be polarised by dynamic nuclear polarisation, and the samples studied with polarised neutrons. This makes it possible to change the contrast within one sample without modification of the solvent. The levels of contrast that can be reached are substantially higher than those obtained in classical solvent or solute contrast variation.

Multiphase SAXS/SANS model of the ribosome. In a comprehensive study of the ribosome (Svergun *et al.* 1997, Svergun and Nierhaus 2000), a set of forty-two synchrotron X-ray and neutron solution scattering curves from hybrid *E. coli* ribosomes was obtained (Fig. 6.13), where the protein and rRNA moieties in the subunits were either protonated or deuterated in all possible combinations. This is probably the most extensive set of consistent X-ray and neutron contrast-variation data collected on a single object.

Fig. 6.13 Low-resolution model of the 70S ribosome. (A) and (B), neutron and X-ray scattering data, respectively, from hybrid 70S ribosomes and free subunits fitted by a four component dummy atoms model. (A), H and D denote protonated and deuterated components, respectively, whereby the first letter is related to proteins, the second to RNA (for example, HH30 + DH50 describes a particle with fully protonated 30S subunit and the 50S subunit with proteins deuterated, proteins protonated). 'Spin contrast' denotes the data obtained by spin-dependent contrast variation (Stuhrmann and Nierhaus 1996); the upper six curves were collected on free ribosomal subunits. The experimental data are presented as dots with error bars, the fits as solid lines. Successive curves are displaced up by one logarithmic unit corresponding to the distance between the ordinate tick marks (in panel A, also by $\Delta q = 0.05\,\text{nm}^{-1}$ along the abscissa) for better visualisation. The plots are adapted from Svergun and Nierhaus (2000), © American Society for Biochemistry and Molecular Biology. (C) The multiphase bead model of the 70S ribosome compared with later high-resolution crystallographic models. Top panel, 30S, bottom panel, 50S. Proteins are depicted in dark grey, ribosomal RNA in light grey. The bead radius in the SANS/SAXS models is $r_0 = 0.5\,\text{nm}$. The crystallographic models are those of *Th. thermophilus* 30S subunit, resolution 0.33 nm (Schluenzen et al. 2000), and of 50S subunit from *H. marismortui* (Nissen et al. 2000), resolution 0.24 nm (note that the peripheral proteins L1 and L7/L12 are not seen in the crystal). In each panel, upper row displays the solvent view; lower row displays the interface view.

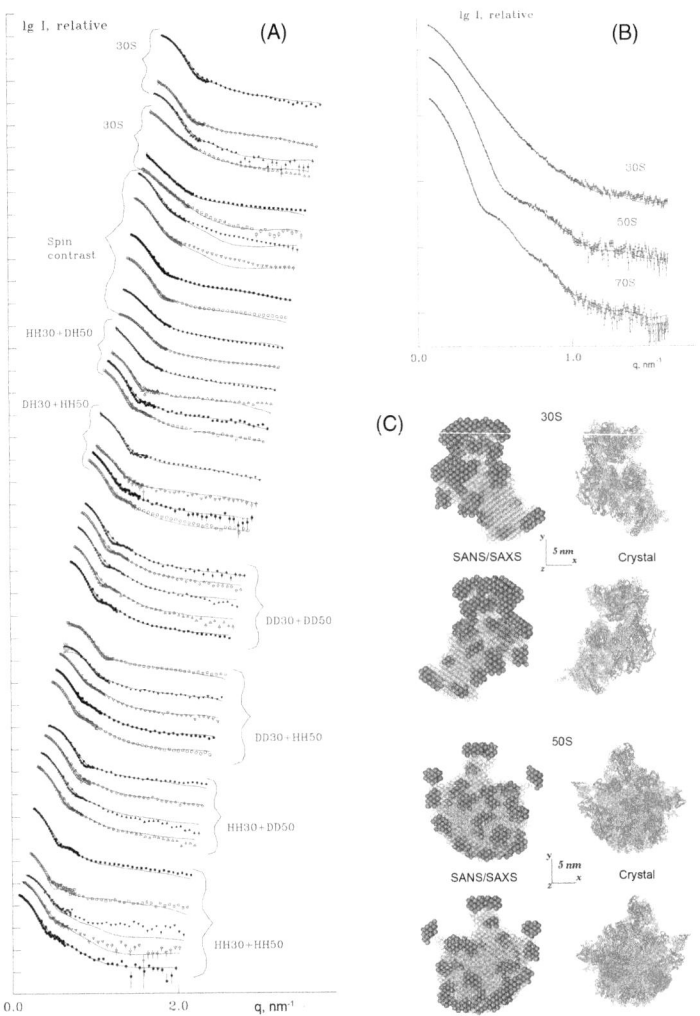

This data set was analysed using the multiphase dummy-atoms technique (program MONSA) described in Section 4.3.2. The search volume of the entire 70S ribosome defined by the cryo-EM model (Frank et al. 1995) was divided into 7890 densely packed spheres of radius 0.5 nm. MONSA was employed to assign each sphere to solvent, protein or RNA to simultaneously fit all scattering curves (Fig. 6.13A, B). Twelve independent reconstructions starting from random approximations yielded reproducible results, and were averaged to yield the models displaying fifteen and twenty protein sub-volumes in the 30S and 50S subunit, respectively, connected by RNA. Figure 6.13C illustrates the comparison of the map obtained from solution scattering with the high-resolution X-ray crystallographic maps of the ribosomal subunits, which became available later (Ban et al. 2000,

Schluenzen *et al.* 2000, Wimberly *et al.* 2000). The positions of protein globules predicted from solution scattering agree astonishingly well with the crystallographic results—especially given that the resolution of the neutron-derived map was only 3 nm. This *a posteriori* comparison further underlines the potential of contrast variation and the joint use of X-ray and neutron scattering in the study of large macromolecular complexes.

6.2.7 Chaperones

Chaperones are protein complexes that provide an environment assisting non-native proteins to find their native structures. The best-known chaperone, GroEL of *E. coli*, has the form of a cylinder composed of two 7-rings. Protein folding is powered by cycles of ATP binding and hydrolysis. ATP binding to GroEL triggers encapsulation and folding of the substrate in the central cavity sealed off by GroES.

Structural studies of GroEL/GroES complexes. In parallel to an X-ray crystallographic structural determination of the asymmetric 1:1 ('bullet') complex of GroEL and GroES (Xu *et al.* 1997), the formation of complexes of GroEL, GroES and substrate proteins, as well as that of the 1:2 ('football') complex of GroEL and GroES, was observed by SANS (Stegmann *et al.* 1998). The 1:1 complex in solution tallies with that observed in the crystal. The amino acids that are missing in the crystal structure could be placed in the centre of GroES, and the molecule in solution appears to be slightly more compact than in the crystal. The solution structure of GroEL single-rings and their complex with GroES was determined by Holzinger (2002). In a titration study, the formation of the symmetrical GroES-GroEL-GroES complex was also followed using deuterated GroES and H-GroEL matched by 40% D_2O, thus only observing the GroES–GroES interferences. Kinetic studies of GroEL/GroES complexes are described in Chapter 7.

Thermosome structure. The chaperonin 'thermosome' of the thermophilic archaeon *Thermoplasma acidophilum* works without a co-chaperonin. It had been observed in a GroEL-like cylindrical conformation by EM (Nitsch *et al.* 1998), and in a spherical conformation by X-ray crystallography (Ditzel *et al.* 1998). Both structures are compared and their theoretical scattering curves compared to the measured data in Fig. 6.14.

In a study of the influence of buffer conditions, nucleotide composition, and temperature on the thermosome structure, it was found that the spherical shape of the thermosome in the crystal structure is induced by sulphate in the crystallisation buffer (Gutsche *et al.* 2000). In physiological conditions, Adenosine diphosphate (ADP) induces a closed thermosome conformation in the presence of inorganic phosphate; ATP induces this conformation only at high temperatures, but not at room temperature (Gutsche *et al.* 2001).

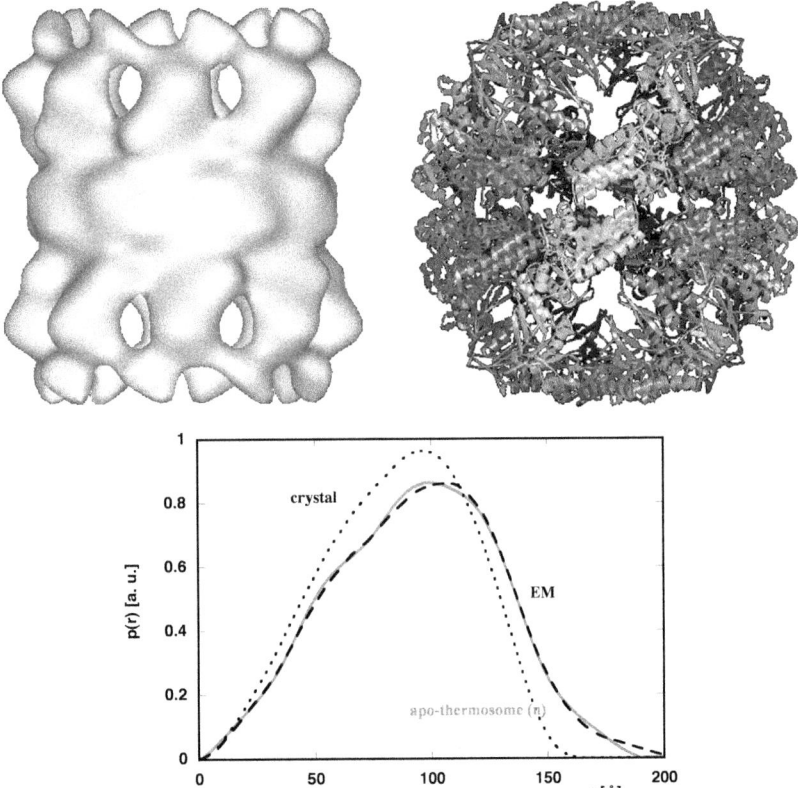

Fig. 6.14 The thermosome as seen by EM (left) (reproduced from Nitsch *et al.* (1998), with permission from Macmillan Publishers Ltd, © 1998) and by X-ray crystallography (reproduced from Ditzel *et al.* (1998), with permission from Elsevier, © 1998). Bottom: The experimental neutron distribution function (grey) reveals that the solution scattering resembles the open (EM) structure in most buffer and temperature conditions. Adapted from Gutsche *et al.* (2000), with permission from Elsevier, © 2000.

6.3 Equilibrium mixtures and oligomeric composition

In the previous sections we considered applications of SAXS and SANS to analyse the scattering from monodisperse systems, with the major aim of reconstructing the low-resolution structure of macromolecules and complexes. SAS is also extremely useful in characterisation of polydisperse systems, especially mixtures, described in terms of the volume fractions of their components. The data analysis methods depend on the kind of the heterogeneity in the system as presented in Chapter 5, and below we shall review typical applications of SAS to this type of problem.

6.3.1 Simple oligomeric equilibrium

Perhaps the most common and practically very important applications are related to equilibrium oligomeric mixtures. One of the typical scenarios is that the high-resolution structure of a monomer is available from crystallography, but the biologically active species are higher oligomers. These may or may not exist in the crystal, and can coexist in solution together with the monomers and/or other species.

The composition—that is, the volume fractions of components in the mixture—often changes with the solute concentration, but may also depend on other conditions (temperature, ionic strength, pH and so on). SAS provides a reliable way of characterising quantitatively the behaviour of such mixtures.

There are numerous SAS studies of the simplest case of a monomer–dimer equilibrium, when the structures of the monomer and of the dimer are known. An example of a comprehensive analysis is given by a study of several monomeric and dimeric kinesin constructs from *Homo sapiens* and *Drosophila melanogaster* (Kozielski *et al.* 2001). The kinesin monomers consist of a globular head and a helical tail responsible for dimerisation. In the crystal, kinesin from *Rattus norvegicus* (Kozielski *et al.* 1997) revealed a dimer with the two heads related by a rotation of about 120° around the axis of the α-helical coiled-coil, and not, as expected, by a two-fold symmetry. SAS analysis demonstrated that the organisation of the *H. sapiens* and *D. melanogaster* dimers in solution is similar to that of the rat kinesin in the crystal. Moreover, the propensity to dimerise was directly related to the length of the helical tail and to the protein concentration. Thus, for the intermediate length *Drosophila* construct (365 residues), the several data sets collected in the concentration range 1–10 mg/mL could be fitted by a mixture of monomers and dimers (see the models and the scattering patterns in Fig. 5.3). The construct was mostly monomeric at 1 mg/mL and fully dimeric above 11 mg/mL, with a tendency to form higher oligomers at yet higher concentrations.

6.3.2 Multicomponent oligomeric mixtures

Situations where several types of oligomers may coexist, and where their relative amounts should be established, are more challenging. A good example is provided by the oligomerisation of the 6.5 kDa protein Bovine Pancreatic Trypsin Inhibitor (BPTI) (Hamiaux *et al.* 2000). Crystallisation of BPTI at acidic pH in the presence of thiocyanate, chloride and sulphate ions leads to three different polymorphs in P21, P6422 and P6322 space groups, all containing a compact decamer with ten BPTI molecules organised through two perpendicular two-fold and five-fold axes. In contrast, only monomeric crystal forms are observed at basic pH. To check whether smaller oligomers also exist in solution and whether the oligomers are stable species and not just crystallisation intermediates, SAXS measurements were performed on undersaturated and supersaturated BPTI solutions at pH 4.5 in the presence of the three anions (Fig. 6.15). The data were analysed in terms of mixtures of monomer, dimers, pentamers and decamers, as identified within the crystal structures. A non-linear fitting procedure to obtain the best global fit to the experimental data demonstrated that only monomers and decamers coexist in solution, as subsequently confirmed by gel filtration. The volume fraction of decamers was found to increase with salt concentration; that is, when reaching and crossing the solubility curve. This suggests a two-step process for the crystallisation of BPTI

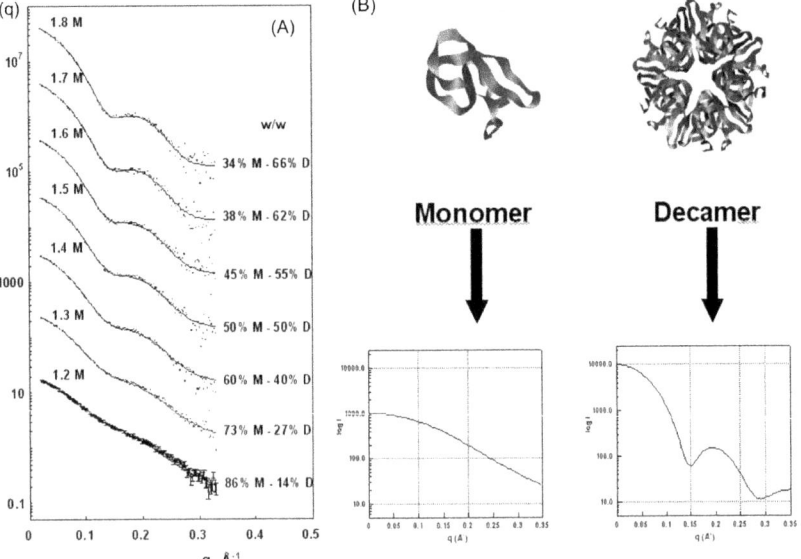

Fig. 6.15 BPTI decamer formation. (A) Scattering patterns of 30 mg/mL BPTI solutions containing at NaCl concentrations from 1.2 to 1.8 M (dots, experimental intensities, continuous lines, best fits by linear combinations of the scattering from monomer and decamer. (B) Ribbon representation of monomeric and decameric BPTI and their calculated scattering patterns. Adapted from Hamiaux *et al.* (2000), with permission from Elsevier, © 2000.

at acidic pH whereby decamers first form in under- and supersaturated solutions followed by the growth of a kind of 'BPTI decamer' crystal. Interestingly, the same binding for all three anions is located at a dimer–dimer interface in the decamer, and the strongest dimer–dimer interaction is found between the two BPTI molecules involved in an inter-pentamer dimer. Such dimers are thus likely to form in solution as protein–protein attractions increase with anion concentration, followed by association of two such dimers through anion binding and cooperative formation of the decamer by association of five dimers around a five-fold symmetry axis. Given that no dimers are present in solution, the decamer assembly must be a very cooperative process, whereby the dimers interact with anions immediately upon formation, to associate into a decamer.

A search for the possible oligomers in mixtures is also possible if crystal structures are not available, and such a procedure is significantly facilitated if supported by data from independent methods. An example is given by the study of chemokine CCL5 (Wang *et al.* 2011)—an 8.4 kDa protein known to activate leukocytes. The monomeric form of CCL5 suffices to cause cell migration *in vitro*, but its propensity for aggregation is essential for important *in vivo* functions including T-cell activation and apoptosis, and HIV entry into cells. The solution structure of a CCL5 oligomer was investigated using a combination of techniques. NMR RDC data were used to define allowed relative orientations of the monomers, and interface residues were identified by hydroxyl radical footprinting and NMR cross-saturation experiments. Based on these data, CCL5 dimers were constructed, and the overall shape from SAXS was employed to find the best translational placement of the dimers within the tetramer. The

SAXS data further revealed an equilibrium between tetramers and higher linear aggregates, mostly hexamers. This led to the hypothesis that the observed aggregation of CCL5 in a linear array probably facilitates movement of leukocytes along cell surfaces and pathogen entry *in vivo*.

An interesting example is given by a study of a dynamic complex between two redox proteins: adrenoxin and cytochrome C (Xu *et al.* 2008). The paramagnetic NMR spectroscopy data reveal a rather broad contact area between the two proteins, and suggest that the complex may be purely stochastic. Using SAXS, no complex formation is observed at low concentrations (below 5 mg/mL), where the two proteins stay monomeric. Above 20 mg/mL, distinct tetramers are formed, whereas the solutions at intermediate concentrations clearly contain lower-orger oligomers. Very interestingly, the intermediate concentration data can only be fitted if one accepts not only heterodimers, but also trimers in solution. Given that for specific dimers no trimer formation is possible (the binding sites are blocked), the very fact of the existence of trimers indicates that one deals with an encounter complex without specific contact interface. The conclusions from the SAXS analysis provide independent and conclusive evidence corroborating the NMR results.

6.3.3 Equilibria between defined conformational states

Decomposition into individual components is also applicable in the analysis of equilibria between different conformational states of the same macromolecule. A typical example is the characterisation of an equilibrium between open and closed states of an enzyme, as in the case of domain movements in phosphoglycerate kinase (PGK) (Zerrad *et al.* 2011)—an enzyme responsible for the first ATP-generating step of glycolysis. The crystal structures of PGK trapped in several conformational states are available, and the SAXS data collected on the enzyme in complex with substrate and product analogues were interpreted in terms of mixtures of these different conformers. It was found that the mixtures mostly contained open enzymes indicating that PGK appears to spend most of the time in a fully open conformation and only short periods in the closed state. Based on these results the authors suggested a spring-loaded release mechanism regulating the domain movement, catalysis and efficient product release.

Equilibria between open and closed states can be meaningfully addressed by SAS revealing both quaternary structure changes and shifts in the volume fractions of the conformers, as illustrated by a study of conformational changes in the Hsp90 molecular chaperone upon substrate binding (Street *et al.* 2011). The authors investigated a model system in which Hsp90 interacts with a partially folded fragment (Δ131Δ) of staphylococcal nuclease. Under apo conditions Hsp90 was found to partially close around Δ131Δ while in the presence of a non-hydrolysable ATP analog, AMP-PNP, Δ131Δ was binding with an increased affinity to Hsp90's fully closed state. For the examples of the analysis of multiple conformational states, see Section 6.5.1.

6.3.4 Complex mixtures and assembly intermediates

SAS is also applicable in much more complicated cases, when neither the scattering from all the components nor the number of components are available *a priori*. These scenarios are typical, for example, for assembly processes, where the initial and final states may be known but the pathway could contain different intermediates. Usually, multiple scattering patterns are available corresponding to mixtures with varying contributions of the components, and the number of components can be estimated by model-independent SVD analysis of the data (see Section 5.3). An example of multiple components detected during the assembly process is given by the study of *Bacillus subtilis* and *Aquifex aeolicus* lumazine synthase (Zhang *et al.* 2006). These proteins are known to form multiple assemblies in solution, whereby some species can be pentameric or decameric, and some may form icosahedral capsids with sixty subunits or more. The lumazine synthase samples from the two species, including wild-type proteins and several point mutants with reduced catalytic activity, were studied in solution in different buffers (borate, phosphate and Tris buffer) at varying pH. Capsid formation was found to be a pH-dependent process strongly influenced by mutations and by the buffer composition. Interestingly though, the SVD analysis of the bulk of the scattering patterns revealed that there are only five major components (see Fig. 5.4). These were identified as complete and incomplete (or deformed) icosahedral capsids of two types, with diameters of about 16 nm and 30 nm, respectively, and much smaller species modelled as pentamers of lumazine synthase. The relative abundance of smaller and larger capsids co-existing in the same solutions was determined by a non-linear fitting procedure accounting also for moderate polydispersity of the capsids. Several mutants with inhibited enzymatic activity had a tendency towards forming the larger capsids with a diameter of about 30 nm. The SAXS results, which were corroborated by EM data, demonstrated that multiple assembly forms are a general feature of lumazine synthases.

The usefulness of SAS for the structural characterisation of biological objects in an extremely broad range of sizes is further illustrated by a comprehensive study of insulin fibrillation (Vestergaard *et al.* 2007). Amyloid fibril formation is a cause of many critical diseases, and is also often observed for protein drugs, such as insulin, posing severe problems for their storage. The time course of insulin fibrillation was monitored by a series of SAXS experiments in solutions with a low pH and temperature of 45°C. Formally, this was a time-resolved study, but as the fibrillation occurred on a long time-scale (within a few hours), and the applied methodology is of general value for the analysis of equilibrium mixtures, the example is presented in this section. The major question addressed was whether the individual insulin monomers are the building elements of the fibrils or whether some intermediate structures exist on the pathway. The SAXS data collected during the fibrillation process (over the time interval of 9 hours) are presented in Fig. 6.16A. The initial solution (curve 1) contains only

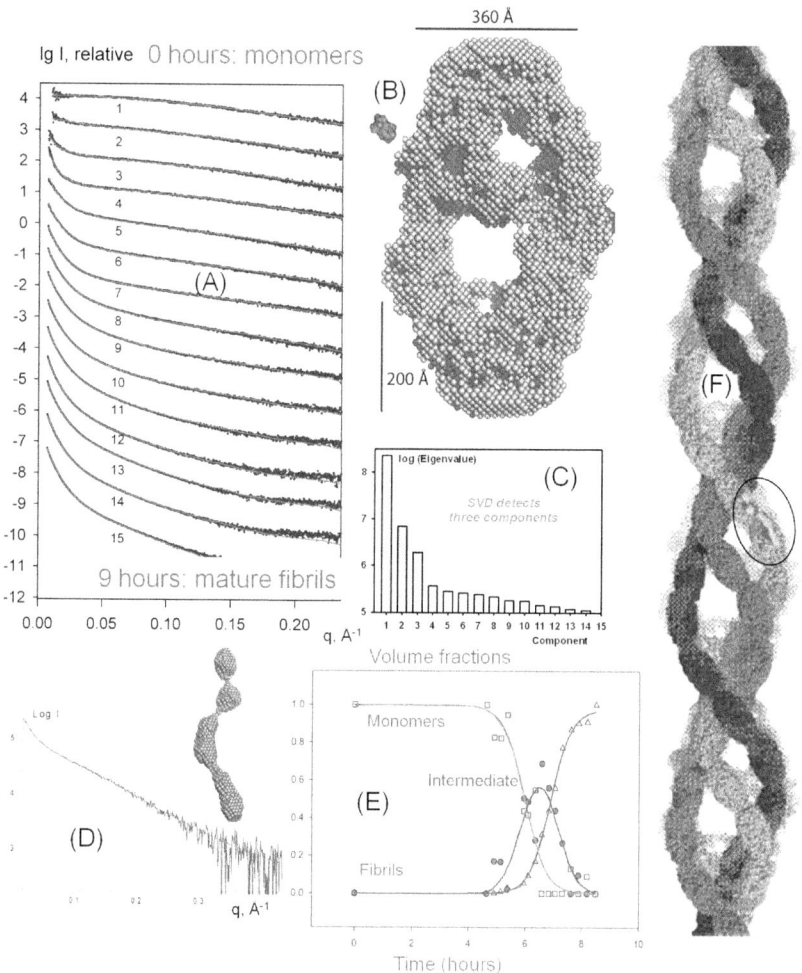

Fig. 6.16 A SAXS study of insulin fibrillation (Vestergaard *et al.* 2007). (A) Experimental scattering patterns (dots) and the fits computed from the three-component mixtures (solid lines); (B) *ab initio* shapes of the monomeric insulin and of the repeating unit of the mature fibril obtained from curves 1 and 15, respectively; (C) singular values of the experimental data showing three major components; (D) the scattering pattern from the third (intermediate) component extracted from the residuals of the two-component fit (noisy line) and the reconstructed shape of the intermediate oligomer (displayed in beads) fitting this pattern (smooth line); (E) volume fractions of the three major components computed by fitting all experimental data by the linear combinations of the scattering from the monomer, oligomer and matyre fibril; (F) a tentative model of the mature fibril formed by three intertwining protofibrils (four repeats are shown superposed with the semitransparent *ab initio* shape in (B). The oligomeric precursor (D) is displayed in darker beads and identified by an oval). Adapted from Vestergaard *et al.* (2007).

monomeric insulin molecules with $M = 5.8\,\text{kDa}$, and the shape of the insulin can be well reconstructed (top of Fig. 6.16A). The *ab initio* low-resolution shape of the sample after 9 hours incubation (curve 15) corresponds to the shape of the repeating unit of the mature fibril (Fig. 6.16B). The scattering patterns collected during the fibrillation process allow one to detect possible intermediates. Indeed, if the fibrils were elongated by association of monomeric insulin molecules, one should be able to fit all the observed SAXS patterns by linear combinations of the scattering from insulin monomers and mature fibrils using eq. (5.6). This was, however, not the case, and the fits by the two-component mixtures displayed significant systematic deviations. The SVD analysis indicated that the mixtures contained three rather than two major components (Fig. 6.16C). The scattering pattern from the third component can be determined by the analysis of the residuals of the two-component fits (Fig. 6.16D). The *ab initio* shape reconstructed from this scattering pattern depicted the third component as a helical oligomeric unit composed of about five or six insulin monomers (Fig. 6.16D). Decomposition of the experimental scattering patterns using the three components in eq. (5.6) yielded the volume fractions in Fig. 6.16E, and the computed scattering from the mixtures fitted all observed experimental data well (Fig. 6.16A). Remarkably, the derivative of the volume fraction of the fibrils (that is, their growth rate) was proportional to the amount of the oligomers in solution, suggesting that the oligomers (and not monomers) elongate the fibrils. Given the shape and size of the oligomeric precursor and the shape of the repeating unit, a structural model of the mature fibril was constructed (Fig. 6.16F) consisting of three intertwining protofibrils, in agreement with the available EM data. A similar pathway was later reported for glucagon fibrillation (Oliveira *et al.* 2009) and, more recently, also for the formation by α-synuclein of amyloid fibrils, that are one of the major hallmarks of Parkinson's disease (Giehm *et al.* 2011). In the latter case the α-synuclein oligomers had a wreath-shaped (not helical) appearance, but they were clearly the building blocks of the oligomeric fibrils, and, moreover, the purified oligomers were shown to disrupt liposomes indicating their potential cytotoxicity. These SAXS results suggesting a novel elongation pathway of amyloid fibrillation provide a conceptually new basis of structure-based drug design against amyloid diseases.

The examples above illustrate that in many cases the SAS results for mixtures are unique inasmuch as they can hardly be obtained by other methods. Additionally, SAS provides the capacity to monitor the changes in real time, as further discussed in Chapter 7.

6.4 Membrane proteins and lipoproteins

Membrane proteins and lipoproteins are potentially ideal objects for study by neutron contrast variation as they have components of widely differing scattering length densities, which can be matched by H/D exchange. Detergents and lipids being predominantly hydrocarbons have scattering densities close to that of H_2O both for neutrons and for X-rays whilst for neutrons they have a very high contrast when in D_2O solution. Note that the electron density of lipids and detergents is lower than that of water (see Table 1.2), which makes the X-ray contrast variation much less feasible for these objects.

6.4.1 Solubilised membrane proteins

Membrane proteins are particularly difficult to study in solution due to their insolubility in aqueous media. In order to solubilise them they have to be associated with lipophilic agents such as detergents, lipids or special constructs such as amphipols (Popot *et al.* 2003, Popot *et al.* 2010). These agents, as well as rendering the protein molecules water soluble, must also preserve their native conformation and hence their biological activity.

Detergents. Some early experiments on detergents and membrane protein/detergent complexes are described in (Timmins *et al.* 1994). Extensive studies on detergent micelles alone have been carried out using both SAXS and SANS (Timmins *et al.* 1991, Thiyagarajan and Tiede 1994, Marone *et al.* 1999). Efficient solubilisation of membrane proteins requires the detergent to be above its critical micellar concentration (cmc) (Le Maire *et al.* 2008) and membrane protein/detergent solutions thus contain also free detergent micelles. In order to obtain information on the protein and its bound detergent it is therefore necessary to eliminate contributions to the scattering from these free micelles. There are two principal ways to do this.

1. As with all macromolecular SAS experiments scattering is measured from a solution of the membrane protein complex and from its solvent. As mentioned above this will comprise contributions from the protein detergent complexes as well as scattering from the solvent which is itself an aqueous solution of free detergent micelles. In principle subtraction of the scattering of the micellar solution from that of the membrane protein solution will result in a scattering curve of the membrane protein/detergent complex alone. However, in practice this is only the case if the concentration of free micelles in the solvent and in the membrane protein solution is identical. The best way to prepare such samples is by extensive dialysis, but equilibration of free micelle concentration in solvent and protein solution will only be attained if the detergent is dialysable and will depend on dialysis time, temperature and so on. The conditions required to attain equilibrium are not easy to find and it is not easy to ensure that equilibrium has been reached, so there is always a risk of there being excess micelles in the solution or the solvent. Ways in which it is possible to at least place limits on the difference have been described (Lipfert *et al.* 2007). If the free micelle background can be subtracted satisfactorily then the resulting scattering is from the protein/detergent complex and will depend on the relative contrasts of the two components with the solvent. A method for doing this has been demonstrated in an extensive SAXS study of the detergent organisation around aquaporin-0 by (Berthaud *et al.* 2012). Here the detergent used for purification, β-octyl glucoside, was replaced by dodecyl maltoside (DDM), which has a considerably lower cmc. The aquaporin/DDM solution was then eluted on a size exclusion chromatography column and passed through a UV-vis cell before entering a SAXS flow cell and finally a refractometer. In this way it was possible to measure the scattering from pure detergent in equilibrium with the membrane protein solution, which eluted first, followed by the aquaporin/DDM whose concentration was monitored by the UV-vis cell. The scattering of the aquaporin/DDM complex (Fig. 6.17) was obtained by subtracting the scattering of detergent from membrane protein. The detergent corona surrounding the aquaporin molecule was then modelled using the known atomic structure of the aquaporin and the calculated electron density of the head and tail of the DDM molecule.

2. If information is required about the membrane protein alone (or indeed about the detergent alone) then one must resort to contrast

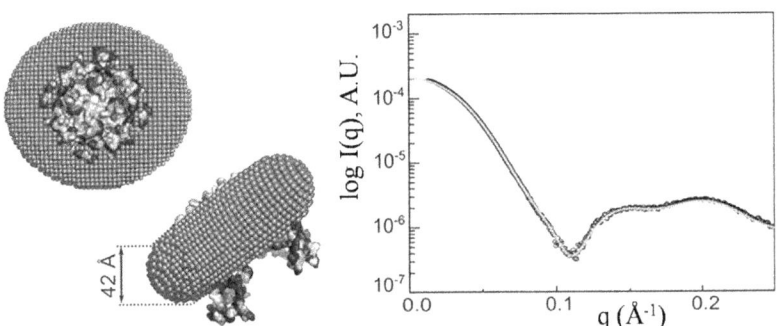

Fig. 6.17 Scattering curve of aquaporin/DDM complex with derived model of the molecule plus its detergent corona. Reproduced, with permission, from Berthaud *et al.* (2012) © American Chemical Society, 2012.

variation so that the detergent micelles and the detergent bound to the protein are effectively invisible. For SAXS this is rather difficult, as there are few non-perturbing solvents which can match the electron density of detergents. This has however been done in a study of the α-helical transmembrane domain of the human erythrocyte glycophorinA (GpA) fused to the carboxyl terminus of monomeric staphylococcal nuclease (SN/GpA) which was solubilised in a zwitterionic detergent micelle, *N*-dodecyl-*N*,*N*-(dimethylammonio)butyrate (DDMAB), having the same electron density as water and therefore rendering the detergent invisible (Bu and Engelman 1999). Contrast variation is, however, best carried out with SANS, where the H_2O/D_2O ratio can be adjusted to match the scattering length density of any detergent. In a SANS experiment on a membrane protein complex the first step is to investigate the contrast behaviour of the detergent, and from this to determine the H_2O/D_2O ratio at which the scattering is matched out. The calculated scattering length densities and match points of a number of more or less common detergents used for solubilising membrane proteins is given in Table 6.1. Note that in cases where the head group contains exchangeable hydrogens, then the scattering length density of the head varies with the deuterium content of the aqueous solvent. Moreover, the head group and tail can have very different scattering length densities, which means that care must be applied when using contrast variation to match out the detergent contribution, as the average scattering length density of the detergent may be matched by that of the solvent but head and tail may each have significant contrast—one being positive and the other negative. If the detergent can, for example, be deuterated in the tail then it is in principle possible by mixing deuterated and hydrogenated molecules to produce a micelle having an homogeneous scattering density.

Once the match point of the detergent has been determined the scattering from the membrane protein/detergent complex is measured in the corresponding H_2O/D_2O mixture along with the H_2O/D_2O buffer, and the resultant difference yields the scattering of the protein alone. An even more rigorous way of conducting the experiment is to measure the protein/detergent complex in a number of different H_2O/D_2O mixtures and to interpolate the scattering at the calculated match point. This

Table 6.1 Neutron scattering length densities for some detergents used in SANS studies of various membrane proteins. From Timmins *et al.* (1994), with corrections. Molecular volumes are calculated from experimentally determined partial specific volumes. For head groups and tails, volumes are reasonable estimates based in their chemical nature except in the case of alkyl chains where they are calculated according Tanford's formula (Tanford 1972): $v = 27.4 + 26.9\,n_c$, where n_c is the number of carbon atoms. x = mole fraction D_2O in solvent.

Detergent	Volume (10^{-3} nm^3)	Scattering length density (10^{10} cm^{-2})	Match point %D$_2$O
DDAO (*N*-decyl-*N,N*-dimethylamine-*N*-oxide)			
protonated head	81.0	0.748	18.8
protonated C$_{10}$ tail	296.4	−0.406	2.6
whole molecule protonated	377.4	−0.158	5.8
deuterated C$_{10}$ tail	296.4	6.972	108.1
whole molecule (protonated head group, deuterated tail)	377.4	5.636	88.9
LDAO (*N*-dodecyl-*N,N*-dimethylamine-*N*-oxide)			
protonated head	81.0	0.748	18.8
protonated C$_{12}$ tail	350.2	−0.392	2.4
whole molecule protonated	431.2	−0.179	5.5
deuterated C$_{12}$ tail	350.2	7.041	109.1
whole molecule (protonated head group, deuterated tail)	431.2	5.859	92.1
C12MS (dodecyl-β-D-maltoside)			
protonated head	358.4	$1.820 + 2.030x$	48.2
protonated tail	350.2	−0.391	2.4
whole molecule	708.6	$0.725 + 1.028x$	21.7
ES12H (1-dodecanoylpropanediol-3-phosphoryl-choline)			
protonated head	263.3	0.856	20.3
protonated tail	44.4	0.024	5.4
whole molecule protonated	710.7	0.355	13.2
Triton X-100			
protonated head	633.4	$0.687 + 0.164x$	17.5
protonated tail	378.6	0.385	13.5
whole molecule protonated	1012.0	$0.573 + 0.103x$	16.5
C8E4 (octyl tetraoxyethylene)			
protonated head	267.4	$0.696 + 0.389x$	19.1
protonated C$_8$ tail	236.7	−0.440	1.8
whole molecule protonated	504.1	$0.163 + 0.207x$	9.3
deuterated C$_8$ tail	236.7	7.037	109.0
whole molecule (protonated head group, deuterated tail)	504.1	$3.670 + 0.207x$	68.3
β-OG (octyl-*b*-D-glucopyranoside)			
protonated head	180.3	$1.859 + 2.310x$	52.0
protonated C$_8$ tail	236.7	−0.440	1.8
whole molecule protonated	417.0	$0.554 + 0.999x$	18.6

approach has the advantage of providing data of improved statistical accuracy, particularly if the match point of the detergent is not very different from that of the protein.

A further advantage of contrast variation study is that as in principle each component, detergent and protein, can be seen separately, the amount of detergent bound to the protein can be determined simply from the $I(0)$ values of the scattering curves measured at different contrasts, which yield the match point of the protein/detergent complex:

$$x = \frac{(\rho_P - \rho_S)\,\overline{v_P}}{(\rho_P - \rho_S)\,\overline{v_P} - (\rho_D - \rho_S)\,\overline{v_D}} \qquad (6.1)$$

where

x = mass fraction of detergent in the complex

$\overline{v_p}$ = partial specific volume of the protein

$\overline{v_D}$ = partial specific volume of the detergent

ρ_P = scattering length density of the protein at the match point of the complex

ρ_D = scattering length density of the detergent at the match point of the complex

ρ_S = scattering length density of solvent at the match point of the complex

A typical example of the use of SANS and contrast variation is in the study of the bacterial SLC26 transporter (Compton *et al.* 2011). The protein was solubilised in the detergent Fos-choline-12 the match point of which was determined to be at 11% D_2O in a series of separate experiments on the detergent micelles alone. SANS experiments were then carried out on the protein/detergent complex in a solvent containing 11% D_2O and the data modelled using the program DAMMIN. The molecular weight derived from the SANS data indicated a dimer, and the model was then constrained to have P2 symmetry. The scattering data from detergent and complex are shown in Fig. 6.18 and the derived model in Fig. 6.19. Note that the model and its mirror image always fit the data equally well, and in this case the correct enantiomorph was chosen with reference to the known crystal structure.

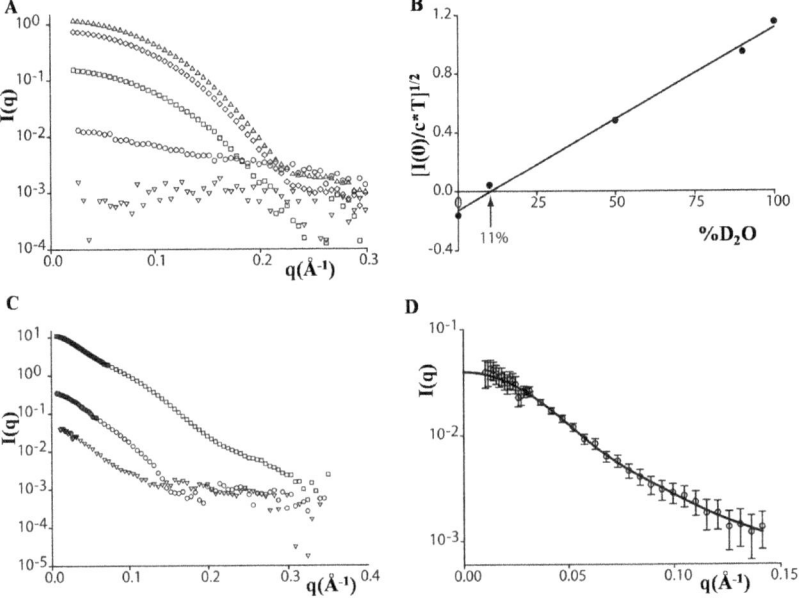

Fig. 6.18 SANS data analysis. A: Fos-choline-12 contrast variation series 0% (o), 10% (∇), 50% ($\square$), 90% ($\diamond$) and 100% ($\triangle$) D_2O. B: Experimental determination of the Fos-choline-12 contrast match point. C: SANS contrast series of the protein detergent complex (o) 0%, (∇) 11% and ($\square$) 100% D_2O. D: fit of the DAMMIN model (line) against the experimental YeSlc26A2 SANS data (open circles). Reproduced from Compton *et al.* (2006), with permission from the American Society for Biochemistry and Molecular Biology, © 2011.

A   No symmetry applied

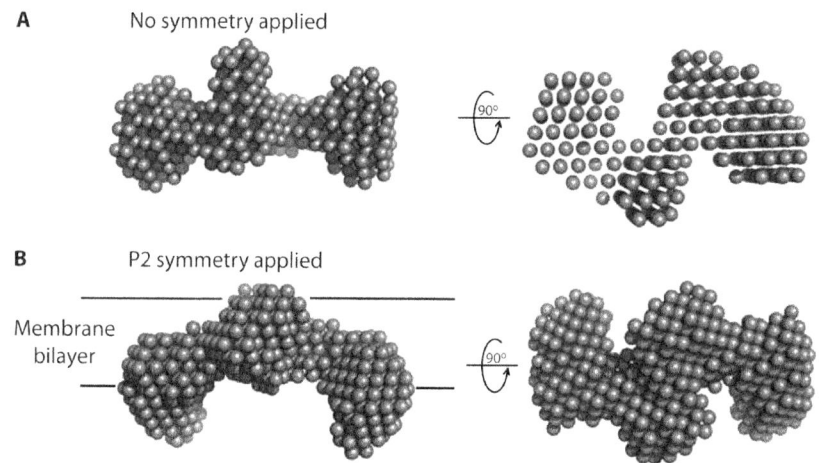

B   P2 symmetry applied

Membrane
bilayer

Fig. 6.19 *Ab initio* envelopes of YeSlc26A2. (A) P1 (no symmetry applied). (B) P1 DAMMIN models of YeSlc26A2 reconstructed using $D_{max} = 120$ Å and $q_{max} = 0.141$ Å^{-1}. The models on the right are rotated by 90° about their long axes. Reproduced from Compton *et al.* (2006), with permission from the American Society for Biochemistry and Molecular Biology, © 2011.

Membrane proteins in vesicles. Lipid vesicles constitute the environment closest to that of the living cell for membrane proteins. They are therefore a tempting choice for studying membrane protein structure in solution by SAS. However, there are a number of difficulties involved in doing so. First, the vesicles are usually rather large objects and will dominate the scattering, particularly at the smallest angles, depending on the contrast conditions. Second, it may be difficult to control the number of membrane protein molecules incorporated in a vesicle. Even when the overall protein concentration is very low if there are more than one protein molecules, the scattering can be influenced by interference effects between the molecules. Third, the difference in scattering length density between lipid headgroups and tails (consider detergents) could lead to incomplete matching of the lipid. These problems were addressed in a SANS study of bacteriorhodopsin adsorbed in small unilamellar vesicles which showed how such effects could be taken into account (Hunt *et al.* 1997). By carrying out experiments at several lipid/protein ratios it could be shown that the interference effects were limited to a *q*-range below that of interest to the protein shape modelling. The effects of inhomogeneity in the lipid scattering could be minimised by judicious deuteration of the lipid tail, and vesicle–vesicle interactions could be modelled by measuring the scattering of empty vesicles in a range of concentrations. The conclusion of the study was that experiments on bacteriorhodopsin ($R_g \sim 1.6$ nm) could be carried out reliably, but for proteins of larger R_g the vesicle size would have to be increased accordingly.

Membrane proteins and amphipols. Amphipols are medium-sized amphiphilic molecules consisting of an hydrophilic backbone to which is attached a large number of hydrophobic chains. They were developed to obviate the problem of solubilising membrane proteins in detergent solution which contains large numbers of free micelles, which can destabilise the membrane proteins (Popot 2010). Extensive studies have been carried out on amphipols alone and on complexes with membrane proteins. Studies on amphipols alone have shown them to be relatively well-defined particles, despite the wide range of molecular masses of the molecules comprising them (Gohon *et al.* 2004). Amphipol/membrane protein complexes can be studied by SAS, just like any other two-component complex and unlike with detergents, if the stoichiometry is correct, there are no free amphipols in solution and certainly no micelles. A SANS study of bacteriorhodopsin solubilised in the amphipol A8-35, which in aqueous solution comprises four amphipol molecules associated into 40kDa globular particles, has been described (Gohon *et al.* 2008). This study used deuterated amphipol in order to increase the contrast and reduce the background, and showed that the bacteriorhodopsin was incorporated predominantly as monomers with its native complement of lipids. Some free amphipol was present as very large aggregates, as evidenced by increased scattering at very small angles.

Membrane proteins and Nanodiscs. Another recently developed method for solubilising membrane proteins is via Nanodiscs®. These are nanometer-sized discs of phospholipid which can be stabilised in solution by an amphipathic protein belt the size of which controls the maximum disc size. SAXS and SANS studies indicate that the discs are flat with an elliptical cross-section and the hydrophobic periphery covered by protein (Skar-Gislinge and Arleth 2011).

6.4.2 Low-density and high-density lipoproteins

The most common lipoprotein to be studied by SANS is low-density lipoprotein (LDL). LDL particles are usually considered to be globular and about 22 nm in diameter. They are highly heterogeneous in composition, and can vary in size from 18 to 22.5 nm, but there is a general consensus that they consist of an apolar core surrounded by an amphipathic shell. This outer shell comprises a phospholipid bilayer, free unesterified cholesterol and a single molecule of protein, apoB100 (Prassl and Laggner 2009). SAXS and SANS studies have indicated that the neutral lipids are organised in a fluid, oil-like state. SANS contrast variation studies have indicated that the apoB100 protein is centred some 0.8 nm further from the centre of mass than the headgroups. Studies of the protein extracted from the LDL particle and solubilised in detergent have shown the molecule to be elongated, as in the native LDL (Johs *et al.* 2006). In general, although LDL would seem an ideal candidate for SAS experiments, the results obtained to date have been limited by the highly dynamic structures of the protein and lipids in the particles.

High-density lipoprotein (HDL)—the carrier of so-called 'good' cholesterol—has also been studied by SAS (Wu *et al.* 2009; Wu *et al.* 2011). In a study of reconstituted nascent HDL (Wu *et al.* 2009), SANS data were measured at four different contrasts, including 12% D_2O at which the lipid contribution is minimised and 42% D_2O at which the protein contribution is minimised. The signal from native reconstituted

LDL at 12% D_2O was very weak, and a sample with this D_2O content was therefore prepared containing recombinant deuterated apolipo-protein AI (apo-AI) in order to increase the contrast. SAXS data were also recorded to increase the number of contrast points. The resultant data were analysed in the Guinier region, which gave reliable radii of gyration for both the protein and lipid components as well as the overall particle, and at the same time confirmed the particle mass of 200 kDa with two apo-AI molecules per particle, as observed in dynamic light scattering, biochemical and cross-linking/mass spectro-metry analyses. The extended scattering curves were then modelled using DAMMIN to give independent structures of the protein and lipid components. These models were then refined by incorporating data from H/D exchange through mass spectrometry with energy minimisation techniques. Additional constraints from chemical cross-linking, fluorescence resonance energy transfer (FRET) and electron spin resonance measurements were also incorporated into the final model. This is a good illustration of how the use of complementary information from several different techniques can lead to a detailed all-atom model of the structure, though the SANS data themselves do not warrant such a detailed model. Another more mature and more abundant form of HDL, spherical high-density lipoprotein (sHDL), has also been studied by SAS. Due to its size (three or four apo-AI molecules per particle) and inhomogeneity it has proved more difficult to study, but recent results (Wu *et al.* 2011) are beginning to throw more light on the structure. SANS data were again collected in 0, 12, 42 and 90% D_2O and analysed by Guinier analysis and DAMMIN modelling. The Stuhrmann plot indicated that the protein lies on the outside of the complex, and the final DAMMIN model shows a rather spherical model of apo-AI surrounding a compact lipid core. Cross-linking and mass spectrometry were used extensively to discriminate between three models having very similar agreement with the SANS data.

6.5 Flexible systems

The ability of SAS to study flexible systems is largely related to the very fact that SAS is a low-resolution method. Indeed, low-resolution crystallographic data still bear information about flexible loops in the crystals, but the information about such loops disappears completely in high-resolution electron density maps. As the solution scattering pat-terns largely contain the information about the overall structure, all parts of the macromolecule, whether flexible or not, contribute to the experimental data. SAS is applicable to a broad variety of systems with different degree and origin of flexibility. Amenable systems include ri-gid structures with flexible loops or domains, multidomain proteins interconnected by flexible hinges or linkers, as well as intrinsically unfolded proteins. In the next section, typical applications to flexible systems will be considered.

6.5.1 Flexibility in multidomain assemblies and complexes

Although *ab initio* methods assume sample monodispersity, these methods may still provide useful information about the overall structure of objects with limited flexibility. *Ab initio* shapes provided the structural bases for the analysis of various potentially flexible macromolecular objects. Examples include solubilised human apolipoprotein B-100 (Johs *et al.* 2006), various cellulosome-like assemblies (Hammel *et al.* 2005), or a complex between the C-terminal domains of the measles virus nucleoprotein and the phosphoprotein (Bourhis *et al.* 2005). In many studies, *ab initio* shapes provide an initial guess of potential flexibility and facilitate further modelling using more elaborate algorithms. An example is provided by a SAXS study of ubiquitination of the proliferating cell nuclear antigen (PCNA), occurring in response to DNA damage and leading to the recruitment of specialised translesion polymerases to the damage locus (Tsutakawa *et al.* 2011). The crystal structure of the yeast PCNA–ubiquitin (PCNA-Ub) complex (Freudenthal *et al.* 2010) displays the Ub on the back face of PCNA, whereas most of the PCNA-binding proteins are located on the front face, where replication occurs. The SAXS data were recorded on yeast PCNA–Ub complexes formed by either split-fusion or by chemical cross-linking to make the results independent on the type of linkage between Ub and PCNA. The experimental scattering curves from the two complexes were nearly identical, and neither of them can be fitted by the scattering computed from the crystal structure (Fig. 6.20, left panel). The *ab initio* shapes reconstructed from the experimental data (Fig. 6.20, right panel) had a toroidal appearance similar to that of the crystal structure but displayed asymmetric protrusions suggesting different positions of Ub compared to the crystal structure. To further explore the conformational space of the Ub molecules in the complex, computational modelling was performed combining tethered Brownian dynamics, protein–protein docking, flexible loop modelling and molecular dynamics algorithms. The generated structures were analysed using the MES algorithm (Pelikan *et al.* 2009), yielding good fits to the experimental data for the mixtures containing

Fig. 6.20 SAXS analysis of PCNA–Ub in solution suggests that ubiquitin is not exclusively oriented in the position determined by crystallography. (A) SAXS curves of split-fusion and cross-linked PCNA–Ub and the pattern calculated from the PCNA–Ub trimer in the crystal (PDB entry 3L10, dashed grey line); (B) Guinier plots of the two constructs are linear suggesting lack of aggregation in the samples. (C) *Ab initio* shapes (semi-transparent envelopes) calculated from the scattering curves of split-fusion or cross-linked PCNA–Ub superimposed with the crystal structure (ribbons). The SAXS shapes suggest that only one or two ubiquitin positions are extending away from the PCNA ring. Adapted by permission from Tsutakawa *et al.* (2011), © National Academy of Sciences, USA, 2011.

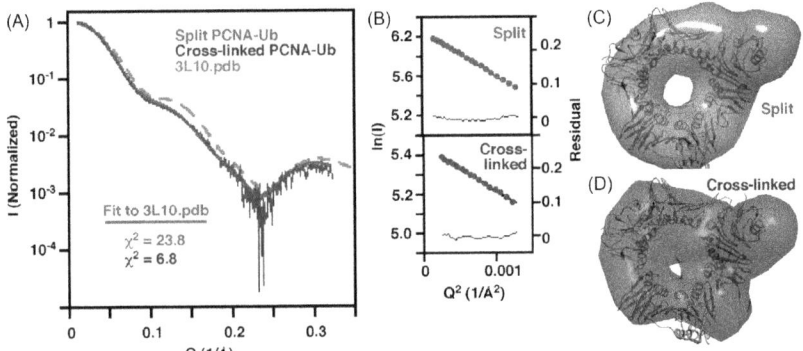

three types of complex. The complexes with the ubiquitin in the crystallographic position were populated to about 25–30%, those with ubiquitin in the computationally determined positions to about 40–50%, and about 25–30% of population were models with the ubiquitin flexible and loosely attached to PCNA. These results indicate that ubiquitin is dynamically associated with PCNA and capable of transitioning between a few discrete sites on the PCNA surface. A similar modelling approach was employed to analyse conformational flexibility of a complex between DNA repair protein RAD50 with Mre11 nuclease essential for dsDNA break repair (Williams *et al.* 2011). The SAXS data also allowed monitoring of the transition from a conformationally flexible complex to a globular, closed complex upon ATP binding.

The study of a pH-induced fusogenic structural transition of a soluble form of vesicular stomatitis virus (VSV) glycoprotein G ectodomain (G_{th}) provides an interesting example of accounting for conformational flexibility (Albertini *et al.* 2012). This protein catalyses fusion between viral and endosomal membranes at low pH, by going from a pre-fusion to a post-fusion conformation. Both these conformations are trimeric, and their high-resolution crystal structures are available. There are, however, topological issues concerning the transition pathway, as the observed structural rearrangements cannot occur without breaking the three-fold symmetry. Structural studies using multiple methods including AUC, CD, EM and SAXS indicated that the post-fusion trimer is the major species at low pH, but the pre-fusion trimer is not detected in solution. SAXS demonstrated that in solutions with high pH, G_{th} is a flexible monomer exploring a large conformational space. The scattering data at pH 8.8 and 7.5 could be interpreted in terms of mixtures of tentative G_{th} models of different compactness, generated from the high-resolution structures. Interestingly, lower pH values lead to more elongated conformations in the mixture (Fig. 6.21). EM data on negatively stained particles revealed the presence of monomeric G_{th} at the viral surface both at pH 7.5 and pH 6.7. The authors propose a mechanism where the monomers are intermediates during the fusion-associated conformational change, and the trimeric VSV G ectodomain fully dissociates at the viral surface during the structural transition.

Fig. 6.21 Pathway for the pH-dependent structural transition of G glycorotein in solution and at the viral surface. (A) G_{th} species detected in solution at various pH. From pH 8.8 to pH 7.5 only monomers of Gth are present. At pH 6.7, at low protein concentration, trimerisation does not occur, and Gth monomers associate to form rosettes through interaction via their fusion loops. At yet lower pH, monomers reassociate to form post-fusion trimers, which interact through their fusion loops to form dimers of trimers and rosettes. (B) Suggested structural transition pathway of G at the viral surface. At pH 7.5, pre-fusion trimers and flexible monomers are in equilibrium at the viral surface. Lowering the pH to 6.7 favours the formation of elongated monomers oblique to the viral membrane, and some post-fusion trimers appear. At pH 6.0, all monomers are reassociated to form post-fusion trimers regularly spaced at the viral surface. Adapted from Albertini *et al.* (2012).

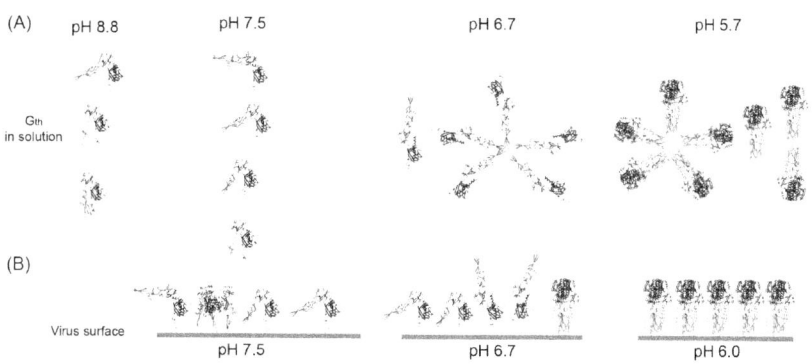

In some cases, results from other methods may assist in the definition of the configurational space occupied by a macromolecule. An example is given by a combined MX, NMR and SAXS analysis of flexibility of the two Ig fragment from the N-terminus of titin, Z1Z2 (Marino *et al.* 2006). The low-resolution shape analysis of this construct was considered in Section 6.1, and the SAXS shape of the tandem depicts an extended but still somewhat bent conformation (see Fig. 6.3). In the crystal, Z1Z2 shows two distinct conformations in the unit cell. One of them has a nearly linear domain arrangement where the two Ig's are in almost coaxial orientation, whereas the other reveals a compact V-shaped conformation. Neither of the crystallographic models was able to fit the experimental SAXS data. The NMR RDC data recorded in a partially oriented alignment medium provide information about relative orientations on the two domains. Due to a four-fold angular degeneracy in the interpretation of the RDCs, four possible clusters of Z1Z2 are compatible with the experimental RDC data, and the models in one of these clusters are similar to the *ab initio* SAXS model in Fig. 6.3. The average over the configurations in this RDC cluster yields an excellent fit to the experimental SAXS data. The fit from the NMR ensemble is significantly better than that obtained from the best mixture of the two (fully opened and compact) crystal structures. These results suggest a limited modular flexibility around a hinge region between the two domains in Z1Z2, whereby the species that greatly deviate from the observed equilibrium, if they exist, are likely to comprise only a minor fraction of the population.

A systematic approach to generation and analysis of multiple conformations was employed in the study of flexibility of Hck tyrosine kinase—a multidomain allosteric enzyme of the Src kinase family implicated in the signalling pathways regulating cell growth and proliferation (Yang *et al.* 2010b). The Src kinases possess the SH3 and SH2 binding domains followed by a highly conserved catalytic domain connected by flexible linkers. There was evidence that the Src inactivation/activation might not be a simple two-state process involving an assembled (inactive) and a disassembled (active) conformation, both available as crystal structures. In particular, different signals could be expected to promote different assembly states, leading to differently configured activated kinases. SAXS was employed to probe the spatial organisation of the wild-type Hck, its high-affinity mutant, and in the presence of external peptide ligands. A simplified coarse-grained model was employed to explore and sample the accessible conformational space of the multidomain Hck complex. These configurations are clustered into a small number of distinct putative assembly states (Fig. 6.22), and fitting the SAXS data by their mixtures yields an estimate of the volume fractions using eq. (5.6). The analysis indicated that the Hck indeed adopts multiple conformations in solution. The wild-type Hck is predominantly (82%) in the assembled state 1, and the other states are the fully disassembled state 6 and the partially disassembled state 8 resembling the partially disassembled crystal structure

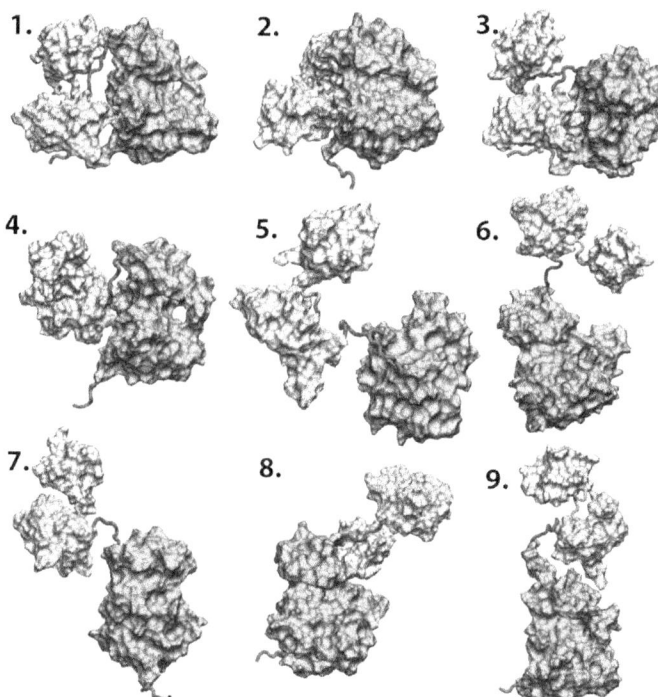

Fig. 6.22 Representatives of the conformational states adopted by the multidomain Hck from selected from a set of coarse grained modelling. The nine states, range in size from compact to extended forms and in architecture from fully to partially assembled to disassembled states. The catalytic domain is depicted in dark grey, SH2 and SH3 domains light grey. States 1 and 8 are similar to the assembled Hck (PDB entry 1QCF (Schindler *et al.* 1999)) and to the partially active c-Src (PDB entry 1Y57 (Cowan-Jacob *et al.* 2005)) structures, respectively. Adapted from Yang *et al.* (2010b), © National Academy of Sciences, USA, 2011.

of an homologous c-Src. The assembled state is also dominating for the high-affinity Ctail mutant, whereas ligand binding, either to SH2 or SH3 domains, significantly diminishes the population of the assembled Hck and increases the fraction of disassembled states. Simultaneous binding of the two ligands completely removes the assembled stated, and fully disassembled states 5 and 6 prevail. For the ligand-bound constructs differences are observed between the wild type and the mutant Hck. The study clearly demonstrates that the available crystal structures are not always sufficient to describe the conformational states of complicated assemblies in solution, and the strategy employed provides a tool to quantify and visualise the behaviour of potentially flexible multidomain proteins.

6.5.2 Intrinsically disordered proteins

The examples above described applications to folded proteins and complexes with limited flexibility. Notoriously flexible objects such as multidomain proteins with long flexible linkers and IDPs are even more challenging. Recent methodological developments have made it possible to advance from qualitative assessments of disorder with Kratky plots and R_g values to a quantitative analysis. The modern possibilities for characterisation of IDPs are illustrated by a comprehensive SAXS study of tau—a microtubule-associated protein occurring mainly in the axons (Mylonas *et al.* 2008).

Tau is involved in Alzheimer's disease via formation of β-sheet rich intracellular aggregates (paired helical filaments, PHFs). In the human central nervous system, tau is found in six alternatively spliced isoforms ranging from 352 to 441 residues. The protein is known to be an IDP even upon binding to physiological partners such as microtubules. Tau contains four semiconserved sequences of 31 or 32 residues, so-called 'repeats', whereby the second repeat may be absent in some of the isoforms. The repeat domain is essential both for the binding to microtubules and for the aggregation of tau into PHFs. Two tau isoforms—a three-repeat ht23 (predominant in the fetal brain)—and the full-length protein (ht40) were studied (Fig. 6.23A), and for both, SAXS data from several deletion mutants were collected (Fig. 6.23B).

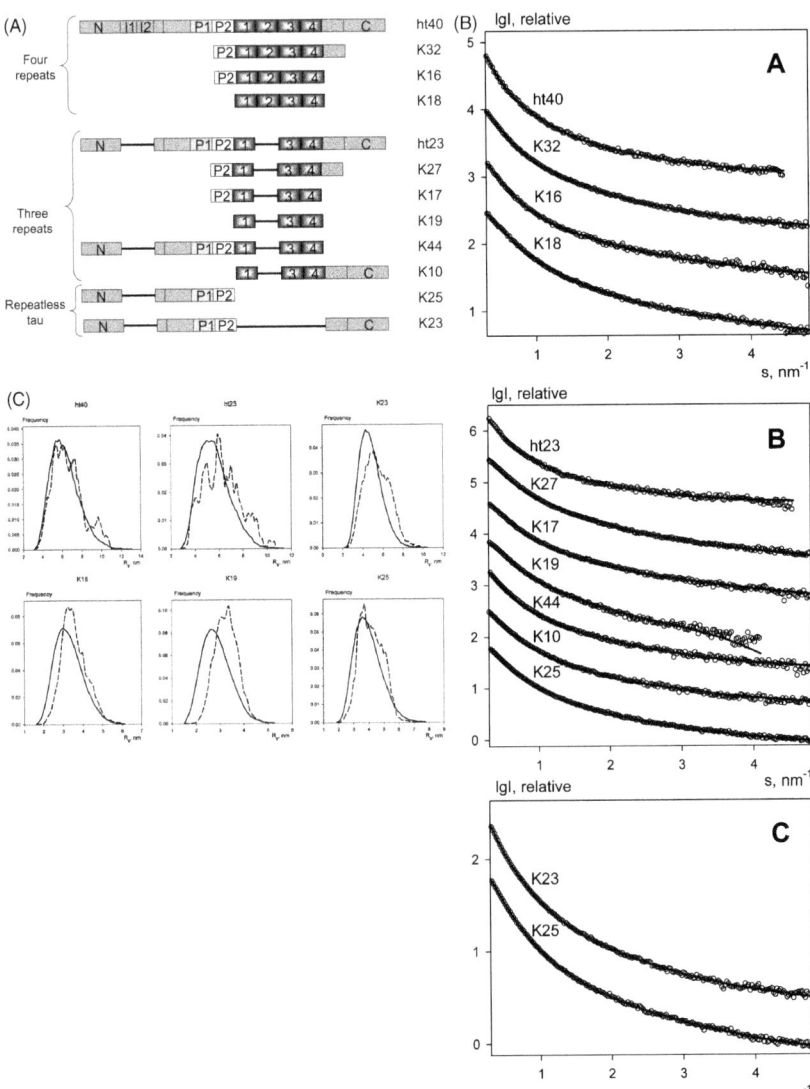

Fig. 6.23 SAXS analysis of tau protein. (A) Bar diagrams of isoforms and mutants of tau protein. All exons refer to the sequence of ht40, the longest of human tau isoforms with 441 residues. Constructs with four repeats are denoted as ht40, K32, K16, and K18; those with three repeats are ht23, K27, K17, K19, K44 and K10. For repeatless tau, K25 contains the N-terminal domain of ht23 and K23 represents the ht23 molecule without repeats. (B) Experimental SAXS data (circles) with the corresponding ensemble fit (lines). Plots A, B and C correspond to the four-repeats, three-repeats and repeatless constructs. (C) R_g distributions of the pools (solid lines) *versus* the ensembles selected by EOM. The integral of the area defined by the histograms equals to 1. Reproduced, with permission, from Mylonas *et al.* (2008), © American Chemical Society, 2008.

All constructs displayed characteristic features of IDPs and were analysed using a multiple-curve fitting option of EOM (Bernado *et al.* 2007), where several scattering patterns, from a full-length protein and from deletion mutants, are simultaneously fitted using one and the same pool. Interestingly, the repeatless and the full-length constructs behaved like random chains, whereas the deletion mutants containing mostly the repeats, although disordered, showed the reconstructed distributions shifted to larger R_gs (Fig. 6.23C). This finding was further confirmed by the analysis of the averaged C_α–C_α inter-residue distance matrix in the selected ensembles. This matrix displayed a distinct conformational behaviour depending on the number of repeats present in the isoforms. For ht23 with three repeats, the maximum separation was found within the repeat domain itself. The full-length ht40 with four repeat domains revealed an enhanced separation between the repeat domain and the preceding region. These results suggest that the different number of turns (one per repeat) may lead to different global arrangements of the chain in that region increasing or shortening the average interdomain distances compared to the random coil. This analysis unambiguously identified the repeat region as the source of residual secondary structure in tau, corroborating the earlier NMR data indicating the presence of turns and extended fragments in this region. The SAXS results lend support to a 'paper-clip' hypothesis of tau organisation, which postulated proximity of the N- and C-termini of the IDP (Jeganathan *et al.* 2006), and they were further confirmed by independent studies by other groups employing different techniques; for example, FRET, Elbaum-Garfinkle and Rhoades (2012).

There are numerous studies employing ensemble fitting to characterise the structure of IDPs and of multidomain proteins and complexes containing long flexible linkers. An interesting example is protein kinase R (PKR), a key component of the interferon antiviral pathway (VanOudenhove *et al.* 2009). The protein is composed of three folded domains with known structures separated by two long and presumably disordered linkers. Here, attempts to describe the structure with a single model using BUNCH did not provide adequate fits to experimental SAXS data, whereas using ensemble fitting did fit the data well. The results indicated that the protein explores a significant conformational space with little or no interactions between the individual domains in solution. Flexible fitting revealed multiple conformational states of two-domain metalloproteases MMP1 and MMP12 (Bertini *et al.* 2008, Bertini *et al.* 2009), specific behaviour of various constructs of multidomain complement regulator, factor H (Schmidt *et al.* 2010, Morgan *et al.* 2011, Morgan *et al.* 2012) and co-existence of compact and extended populations of a disordered F-actin-binding protein juxtanodin (Ruskamo *et al.* 2012). Accounting for flexibility also allows one to meaningfully analyse structural changes between the proteins in the free state and upon complex formation, as in a combined MX-EM-SAXS study of allosteric competitive inactivation of human CSF-1 protein by the Epstein–Barr viral decoy receptor BARF1 (Elegheert *et al.* 2012).

6.5.3 Flexibility of RNA molecules

The ensemble fitting approach is also applicable to characterise flexibility of RNA, as demonstrated by the study *E. coli* RNA chaperone Hfq. After providing insights into the flexibility and function of the C-terminus of the Hfq (Beich-Frandsen *et al.* 2011), its complex with a 34-nucleotides-long natural substrate small RNA, DsrA(34) was analysed in a subsequent study (de Almeida Ribeiro *et al.* 2012). In the complex, the protein maintained its doughnut-like structure, whereas the extended DsrA(34) was found to explore one hemisphere in conformational space contacting a broad area of the protein structure in agreement with the NMR chemical shifts data. It is speculated that the structural flexibility of RNA ligands bound to Hfq stochastically may facilitate base pairing providing the foundation for the RNA chaperone function inherent to Hfq.

Another example of accounting for flexibility in the RNA modelling is the study of solution structures of three phylogenetically distinct bacterial RNase P RNAs from *E. coli*, *A. tumefaciens* and *B. stearothermophilus* (Kazantsev *et al.* 2011). The RNA component of the ribonucleoprotein enzyme ribonuclease P (RNase P) processes tRNAs by cleavage of precursor-tRNAs. The crystal structures of the bacterial RNase P, including full-length RNAs and a ternary complex with substrate are available but the solution structures of free RNAs are unknown. The authors employed SAXS using a combination of homology modelling, normal mode analysis and molecular dynamics with selective 2'-hydroxyl acylation analysed by primer extension. The approach was used to refine the available RNA structures in solution under the high ionic strength conditions required for catalytic activity.

Overall, the novel approaches for the characterisation of flexible macromolecular objects from SAS data—specifically those operating in terms of ensemble distributions—allow one to conduct meaningful quantitative analysis of these challenging systems. It should, however, always be kept in mind that such an analysis is even more susceptible to overinterpretation and overfitting than that of rigid monodisperse systems. For an interesting analysis of structural diversity intrinsic to protein ensembles reflected in SAXS and WAXS, see, for example Makowski *et al.* (2011). To increase the information content the SAS studies of flexible systems are therefore in most cases combined with complementary techniques, most often with NMR but also with FRET, CD, AFM and other methods (see Chapter 9).

References

Albertini, A. A., Merigoux, C., Libersou, S. *et al.* (2012). 'Characterisation of monomeric intermediates during VSV glycoprotein structural transition', *PLoS Pathog* 8: e1002556.

Albesa-Jove, D., Bertrand, T., Carpenter, E. P. *et al.* (2010). 'Four distinct structural domains in Clostridium difficile toxin B visualised using SAXS', *J. Mol. Biol.* 396: 1260–70.

Armbruster, A., Svergun, D. I., Coskun, U., Juliano, S., Bailer, S. M. and Gruber, G. (2004). 'Structural analysis of the stalk subunit Vma5p of the yeast V-ATPase in solution', *FEBS Lett.* 570: 119–25.

Ashton, A. W., Boehm, M. K., Gallimore, J. R., Pepys, M. B. and Perkins, S. J. (1997). 'Pentameric and decameric structures in solution of serum amyloid P component by X-ray and neutron scattering and molecular modelling analyses', *J. Mol. Biol.* 272: 408–22.

Baker, N. M., Weigand, S., Maar-Mathias, S. and Mondragon, A. (2010). 'Solution structures of DNA-bound gyrase', *Nucleic Acids Res.* 39: 755–66.

Baldock, C., Oberhauser, A. F., Ma, L. *et al.* (2011). 'Shape of tropoelastin, the highly extensible protein that controls human tissue elasticity', *P. Natl. Acad. Sci. USA* 108: 4322–7.

Ban, N., Nissen, P., Hansen, J., Moore, P. B. and Steitz, T. A. (2000). 'The complete atomic structure of the large ribosomal subunit at 2.4 Å resolution', *Science* 289: 905–20.

Bang, M. L., Centner, T., Fornoff, F. *et al.* (2001). 'The complete gene sequence of titin, expression of an unusual approximately 700-kDa titin isoform, and its interaction with obscurin identify a novel Z-line to I-band linking system', *Circ. Res.* 89: 1065–72.

Beich-Frandsen, M., Vecerek, B., Konarev, P. V. *et al.* (2011). 'Structural insights into the dynamics and function of the C-terminus of the E-coli RNA chaperone Hfq', *Nucleic Acids Res.* 39: 4900–15.

Beltramini, M., Borghi, E., Dimuro, P., Lamonaca, A., Salvato, B. and Santini, C. (1996). 'The use of small-angle X-ray scattering in the study of quaternary organisation of giant proteins', *J. Mol. Struct.* 383: 231–6.

Bernado, P., Mylonas, E., Petoukhov, M. V., Blackledge, M. and Svergun, D. I. (2007). 'Structural characterization of flexible proteins using small-angle X-ray scattering', *J. Am. Chem. Soc.* 129: 5656–64.

Berthaud, A., Manzi, J., Perez, J. and Mangenot, S. (2012). 'Modeling detergent organization around aquaporin-0 using small-angle X-ray scattering', *J. Am. Chem. Soc.* 134: 10080–8.

Bertini, I., Fragai, M., Luchinat, C. *et al.* (2009). 'Interdomain flexibility in full-length matrix metalloproteinase-1 (MMP-1)', *J. Biol. Chem.* 284: 12821–8.

Bertini, I., Calderone, V., Fragai, M. *et al.* (2008). 'Evidence of reciprocal reorientation of the catalytic and hemopexin-like domains of full-length MMP-12', *J. Am. Chem. Soc.* 130: 7011–21.

Bonner, A., Almogen, A., Furtado, B. P., Kerr, M. A. and Perkins, S. J. (2009). 'Location of secretory component of the Fc edge of dimeric IgA1 reveals insight into the role of secretory IgA1 in mucosal immunity', *Mucosal Immunol.* 2: 74–84.

Bordas, J., Mandelkow, E. M. and Mandelkow, E. (1983). 'Stages of tubulin assembly and disassembly studied by time-resolved synchrotron X-ray scattering', *J. Mol. Biol.* 164: 80–135.

Bourhis, J. M., Receveur-Brechot, V., Oglesbee, M. *et al.* (2005). 'The intrinsically disordered C-terminal domain of the measles virus nucleoprotein interacts with the C-terminal domain of the phosphoprotein via two distinct sites and remains predominantly unfolded', *Protein Sci.* 14: 1975–92.

Brodersen, D. E., Clemons Jr, W. M., Carter, A. P., Wimberly, B. T. and Ramakrishnan, V. (2002). 'Crystal structure of the 30 S ribosomal subunit from Thermus thermophilus: Structure of the proteins and their interactions with 16 S RNA', *J. Mol. Biol.* 316: 725–68.

Bu, Z. M. and Engelman, D. M. (1999). 'A method for determining transmembrane helix association and orientation in detergent micelles using small angle X-ray scattering', *Biophys. J.* 77: 1064–73.

Capel, M. S., Engelman, D. M., Freeborn, B. R. *et al.* (1987). 'A complete mapping of the proteins in the small ribosomal subunit of *Escherichia coli*', *Science* 238: 1403–6.

Chacon, P., Diaz, J. F., Moran, F. and Andreu, J. M. (2000). 'Reconstruction of protein form with X-ray solution scattering and a genetic algorithm', *J. Mol. Biol.* 299: 1289–302.

Chao, L. H., Stratton, M. M., Lee, I. H. *et al.* (2011). 'A mechanism for tunable autoinhibition in the structure of a human Ca2+/calmodulin-dependent kinase II holoenzyme', *Cell* 146: 732–45.

Chauvin, C., Witz, J. and Jacrot, B. (1978). 'Structure of tomato bushy stunt virus: model for protein-RNA interaction', *J. Mol. Biol.* 124: 641–51.

Civril, F., Bennett, M., Moldt, M. *et al.* (2011). 'The RIG-I ATPase domain structure reveals insights into ATP-dependent antiviral signalling', *EMBO Rep.* 12: 1127–34.

Compton, E. L. R., Karinou, E., Naismith, J. H., Gabel, F. and Javelle, A. (2011). 'Low resolution structure of a bacterial SLC26 transporter reveals dimeric stoichiometry and mobile intracellular domains', *J. Biol. Chem.* 286: 27058–67.

Cowan-Jacob, S. W., Fendrich, G., Manley, P. W. *et al.* (2005). 'The crystal structure of a c-Src complex in an active conformation suggests possible steps in c-Src activation', *Structure* 13: 861–71.

Crichton, R. R., Engelman, D. M., Haas, J. *et al.* (1977). 'Contrast variation study of specifically deuterated Escherichia-coli ribosomal-subunits', *P. Natl. Acad. Sci. USA* 74: 5547–50.

Cusack, S., Ruigrok, R. W. H., Krygsman, P. C. J. and Mellema, J. E. (1985). 'Structure and composition of influenza virus: a small-angle neutron-scattering study', *J. Mol. Biol.* 186: 565–82.

de Almeida Ribeiro, E., Jr., Beich-Frandsen, M., Konarev, P. V. *et al.* (2012). 'Structural flexibility of RNA as molecular basis for Hfq chaperone function', *Nucleic Acids Res.* 40: 8072–84.

Decker, H., Hartmann, H., Sterner, R., Schwarz, E. and Pilz, I. (1996). 'Small-angle X-ray scattering reveals differences between the quaternary structures of oxygenated and deoxygenated tarantula hemocyanin', *FEBS Lett.* 393: 226–30.

Devaux, C., Timmins, P. A. and Berthet-Colominas, C. (1983). 'Structural studies of adenovirus type 2 by neutron and X-ray scattering', *J. Mol. Biol.* 167: 119–32.

Didry, D., Cantrelle, F.-X., Husson, C. *et al.* (2012). 'How a single residue in individual β-thymosin/WH2 domains controls their function in actin assembly', *EMBO J.* 31: 1000–13.

Diepholz, M., Venzke, D., Prinz, S. *et al.* (2008). 'A different conformation for EGC stator subcomplex in solution and in the assembled yeast V-ATPase: Possible implications for regulatory disassembly', *Structure* 16: 1789–98.

Ditzel, L., Löwe, J., Stock, D. *et al.* (1998). 'Crystal structure of the thermosome, the archaeal chaperonin and homolog of CCT', *Cell* 93: 125–38.

Drory, O., Frolow, F. and Nelson, N. (2004). 'Crystal structure of yeast V-ATPase subunit C reveals its stator function', *EMBO Rep.* 5: 1148–52.

Elbaum-Garfinkle, S. and Rhoades, E. (2012). 'Identification of an aggregation-prone structure of tau', *J. Am. Chem. Soc.* 134: 16607–13.

Elegheert, J., Bracke, N., Pouliot, P. *et al.* (2012). 'Allosteric competitive inactivation of hematopoietic CSF-1 signaling by the viral decoy receptor BARF1', *Nat. Struct. Mol. Biol.* 19: 938–47.

Fagan, R. P., Albesa-Jove, D., Qazi, O., Svergun, D. I., Brown, K. A. and Fairweather, N. F. (2009). 'Structural insights into the molecular organization of the S-layer from Clostridium difficile', *Mol. Microbiol.* 71: 1308–22.

Fan, L. X., Svergun, D. I., Volkov, V. V. *et al.* (2000). 'Structural studies of the 30S subunit of ribosomes *Thermus thermophilus* by small-angle neutron and X-ray scattering', *J. Appl. Crystallogr.* 33: 515–8.

Foerster, F., Webb, B., Krukenberg, K. A., Tsuruta, H., Agard, D. A. and Sali, A. (2008). 'Integration of small-angle X-ray scattering data into structural modeling of proteins and their assemblies', *J. Mol. Biol.* 382: 1089–1106.

Frank, J., Zhu, J., Penczek, P. *et al.* (1995). 'A model of protein-synthesis based on cryoelectron microscopy of the E-coli ribosome', *Nature* 376: 441–4.

Freudenthal, B. D., Gakhar, L., Ramaswamy, S. and Washington, M. T. (2010). 'Structure of monoubiquitinated PCNA and implications for translesion synthesis and DNA polymerase exchange', *Nat. Struct. Mol. Biol.* 17: 479–84.

Fujisawa, T., Ueki, T. and Iida, S. (1990). 'Structural change of troponin c molecule and its domains upon Ca2+ binding in the presence of Mg2+ ions measured by a solution X-ray scattering technique', *J. Biochem (Tokyo)* 107: 343–51.

Funari, S. S., Rapp, G., Perbandt, M. *et al.* (2000). 'Structure of free Thermus flavus 5 s rRNA at 1.3 nm resolution from synchrotron X-ray solution scattering', *J. Biol. Chem.* 275: 31283–8.

Garcia de la Torre, J., Huertas, M. L. and Carrasco, B. (2000). 'Calculation of hydrodynamic properties of globular proteins from their atomic-level structure', *Biophys. J.* 78: 719–30.

Giehm, L., Svergun, D. I., Otzen, D. E. and Vestergaard, B. (2011). 'Low-resolution structure of a vesicle disrupting α-synuclein oligomer that accumulates during fibrillation', *P. Natl. Acad. Sci. USA* 108: 3246–51.

Gohon, Y., Pavlov, G., Timmins, P., Tribet, C., Popot, J.-L. and Ebel, C. (2004). 'Partial specific volume and solvent interactions of amphipol A8-35', *Anal. Biochem.* 334: 318–334.

Gohon, Y., Dahmane, T., Ruigrok, R. W. *et al.* (2008). 'Bacteriorhodopsin/amphipol complexes: Structural and functional properties', *Biophys. J.* 94: 3523–37.

Gorynia, S., Banderas, T. M., Pinho, F. G. *et al.* (2011). 'Structural and functional insights into a dodecameric molecular machine: the RuvBL1/RuvBL2 complex', *J. Struct. Biol.* 176: 279–91.

Graebsch, A., Roche, S. and Niessing, D. (2009). 'X-ray structure of Pur-alpha reveals a Whirly-like fold and an unusual nucleic-acid binding surface', *P. Natl. Acad. Sci. USA* 106: 18521–6.

Grishaev, A., Wu, J., Trewhella, J. and Bax, A. (2005). 'Refinement of multidomain protein structures by combination of solution small-angle X-ray scattering and NMR data', *J. Am. Chem. Soc.* 127: 16621–8.

Gutsche, I., Holzinger, J., Rauh, N., Baumeister, W. and May, R. P. (2001). 'ATP-induced structural change of the thermosome is temperature-dependent', *J. Struct. Biol.* 135: 139–46.

Gutsche, I., Holzinger, J., Rossle, M., Heumann, H., Baumeister, W. and May, R. P. (2000). 'Conformational rearrangements of an archaeal chaperonin upon ATPase cycling', *Current Biology* 10: 405–8.

Hamiaux, C., Perez, J., Prange, T., Veesler, S., Ries-Kautt, M. and Vachette, P. (2000). 'The BPTI decamer observed in acidic pH crystal forms pre-exists as a stable species in solution', *J. Mol. Biol.* 297: 697–712.

Hammel, M., Fierobe, H. P., Czjzek, M. *et al.* (2005). 'Structural basis of cellulosome efficiency explored by small angle X-ray scattering', *J. Biol. Chem.* 280: 38562–8.

Hartmann, H. and Decker, H. (2002). 'All hierarchical levels are involved in conformational transitions of the 4 x 6-meric tarantula hemocyanin upon oxygenation', *Biochim. Biophys. Acta* 1601: 132–7.

He, L., Piper, A., Meilleur, F. *et al.* (2010). 'The structure of Sindbis virus produced from vertebrate and invertebrate hosts as determined by small-angle neutron scattering', *J. Virol.* 84: 5270–6.

He, L., Piper, A., Meilleur, F., Hernandez, R., Heller, W. T. and Brown, D. T. (2012). 'Conformational changes in Sindbis virus induced by decreased pH are revealed by small-angle neutron scattering', *J. Virol.* 86: 1982–7.

Hjelm, R. P., Kneale, G. G., Suau, P., Baldwin, J. P., Bradbury, E. M. and Ibel, K. (1977). 'Small-angle neutron-scattering studies of chromatin subunits in solution', *Cell* 10: 139–51.

Holzinger, J. (2002) 'Untersuchung der Reaktionszyklen von Chaperoninen aus Escherichia coli und Thermoplasma acidophilum mit Hilfe der Neutronenkleinwinkelstreuung', Thesis, Ludwig-Maximilians Universität München.

Hunt, J. F., McCrea, P. D., Zaccai, G. and Engelman, D. M. (1997). 'Assessment of the aggregation state of integral membrane proteins in reconstituted phospholipid vesicles using small angle neutron scattering', *J. Mol. Biol.* 273: 1004–19.

Ibel, K. (1976). 'The neutron small-angle camera D11 at the high-flux reactor, Grenoble', *J. Appl. Crystallogr.* 9: 296–309.

Jacques, D. A., Guss, J. M., Svergun, D. I. and Trewhella, J. (2012). 'Publication guidelines for structural modelling of small-angle scattering data from biomolecules in solution', *Acta Crystallogr. D* 68: 620–6.

Jacrot, B. and Zaccai, G. (1981). 'Determination of molecular weight by neutron scattering', *Biopolymers* 20: 2413–26.

Jacrot, B., Pfeiffer, P. and Witz, J. (1976). 'The structure of a spherical plant virus (bromegrass mosaic virus) established by neutron diffraction', *Philos. T. Roy. Soc. B* 276: 109–12.

Jacrot, B., Chauvin, C. and Witz, J. (1977). 'Comparative neutron small-angle scattering study of small spherical RNA viruses', *Nature* 266: 417–21.

Jaenicke, E., Pairet, B., Hartmann, H. and Decker, H. (2012). 'Crystallization and preliminary analysis of crystals of the 24-meric hemocyanin of the emperor scorpion (Pandinus imperator)', *PLoS One* 7: e32548.

Jeganathan, S., Von Bergen, M., Brutlach, H., Steinhoff, H. J. and Mandelkow, E. (2006). 'Global hairpin folding of tau in solution', *Biochemistry* 45: 2283–93.

Johs, A., Hammel, M., Waldner, I., May, R. P., Laggner, P. and Prassl, R. (2006). 'Modular structure of solubilized human apolipoprotein B-100. Low resolution model revealed by small angle neutron scattering', *J. Biol. Chem.* 281: 19732–9.

Kazantsev, A. V., Rambo, R. P., Karimpour, S., Santalucia, J., Jr., Tainer, J. A. and Pace, N. R. (2011). 'Solution structure of RNase P RNA', *RNA* 17: 1159–71.

Kennaway, C. K., Taylor, J. E., Song, C. F. *et al.* (2012). 'Structure and operation of the DNA-translocating type I DNA restriction enzymes', *Gene Dev.* 26: 92–104.

Knop, W., Hirai, M., Schink, H. J. *et al.* (1992). 'A new polarized target for neutron-scattering studies on biomolecules: 1st results from apoferritin and the deuterated 50S subunit of ribosomes', *J. Appl. Crystallogr.* 25: 155–65.

Koch, M. H. J. (1989). 'The structure of chromatin and its condensation mechanism', in Heinemann, U. and Saenger, W. (eds.), *Protein–Nucleic Acid Interaction*. London: McMillan.

Kornberg, R. D. (1974). 'Chromatin structure: A repeating unit of histones and DNA', *Science* 184: 868–71.

Kozielski, F., Svergun, D., Zaccai, G., Wade, R. H. and Koch, M. H. J. (2001). 'The overall conformation of conventional kinesins studied by small angle X-ray and neutron scattering', *J. Biol. Chem.* 276: 1267–75.

Kozielski, F., Sack, S., Marx, A. *et al.* (1997). 'The crystal structure of dimeric kinesin and implications for microtubule-dependent motility', *Cell* 91: 985–94.

Kruse, J., Timmins, P. A. and Witz, J. (1982). 'A neutron-scattering study of the structure of compact and swollen forms of southern bean mosaic virus', *Virology* 119: 42–50.

Labeit, S. and Kolmerer, B. (1995). 'Titins: Giant proteins in charge of muscle ultrastructure and elasticity [see comments]', *Science* 270: 293–6.

Le Maire, M., Arnou, B., Olesen, C., Georgin, D., Ebel, C. and Moller, J. V. (2008). 'Gel chromatography and analytical ultracentrifugation to determine the extent of detergent binding and aggregation, and Stokes radius of membrane proteins using sarcoplasmic reticulum Ca^{2+}-ATPase as an example', *Nature Protocols* 3: 1782–95.

Li, W., Kamtekar, S., Xiong, Y., Sarkis, G. J., Grindley, N. D. F. and Steitz, T. A. (2005). 'Structure of a synaptic γδ resolvase tetramer covalently linked to two cleaved DNAs', *Science* 309: 1210–5.

Lipfert, J., Columbus, L., Chu, V. B., Lesley, S. A. and Doniach, S. (2007). 'Size and shape of detergent micelles determined by small-angle X-ray scattering', *J. Phys. Chem. B* 111: 12427–38.

Luger, K., Mader, A. W., Richmond, R. K., Sargent, D. F. and Richmond, T. J. (1997). 'Crystal structure of the nucleosome core particle at 2.8 angstrom resolution', *Nature* 389: 251–60.

Madl, T., Gabel, F. and Sattler, M. (2011). 'NMR and small-angle scattering-based structural analysis of protein complexes in solution', *J. Struct.l Biol.* 173: 472–82.

Makowski, L., Gore, D., Mandava, S. *et al.* (2011). 'X-ray solution scattering studies of the structural diversity intrinsic to protein ensembles', *Biopolymers* 95: 531–42.

Marino, M., Zou, P., Svergun, D., *et al.* (2006). 'The Ig doublet Z1Z2: A model system for the hybrid analysis of conformational dynamics in Ig tandems from titin', *Structure* 14: 1437–47.

Marone, P. A., Thiyagarajan, P., Wagner, A. M. and Tiede, D. M. (1999). 'Effect of the alkyl chain length on crystallization of a detergent-solubilized membrane protein: Correlation of protein-detergent particle size and particle-particle interaction with crystallization of the photosynthetic reaction center from Rhodobacter sphaeroides', *J. Crystal. Growth* 207: 214–25.

Mattinen, M. L., Paakkonen, K., Ikonen, T., *et al.* (2002). 'Quaternary structure built from subunits combining NMR and small-angle X-ray scattering data', *Biophys. J.* 83: 1177–83.

May, R. P. and Nowotny, V. (1989). 'Distance information derived from neutron low-q scattering', *J. Appl. Crystallogr.* 22: 231–7.

May, R. P., Nowotny, V., Nowotny, P., Voß, H. and Nierhaus, K. H. (1992). 'Inter-protein distances within the large subunit from *Escherichia coli* ribosomes', *EMBO Journal* 11: 373–8.

McDevitt, C. A., Shintre, C. A., Grossmann, J. G. *et al.* (2008). 'Structural insights into P-glycoprotein (ABCB1) by small angle X-ray scattering and electron crystallography', *FEBS Lett.* 582: 2950–6.

Mellema, J. E., Andree, P. J., Krygsman, P. C. J. *et al.* (1981). 'Structural investigations of influenza B virus', *J. Mol. Biol.* 151: 329–36.

Mertens, H. D. T. and Svergun, D. I. (2010). 'Structural characterization of proteins and complexes using small-angle X-ray solution scattering', *J. Struct. Biol.* 172: 128–41.

Moore, P. B., Engelman, D. M. and Schoenborn, B. P. (1974). 'Asymmetry in 50S ribosomal subunit of Escherichia coli', *P. Natl. Acad. Sci. USA* 71: 172–6.

Morgan, H. P., Mertens, H. D., Guariento, M. *et al.* (2012). 'Structural analysis of the C-terminal region (modules 18–20) of complement regulator factor H (fh)', *PloS One* 7: e32187.

Morgan, H. P., Schmidt, C. Q., Guariento, M. *et al.* (2011). 'Structural basis for engagement by complement factor H of c3b on a self surface', *Nat. Struct. Mol. Biol.* 18: 463–70.

Mylonas, E., Hascher, A., Bernado, P., Blackledge, M., Mandelkow, E. and Svergun, D. I. (2008). 'Domain conformation of tau protein studied by solution small-angle X-ray scattering', *Biochemistry* 47: 10345–53.

Nierhaus, K. H., Lietzke, R., May, R. P. *et al.* (1983). 'Shape determinations of ribosomal proteins in situ', *P. Natl. Acad. Sci. USA* 80: 2889–93.

Nissen, P., Hansen, J., Ban, N., Moore, P. B. and Steitz, T. A. (2000). 'The structural basis of ribosome activity in peptide bond synthesis', *Science* 289: 920–30.

Nitsch, M., Walz, J., Typke, D., Klumpp, M., Essen, L. O. and Baumeister, W. (1998). 'Group II chaperonin in an open conformation examined by electron tomography', *Nat. Struct. Biol.* 5: 855–7.

Nöllmann, M., He, J., Byron, O. and Stark, W. M. (2004). 'Solution structure of the tn3 resolvase-crossover site synaptic complex', *Mol. Cell.* 16: 127–37.

Nöllmann, M., Stark, W. M. and Byron, O. (2005). 'A global multi-technique approach to study low-resolution solution structures', *J. Appl. Crystallogr.* 38: 874–87.

Nowotny, P., Rühl, M., Nowotny, V. *et al.* (1994). 'Direct shape determination of ribosomal proteins in solution and within the ribosome by means of neutron scattering', *Biophys. Chem.* 53: 115–22.

Oliveira, C. L., Behrens, M. A., Pedersen, J. S., Erlacher, K. and Otzen, D. (2009). 'A SAXS study of glucagon fibrillation', *J. Mol. Biol.* 387: 147–61.

Pardon, J. F., Worcester, D. L., Wooley, J. C., Cotter, R. I., Lilley, D. M. J. and Richards, B. M. (1977). 'The structure of the chromatin core particle in solution', *Nucleic Acids Res.* 4: 3199–214.

Pattanayek, R., Williams, D. R., Gian Rossi, G. *et al.* (2011). 'Combined SAXS/em based models of the S. elongatus post-translational circadian oscillator and its interactions with the output His-kinase sasA', *PLoS One* 6: e23697.

Pelikan, M., Hura, G. L. and Hammel, M. (2009). 'Structure and flexibility within proteins as identified through small angle X-ray scattering', *Gen. Physiol. Biophys.* 28: 174–89.

Perkins, S. J., Nan, R., Li, K., Khan, S. and Abe, Y. (2011). 'Analytical ultracentrifugation combined with X-ray and neutron scattering: Experiment and modelling', *Methods* 54: 181–99.

Popot, J.-L. (2010). 'Amphipols, nanodiscs, and fluorinated surfactants: Three nonconventional approaches to studying membrane proteins in aqueous solutions'. In Kornberg, R. D., Raetz, C. R. H., Rothman, J. E. and Thorner, J. W. (eds.), *Annual Review of Biochemistry*, vol. 79.

Popot, J.-L., Berry, E. A., Charvolin, D. *et al.* (2003). 'Amphipols: Polymeric surfactants for membrane biology research', *Cell. Mol. Life Sci.* 60: 1559–74.

Prassl, R. and Laggner, P. (2009). 'Molecular structure of low density lipoprotein: Current status and future challenges', *Eur. Biophys. J. Biophys. Lett.* 38: 145–58.

Rai, N., Nöllmann, M., Spotorno, B., Tassara, G., Byron, O. and Rocco, M. (2005). 'SOMO(SOlution MOdeler): Differences between X-ray- and NMR-derived bead models suggest a role for side chain flexibility in protein hydrodynamics', *Structure* 13: 723–34.

Rambo, R. P. and Tainer, J. A. (2010). 'Bridging the solution divide: Comprehensive structural analyses of dynamic RNA, DNA, and protein assemblies by small-angle X-ray scattering', *Curr. Opin. Struc. Biol.* 20: 128–37.

Rhys, N. H., Wang, M.-C., Jowitt, T. A., Helliwell, J. R., Grossmann, J. G. and Baldock, C. (2011). 'Deriving the ultrastructure of α-crustacyanin using lower-resolution structural and biophysical methods', *J. Synchrotron Rad.* 18: 79–83.

Rice, P. A. and Steitz, T. A. (1994). 'Model for a DNA-mediated synaptic complex suggested by crystal packing of γδ resolvase subunits', *EMBO J.* 13: 1514–24.

Rodnina, M., Wintermeyer, W. and Green, R. (eds.) (2011). *Ribosomes Structure, Function, and Dynamics.* Vienna: Springer.

Ruskamo, S., Chukhlieb, M., Vahokoski, J. *et al.* (2012). 'Juxtanodin is an intrinsically disordered F-actin-binding protein', *Sci. Rep.* 2: 899.

Sabatucci, A., Vachette, P., Beltramini, M., Salvato, B. and Dainese, E. (2005). 'Comparative structural analysis of low-molecular mass fragments of *Rapana venosa* hemocyanin obtained using two different procedures', *J. Struct. Biol.* 149: 127–37.

Sabatucci, A., Vachette, P., Vasilyev, V. B. *et al.* (2007). 'Structural characterization of the ceruloplasmin:lactoferrin complex in solution', *J. Mol. Biol.* 371: 1038–46.

Satyshur, K. A., Pyzalska, D., Greaser, M., Rao, S. T. and Sundaralingam, M. (1994). 'Structure of chicken skeletal muscle troponin c at 1.78 a resolution', *Acta Crystallogr. D* 50: 40–49.

Schindler, T., Sicheri, F., Pico, A., Gazit, A., Levitzki, A. and Kuriyan, J. (1999). 'Crystal structure of Hck in complex with a Src family-selective tyrosine kinase inhibitor', *Mol. Cell* 3: 639–48.

Schluenzen, F., Tocilj, A., Zarivach, R. *et al.* (2000). 'Structure of functionally activated small ribosomal subunit at 3.3 Å resolution', *Cell* 102: 615–23.

Schmid, M. F., Sherman, M. B., Matsudaira, P., Tsuruta, H. and Chiu, W. (1999). 'Scaling structure factor amplitudes in electron cryomicroscopy using X-ray solution scattering', *J. Struct. Biol.* 128: 51–7.

Schmidt, C. Q., Herbert, A. P., Mertens, H. D. *et al.* (2010). 'The central portion of factor H (modules 10–15) is compact and contains a structurally deviant CCP module', *J. Mol. Biol.* 395: 105–22.

Schneidman-Duhovny, D., Kim, S. J. and Sali, A. (2012). 'Integrative structural modeling with small angle X-ray scattering profiles', *BMC Struct. Biol.* 12: 17.

Silva, J. C., Borges, J. C., Cyr, D. M., Ramos, C. H. and Torriani, I. L. (2011). 'Central domain deletions affect the SAXS solution structure and function of yeast Hsp40 proteins Sis1 and Ydj1', *BMC Struct Biol.* 11: 40.

Singh, R. K., Larson, J. D., Zhu, W. *et al.* (2011). 'Small-angle X-ray scattering studies of the oligomeric state and quaternary structure of the trifunctional proline utilization A (PutA) flavoprotein from Escherichia coli', *J. Biol. Chem.* 286: 43144–53.

Skar-Gislinge, N. and Arleth, L. (2011). 'Small-angle scattering from phospholipid nanodiscs: Derivation and refinement of a molecular constrained analytical model form factor', *Phys. Chem. Chem. Phys.* 13: 3161–70.

Spirin, A. S., Serdyuk, I. N., Shpungin, J. L. and Vasiliev, V. D. (1979). 'Quaternary structure of the ribosomal 30S subunit: Model and its experimental testing', *P. Natl. Acad. Sci. USA* 76: 4867–71.

Stegmann, R., Manakova, E., Rössle, M. *et al.* (1998). 'Structural changes of the Escherichia coli GroEL–GroES chaperonins upon complex formation in solution: A neutron small angle scattering study', *J. Struct. Biol.* 121: 30–40.

Stöffler, G. and Stöffler-Meilicke, M. (1984). 'Immunoelectron microscopy of ribosomes', *Annu. Rev. Biophys. Bio.* 13: 303–30.

Street, T. O., Lavery, L. A. and Agard, D. A. (2011). 'Substrate binding drives large-scale conformational changes in the hsp90 molecular chaperone', *Mol. Cell* 42: 96–105.

Stubbs, M. J., Miller, A., Sizer, P. J. H., Stephenson, J. R. and Crooks, A. J. (1991). 'X-ray solution scattering of Sindbis virus: Changes in conformation induced at low pH', *J. Mol. Biol.* 221: 39–42.

Stuhrmann, H. B. and Nierhaus, K. H. (1996). 'The determination of the *in situ* structure by nuclear spin contrast variation', *Basic Life Sci.* 64: 397–413.

Stuhrmann, H. B., Haas, J., Ibel, K. *et al.* (1976). 'New low resolution model for 50S subunit of Escherichia coli ribosomes', *P. Natl. Acad. Sci. USA* 73: 2379–83.

Svergun, D. I. (1999). 'Restoring low resolution structure of biological macromolecules from solution scattering using simulated annealing', *Biophys. J.* 76: 2879–86.

Svergun, D. I., Koch, M. H. J. and Serdyuk, I. N. (1994). 'Structure of the 50S subunit of E. coli ribosomes from solution scattering 1) synchrotron radiation study', *J. Mol. Biol.* 240: 66–77.

Svergun, D. I., Burkhardt, N., Pedersen, J. S. *et al.* (1997). 'Solution scattering structural analysis of the 70 S Escherichia coli ribosome by contrast variation. 2. A model of the ribosome and its RNA at 3.5 nm resolution', *J. Mol. Biol.* 271: 602–18.

Svergun, D. I. and Nierhaus, K. H. (2000). 'A map of protein-rRNA distribution in the 70S Escherichia coli ribosome', *J. Biol. Chem.* 275: 14432–9.

Tabarani, G., Thépaut, M., Stroebel, D. *et al.* (2009). 'DC-SIGN neck domain is a pH-sensor controlling oligomerization', *J. Biol. Chem.* 284: 21229–40.

Tanford, C. (1972). 'Micelle shape and size', *J. Phys. Chem.* 76: 3020–4.

Thiyagarajan, P. and Tiede, D. M. (1994). 'Detergent micelle structure and micelle–micelle interactions determined by small-angle neutron-scattering under solution conditions used for membrane-protein crystallization', *J. Phys. Chem.* 98: 10343–51.

Thuman-Commike, P. A., Tsuruta, H., Greene, B., Prevelige, J., P.E., King, J. and Chiu, W. (1999). 'Solution X-ray scattering-based estimation of electron cryomicroscopy imaging parameters for reconstruction of virus particles', *Biophys. J.* 76: 2249–61.

Timmins, P. A., Hauk, J., Wacker, T. and Welte, W. (1991). 'The influence of heptane-1,2,3-triol on the size and shape of LDAO micelles. Implications for the crystallisation of membrane proteins', *FEBS Lett.* 280: 115–20.

Timmins, P. A., Pebay-Peyroula, E. and Welte, W. (1994). 'Detergent organisation in solutions and in crystals of membrane proteins', *Biophys. Chem.* 53: 27–36.

Tskhovrebova, L., Walker, M. L., Grossmann, J. G., Khan, G. N., Baron, A. and Trinick, J. (2010). 'Shape and flexibility in the titin 11-domain super-repeat', *J. Mol. Biol.* 397: 1092–105.

Tsutakawa, S. E., Van Wynsberghe, A. W., Freudenthal, B. D. *et al.* (2011). 'Solution X-ray scattering combined with computational modeling reveals multiple conformations of covalently bound ubiquitin on PCNA', *P. Natl. Acad. Sci. USA* 108: 17672–7.

VanOudenhove, J., Anderson, E., Krueger, S. and Cole, J. L. (2009). 'Analysis of PKR structure by small-angle scattering', *J. Mol. Biol.* 387: 910–20.

Vestergaard, B., Groenning, M., Roessle, M. *et al.* (2007). 'A helical structural nucleus is the primary elongating unit of insulin amyloid fibrils', *PLoS Biol.* 5: e134.

Vijayakrishnan, S., Kelly, S. M., Gilbert, R. J. C. *et al.* (2010). 'Solution structure and characterisation of the human pyruvate dehydrogenase complex core assembly', *J. Mol. Biol.* 399: 71–93.

Vinogradova, M. V., Stone, D. B., Malanina, G. G. *et al.* (2005). 'Ca(2+)-regulated structural changes in troponin', *P. Natl. Acad. Sci. USA* 102: 5038–43.

Volkov, V. V. and Svergun, D. I. (2003). 'Uniqueness of *ab initio* shape determination in small angle scattering', *J. Appl. Crystallogr.* 36: 860–4.

Wang, X., Watson, C., Sharp, J. S., Handel, T. M. and Prestegard, J. H. (2011). 'Oligomeric structure of the chemokine CCL5/RANTES from NMR, MS, and SAXS data', *Structure* 19: 1138–48.

Whelan, F., Stead, J. A., Shkumatov, A. V., Svergun, D. I., Sanders, C. M. and Antson, A. A. (2012). 'A flexible brace maintains the assembly of a hexameric replicative helicase during DNA unwinding', *Nucleic Acids Res.* 40: 2271–83.

Williams, G. J., Williams, R. S., Williams, J. S. *et al.* (2011). 'ABC ATPase signature helices in Rad50 link nucleotide state to Mre11 interface for DNA repair', *Nat. Struct. Mol. Biol.* 18: 423–31.

Willumeit, R., Diedrich, G., Forthmann, S. *et al.* (2001). 'Mapping proteins of the 50S subunit from Escherichia coli ribosomes', *BBA-Gene Struct. Expr.* 1520: 7–20.

Wimberly, B. T., Brodersen, D. E., Clemons, W. M. *et al.* (2000). 'Structure of the 30S ribosomal subunit', *Nature* 407: 327–39.

Wu, Z. P., Gogonea, V., Lee, X. *et al.* (2009). 'Double super-helix model of high density lipoprotein', *J. Biol. Chem.* 284: 36605–19.

Wu, Z. P., Gogonea, V., Lee, X. *et al.* (2011). 'The low resolution structure of ApoA1 in spherical high density lipoprotein revealed by small angle neutron scattering', *J. Biol. Chem.* 286: 12495–508.

Xu, Z., Horwich, A. L. and Sigler, P. B. (1997). 'The crystal structure of the asymmetric GroEL-GroES-(ADP)7 chaperonin complex', *Nature* 388: 741–50.

Xu, X., Reinle, W., Hannemann, F. *et al.* (2008). 'Dynamics in a pure encounter complex of two proteins studied by solution scattering and paramagnetic NMR spectroscopy', *J. Am. Chem. Soc.* 130: 6395–403.

Yang, S., Park, S., Makowski, L. and Roux, B. (2009). 'A rapid coarse residue-based computational method for X-ray solution scattering characterization of protein folds and multiple conformational states of large protein complexes', *Biophys. J.* 96: 4449–63.

Yang, S., Parisien, M., Major, F. and Roux, B. (2010a). 'RNA structure determination using SAXS data', *J. Phys. Chem. B* 114: 10039–48.

Yang, S., Blachowicz, L., Makowski, L. and Roux, B. (2010b). 'Multidomain assembled states of Hck tyrosine kinase in solution', *P. Natl. Acad. Sci. USA* 107: 15757–62.

Yang, W. and Steitz, T. A. (1995). 'Crystal structure of the site-specific recombinase γδ resolvase complexed with a 34 bp cleavage site', *Cell* 82: 193–207.

Zerrad, L., Merli, A., Schroder, G. F. *et al.* (2011). 'A spring-loaded release mechanism regulates domain movement and catalysis in phosphoglycerate kinase', *J. Biol. Chem.* 286: 14040–8.

Zhang, X., Konarev, P. V., Petoukhov, M. V. *et al.* (2006). 'Multiple assembly states of lumazine synthase: A model relating catalytic function and molecular assembly', *J. Mol. Biol.* 362: 753–70.

Zou, P., Gautel, M., Geerlof, A., Wilmanns, M., Koch, M. H. and Svergun, D. I. (2003). 'Solution scattering suggests cross-linking function of telethonin in the complex with titin', *J. Biol. Chem.* 278: 2636–44.

Zou, P., Pinotsis, N., Lange, S. *et al.* (2006). 'Palindromic assembly of the giant muscle protein titin in the sarcomeric Z-disk', *Nature* 439: 229–33.

Kinetic and perturbation studies

7

7.1 Dynamics and kinetics   221

7.2 Perturbation methods   223

7.3 Temperature scans and
T-jumps   230

7.4 High-pressure experiments   233

7.5 Stopped-flow and
continuous-flow mixing   237

7.6 Light-triggered processes   247

References   250

Biological processes, which involve chemical reactions as well as phase transitions, cover a wide range of times from femtoseconds to years. Clearly, the higher brilliance of X-ray sources is a decisive advantage in the study of faster processes, as the time resolution of SANS is currently limited in the most favourable cases to about 100 ms for stopped-flow processes (Grillo, 2009) and 20 ms for light-induced processes. This may change with the advent of pulsed-source instrumentation.

Early time-resolved SAXS experiments on solutions made on second-generation SR sources using gas detectors were limited by the available flux, and therefore concentrated with few exceptions on relatively slow processes (s to min) such as the transitions associated with the (dis)assembly of multisubunit proteins (Moody *et al.* 1980, Inoko *et al.* 1983), components of the cytoskeleton (such as microtubules (Mandelkow 1980), actin (Sayers *et al.* 1985, Matsudaira *et al.* 1987), collagen (Suarez *et al.* 1985)) or virus capsids (Cuillel *et al.* 1983). The higher brilliance of third-generation sources made it possible to study faster (100 μs to s) transitions such as protein (un)folding and to use integrating detectors (CCDs) despite their higher intrinsic noise. To reduce this noise, CCDs are in turn being replaced by pixel area detectors (PADs).

In recent years a number of fast pump–probe SAXS/WAXS measurements (20 ps to s) have also been made to monitor intramolecular structural changes in proteins during reactions. On present storage rings the time resolution is limited by the bunch length, which cannot be reduced much below a few ps without drastically reducing the circulating current. It is expected, however, that in the not too distant future even faster measurements will become possible using coherent radiation from X-ray free electron lasers where the minimum observation time is defined by the duration of a cycle of the radiation, which in the case of X-rays may be as short as 10^{-18} s. These very fast measurements are, however, limited to processes that can be triggered by a light pulse.

Slow processes are best discussed in terms of thermodynamic forces ($1/T$, p/T and μ/T) and their associated flows (Kondepudi and Prigogine 1998, Dill and Bromberg 2003) whereas for the very fast reactions monitored with higher time resolution molecular dynamics modelling may be attempted.

7.1 Dynamics and kinetics

Dynamics implies the study of changes in motion and their causes. In contrast, kinetics refers to the results of the action of (thermodynamic) forces on matter. Elastic scattering experiments on solutions, which yield a time and spatial average, can thus give only kinetic information.

The rate (k) at which the equilibrium between the initial (reactants) and final state (products) is established in thermal reactions is usually represented by the empirical Arrhenius equation $k(T) = A \exp(-E_a/RT)$, where E_a is the activation energy. This equation is well explained by transition state theory, which assumes that there is a static barrier of higher free energy between reactants and products; it is applicable to situations where the energy landscape of the reaction is simple, which is usually valid for slow reactions like assembly. According to transition state theory, $A_{\text{theor}} = \kappa(k_B T/h) \exp(\Delta S^\ddagger/R)$, with $\Delta S^\ddagger$ the activation entropy and κ the transmission coefficient that takes into account that some of the activated molecules may return to the initial state (see, for example, Pilling and Seakins (1995)).

As shown by Frauenfelder and co-workers, transition state theory is not applicable to protein kinetics, which corresponds to a very complex energy landscape with fluctuating barriers where the hydration sphere, bulk solvent, reactants and products all play significant roles (Beece *et al.* 1980, Frauenfelder *et al.* 2010). The rates of individual reaction steps in these complex mechanisms can, however, usually still be approximated by the Arrhenius equation or the closely related theoretical Eyring equation (Beece *et al.* 1980).

Note that in transition state theory the actual time it takes for a molecule to cross the barrier does not appear in the expression for the rate constant, which can be rewritten as $k(T) = (\kappa k_B T/h) \exp(-\Delta G^\ddagger/RT)$. The first factor indicates that the molecules cross the barrier very fast, whereas the second one presents the probability that a molecule would have enough energy to do so. Time enters indirectly through the inverse of the ratio $h/k_B = 4.799237 \times 10^{-11}$ [s $\times$ K]. Since $\Delta S = k\Delta \ln W$, where W is the multiplicity of microscopic degrees of freedom, this represents the cost (h) in terms of action (energy $\times$ time) associated with a change of one natural log unit in W (Cohen-Tannoudji *et al.* 1998). For $\Delta S^\ddagger = 0$ and $\kappa = 1$ the value of the pre-exponential factor at room temperature (300 K) is expected to be around 10^{13} s^{-1} for first-order reactions, which corresponds to interatomic vibration frequencies. Although the elementary steps in chemical reactions are thus expected to occur in tens to hundreds of femtoseconds (10^{-15} s), electron rearrangements in the transition state may be even faster, in the attosecond (10^{-18} s) range (Bucksbaum 2007).

As molecules with different energies have different trajectories between two successive states and the times involved are much too short for current structural X-ray and even more neutron techniques, it is not justified to present the results of time-resolved X-ray structure analysis as the equivalent of a movie, as is unfortunately done too often. All one observes are the changes in the populations of intermediate states containing a measurable fraction of the molecules. Even in the case of intramolecular reactions in proteins during pump–probe experiments one does not observe the movement of atoms, only the result of these movements. The approaches used for the analysis of time-resolved scattering patterns are thus the same as for mixtures (least-squares fitting, SVD analysis and so on). DLS is probably the most useful complementary technique, as it yields the number of species in solution. Shape reconstruction from time-resolved data is meaningful only if the individual scattering curves of intermediates can be extracted. If this is not the case, reconstruction software will generate objects along the reaction pathway which may not correspond to any physical reality (Lamb 2008).

The nature of the structural changes—work or heat—occurring during fast reactions of macromolecules also matters. Time does not appear in any of the laws of thermodynamics, but as pointed out by McClare forty years ago: 'Work implies the transformation of one form of stored energy into another in a time interval which does not allow exchange with the intra- and intermolecular translational, rotational, and vibrational energies which constitute heat' (McClare 1971).

This sets a lower limit to the rates of the reactions in which work is done in molecular machines, and implies that most of these reactions must have low activation energies and occur close to equilibrium where energy-transfer efficiency is highest. Molecules that are close to equilibrium and have the same activation barriers relax exponentially, but when there is a distribution of barriers, as is the case in proteins, relaxation is non-exponential (Frauenfelder *et al.* 1999).

If to be efficient the steps involving energy transfer in biological macromolecules must occur close to equilibrium and must be fast, the return to the initial state, which involves a hierarchy of vibrational, rotational and translational energy transfer processes dissipating the excess energy, can be slow. This also explains that after a reaction there is a refractory period during which enzymes are inactive, as observed in single-molecule fluorescence experiments on cholesteryl oxidase (Lu *et al.* 1998), or during the slow return to the initial state at the end of the photocycle of bacteriorhodopsin (Váró and Lanyi 1991).

As work is a constrained release of energy over a few degrees of freedom, in contrast with heat which is distributed over many degrees of freedom (see, for example, Atkins (1984)), the view that biological systems are molecular machines implies that one should be looking for coherent displacements associated with work and for structures which can select coherent motion from uncorrelated motion (as, for instance, in a piston). These structures may extend across several length scales, as in the case of muscle. It should be noted, however, that at the molecular level work is always associated with some form of charge separation that may take place faster than chemical reactions.

The ensuing overview of different perturbation methods is followed by a few representative examples of applications illustrating the complementarity of SAS/WAS with other, mainly hydrodynamic, spectroscopic and imaging methods. The separation of these applications in terms of the thermodynamic forces is somewhat arbitrary as perturbation methods are sometimes combined (for example, temperature and pressure, temperature and chemical denaturation or pressure and chemical denaturation).

7.2 Perturbation methods

7.2.1 Temperature scans and jumps

Temperature is the most easily controllable thermodynamic variable; a programmable water bath and a temperature-controlled sample cell are usually all that is required for experiments involving continuous or step scans or static measurements at different temperatures.

Assembly reactions and other phase transitions including protein (un)folding are preferably triggered by rapid temperature changes. A number of devices based on heat exchange with water baths having dead times of the order of a few hundred ms have been described (Renner *et al.* 1983, Hiragi *et al.* 1988b). More recent instruments are based on Peltier elements controlling the temperature of the syringes of a stopped-flow device. Mixing small volumes of solutions with different temperatures yields reversible jumps of $\pm 40°$ C with ms dead times but at the cost of dilution. Continuous mixing devices such as those described here (Kathuria *et al.* 2011), which should be able to reach the $100\,\mu$s range, appear not to have been used for this purpose yet.

There are few recent T-jump SAXS or SANS studies on protein solutions, but the method has remained popular for lipid systems. In a recent application the reversible transition from a thermotropic crystalline to an isotropic state of the lipid core in LDLs near physiological temperature was monitored with a time resolution of 10 ms limited by the available X-ray flux (Prassl *et al.* 2008). A commercial infrared erbium laser ($\lambda = 1.5\,\mu$m) with an output of 2 J was used to increase the temperature of LDL solutions by 10–12° C with a single 2 ms flash. Laser T-jumps used in optical experiments usually have shorter pulses (35 ps) with lower power (5 mJ), which is compensated by inert absorbing dyes with fast internal conversion (Phillips *et al.* 1995). This technique, which can produce T-jumps of tens of degrees in 10 ps with thin (50 μm) samples, seems not to have been used yet in combination with SAXS/WAXS.

For fast (ps–μs) measurements the effects of impulsive heating on the scattering of the solvent must be taken into account, and this was studied by X-ray scattering using short (100–150 fs) near-infrared pulses with organic solvents (Cammarata *et al.* 2006) and water (Lindenberg *et al.* 2005). As the scattering volume is always an open system, which can exchange energy and matter with its immediate surroundings, a short period of heating at nearly constant volume (≈ 10 ns) is followed by solvent expansion and pressure relaxation over about 100 ns. The fundamental time-scale for water thermalisation through hydrogen bonds is of the order of 5 ps, and can presently be reached only in optical measurements, which show that significant temperature-dependent changes occur during the melting of a DNA hairpin, for example, in less than 100 ps (Ma *et al.* 2006).

7.2.2 Pressure and pressure jumps

Although pressure studies are the natural complement of temperature studies they have lagged behind due to greater technical difficulties and have often not been pursued beyond feasibility studies. The hydrostatic pressures that can be applied to solutions are limited by ice formation at pressures around 1–1.5 GPa ($1-1.5 \times 10^4$ bars), which is an order of magnitude higher than the maximum pressure to which organisms are subjected at the deepest point in the ocean.

Pressure cells are made of low-Z materials, which have a low absorption for X-rays, such as beryllium, or of materials such as steel that can withstand high pressures but are highly absorbing. In this case, windows made of light elements, such as beryllium or diamond, are required along the path of the beam. Some devices are optimised to achieve high pressures in static measurements, and others to provide precise control of fast pressure jumps. A variety of cells have been developed in different areas such as lipid research or polymers, and early feasibility studies on solutions of biological macromolecules were based on diamond anvil cells (Czeslik *et al.* 1996) or on a piston-cylinder-type cell designed for the study of polymers (Lorenzen *et al.* 1993, Kleppinger *et al.* 1997). One of the requirements for SAXS/WAXS on solutions is that the sample thickness should not vary with pressure or between successive experiments, as is often the case with diamond anvil cells, which are therefore no longer used for solution work. The characteristics of some recent X-ray pressure cells for solutions are given in Table 7.1. Diamond windows give Kossel lines (Nishikawa *et al.* 2001), and it is therefore useful to have cell designs that do not require the removal of the windows when the sample is changed (Krywka *et al.* 2008).

Large submicrosecond pressure drops (0.2 GPa in 0.7 μs) have been achieved with optical measurements using a technique that should be adaptable to SAXS with small beams (Dumont *et al.* 2009).

There are also several designs of high pressure cells for SANS (Table 7.2). These are larger as required by the size of the neutron beams and have sapphire windows. Their necessary thickness is not a problem, due to the low absorption of neutrons. Sample volumes are about a hundred times larger than for SAXS.

Table 7.1 Characteristics of some recent pressure cells for solution work with X-rays. Note that the window thickness (D_{window}) and the path length (D) of the cells have been optimised for different X-ray wavelengths. All cells can be thermostated in the range $-20°C$ to $+80°C$ or more. (* Still under development for solutions).

Window	D_{window} (mm)	Sample (μL)	D (mm)	2θ max. (°)	Max. pressure (GPa = 10^5 bars)	P-jump	Reference
Diamond	1	50	4.0	15	0.5	–	(Kato and Fujisawa 1998, Nishikawa et al. 2001)
Be	1.5	15	1.0	30	0.3	60 s	(Pressl et al. 1997)
Diamond	0.5	12	2.0	15	0.4	–	(Ando et al. 2008b)
Diamond	0.8	19	1.5	12	0.7	5 ms	(Woenckhaus et al. 2000)
Diamond	1	< 100	10	12	0.7	–	(Skouri-Panet et al. 2006)
Diamond	1	< 10	2	21	0.5	5 ms	(Brooks et al. 2010)*

Table 7.2 Characteristics of some high-pressure cells for neutron scattering. Note the larger sample volume required. The cells can be thermostated in the range $-20°C$ to $+80°C$.

D_{window} (mm)	Sample (mL)	D (mm)	2θ range (°)	Max. pressure (GPa = 10^5 bars)	Reference
10	~2	5	17	0.4	(Lechner et al. 1985)
20	~2	2	9.5	0.5	(Bonetti and Calmettes 2005)
35	1.3	2	20	0.5	(Kohlbrecher et al. 2007)*

*This cell can be used for simultaneous DLS measurements.

7.2.3 Stopped-flow and continuous-flow mixing

The stopped-flow devices used with X-rays and neutrons are similar to those initially developed for optical experiments. As illustrated in Fig. 7.1A, they consist of two or three motor- or pneumatically-driven syringes and one or two mixing chambers where narrow channels result in turbulent mixing (Panine et al. 2006). Their performance depends essentially on the hydraulic diameter of the flow channel (L) determined by its exact shape, the mean flow rate (v), density (ρ) and dynamic viscosity (η) of the liquid. These variables determine the Reynolds number ($Re = vL\rho/\eta$), which must have values above 10^3 for useful devices. At length scales smaller than the viscous dissipation or Kolmogorov scale $I_d \approx LRe^{-3/4}$, diffusion sets the ultimate limit to the mixing rate.

With neutrons the main limitation for stopped-flow experiments is the amount of material needed due to the number of repetitions (10–100) required to collect enough counts, particularly in the sub-second time range. For this reason, stopped-flow neutron experiments are more frequent in the field of soft matter than in biology (Grillo 2009).

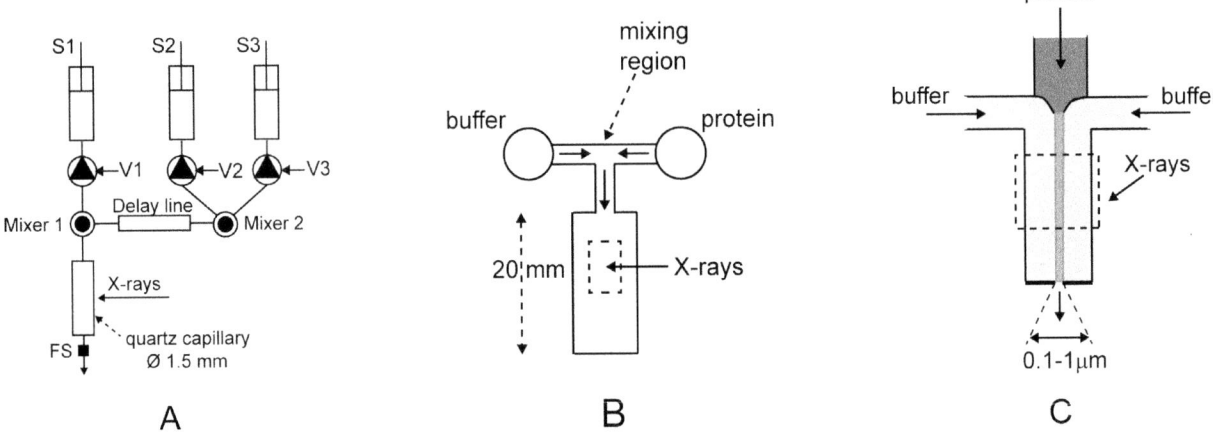

Fig. 7.1 (A) Schematic diagram of a conventional stopped-flow instrument. (Adapted from Panine *et al.* (2006), with the permission of Elsevier, © 2006.) The reagents are contained in the syringes S1 and S2, whereas S3 contains buffer to clean the capillary between shots. The syringes are manually filled using the valves (V). When the syringes stop the fast stopper (FS) immediately interrupts the flow. Mixing can be optimised by varying the volume of the delay line to let the solution age. These devices typically have dead times of 2.5 ms. The continuous flow mixing devices (adapted from Kathuria *et al.* (2011), and reproduced with the permission of John Wiley & Sons, Inc., © Wiley Periodicals, Inc., 2011) based on turbulent flow (B) or laminar flow (C) have dead times of 100–150 μs, but especially laminar flow devices where the focused central channel has dimensions of a few μm at most require much smaller and intense stable X-ray beams than usually available. Stopped-flow and continuous-flow techniques have been combined with many observational techniques (for a review, see Roder *et al.* 2006), especially to study folding of proteins and RNA, and it is useful to acquire some experience with optical methods before embarking on X-ray experiments.

Early X-ray experiments were done with custom-built instruments (Renner *et al.* 1983, Fowler *et al.* 1983, Berthet-Colominas *et al.* 1984, Nagamura *et al.* 1985, Díaz *et al.* 1998). One of the most successful early models operates in the −20°C to +40°C range, with a dead time of 10 ms even with viscous solutions (20 cP), and uses 180 μl per shot (Tsuruta *et al.* 1989). On third-generation sources, commercial devices primarily designed for optical measurements can be used (Panine *et al.* 2006). These typically have dead times of 2.5 ms and dead volumes of 25 μL.

Stopped-flow experiments with neutrons require high fluxes, and were therefore pioneered at the SANS instrument D22 of the high-flux reactor of the ILL. Initially, an improvised device consisting of step-motor-controlled syringes and a 137-QS flow-through quartz cell (Hellma, Müllheim, Germany) with 1 mm path length was used (Holzinger 2002). Subsequently, it was replaced by a commercial stopped-flow instrument (Bio-Logic, Claix/Isère, France) that was modified from an optical device with a 1 mm quartz cell (Grillo *et al.* 2003).

The advantage of stopped-flow methods, which can be carried out routinely at several facilities, lies in the ability to record a complete kinetic trace over a long period of time with a small sample volume and in the flexibility to modify conditions by altering the ratio of the volumes injected by the different syringes. Their main limitations are the dead time, and for longer observation times the possibility of radiation damage, which can be circumvented partly by delaying the start of the X-ray exposure relative to the mixing time and/or protecting the sample between frames using a fast X-ray shutter.

More recent mixing devices aiming at higher time resolutions and a more efficient use of the sample are produced by micromachining or lithographic techniques to take full advantage of small X-ray beams, as illustrated in Fig. 7.1B, C. They are based on continuous-flow mixing with turbulent or laminar flow (Kathuria *et al.* 2011).

Using a micromachined T-shaped turbulent flow mixer, relying like stopped-flow devices on high Reynolds numbers, a mixing time of $70\,\mu s$ at a flow rate of $4\,ms^{-1}$ and a dead time of $160\,\mu s$ were achieved (Akiyama *et al.* 2002). The device is $400\,\mu m$ thick and has a $33\,\mu m$ wide and $100\,\mu m$ long mixing region followed by a $200\,\mu m$ wide observation channel, resulting in a sample consumption of about $0.5\,mL\,s^{-1}$.

In the laminar-flow devices a thin stream of solution of macromolecules is focused in the plane of the channels by faster-flowing solution from the side channels. This solution contains small molecules or ions that rapidly diffuse to the central channel and trigger the desired reaction. As macromolecules have diffusion coefficients that are two orders of magnitude smaller than those of small molecules, they remain in the central channel during the short periods of time involved. The width of the central stream is determined by the value of the ratio ($\sim$1.3) between the pressure on the sides and the inlet pressure of the solution of macromolecules (Knight *et al.* 1998, Brennich *et al.* 2011). Mixing uniformity can be improved by hydrodynamic focusing prior to mixing (Park *et al.* 2006). Mixing times of $10\,\mu s$ can be achieved, but with the devices used for SAXS the highest time-resolution reported was $100\,\mu s$ (Pollack *et al.* 1999, Pollack *et al.* 2001). Given the flow rate of the central stream ($\approx$0.3 m s^{-1}) and its small dimensions, 2.5–5 μm width, and $390\,\mu m$ depth, the device does not require much sample but much smaller and more intense beams than available on most SAXS instruments. The lower flow-rates than in turbulent mixers require smaller beams in the direction of flow to achieve good time-resolution. Diffusive broadening may lead to changes in the concentration of the macromolecule in the central jet for observation times longer than 1s (Pollack 2011). Three-dimensional laminar flow focusing has recently been achieved with optical measurements and mixing times of $10\,\mu s$ (Gambin *et al.* 2010).

With continuous-flow methods, successive time-points in the kinetic trace are measured by increasing the distance between the mixing and observation points. This approach obviously requires relatively larger total sample volumes, which can, however, be compensated by the small size of the devices. The radiation damage in continuous-flow experiments is usually negligible, even at high brilliance sources.

A free-jet micromixer with an exit speed at the nozzle of $13\,m\,s^{-1}$ and a dead time of $75\,\mu s$ has also been successfully tested for inorganic reactions. The internal pressure, which is of the order of $1\,MPa$, may, however, affect the outcome of some experiments on biological macromolecules (Marmiroli *et al.* 2009).

7.2.4 Removal of components from the solution

Most self-assembly or aggregation processes take place in seconds or minutes, and may even continue for days. In technology, such phenomena are important for the stability of preparations and their safe delivery that often implies the removal of a small molecule component, or of ions from the solution as might occur, for example, after an injection. Small molecules can be removed from solutions by placing a short HPLC column in front of the sample cell, and this was applied successfully to study the aggregation of insulin analogues, which started without measurable delay after removal of phenol from the solution (Hillerup Jensen *et al.* 2010). Divalent ions that play a role in the structure and stability of nucleic acids or nucleoprotein complexes such as viruses can be removed easily by mixing with complexants such as EDTA (Aramayo *et al.* 2005).

7.2.5 Triggering with light

Light-triggered processes are usually studied with SAXS/WAXS, though they are in some cases also amenable to SANS, as they do not have the same limitations as mixing experiments. Since cold neutrons do not damage the sample, thousands of repetitions are, in principle, possible, depending on the relaxation time to the initial state and possible photo-induced damage.

When using light as a trigger it must be taken into account that absorbance in the infrared to UV range is larger than for X-rays and even more for neutrons, and relatively thin samples must be used to ensure that they are irradiated over their entire thickness.

To fulfil these conditions a 0.3 mm thick high-pressure sapphire slit nozzle with a $3 \, \text{m s}^{-1}$ jet speed has been used for most fast chemical reactions at the ESRF (Kim *et al.* 2006). Capillaries containing concentrated protein solutions may not always guarantee homogeneous irradiation. Moreover, unwanted heating effects accompanying intense light flashes must be taken into account. In the case of neutrons one can place an aluminium reflector behind the sample to reflect the residual light, and 1 mm quartz cells of 100–200 μL are employed rather than capillaries.

Fast (ps–μs) time-resolved measurements can be used to study structural changes during reactions, including the intramolecular reactions accompanying the working cycle of molecular machines. The accompanying structural changes mainly affect the scattering pattern at higher angles in the range of resolution of 3–0.5 nm ($2 \, \text{nm}^{-1} < q < 15 \, \text{nm}^{-1}$). The resolution of the useable data is typically limited to $q \leq 15 \, \text{nm}^{-1}$ by the heating of the aqueous solvent due to the laser pulse, though with sufficiently concentrated protein solutions it is possible to obtain accurate data up to at least $q = 25 \, \text{nm}^{-1}$.

When the laser pulse is much shorter than the X-ray pulse ($\approx 100 \, \text{ps}$) and the reaction or transition being studied, the signal corresponds to that of a simple mixture of the different molecular species in the solution averaged over the duration of the X-ray pulse. In most cases the laser pulse is much longer (≈ 250 ns) than the X-ray pulses. There must then be a sufficiently long delay between the light and X-ray pulses to avoid any interaction between the two beams or orientation effects due to the laser beam (see, for example, Kim *et al.* (2011)) and to guarantee that the solution is isotropic.

Most fast experiments to date have been carried out at the ID09B beamline at the ESRF using quasi-monochromatic radiation from an undulator selecting a variable number of single X-ray pulses with 100-ps duration from the storage ring pulse train, using a high-speed mechanical chopper and a millisecond shutter. The scattering patterns are collected with a CCD detector (Cammarata *et al.* 2008). Experiments were also carried out with a Pilatus detector and monochromatic radiation (Westenhoff *et al.* 2010). The more recent BioCARS facility at the APS has similar capabilities (Graber *et al.* 2011).

The use of caged compounds is often suggested as a possibility to extend the applicability of light activation as a variety of caged compounds (nucleotides, ions, peptides, neurotransmitters and so on) are available (Giovannardi *et al.* 1998). In each case the possible side reactions and effects of the X-ray beam must be considered. Some reactions, such as microtubule assembly and oscillations, were studied following flash photolysis of caged-GTP to establish assembly conditions in a much shorter time than during T-jumps (Marx *et al.* 1990). The absence of effect of the rate of perturbation on the rate of the structural transitions indicates that either the latter is determined by the protein itself or it is diffusion-limited. An X-ray-dose-dependent but concentration-independent (in the mM concentration range) slow breakdown of caged-GTP, possibly due to free radicals, induced assembly even in the absence of UV photolysis. Such side reactions and the propensity of caged compounds to produce or react with free radicals may be limiting factors in the application of the technique.

7.2.6 Orientation by external fields and forces

Biological macromolecules can be oriented in various fields, but these effects have hitherto hardly been investigated by SAXS or SANS. Among the few applications, solutions of microtubules at concentrations of 4–8 mg/mL have been oriented successfully for SAXS in magnetic fields of 11–17 Tesla (Bras *et al.* 1998). Magnetic fields have been used to orient liquid crystalline systems such as concentrated DNA, or to follow the nematic to smectic transition in tobacco mosaic virus suspensions (Hirai *et al.* 1997). Left–right circular orientation of tobacco mosaic virus (TMV) was observed by SAXS using magnetic field gradients (Hirai 2003). The orientation of photosynthetic membranes and investigation of their light-induced structural changes has been observed by SANS (Nagy *et al.* 2011).

Orientation of TMV (Koch *et al.* 1988) and DNA (Koch 1995) in electric fields has been observed by SAXS with a time resolution of 2 ms applying 1 or 2 ms DC pulses of 0.5–5 kVcm^{-1}. Complete orientation resulting from the field-induced dipole due to displacement of ions can be achieved for short periods of time, even with relatively dilute systems.

Linearly polarised light was used to orient myoglobin in solution via its transition dipole in the 100 ps–1 μs time-range. Although it is unclear whether full orientation was reached, a rotational diffusion time of about 15 ns, in agreement with independent NMR measurements, was obtained (Kim *et al.* 2011). Note that this mechanism of orientation differs fundamentally from that of electric field alignment.

Rheometers with shear–flow alignment have been used sometimes with spectacular results with elongated molecules and fibres (Nordén *et al.* 1998, Castelletto 2006, Sugiyama *et al.* 2009). Unintended flow alignment should be a concern when using microfluidic devices with elongated molecules.

7.2.7 Data collection

For experiments with time resolutions down to the ms, scattering and ancillary data (*T*, *p*, pH ...) can be collected in a series of time frames possibly separated by short pauses. The duration of the frames and pauses can be equal or have a structure that is more appropriate to the particular experiment (for example, exponential; Aramayo *et al.* 2005). Adequate synchronisation of triggering and data-acquisition systems must also be available—a requirement which becomes, of course, more stringent at higher time-resolutions. If necessary, the data acquisition cycle is repeated as often as needed to achieve the required statistics.

For measurements where the detector readout time is limiting and/or the sample flows, as in continuous flow mixing or some flash photolysis experiments, a single time frame is recorded repeatedly until sufficient statistics are obtained. After that, the delay between the pump and probe pulse, or the distance between mixing and observation points, must be adjusted before recording the next time-frame.

7.3 Temperature scans and T-jumps

The effects of heat on solutions of biological macromolecules are essentially associated with entropy-driven processes, and one of the main challenges is to relate structural changes as monitored by different methods to changes in heat capacity ($C_p = (\partial H/\partial T)_p = T(\partial S/\partial T)_p$). Whereas the combination of differential scanning calorimetry (DSC) and SAXS is very frequent for synthetic polymers, there are astonishingly few combined studies of solutions of biological macromolecules.

Equilibrium measurements can provide information about stable intermediates and the conditions or forces that stabilise them. They do not inform about the forces driving transitions between equilibrium structures, nor about the changes of these forces during transitions. This information can usually be obtained only by a combination of methods probing the local structure (such as fluorescence, CD, Fourier transform infrared (FTIR) spectroscopy and NMR) and the overall structure (SAXS/SANS, NMR and hydrodynamic methods) and modelling.

Slow continuous or step scans of the temperature or series of static measurements at different temperatures have been carried out on a number of proteins mainly to follow assembly, and unfolding, or on DNA to follow melting (Puigdomenech *et al.* 1989, Barone *et al.* 1999). Some studies on the unfolding of lysozyme (Hirai 1998, Arai and Hirai 1999, Meersman *et al.* 2010, Koizumi *et al.* 2007, Shiu *et al.* 2008), neocarzinostatin (Pérez *et al.* 2001) and apoflavodoxin (Ayuso-Tejedor *et al.* 2011) provide useful examples of data analysis of conformational ensembles and of combinations of SAXS with other methods. The choice of models to account for the scattering of the unfolded protein—such as the thick persistence length chain and the infinitely thin Gaussian chain without persistence, and their relationship to polymers—is described in detail for neocarzinostatin (Pérez *et al.* 2001).

During thermal unfolding of lysozyme and neocarzinostatin, small but significant non-linear changes in the forward scattering are observed at the transition, which have hitherto not been satisfactorily explained. Similar effects have been observed in the integrated low-angle intensity during DNA melting (Puigdomenech *et al.* 1989).

A number of time-resolved SAXS studies have been carried out on the heat-induced assembly of components of the cytoskeleton (profilactin; Sayers *et al.* 1985), collagen (Suarez *et al.* 1985), microtubules (Mandelkow 1980, Bordas *et al.* 1983, Spann *et al.* 1987) and tobacco mosaic virus protein in equilibrium conditions (Hiragi *et al.* 1988a) or after T-jumps (Potschka *et al.* 1988, Hiragi 1990) or full virus (tobacco mosaic virus (Sano *et al.* 1995), and cucumber green mottle mosaic virus (Sano 1999)) using low-temperature quenching. An equilibrium study of the temperature dependence of actin polymerisation was carried out by SANS (Norman *et al.* 2005). A detailed account of the data analysis used for microtubules, which can be extended readily to similar systems, is given in (Bordas *et al.* 1983), together with a useful method for detecting intermediates based on changes in the slope of plots of the intensity in one part of the pattern against that in another part (intensity correlation plots). Today, such analysis would probably rely rather on the SVD technique.

Most assembly processes can be described in the frame of the nucleation and growth model of Oosawa (Oosawa and Asakura 1975), even if polymerisation is sometimes only partly reversible, as illustrated in Fig. 7.2A for collagen. Nucleation is probably not physiologically relevant, as growth of filaments *in vivo* starts from specific initiation centres. SAXS is particularly useful for studying the early stages of assembly, but becomes rapidly insensitive to the length of filaments to which light scattering with its longer wavelength ($\lambda \approx 60\,\text{nm}$ for a He–Ne

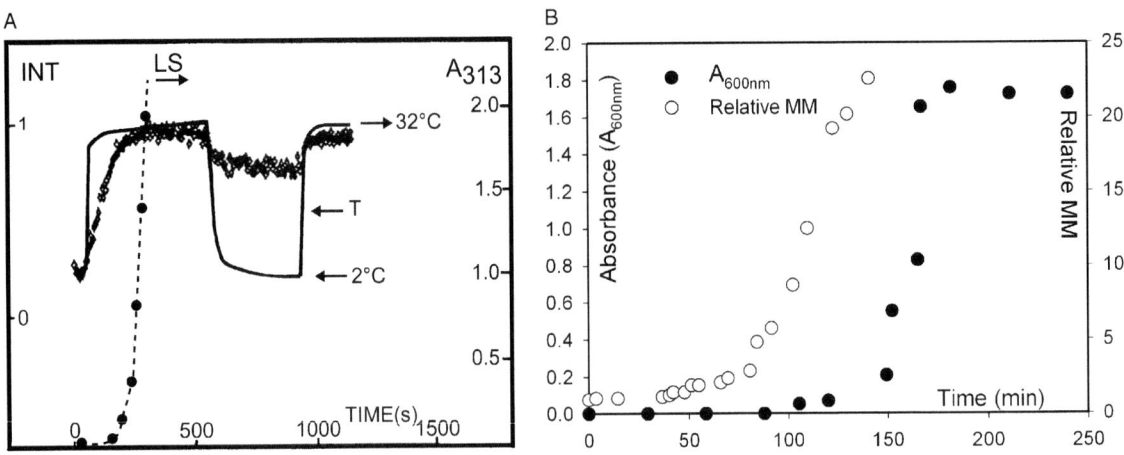

Fig. 7.2 (A) Time course of the temperature (*T*), the integrated scattered intensity (INT($I(q)$ for $1.0 \times 10^{-1}\,\mathrm{nm}^{-1} < q < 1.8 \times 10^{-1}\,\mathrm{nm}^{-1}$) of a collagen solution (0.56 mg/mL, 0.0327 M phosphate buffer, pH 7.0), during heating and cooling T-jumps, and of the absorbance signal (LS) after a T-jump from 2°C to 32°C. The SAXS signal increases immediately when the temperature is raised, whereas absorbance displays a lag phase, though it was measured at a higher concentration (1.05 mg/mL). (Adapted from Suarez *et al.* (1985), © National Academy of Sciences, USA, 1985.) (B) Time course of the absorbance ($A_{600\,\mathrm{nm}}$, •) and of the relative average molecular mass (○) measured by SANS for insulin solutions (10 mg/mL, pH 1.6) in D_2O after heating to 65°C. Note that the onset of the transition is observed much earlier in SANS, whereas the absorbance displays a long lag phase due to lack of sensitivity to smaller aggregates. (Adapted from Nayak *et al.* 2009, and reproduced with the permission of John Wiley and Sons, Inc., © Wiley-Liss, Inc., 2008.) Note the difference in time-scale between Figs. A and B.

laser) is sensitive. As a result, the lag phases observed in SAXS/SANS are usually much shorter than in light scattering, as illustrated in Fig. 7.2A, B.

It is always useful, if not indispensable, to compare the structural results of assembly experiments with those of EM under conditions as close as possible to those of the scattering experiments. There are, however, some difficulties in comparing results from fast perturbations with those of equilibrium measurements, as the systems involved may give rise to heating-rate-dependent overshoots or undershoots, resulting from transient structures. Moreover, SAXS/SANS sample the complete distribution of structures in the solution, whereas in EM the structures that are imaged tend to be selected on the basis of more subjective criteria. This occasionally leads to differences in interpretation, as in the case of the assembly of tobacco mosaic virus protein (see Potschka *et al.* 1988, Hiragi 1990, Butler *et al.* 1992). This is less likely to occur if the scattering and EM studies are performed in parallel, as in the case of profilactin (Sayers *et al.* 1985) or tubulin (Bordas *et al.* 1983).

The importance of the heating rate, even for smaller structures, is illustrated by the different forms of tau protein following slow heating/cooling, which does not lead to any changes in overall size, and following fast heating/cooling, which leads to a more compact form that is stable for 3 hours (Shkumatov *et al.* 2011).

Phenomena such as protein (un)folding or the assembly of collagen or virus capsids are transitions that do not involve any chemical reactions. In contrast, the assembly of tubulin depends on the presence of GTP as source of chemical energy. This and other features of the system lead to oscillations at high protein concentrations (>10 mg/mL). The mechanism of oscillations based on SAXS (Marx and Mandelkow 1994) explains a number of experimental observations (Lange *et al.* 1988, Mandelkow *et al.* 1988), but contains concentration-dependent rates and requires an additional synchronisation mechanism, suggesting that diffusion plays a significant role. This view is supported by the fact that upon polymerisation the relative viscosity of an 8.25 mg/mL tubulin solution increases by a factor of 80 (Wagner *et al.* 1999). Several studies on microtubule assembly using different methods and different conditions emphasise the importance of diffusion, but there is hitherto no generally accepted kinetic mechanism.

In recent years considerable effort has been expended on the study of amyloid fibril formation, and fibrillation of various proteins *in vitro* using SAXS and SANS, besides other techniques. These processes may take several hours, and are in general much slower than assembly of cytoskeletal components or cylindrical virus capsids, and irreversible. It is, however, important to verify the physiological relevance of these phenomena. Whereas recombinant α-synuclein forms fibrils with a lag phase of 5 h at a concentration of 12 mg/mL (Giehm *et al.* 2011), the physiological form is a helically folded tetramer that resists aggregation (Bartels *et al.* 2011). Even longer lag phases (10 hours) have been observed by SANS during assembly into oligomers of glucagon at pH 2.5 and room temperature. These oligomers in turn associate into protofibrils and fibrils by a mechanism that has been modelled in detail (Oliveira *et al.* 2009). The heat-induced fibrillar aggregation of bovine β-lactoglobulin at pH 2.0 is another process extending over several hours that was monitored by SANS (Arnaudov *et al.* 2006). The technique consisting of quenching a solution prior to SANS measurements after a known residence time at high temperature used for multistranded ribbon-like β-lactoglobulin fibrils is, of course, applicable only with completely irreversible processes (Bolisetty *et al.* 2011).

Some of the interest in assembly phenomena has now shifted to applications in nanotechnology that may require more detailed control of the mechanisms.

7.4 High-pressure experiments

The effects of pressure on intra- and intermolecular interactions in solutions of biological macromolecules are not only of fundamental interest but also important for applications ranging from food processing to vaccine development (Heremans 1997, Aertsen *et al.* 2009). The fundamental aspects have recently been reviewed from different perspectives (Heremans and Smeller 1998, Boonyaratanakornkit *et al.* 2002, Ravindra and Winter 2003, Meersman 2006). Changes in the SAXS pattern of water with increasing pressure reveal modifications in its large-scale structure similar to some of those observed by increasing temperature (Cunsolo *et al.* 2009).

As a consequence of Le Chatelier's principle an increase in pressure will always drive the system towards the side of the equilibrium where it occupies the smallest volume. For single-chain proteins where pressure and temperature changes result in a reversible transition between two equilibrium states—a folded near-native state and an unfolded one—it can be assumed that the internal energy (U) of the system is constant in the useful range of pressure and hence the change in enthalpy (H) proportional to the change in volume (V) ($\Delta H \approx p\Delta V$).

Using partial molar quantities, the difference (ΔG) between the free energies of the two states is given by eq. (7.1), where ΔV refers to the change in partial molar volume ($V_i \equiv (\partial V/\partial N_i)_{p,T,N_{j\neq i}}$) of the macromolecule, and ΔS is the change in its partial molar entropy.

$$d(\Delta G(p,T)) = \Delta V(p,T)dp - \Delta S(p,T)dT \qquad (7.1)$$

This suggests that the effects of temperature and density and hence also of contrast can be separated in pressure experiments. When eq. (7.1) is integrated, starting from a reference state (p_0, T_0) and neglecting terms above the second order in the Taylor expansion, one obtains eq. (7.2) (Bridgman 1915), which corresponds to the elliptical stability diagram shown in Fig. 7.3A (Hawley 1971, Meersman 2006). This diagram describes the main observations, though actual diagrams may be more complex, as shown by Fourier transform infrared (FTIR) spectroscopy in the case of myoglobin (Meersman *et al.* 2005).

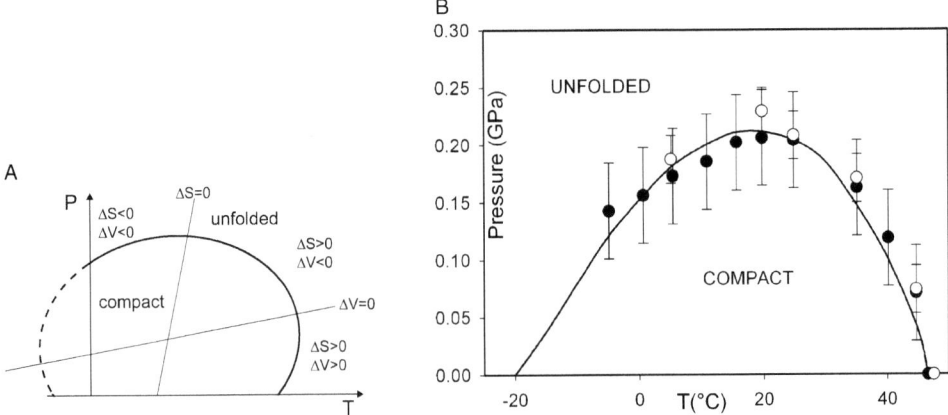

Fig. 7.3 (A) p,T stability diagram for a single-chain protein undergoing a reversible transition from native to unfolded without intermediates and without aggregation. The dotted part of the ellipse represents the region of cold denaturation. (Adapted from Meersman 2006, and reproduced by permission of the Royal Society of Chemistry.) (B) p,T stability diagram of staphylococcal nuclease (10 mg/mL) at pH 5.5 measured by FTIR (•) and SAXS (○). (Adapted from Winter and Dzwolak 2005, with permission from Royal Society.)

$$\Delta G(p, T) = G(p, T)_{\text{unfolded}} - G(p, T)_{\text{native}}$$

$$= \left\{ \Delta G^0 + \Delta V^0(p - p_0) - \frac{\Delta \beta}{2}(p - p_0)^2 \right\} + \Delta \alpha (T - T_0)(p - p_0)$$

$$- \Delta S^0(T - T_0) + \Delta C_p \left[(T - T_0) - T \ln(T/T_0) \right] \qquad (7.2)$$

$\Delta G^0 = G(p_0, T_0)_{\text{unfolded}} - G(p_0, T_0)_{\text{folded}}$, and similarly ΔV^0 and ΔS^0 correspond to the reference state. $\Delta \alpha = (\partial \Delta V / \partial T)_p$, $\Delta \beta = -(\partial \Delta V / \partial p)_T$ and $\Delta C_p = T(\partial \Delta S / \partial T)_p$ refer, respectively, to the differences between the values of the thermal expansion and compressibility factors and of the heat capacities of the unfolded and folded states. These differences are assumed to be independent of temperature and pressure.

The pressure dependence of ΔG at constant temperature represented by the terms in the curly brackets in eq. (7.2) is related to the compressibility factor $\Delta \beta = V \Delta \beta_T$, which in turn is related to the isothermal compressibility $\beta_T = -V^{-1}(\partial V / \partial p)_T$ and the volume fluctuations, and hence indirectly also to the dynamics of the system (Meersman 2006). As a result, the magnitude and sign of the changes in partial molar volume, which have essentially two contributions, one from the change in the volume of cavities of the macromolecule and the second one from hydration effects, cannot be predicted from ΔV^0 alone. This can lead to complex behaviour, especially with assemblies such as lactate dehydrogenase, where the tetramers dissociate into dimers around 0.12 GPa at 25°C. At higher pressures the dimers do not dissociate into monomers but irreversibly aggregate (Fujisawa *et al.* 1999). In the case of urate oxidase the tetramers dissociate into monomers that aggregate (Girard *et al.* 2010). For small heat-shock proteins (sHSPs) it was found that the size of αN- and αB-crystallins, which form large oligomers, reversibly increases between atmospheric pressure and 0.3 GPa, whereas temperature-induced denaturation is irreversible (Skouri-Panet *et al.* 2006). In contrast, yeast HSP126, a monodisperse 24-mer at atmospheric pressure, dissociates reversibly into monomers between 0.2 and 0.3 GPa. Equation (7.2), which is based on the assumption that the number of particles is constant, can thus not be applied to such oligomeric systems.

A study of the effect of cavity size on pressure denaturation for a series of lysozyme T4 mutants suggested that the small volume changes upon denaturation and the compactness of the pressure-denatured state is due to preferential filling of large cavities (Ando *et al.* 2008a). It also confirmed that the pressure-denatured state differs significantly from the heat-denatured or chemically denatured states.

The p,T-values for the mid-point of the native to unfolded transition ($\Delta G = 0$) of staphylococcal nuclease (SNase) obtained by FTIR spectroscopy and SAXS pressure jump measurements are in good agreement with the elliptic diagram, as illustrated in Fig. 7.3B (Panick *et al.* 1998). Between atmospheric pressure and about 0.2 GPa the radius of gyration of SNase in aqueous buffer increases from 1.7 to 3.5 nm at 25°C. The mid-point of the transition is shifted from 0.21 GPa in H_2O to 0.25 GPa in D_2O (Paliwal *et al.* 2004), where unfolded SNase also aggregates slightly. The aggregation was not observed in H_2O, pointing to the different properties of the two solvents (see Section 8.9). It should be noted that the pathways of heat and pressure unfolding are different, with pressure unfolding involving states with a significant fraction of secondary structure. In contrast, thermal denaturation leads to an unfolded structure with a radius of gyration of 4.5 nm at 50°C and atmospheric pressure corresponding to a random coil with little secondary structure as indicated by FTIR. The bimodal distance distribution function $P(r)$ for the pressure unfolded state corresponds to the existence of two domains, in agreement with the results of molecular dynamics (Paliwal *et al.* 2004). Similar observations have been made for a number of SNase mutants (Schroer *et al.* 2010). No changes in the distance distribution of the ankyrin repeat domain of the Notch receptor, Nank1-7, were observed at high pressure and with pressure jump SAXS up to 0.3 GPa at 24°C in the absence of urea, though under those conditions large changes are already observed in the tryptophan fluorescence (Rouget *et al.* 2010). Addition of 2.0–2.2 M urea led to a large increase in the maximum dimension and in the fluorescence signal. The experiments show that the unfolded state ensemble and folding transition state ensemble have a higher degree of plasticity than those of SNase. The variety of behaviours is also illustrated by proteins such as the extracellular peroxidase of *Marasmius scorodonius* (MSp1), which is highly resistant to pressure and temperature (65°C and 1 GPa at RT). Its enzymatic activity increases up to 50 MPa—possibly due to stabilisation of the transition state. No structural changes were observed either in fluorescence or SAXS up to 0.8–1.0 GPa (Pühse *et al.* 2009).

A SANS study of horse-heart metmyoglobin in D_2O showed that the R_g of the protein is constant up to 0.3 GPa, but changes in hydration resulting in lower repulsive intermolecular interactions with pressure were detected (Loupiac *et al.* 2002). Spectroscopic methods had indicated structural rearrangements at such pressures, leading to a molten globule state, which precedes denaturation around 1 GPa. The (p, T) diagram of myoglobin is more complex with transient aggregated states and pressure induced thermostabilisation at higher temperatures (Meersman *et al.* 2005), but no scattering data have yet been obtained on these phenomena.

Some compact proteins like glucose/xylose isomerase from *Streptomyces rubiginosus* retain their tertiary and quaternary structure at pressures at least up to 0.15 GPa as shown by SANS (Banachowicz *et al.* 2009). The only significant difference was in the forward scattering, which increased by 17%/0.1 GPa, as expected from the 4.3%/0.1 GPa change in density of D_2O. These authors also found no change in the scattering curves of lysozyme up to 0.35 GPa and of BSA up to 0.45 GPa. In the presence of 4 M urea BSA already unfolds at 0.2 GPa, as shown by SANS (Aswal *et al.* 2009).

The effect of osmolytes on the pressure stability of proteins was investigated in the case of SNase (Krywka *et al.* 2008), and it was found that substances such as trimethylammonium N-oxide (TMAO), which stabilise biological macromolecules against thermal denaturation, also protect them against the effects of pressure, whereas urea increases the effects of pressure. This confirms the major role of the properties of the solvent on the stability of biological macromolecules in solution (Bolen and Rose 2008), as also revealed by the effects of pressure on the intermolecular interactions in concentrated solution observed by SAXS in the case of lysozyme (Ortore *et al.* 2009, Schroer *et al.* 2011). The short-range attractive part of the interaction potential decreases above atmospheric pressure, reaching a minimum around 0.2 GPa, and then starts increasing, but in concentrated solutions the protein does not denature at pressures at least up to 0.4 GPa.

Solvent properties also play a major role in the stability of oligomeric assemblies such as casein, which dissociates at high pressure (Jackson and McGillivray 2011). Protein–nucleic acid assemblies such as viruses also tend to decapsidate irreversibly in a pH-dependent manner at moderate pressures following conformational changes, as observed by SANS for brome grass mosaic virus (Leimkühler *et al.* 2000) and turnip yellow mosaic virus (Leimkühler 2001), even if in this last case the capsids do not dissociate into smaller aggregates.

7.5 Stopped-flow and continuous-flow mixing

Mixing experiments are among the most popular ones, partly because they enable rapid perturbations for experiments on conformational changes of proteins and nucleic acids.

Early buffer-exchange experiments were related to assembly of profilactin (Sayers *et al.* 1985) and actin (Matsudaira *et al.* 1987) or chromatin condensation (Bordas *et al.* 1986). The previous remarks concerning the lag phase for heat driven assembly are equally valid for mixing experiments. Assembly of filaments is often accompanied by lateral aggregation or bundling, as is the case of microtubules or actin. This reduces the axial ratio of the objects in solution, and in the absence of X-ray measurements or EM this leads easily to misinterpretation of viscosity measurements, as illustrated in Fig. 7.4.

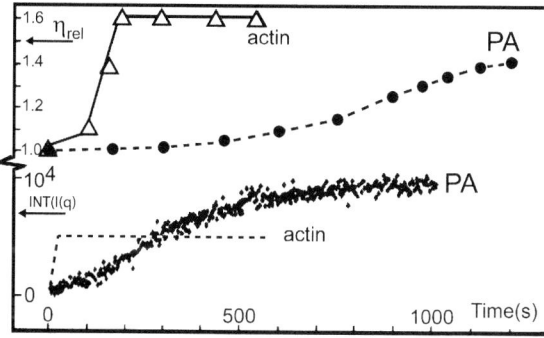

Fig. 7.4 Polymerisation of actin and profilactin (PA). Actin polymerises faster and reaches a higher relative viscosity than profilactin, which reaches a higher level of scattering intensity due to lateral assembly. (Adapted from Sayers *et al.* (1985), © Springer-Verlag, 1985, with kind permission from Springer Science and Business Media.) Taken separately, the viscosimetry results do not allow one to decide between a reduction in filament length or lateral aggregation.

A phenomenon related to assembly is that of crystal nucleation and growth in protein solutions, where the influence of physicochemical parameters (temperature, pH, additives) on the kinetics has been studied by time-resolved SAXS or SANS for a number of cases such as BPTI (Budayova-Spano *et al.* 2002) or brome mosaic virus (Casselyn *et al.* 2004). Although the nucleation theory explains many of these observations there is no consensus on the structure of the critical cluster, which may be crystalline from the onset or initially amorphous and rearrange into a crystal lattice (Garcia-Ruiz 2003). A combination of scattering and direct imaging may clarify this problem in future.

Slow processes extending over hours, such as the formation of amyloid protofilaments of hen egg-white lysozyme (Fujiwara *et al.* 2003), aggregation in supersatured lysozyme solution following addition of salt (Niimura *et al.* 1995), or the pH and temperature induced gelation of proteins (Chodankar *et al.* 2009) can also easily be followed by SANS.

7.5.1 Protein folding

Protein and nucleic acid folding has provided challenging problems during the last decades, and in this context SAXS has yielded useful information about the overall dimensions of intermediates that is difficult to obtain by other methods. Time-resolved experiments on folding are obviously more demanding than those on assembly, and to compensate for the intensity limitations on second-generation SR sources different methods were used to improve the signal-to-noise ratio, such as multilayer monochromators with a wider band pass, or early CCD detectors, as illustrated by some of the examples to follow. On third-generation sources it was also attempted to use pink beams from undulators with continuous-flow mixers.

Acid and chemical (urea, guanidinium chloride (GdmCl)) denaturation are the most common techniques for protein (un)folding experiments. With few exceptions, after reduction of any disulphide bridges, chemically unfolded proteins behave like random coils in SAS (Kohn *et al.* 2004), but even for single-domain proteins the extent of residual secondary structure suggested by spectroscopic methods is still a matter of debate (Sosnick and Barrick 2011).

Different classes of proteins (all β, all α and α+β) behave slightly differently, especially during the early stages of refolding, as summarised in Table 7.3 and recently reviewed (Kathuria *et al.* 2011). Comparisons are, however, complicated by the fact that most of the recent results for all-α proteins have been obtained by pH-jump, where solvent exchange may be faster due to the anomalous diffusion properties of protons, and for the other two classes mainly by urea or GdmCl denaturation. Moreover, in several cases it appears that unfolding may not have been complete, whereas in others (such as ubiquitin and acylphosphatase) the early data points suggest the possibility of minor aggregation.

Table 7.3 Values of the radius of gyration (R_g) in nm for proteins belonging to different classes during refolding. N is the number of residues in the chain. (eq) are values determined in equilibrium conditions, (ex) are values extrapolated to zero time in time resolved experiments and * are equilibrium values taken from Kohn *et al.* (2004).

Class/name	N	R_g (nm)	Solvent	Comment/reference
All β				
Neocarzinostatin	113	2.61 →? → 1.40 3.3 →? → 1.29	$T_{max} = 76°C$ 5M GdmCl	(Pérez *et al.* 2001) (Russo *et al.* 2001) Equilibrium only
β-Lactoglobulin	162	(4.0)eq→2.2 (100 ms)→1.97	8M→3 M urea	Non-native α-helix in intermediate (Arai *et al.* 1998)
Human tear lipocalin C101A	158	(4.3)eq→2.55(6 ms)→ 2.23(15 s)→(2.10)eq	8M→1.1 M urea	(Tsukamoto *et al.* 2009)
All α				
cytochrome C	104	(2.43)eq→2.0 (160 μs)→1.8→1.39	pH jump 2→4.5	(Akiyama *et al.* 2002)
myoglobin	154	2.9→3.2 (dimer, 100 ms)→1.8	7 M→1.9 M urea	(Eliezer *et al.* 1993)
apomyoglobin	154	(2.97)eq→2.37 (300 μs)→1.82	pH jump 2.2→6	(Uzawa *et al.* 2004) Rg = 4.0 nm in 6 M GdmCl*
heme oxygenase	263	(3.78)eq→2.63→3.55 (transient oligomer)→2.61→2.3	pH jump 2→5	(Uzawa *et al.* 2006)
λ-repressor fragment $λ_{6-85}$	80	2.67→2.40 (100 ms)→1.42	5 M→0.7 M GdmCl	(Kim *et al.* 2009)
α and β				
apo-superoxide dismutase	153	(?)→2.7 (23 s)→2.2 (dimer)	7 M→1.5 M urea	(Svensson *et al.* 2006) Manual mixing
dihydrofolate reductase	167	3.07→2.32 (300 μs)→1.66	4.5 M→0.45 M urea	(Arai *et al.* 2007)
ribonuclease A	124	2.56eq→2.0 (22 ms)→ 1.58→1.56eq	4.2 M→0.7 M	(Kimura *et al.* 2005a)
r-RNase		2.82eq→2.46(22 ms)→ 1.80→1.61eq		Rg = 3.32 nm in 3.25-6M GdmCl*
single chain monellin	94	2.55eq→1.82 (300 μs)→1.58	pH 13→9.4	(Kimura *et al.* 2005b)
lysozyme	127	2.2eq→1.9→1.5 2.35→1.96 (1 ms)→1.7→1.6		(Chen *et al.* 1996, Chen 1998, Segel *et al.* 1999)
α-lactalbumin	124	2.07eq→1.75ex (10 ms)→1.58	4M→0.75M GdmCl	(Arai *et al.* 2002)
ubiquitin	79	2.59eq→2.76 (2.5 ms)→2.36→1.44 (1.39)eq	6 M−0.75 GdmCl	(Jacob *et al.* 2004) Rg = 2.52 nm in 4.9-6M GdmCl*
acylphosphatase	99	3.02.eq→3.13 (2.5 ms)→2.66→1.46	6 M→0.85 M urea	(Jacob *et al.* 2004)
barnase	110	2.69eq 2.43ex→2.35 (100 ms)→1.6 (2s)→1.38	5 M→0.7 M GdmCl	(Konuma *et al.* 2011)
tryptophan synthase α-subunit	268	4.3ex→3.4 (150 μs–3 ms)→2.7 (1.8 native)	6 M→0.6 M urea	(Wu *et al.* 2008) R_g = 4.88 nm in 6M GdmCl*

Nevertheless, protein refolding follows quite closely Flory's theory for a coil-to-globule transition (Flory 1953) if one considers the first intermediate, which has a very flexible structure, as the globule. The equilibrium R_g of unfolded chains follows the scaling law $R_g = aN^\nu$, where N is the number of residues in the chain, and ν is close to 3/5, whereas both the globules and the compact native proteins have exponents close to 1/3 despite their very different structures (Uzawa *et al.* 2006). Interesting attempts were made at relating equilibrium parameters with kinetic properties, but these have not been successful in the prediction of relative folding rates (Millet *et al.* 2002).

Formation of the globule by collapse of the unfolded chain is very rapid ($<5\,\mu$s) and thus not accessible in mixing experiments. This has been studied in some detail in the case of downhill folding of the λ-repressor fragment λ_{6-85} by CD, fluorescence and SAXS at low ($-30°$C) temperatures (Kim *et al.* 2009).

The transition between unfolded and globular forms can be followed using plots of the integrated intensity in different q-ranges or Kratky plots and calculating the radius of gyration with the Debye formula for the unfolded forms and the Guinier approximation for the globular forms (Semisotnov *et al.* 1996). Intensity correlation plots should also be useful, but do not seem to be common in this field.

Heat and GdmCl unfolding of neocarzinostatin was studied in equilibrium conditions by SAXS and other methods (Pérez *et al.* 2001, Russo *et al.* 2001). SVD analysis of thermal unfolding in the absence of GdmCl indicates that there is at least one weakly structured intermediate but a strict three-state model with a well-defined intermediate still fails to explain the results. This can occur when the scattering curves of putative intermediates can be described by linear combinations of the components found by SVD, and suggests that there is either a set of distinct intermediates or an ensemble of conformations with temperature-dependent occupancy. The results of independent methods (SANS, fluorescence and CD) for unfolding in presence of denaturant taken separately can in first approximation be interpreted in terms of a two-state (folded and unfolded) process. Discrepancies between the Gibbs free energy of unfolding obtained by the different methods and DSC measurements also clearly indicate that more than one intermediate is required to describe the process.

For β-lactoglobulin, which belongs to the lipocalin superfamily, six distinct phases were detected by SAXS, Trp absorption and fluorescence, and CD during the refolding by dilution following urea denaturation (Arai *et al.* 1998). Three of these phases involving local conformational and hydration changes are not detected by SAXS. Interestingly for a predominantly β-sheet protein, the first (burst-phase) intermediate, which has a compact globular shape with little tertiary structure, forms a non-native transient α-helical structure. The value of R_g is only 10% larger than that of the native structure after 100 ms. Transient α-helical intermediates during refolding have also been found by combining time-resolved SAXS and spectroscopy in a number of other proteins (Matsumura *et al.* 2010), but not in a tear lipocalin with similar structure but a low sequence-similarity with β-lactoglobulin (Tsukamoto *et al.* 2009). It was therefore suggested that formation of non-native helices depends essentially on the helical propensities of the amino acid sequence. The collapse of urea-denatured β-lactoglobulin was also followed with ms time-resolution using a microfabricated diffusive mixer and a pink beam, and it was found that 70% of the protein remained in an expanded state after 8 ms, indicating that the collapse is much slower than in the case of the all-α cytochrome C, for example (Pollack *et al.* 2001).

The early folding events for cytochrome C after acid denaturation have been followed with μs time-resolution by SAXS using continuous-flow methods, and the results combined with CD and Trp-fluorescence data taken from the literature to characterise the conformational landscape in terms of R_g and helical content (Akiyama *et al.* 2002). SAXS experiments showing the rapid collapse had been made earlier using a diffusive mixer and also indicated a very rapid collapse ($<150\,\mu s$), as observed already by fluorescence (Pollack *et al.* 1999). The results were interpreted in terms of a sequential mechanism with two monomeric intermediates—an extended one and a kinetic molten globule state with a structured core. The collapse of the initial structure ($R_g = 2.4\,nm$) to an early intermediate with $R_g = 2.0\,nm$ occurs in less than $150\,\mu s$, followed by a second intermediate ($R_g = 1.8\,nm$) which evolves more slowly to the native structure ($R_g = 1.4\,nm$). Note that the value of R_g extrapolated to zero time (2.1 nm) is significantly smaller than that of the initial structure.

Acid-denatured apomyoglobin has a much smaller R_g than the Gdm-Cl denatured form, indicating that the molecules are not completely unfolded (Uzawa *et al.* 2004). Moreover, in this experiment a drop of 33% of the forward scattering, which parallels the decrease in R_g, was observed. Although not consistent with the formation of dimers, this effect, which is also unlikely to result from changes in hydration, remains unexplained. In the case of myoglobin denatured in 7 M urea, a transient mainly dimeric intermediate had been found previously (Eliezer *et al.* 1993).

Refolding of the highly helical heme oxygenase follows a complex mechanism, with an oligomeric intermediate and two distinct reaction pathways probably involving fractions with different levels of *cis-trans* isomerisation of the X-proline peptide bonds (Uzawa *et al.* 2006).

The refolding of metal-free human Cu, Zn superoxide dismutase after urea denaturation proceeds from unfolded monomer over a short-lived folded monomeric intermediate to a native dimer in the presence of urea, as concluded by combining the results of time-resolved CD, fluorescence and SAXS measurements. In contrast, the combined equilibrium measurements are well explained by a two-state mechanism (Svensson *et al.* 2006).

The folding of an *Escherichia coli* dihydrofolate reductase mutant (AS-DHFR) after urea dilution is accompanied by an extensive collapse (from $R_g = 3.07$ nm to $R_g = 2.32$ nm) within 300 μs followed by a much longer evolution towards the native state (Arai *et al.* 2007). Note that the equilibrium value of the R_g of the unfolded form in 8 M urea is significantly larger (3.5 nm) than in the initial conditions of the continuous-flow experiment (4 M urea), suggesting that the molecules are still partially folded.

In the early stages of folding of ribonuclease A (RNAse A) and its reduced form (r-RNase A), where the disulphide bonds are broken, SAXS indicates that the burst-phase intermediates of RNase are significantly smaller than those of r-RNase, whereas their CD signals are identical (Kimura *et al.* 2005a). The R_g value of r-RNase in this study is, however, 0.5 nm smaller than observed by others in the same conditions (Kohn *et al.* 2004), illustrating once more the difficulty in comparing studies where the R_g and estimated molecular mass are not available simultaneously.

Refolding of single-chain monellin, which has a five-stranded β-sheet and a helix, after a pH jump involves two intermediates with slow hydrogen-bond formation in the β-sheets. In this case too the R_g of the burst-phase intermediate or the value extrapolated to zero time are much smaller than that of the equilibrium unfolded form (Kimura *et al.* 2005b).

Like thermal unfolding, chemical unfolding leaves the four disulphide bridges in lysozyme intact. When lysozyme is refolded by lowering the urea concentration from 8 M to 1.1 M at pH 2.9, where the process is much slower than at pH 5.2, the unfolded state ($R_g = 2.2$ nm) rapidly (<1 ms) compacts to a non-specifically collapsed ensemble with $R_g = 1.9$ nm (Chen 1998). This is close to the maximum value reached in reversible thermal unfolding. This collapsed ensemble then evolves much more slowly (~10 s) to the native form ($R_g = 1.5$ nm). A similar intermediate with extended β-domain and a more compact α-domain had been found in a previous equilibrium SAXS study of urea denaturation (Chen *et al.* 1996).

In contrast, analysis of the refolding of GdmCl-denatured lysozyme at pH 5.2 followed by stopped-flow and continuous flow showed that most of the collapse of the unfolded structure ($R_g = 2.35$ nm) occurs through a cooperative two-state process (Bachmann *et al.* 2002) within the dead time to yield a globular intermediate with ($R_g = 1.96$ nm), where the hydrophobic chains are still accessible to the solvent according to fluorescence measurements (Segel *et al.* 1999). This is followed by formation of an intermediate with helical structure in the α-domain and a solvent inaccessible core ($R_g = 1.7$ nm). The R_g of the folded structure (1.6 nm) after 2 s is still somewhat higher than that of native lysozyme (1.5 nm), due to the fact that a small fraction of protein (~10%) folds very slowly (~20 s).

Compared to the previous experiments where a linear gas detector was used, the signal-to-noise ratio in refolding experiments on α-lactalbumin was considerably improved by using an area detector (Arai *et al.* 2002). Here the kinetic folding intermediate formed by rapid collapse in the burst-phase and the equilibrium-molten globules are identical, which also suggests that further evolution to the specific side-chain packing of the native state involves progressive dehydration.

Whereas, as described above, the folding of proteins with more than 120 residues appears to involve a very rapid initial collapse (1 ms) as soon as the solvent conditions are changed, this does not seem to hold for smaller proteins (<100 residues), which fold by a two-state mechanism and remain expanded after the change in solvent conditions. Rapid chain collapse is thus not a necessary step in refolding, as demonstrated by time-resolved SAXS in a number of cases such as protein L (Plaxco *et al.* 1999), acyl-phosphatase and ubiquitin (Jacob *et al.* 2004). With barnase, the initial intermediate after dilution of GdmCl is also expanded, in contradiction with FRET measurements (Konuma *et al.* 2011). The systematic inconsistencies between the molecular dimensions determined by FRET and SAXS for small proteins at low denaturant concentrations, which may be due partly to the effects of protein dynamics on the efficiency of resonance energy transfer, have led to an hitherto not entirely resolved controversy, especially in the case of protein L (Merchant *et al.* 2007, Makarov and Plaxco 2009).

All these experiments illustrate the importance of exactly controlling the experimental conditions during the entire process, as well as the need for parallel experiments using other methods. In dilution studies it is particularly important to normalise the scattering patterns on the transmitted intensity rather than on the incident intensity, or to make appropriate corrections for absorption (see, for example, Arai *et al.* 2002). This normalisation is especially important for obtaining correct values of the forward scattering, and though the procedure usually seems to have been applied it is not always indicated in the original papers.

7.5.2 RNA folding

Folding of nucleic acids differs fundamentally from that of proteins due to the much greater importance of electrostatic forces. SAXS experiments on nucleic acids are often complicated by the higher absorption of the molecules, which also tend to be more sensitive to radiation damage than proteins, and by strong repulsive interactions at low ionic strengths. Interpretation of the scattering profiles relies extensively on Kratky plots, which are useful to monitor compaction but should be used with caution at low ionic strength, as this representation may conceal intermolecular interactions affecting the scattering pattern at low q-values.

RNA folding is induced by rapid mixing of RNA solutions at low ionic strength with solutions containing monovalent (Na^+) or divalent (Mg^{++}) ions. First manual mixing experiments with a dead time of about 1 min on the folding of *Tetrahymena* group I ribozyme revealed that similarly to the case of proteins the initial phase is a rapid (electrostatic) collapse of the unfolded ensemble ($R_g = 7.4$ nm) to a compact kinetically trapped intermediate ($R_g = 5.1$ nm) followed by a slow re-arrangement associated with the formation of specific contacts leading to the native structure ($R_g = 4.7$ nm) (Russell *et al.* 2000). Whereas 15–20 mM Mg^{++} or Ca^{++} lead to compact intermediates, compaction is much reduced in the presence of 1 M Na^+ ($R_g = 6.8$ nm). These experiments illustrate the complementarity of SAXS, time-resolved oligonucleotide hybridisation and hydroxyl radical footprinting for studies on nucleic acids. They were repeated in the time range 0.5 ms–1000 s using a combination of a stopped-flow device with a multilayer monochromator as well as of a continuous flow mixer and a pink beam (Russell *et al.* 2002). SVD analysis detected only two forms, folded and unfolded, because of the similarity of the patterns of the long-lived misfolded state and the folded state. The fastest global events are electrostatic relaxation and tertiary collapse (Das *et al.* 2003). Further studies on mutants, including parallel time-resolved hydroxyl radical analyses, confirmed the generality of the rapid ($\approx$10 ms) collapse to an intermediate with little tertiary structure, followed by a slow (100 ms) formation of tertiary contacts within this compact structure (Kwok *et al.* 2006). The importance of chain stiffness for folding was demonstrated in a study on an autonomously folding portion of the *Tetrahymena* ribozyme. Collapse is greatly accelerated at Mg^{++} concentrations above 10 mM, as the limiting step is no longer determined by a stiff hinge between domains but by tertiary contact formation (Schlatterer *et al.* 2008).

The collapse of the *Azoarcus* group I ribozyme, which involves at least three kinetic phases, was followed with 0.6 ms dead time using a stopped-flow mixer (Roh *et al.* 2010). Intermediate ensembles differing in compactness are formed within 1 ms, depending on the Mg^{++} concentration, but the global fold and local interactions are formed simultaneously. Comparison with parallel hydroxyl footprinting experiments suggests that besides a number of parallel non-specific pathways there is a specific collapse pathway explaining the much faster refolding than in the case of the *Tetrahymena* ribozyme.

For reviews of SAXS experiments on RNA folding, see Pollack and Doniach (2009) and Pollack (2011).

7.5.3 Allosteric transitions

The mechanisms of allosteric transitions present a special interest due to their importance in metabolic regulation and signal transduction (for a review, see Changeux 2012). One of the aims of time-resolved SAXS experiments on these systems is to distinguish between the two main models of allostery. The Monod–Wyman–Changeux (MWC) model predicts a concerted symmetry-conserving all-or-none transition between a tense (T), low-affinity and activity state and a relaxed (R), high-affinity and activity state (Monod *et al.* 1965). The Koshland–Némethy–Filmer (KNF) model (Koshland *et al.* 1966) assumes sequential binding and the existence of several intermediates. Although quaternary structural changes have been observed for several allosteric proteins in static measurements, extensive time-resolved SAXS experiments have been carried out only for the bacterial regulatory enzyme aspartate transcarbamoylase (ATCase). These experiments motivated the developments of stopped-flow instruments. The initial studies required integration over 100–250 ms, the substrate analogue acetyl phosphate rather than the natural substrate carbamoyl phosphate (CP) and ethylene glycol (10–30%), and low temperatures ($-5°C$ to $6.5°C$) had to be used to bring the kinetics in a measurable range (Kihara *et al.* 1987, Tsuruta *et al.* 1994, Tsuruta 2005). No intermediates were found between the T and R states of the enzyme in any of these experiments. Use of a wide band-pass ($\Delta\lambda/\lambda = 3\%$) multilayer monochromator (Tsuruta 1998) and an image-intensified CCD detector (Amemiya *et al.* 1995) made it possible to carry out later experiments in the absence of ethylene glycol, which was shown to alter significantly the homotropic and heterotropic kinetics of the enzyme, and with the natural substrates (CP and L-aspartate) and nucleotide effectors at temperatures of 5–22°C with a time resolution of 5 ms (West *et al.* 2008). These experiments suggest that the allosteric transition is triggered by binding of L-aspartate to the T-state rather than due to a shift of the T/R equilibrium resulting from specific binding to the R-state. Since the allosteric transition is not rate-limiting the enzyme remains in the R-state as long as the substrate concentration is saturating, as illustrated in Fig. 7.5.

The biphasic kinetics of the structural transition after stopped-flow mixing with substrates suggested the existence of a heterogeneous T-state with different ligation substrates having different transition rates. The rate of the T→R transition depends not only on the substrate (L-aspartate) concentration but also on the concentration of ATP, which activates the enzyme. The subtle differences between the effects of ATP and Mg-ATP on the structures of the T and R states, which do not, however, affect the quaternary structural transition, have been investigated in detail in equilibrium conditions (Fetler and Vachette 2001). Although

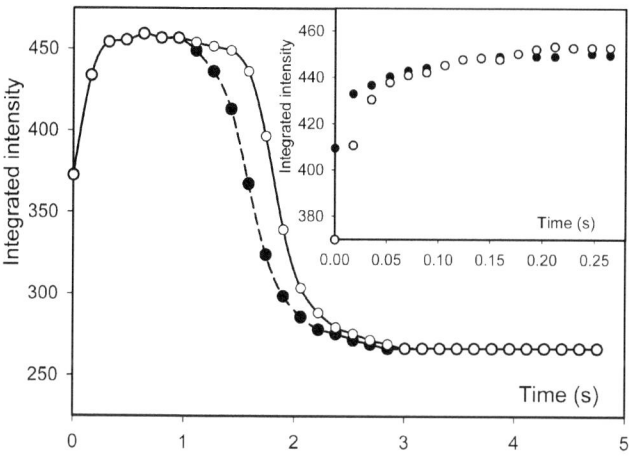

Fig. 7.5 Time course of the integrated intensity in the range $0.63\,\text{nm}^{-1} \leq q \leq 1.26\,\text{nm}^{-1}$ during the structural transition of aspartate carbamoyltransferase, AT-Case, (0.75 mM) following stopped-flow mixing with substrates ([L-Asp] = [CP] = 50 mM), with (•) and without (○) 5 mM ATP. (Adapted from West *et al.* (2008), with permission from Elsevier, © 2008). The inset illustrates the biphasic nature of the transition, suggesting the existence of a heterogeneous T-state.

many observations can be satisfactorily explained in the frame of the MWC model, a number of them also reveal that reality is more complex.

In all likelihood a direct measurement of the rate of the structural transition, which can be expected to occur in the sub-μs range, will be possible only if the reaction can be triggered with light, as described in Section 7.6 for haemoglobin.

Preliminary experiments on the complexation of *E. coli* GroEL and its co-chaperonin GroES in the presence of ADP and the reverse reaction with ATP were carried out with a time-resolution of 300 ms limited by the readout of the CCD detector (Roessle *et al.* 2000). Whereas the complex forms in a single step (5 s) in the presence of ADP, its dissociation seems to be slower (13 s) and may be followed by reassociation on a longer time-scale, so that the reaction appears biphasic. The fact that in the dissociation reaction the final R_g after 35 s is higher than the initial one suggests that radiation damage may have occurred.

Neutrons do not cause radiation damage, and allow one to follow reaction details by specific deuteration and contrast variation. The formation of the symmetric ('football') complex of GroEL with two GroES molecules was studied in D_2O (Roessle *et al.* 2000). Another biologically significant experiment dealt with the chasing of 'visible' GroES bound to 'visible' GroEL by an excess of 'invisible' (partially deuterated) GroES or 'invisible' GP31 (Holzinger, 2002). GP31 is a co-chaperonin present in the phage T4. GP31 yields, when bound to GroEL, a larger volume for misfolded proteins as shown by static SANS (Holzinger, 2002), thus providing an evolutionary advantage for T4. The kinetic SANS measurements indicated a slower dissociation of GP31-GroEL than of GroES-GroEL complexes, showing that the former are more stable.

7.5.4 Viruses

Viruses and virus capsids assemble *in vitro* on a subsecond–minute time-scale, and can undergo large-scale cooperative reorganisations associated with different stages in their cycle. The early time-resolved SAXS/SANS studies on the assembly of capsids of brome mosaic virus (BMV), an icosahedral virus with a triangulation number T = 3 also relied extensively on light scattering (Cuillel *et al.* 1983, Berthet-Colominas 1987). Using a simplified model it was found that several classes of incomplete capsids are formed starting from solutions containing predominantly dimers.

The divalent ion-dependent swelling of the *T* = 3 tomato bushy stunt virus (TBSV) that takes place over minutes, was investigated by a combination of time-resolved SAXS, SANS in 72% D_2O to mask the contribution of RNA, and cryo-EM (Aramayo *et al.* 2005).

Large subsecond time-scale quaternary structural changes resulting in a capsid with a diameter 16% smaller than that of the initial procapsid were observed by stopped-flow pH-jump (pH 7.5→5.0) with a time-resolution of 250–500 ms in virus-like particles of the $T=4$ RNA *Nudaurelia capensis* ω virus (Canady *et al.* 2001). By reducing the pH-jump (pH 7.5→5.5) to slow the transition to the minute-range, a fast-forming intermediate could be detected. The three phases of this process include a size reduction of the particles (<10 ms), continuing for about 5 s and followed by another smaller structural reorganisation after 3–5 min (Matsui *et al.* 2010). The initially reversible conformational transition becomes irreversible after completion of about 15% of the autoproteolytic maturation.

Following a pH drop (pH 7→3.8–4.1), the large-scale cooperative maturation of the icosahedral $T = 7$ capsid of HK97 lacking portal-terminase complex, which has 420 subunits, leads from Prohead II to the expansion intermediate I in minutes (Lee *et al.* 2005). The well-defined isointensity points in the scattering patterns indicate the absence of intermediates during the transition leading to a 6.0 nm increase in diameter of the particles.

Structurally similar end-products such as the icosahedral HK97 and P22 ($T = 7$) capsids may have different assembly mechanisms, as illustrated by the *in vitro* assembly of the coat protein of bacteriophage P22 in the presence of three different scaffolding proteins followed after manual mixing (Tuma *et al.* 2008). Binding of monomeric scaffolding protein, which exists in a monomer–dimer–tetramer equilibrium, activates the coat protein by lowering its critical concentration for assembly. Excess dimeric scaffolding protein leads, however, to kinetic trapping and production of incomplete shells, which can be completed by addition of monomeric wild-type scaffold and excess coat protein.

This illustrates the usefulness of time-resolved measurements in studies aiming at understanding but also controlling the assembly of this type of particle for possible applications.

7.6 Light-triggered processes

Most light-triggered processes require very fast (ps–μs) data acquisition, but some photoreceptors, such as the PAS-LOV (light oxygen voltage sensing) protein, which dimerises in less than 20 ms following exposure to light, remain in the light-adapted state for several hours. Such a process can be followed without need for repeating the photocycle using a flow cell (Lamb *et al.* 2008).

The kinetics of the R–T transition induced by photoactivation of human CO-haemoglobin with a 230-ns (FWHM) laser pulse (527 nm) was followed with 100-ps (FWHM) X-ray pulses delayed between 147 ns and 300 ms relative to the light pulse (Cammarata *et al.* 2010). The signal for $q < 15\,\text{nm}^{-1}$ is associated with changes in the tertiary and quaternary structure, whereas the changes at larger q-values are dominated by heating effects following the laser pulse. Smaller tertiary structural changes may therefore not be detected in the

WAXS pattern. The time-scale of the main structural R–T transitions ($\approx 2\,\mu s$) in these experiments is an order of magnitude faster than that observed by most spectroscopic methods under similar conditions. No structural changes are detected at the $20\,\mu s$ time-scale, which is consistent with the higher sensitivity of spectroscopic methods to local rather than global changes. A step-wise transition with a fast ($2\,\mu s$) and a slow (20–$50\,\mu s$) component had, however, been detected by time-resolved UV resonance Raman spectroscopy (Balakrishnan *et al.* 2004).

After SVD analysis the time course of the reaction can be explained using only two basis patterns—the T-state and a tertiary relaxed R-state—despite the presence of different ligation states. This lack of structural heterogeneity lends support to the MWC model of allostery (reviewed in Changeux 2012). Recent molecular dynamics calculations where quaternary transitions are preceded and followed by phases of tertiary rearrangements explain the WAXS observations (Fischer *et al.* 2011).

Some questions remain open in the interpretation of WAXS patterns of haemoglobin. Concentration-dependent effects on the scattering pattern observed in the concentration range used (0.25–1 mM, 17–68 mg/mL) in slightly different conditions (50 mM Na-phosphate, pH 7, 4°C), and attributed to increasing structural fluctuations (Makowski *et al.* 2008, Makowski *et al.* 2011), were not observed in this case. The extent to which the different states are physiologically relevant is unknown, as the energies involved and the number of spectroscopic states are different for O_2 (four states) and CO (five states) (Beece *et al.* 1980).

A study of the photocycle of detergent-solubilised bacteriorhdopsin and proteorhodopsin at room temperature showed that unlike the case of myoglobin or haemoglobin the intermediate states observed in those conditions did not correspond to any of the intermediates trapped in crystals, usually at low temperatures (Andersson *et al.* 2009). Controversies concerning differences between the crystallographic structures of the BR intermediates have been attributed to radiation damage (Borshchevskiy *et al.* 2011).

The scattering patterns in this work ($0 < q < 12\,\mathrm{nm}^{-1}$) were calculated with CRYSOL, which relies on a continuum solvent model. This approach breaks down for $q > 16\,\mathrm{nm}^{-1}$, beyond which more detailed solvent models are preferable (Park 2009, Virtanen *et al.* 2011). The scattering patterns were approximated using only three models (resting state, early intermediate and late intermediate), which should be compared with up to eight spectroscopic states in the photocycle of bacteriorhodopsin (BR→J→K→L→M1→M2→N→O→) and at least six in proteorhodopsin (PR→K→L→M→N→O→) (Xiao *et al.* 2005). When retinal is substituted by 13-desmethyl-13-iodo retinal the conformational changes in proteorhodopsin are accelerated ten-fold, though the major transient conformations do not seem to be affected (Malmerberg *et al.* 2011).

The use of a limited number of intermediates in the simulations follows an earlier proposal based on electron crystallography, suggesting that the structures of the initial state and the early intermediates (K, L and M(1)) are well-approximated by a conformation in which the Schiff base between retinal and Lys216 is accessible from the extracellular side, and the later intermediates (M(2), N and O) by a different conformation in which it is accessible from the cytoplasmic side. This guarantees that the two sides of the membrane are not in direct contact for a time interval that allows back diffusion of the proton.

It is unclear whether the photocycle of detergent-solubilised BR is the same as in purple membrane, closed vesicles or intact cells. Clearly, proton pumping occurs only between separate compartments, and this requires at least three thermodynamic states, since two-state systems can only oscillate but not do any work. The different states may, of course, not be distinguishable at 0.5 nm resolution.

The photoactive yellow protein (PYP) photoreceptor was studied using combined structural probes. In this case, TR SAXS/WAXS on the native structure in the range $0.05 < q < 0.8 \, \text{nm}^{-1}$ was used with NMR and double electron resonance spectroscopy (DEER) on doubly spin labelled photoreceptor constructs (Ramachandran *et al.* 2011).

The I_2' intermediate was observed 10 ms after a ns laser pulse (460 nm). The results were interpreted in terms of reduced folding, larger radius of gyration and maximum dimension, and surprisingly, a smaller excluded volume in the I_2' intermediate. No X-ray kinetic data or structures of other intermediates in the cycle are reported.

Photoactive yellow protein was also studied by SANS in optimised pH and concentration conditions on the SANS instrument D22 (May *et al.* 2005). The data can be interpreted as showing a slight opening of the protein structure followed by a rapid decay.

Interpretation of these experiments relies on sophisticated modelling techniques by which it is not always easy to determine what the underlying assumptions and simplifications are, and it is therefore useful to acquire the necessary theoretical background before embarking on such projects. Considerable work, both experimental and theoretical, remains for understanding the functioning of molecular machines in terms of changes in the distributions of conformations and multiple transition pathways rather than by analogy with macroscopic machines, where it is legitimate to speak about 'the' structure and 'the' mechanism. An even bigger challenge will be to make use of this knowledge in the modification and/or synthesis of new molecular machines. The present chapter illustrates the progress in instrumentation and modelling made during the last three decades. In light of the progress, it might be useful to revisit some of the earlier TR projects—for example, transitions such as chromatin condensation (Bordas 1986)—which were too fast (<50 ms) for stopped-flow measurements on second-generation SR sources.

References

Aertsen, A., Meersman, F., Hendrickx, M. E. G., Vogel, R. F. and Michiels, C. W. (2009). 'Biotechnology under high pressure: applications and implications', *Trends Biotechnol.* 27: 434–41.

Akiyama, S., Takahashi, S., Kimura, T. *et al.* (2002). 'Conformational landscape of cytochrome C folding studied by microsecond-resolved small-angle X-ray scattering', *P. Natl. Acad. Sci. USA* 99: 1329–34.

Amemiya, Y., Ito, K., Yagi, N. *et al.* (1995). 'Large aperture TV detector with a beryllium-windowed image intensifier for X-ray diffraction', *Rev. Sci. Instrum.* 66: 2290–4.

Andersson, M., Malmerberg, E., Westenhoff, S. *et al.* (2009). 'Structural dynamics of light-driven proton pumps', *Structure* 17: 1265–75.

Ando, N., Barstow, B., Baase, W. A., Fields, A., Matthews, B. W. and Gruner, S. M. (2008a). 'Structural and thermodynamic characterization of T4 lysozyme mutants and the contribution of internal cavities to pressure denaturation', *Biochemistry-US* 47: 11097–109.

Ando, N., Chenevier, P., Novak, M., Tate, M. W. and Gruner, S. M. (2008b). 'High hydrostatic pressure small-angle X-ray scattering cell for protein solution studies featuring diamond windows and disposable sample cells', *J. Appl. Crystallogr.* 41: 167–75.

Arai, M., Ikura, T., G.V., S., Kihara, H., Amemiya, Y. and Kuwajima, K. (1998). 'Kinetic refolding of β-lactoglobulin. Studies by synchrotron X-ray scattering, and circular dichroism, absorption and fluorescence spectroscopy', *J. Mol. Biol.* 275: 149–62.

Arai, M., Ito, K., Inobe, T. *et al.* (2002). 'Fast compaction of α-lactalbumin during folding studied by stopped-flow X-ray scattering', *J. Mol. Biol.* 321: 121–32.

Arai, M., Kondrashkina, E., Kayatekin, C., Matthews, C. R., Iwakura, M. and Bilsel, O. (2007). 'Microsecond hydrophobic collapse in the folding of Escherichia coli dihydrofolate reductase, an α/β-type protein', *J. Mol. Biol.* 368: 219–29.

Arai, S. and Hirai, M. (1999). 'Reversibility and hierarchy of thermal transition of hen egg-white lysozyme studied by small-angle X-ray scattering', *Biophys. J.* 76: 2192–7.

Aramayo, R., Mérigoux, C., Larquet, E. *et al.* (2005). 'Divalent ion-dependent swelling of Tomato Bushy Stunt Virus: A multi-approach study', *Biochim. Biophys. Acta* 1724: 345–54.

Arnaudov, L. N., De Vries, R. and Cohen Stuart, M. A. (2006). 'Time-resolved small-angle neutron scattering during heat-induced fibril formation from bovine β-lactoglobulin', *J. Chem. Phys.* 124: 084701.

Aswal, V. K., Chodankar, S., Kohlbrecher, J., Vavrin, R. and Wagh, A. G. (2009). 'Small-angle neutron scattering study of protein unfolding and refolding', *Phys. Rev. E* 80: 011924.

Atkins, P. W. (1984). *The Second Law*. New York: W. H. Freeman and Company.

Ayuso-Tejedor, S., García-Fandiño, R., Orozco, M., Sancho, J. and Bernadó, P. (2011). 'Structural analysis of an equilibrium folding intermediate in the apoflavodoxin native ensemble by small-angle X-ray scattering', *J. Mol. Biol.* 406: 604–19.

Bachmann, A., Segel, D. and Kiefhaber, T. (2002). 'Test for cooperativity in the early kinetic intermediate in lysozyme folding', *Biophys. Chem.* 96: 141–51.

Balakrishnan, G., Case, M. A., Pevsner, A. *et al.* (2004). 'Time-resolved absorption and UV resonance Raman spectra reveal stepwise formation of T quaternary contacts in the allosteric pathway of hemoglobin', *J. Mol. Biol.* 340: 843–56.

Banachowicz, E., Kozak, M., Patkowski, A., Meier, G. and Kohlbrecher, J. (2009). 'High-pressure small-angle neutron scattering studies of glucose isomerase conformation in solution', *J. Appl. Crystallogr.* 42: 461–8.

Barone, G., Sayers, Z., Svergun, D. I. and Koch, M. H. J. (1999). 'A synchrotron radiation X-ray scattering study of aqueous solutions of native DNA', *J. Synchrotron Radiat.* 6: 1031–4.

Bartels, T., Choi, J. G. and Selkoe, D. J. (2011). 'α-Synuclein occurs physiologically as a helically folded tetramer that resists aggregation', *Nature* 477: 107–10.

Beece, D., Eisenstein, L., Frauenfelder, H. *et al.* (1980). 'Solvent viscosity and protein dynamics', *Biochemistry-US* 19: 5147–57.

Berthet-Colominas, C., Bois, J. M., Cuillel, M., Sedita, J. and Vachette, P. (1984). 'An apparatus for stopped-flow X-ray scattering', *Rev. Phys. Appl.* 19: 769–72.

Berthet-Colominas, C., Cuillel, M., Koch, M.H.J., Vachette, P. and Jacrot B. (1987). 'Kinetic study of the assembly of brome mosaic virus capsid', *Eur. Biophys. J.* 15: 159–68.

Bolen, D. W. and Rose, G. D. (2008). 'Structure and energetics of the hydrogen-bonded backbone in protein folding', *Annu. Rev. Biochem.* 77: 339–62.

Bolisetty, S., Adamcik, J. and Mezzenga, R. (2011). 'Snapshots of fibrillation and aggregation kinetics in multistranded amyloid β-lactoglobulin fibrils', *Soft Matter* 7: 493–9.

Bonetti, M. and Calmettes, P. (2005). 'Sapphire-anvil cell for small-angle neutron scattering measurements in large-volume liquid samples up to 530 MPa', *Rev. Sci. Instrum.* 76: 043903.

Boonyaratanakornkit, B. B., Park, C. B. and Clark, D. S. (2002). 'Pressure effects on intra- and intermolecular interactions within proteins', *Biochim. Biophys. Acta* 1595: 235–49.

Bordas, J., Mandelkow, E. M. and Mandelkow, E. (1983). 'Stages of tubulin assembly and disassembly studied by time-resolved synchrotron X-ray scattering', *J. Mol. Biol.* 164: 80–135.

Bordas, J., Perez-Grau, L., Koch, M. H. J., Vega, M. C. and Nave, C. (1986). 'The superstructure of chromatin and its condensation mechanism. I. Synchrotron radiation X-ray scattering results', *Europ. Journ. Biophys.* 13: 157–74.

Borshchevskiy, V. I., Round, E. S., Popov, A. N., Büldt, G. and Gordeliy, V. I. (2011). 'X-ray-radiation-induced changes in bacteriorhodopsin structure', *J. Mol. Biol.* 409: 813–25.

Bras, W., Diakun, G. P., Diaz, J. F. *et al.* (1998). 'The susceptibility of pure tubulin to high magnetic fields: A magnetic birefringence and X-ray fiber diffraction study', *Biophys. J.* 74: 1509–21.

Brennich, M. E., Nolting, J. F., Dammann, C. *et al.* (2011). 'Dynamics of intermediate filament assembly followed in micro-flow by small angle X-ray scattering', *Lab Chip* 11: 708–16.

Bridgman, P. W. (1915). 'Change of phase under pressure', *Phys. Rev.* 6: 94–112.

Brooks, N. J., Gauthe, B. L. L. E., Terrill, N. J. *et al.* (2010). 'Automated high pressure cell for pressure jump X-ray diffraction', *Rev. Sci. Instrum.* 81: 064103.

Bucksbaum, P. H. (2007). 'The future of attosecond spectroscopy', *Science* 317: 766–9.

Budayova-Spano, M., Bonneté, F., Astier, J. P. and Veesler, S. (2002). 'Investigation of aprotinin (BPTI) solutions during nucleation', *J. Cryst. Growth* 235: 547–54.

Butler, P. J. G., Bloomer, A. C. and Finch, J. T. (1992). 'Direct visualization of the structure of the '20S' aggregate of coat protein of tobacco mosaic virus', *J. Mol. Biol.* 224: 381–94.

Cammarata, M., Levantino, M., Schotte, F. *et al.* (2008). 'Tracking the structural dynamics of proteins in solution using time-resolved wide-angle scattering', *Nat. Methods* 5: 881–6.

Cammarata, M., Levantino, M., Wulff, M. and Cupane, A. (2010). 'Unveiling the timescale of the R–T transition in human hemoglobin', *J. Mol. Biol.* 400: 951–62.

Cammarata, M., Lorenc, M., Kim, T. H. *et al.* (2006). 'Impulsive solvent heating probed by picosecond X-ray diffraction', *J. Chem. Phys.* 124: 124504.

Canady, M. A., Tsuruta, H. and Johnson, J. E. (2001). 'Analysis of rapid, large-scale protein quaternary structural changes: Time-resolved X-ray solution scattering of Nudaurelia capensis ω Virus (NωV) maturation', *J. Mol. Biol.* 311: 803–14.

Casselyn, M., Tardieu, A., Delacroix, H. and Finet, S. (2004). 'Birth and growth kinetics of brome mosaic virus microcrystals', *Biophys. J.* 87: 2737–48.

Castelletto, V. and Hamley, I.W. (2006). 'Capillary flow behavior of worm-like micelles studied by small-angle X-ray scattering and small angle light scattering', *Polym. Adv. Technol.* 17: 137–44.

Changeux, J.-P. (2012). 'Allostery and the Monod–Wyman–Changeux Model after 50 years', *Annu. Rev. Biophys.* 41: 103–33.

Chen, L., Hodgson, K. O. and Doniach, S. (1996). 'A lysozyme folding intermediate revealed by solution X-ray scattering', *J. Mol. Biol.* 261: 658–71.

Chen, L., Wildegger, G., Kiefhaber, T., Hodgson, K. O. and Doniach, S. (1998). 'Kinetics of lysozyme refolding: structural characterization of a non-specifically collapsed state using time-resolved X-ray scattering', *J. Mol. Biol.* 276: 225–37.

Chodankar, S., Aswal, V. K., Kohlbrecher, J., Vavrin, R. and Wagh, A. G. (2009). 'Small-angle neutron scattering study of structure and kinetics of temperature-induced protein gelation', *Phys. Rev. E* 79: 021912.

Cohen-Tannoudji, G. (1998). *Les Constantes Universelles*: Hachette Littératures.

Cuillel, M., Berthet-Colominas, C., Krop, B., Tardieu, A., Vachette, P. and Jacrot, B. (1983). 'Self-assembly of brome mosaic-virus capsids- kinetic study using neutron and X-ray solution scattering', *J. Mol. Biol.* 164: 645–50.

Cunsolo, A., Formisano, F., Ferrero, C., Bencivenga, F. and Finet, S. (2009). 'Pressure dependence of the large-scale structure of water', *J. Chem. Phys.* 131: 194502–5.

Czeslik, C., Malessa, R., Winter, R. and Rapp, G. (1996). 'High pressure synchrotron X-ray diffraction studies of biological molecules using the diamond anvil technique', *Nucl. Instrum. Meth. A* 368: 847–51.

Das, R., Kwok, L. W., Millett, I. S. *et al.* (2003). 'The fastest global events in RNA folding: electrostatic relaxation and tertiary collapse of the Tetrahymena ribozyme', *J. Mol. Biol.* 332: 311–19.

Díaz, J. F., Strobbe, R., Engelborghs, Y., Chacón, P., Andreu, J. M. and Diakun, G. (1998). 'Fast mixing device for time-resolved synchrotron X-ray scattering studies of radiation sensitive proteins', *Rev. Sci. Instrum.* 69: 286–9.

Dill, K. A. and Bromberg, S. (2003). *Molecular Driving Forces: Statistical Thermodynamics in Chemistry and Biology.* New York: Garland Science.

Dumont, C., Emilsson, T. and Gruebele, M. (2009). 'Reaching the protein folding speed limit with large sub-microsecond pressure jumps', *Nat. Methods* 6: 515–9.

Eliezer, E., Chiba, K., Tsuruta, H., Doniach, S., Hodgson, K. O. and Kihara, H. (1993). 'Evidence of an associative intermediate on the myoglobin refolding pathway', *Biophys. J.* 65: 912–7.

Fetler, L. and Vachette, P. (2001). 'The allosteric activator Mg-ATP modifies the quaternary structure of the R-state of Escherichia coli aspartate transcarbamylase without altering the T→R equilibrium', *J. Mol. Biol.* 309: 817–32.

Fischer, S., Olsen, K. W., Nam, K. and Karplus, M. (2011). 'Unsuspected pathway of the allosteric transition in hemoglobin', *P. Natl. Acad. Sci. USA* 108: 5608–13.

Flory, P. J. (1953). *Principles of Polymer Chemistry.* Ithaca, NY: Cornell University Press.

Fowler, A. G., Foote, A. M., Moody, M. F. *et al.* (1983). 'Stopped-flow solution scattering using synchrotron radiation; apparatus, data collection and data analysis', *J. Biochem. Bioph. Meth.* 7: 387–92.

Frauenfelder, H., Chena, G., Berendzen, J. *et al.* (2010). 'A unified model of protein dynamics', *P. Natl. Acad. Sci. USA* 106: 5130–4.

Frauenfelder, H., Wolynes, P. G. and Austin, R. H. (1999). 'Biological physics', *Rev. Mod. Phys.* 71: S419–30.

Fujisawa, T., Kato, M. and Inoko, Y. (1999). 'Structural characterization of lactate dehydrogenase dissociation under high pressure studied by synchrotron high-pressure small-angle X-ray scattering', *Biochemistry-US* 38: 6411–8.

Fujiwara, S., Matsumoto, F. and Yonezawa, Y. (2003). 'Effects of salt concentration on association of the amyloid protofilaments of hen egg white lysozyme studied by time-resolved neutron scattering', *J. Mol. Biol.* 331: 21–8.

Gambin, Y., Simonnet, C., Vandelinder, V., Deniz, A. and Groisman, A. (2010). 'Ultrafast microfluidic mixer with three-dimensional flow focusing for biochemical kinetics', *Lab Chip* 10: 598–609.

Garcia-Ruiz, J. M. (2003). 'Nucleation of protein crystals', *J. Struct. Biol.* 142: 22–31.

Giehm, L., Svergun, D. I., Otzen, D. E. and Vestergaard, B. (2011). 'Low-resolution structure of a vesicle disrupting α-synuclein oligomer that accumulates during fibrillation', *P. Natl. Acad. Sci. USA* 108: 3246–51.

Giovannardi, S., Landò, L. and Peres, A. (1998). 'Flash photolysis of caged compounds: casting light on physiological processes', *News Physiol. Sci.* 13: 251–5.

Girard, E., Marchal, S., Pérez, J. *et al.* (2010). 'Structure-function perturbation and dissociation of tetrameric urate oxidase by high hydrostatic pressure', *Biophys. J.* 98: 2365–73.

Graber, T., Anderson, S., Brewer, H. *et al.* (2011). 'BioCARS: a synchrotron resource for time-resolved X-ray science', *J. Synchrotron Radiat.* 18: 658–70.

Grillo, I. (2009). 'Applications of stopped-flow in SAXS and SANS', *Curr. Opin. Colloid In.* 14: 402–8.

Grillo, I., Kats, E. I. and Muratov, A. R. (2003). 'Formation and growth of anionic vesicles followed by small-angle neutron scattering', *Langmuir* 19: 4573–81.

Hawley, S. A. (1971). 'Reversible pressure–temperature denaturation of chymotrypsinogen', *Biochemistry-US* 10: 2436–42.

Heremans, K. (ed.) (1997). *High Pressure Research in the Biosciences and Biotechnology.* Leuven: Leuven University Press.

Heremans, K. and Smeller, L. (1998). 'Protein structure and dynamics at high pressure', *Biochim. Biophys. Acta* 1386: 353–70.

Hillerup Jensen, M., Nørgaard Toft, K., David, G., Havelund, S., Pérez, J. and Vestergaard, B. (2010). 'Time-resolved SAXS measurements facilitated by online HPLC buffer exchange', *J. Synchrotron Radiat.* 17: 769–73.

Hiragi, Y., Inoue, H., Sano, Y. *et al.* (1988a). 'Temperature-dependence of the structure of aggregates of tobacco mosaic-virus protein at pH 7.2: static synchrotron small-angle X-ray scattering', *J. Mol. Biol.* 204: 129–40.

Hiragi, Y., Inoue, H., Sano, Y., Kajiwara, K., Ueki, T. and Nakatani, H. (1990). 'Dynamic mechanism of the self-assembly process of tobacco mosaic virus protein studied by rapid temperature-jump small-angle X-ray scattering using synchrotron radiation', *J. Mol. Biol.* 213: 495–502.

Hiragi, Y., Nakatani, H., Kajiwara, K., Inoue, H., Sano, Y. and Kataoka, M. (1988b). 'Temperature jump apparatus and measuring system for synchrotron solution X-ray scattering experiments', *Rev. Sci. Instrum.* 59: 64–6.

Hirai, M., Arai, S., Takizawa, T. and Yabuki, Y. (1997). 'Dynamics and phase behavior of a supermacromolecular suspension under a magnetic field studied by time-resolved X-ray scattering', *Phys. Rev. B* 55: 3490–6.

Hirai, M., Arai, S., Iwase, H. and Takizawa, T. (1998). 'Small-angle X-ray scattering and calorimetric studies of thermal conformational change of lysozyme depending on pH', *J. Phys. Chem. B* 102: 1308–13.

Hirai, M., Koizumi, M., Han, R., Hayakawa T. and Sano, Y. (2003). 'Right–left–circular orientation of biological macromolecules under magnetic field gradient', *J. Appl. Crystallogr*. 36: 520–4.

Holzinger, J. (2002). 'Untersuchung der Reaktionszyklen von Chaperoninen aus Escherichia coli und Thermoplasma acidophilum mit Hilfe der Neutronenkleinwinkelstreuung', PhD Thesis, *Faculty of Chemistry and Pharmacy*. München: LMU.

Inoko, Y., Kihara, H. and Koch, M. H. J. (1983). 'Time-resolved small-angle X-ray scattering experiment on association of isolated alpha-chain and beta-chain of human hemoglobin', *Biophys. Chem*. 17: 171–4.

Jackson, A. J. and McGillivray, D. J. (2011). 'Protein aggregate structure under high pressure', *Chem. Commun*. 47: 487–9.

Jacob, J., Krantz, B., Dothager, R. S., Thiyagarajan, P. and Sosnick, T. R. (2004). 'Early collapse is not an obligate step in protein folding', *J. Mol. Biol*. 338: 369–82.

Kathuria, S. V., Guo, L., Graceffa, R. et al. (2011). 'Structural insights into early folding events using continuous-flow time-resolved small-angle X-ray scattering', *Biopolymers* 95: 550–8.

Kato, M. and Fujisawa, T. (1998). 'High-pressure solution X-ray scattering of protein using a hydrostatic cell with diamond windows', *J. Synchrotron Radiat*. 5: 1282–6.

Kihara, H., Takahashi-Ushijima, E., Amemiya, Y. et al. (1987). 'Kinetic of structure and activity changes during the allosteric transition of aspartate transcarbamylase', *J. Mol. Biol*. 198: 745–8.

Kim, J., Kim, K. H., Kim, J. G., Kim, T. W., Kim, Y. and Ihee, H. (2011). 'Anisotropic picosecond X-ray solution scattering from photoselectively aligned protein molecules', *J. Phys. Chem. Lett*. 2: 350–6.

Kim, S. J., Matsumura, Y., Dumont, C., Kihara, H. and Gruebele, M. (2009). 'Slowing down downhill folding: A three-probe study', *Biophys. J*. 97: 295–302.

Kim, T. K., Lorenc, M., Lee, J. H. et al. (2006). 'Spatiotemporal reaction kinetics of an ultrafast photoreaction pathway visualized by time-resolved liquid X-ray diffraction', *P. Natl. Acad. Sci. USA* 103: 9410–5.

Kimura, T., Akiyama, S., Uzawa, T. et al. (2005a). 'Specifically collapsed intermediate in the early stage of the folding of ribonuclease A', *J. Mol. Biol*. 350: 349–62.

Kimura, T., Uzawa, T., Ishimori, K. et al. (2005b). 'Specific collapse followed by slow hydrogen-bond formation of β-sheet in the folding of single-chain monellin', *P. Natl. Acad. Sci. USA* 102: 2748–53.

Kleppinger, R., Goossens, K., Lorenzen, M., Geissler, E. and Heremans, K. (1997). 'Effect of high-pressure treatment of ribonuclease A: small angle X-ray scattering and Fourier transform infrared spectroscopy investigations', in Heremans, K. (ed.) *High Pressure Research in the Biosciences and Biotechnology*. Leuven: Leuven University Press.

Knight, J. B., Vishwanath, A., Brody, J. P. and Austin, R. H. (1998). 'Hydrodynamic focusing on a silicon chip: mixing nanoliters in microseconds', *Phys. Rev. Letters* 80: 3863–6.

Koch, M. H. J., Dorrington, E., Klaering, R. et al. (1988). 'Electric field X-ray scattering measurements on tobacco mosaic virus', *Science* 240: 194–6.

Koch, M. H. J., Sayers, Z., Sicre, P. and Svergun, D. (1995). 'A synchrotron radiation electric field X-ray solution scattering study of DNA at very low ionic strength', *Macromolecules* 28: 4904–7.

Kohlbrecher, J., Bollhalder, A., Vavrin, R. and Meier, G. (2007). 'A high pressure cell for small angle neutron scattering up to 500 MPa in combination with light scattering to investigate liquid samples', *Rev. Sci. Instrum*. 78: 12101.

Kohn, J. E., Millett, I. S., Jacob, J. et al. (2004). 'Random-coil behavior and the dimensions of chemically unfolded proteins', *P. Natl. Acad. Sci. USA* 101: 12491–6.

Koizumi, M., Hirai, H., Onai, T., Inoue, K. and Hirai, M. (2007). 'Collapse of the hydration shell of a protein prior to thermal unfolding', *J. Appl. Crystallogr*. 40: s175–8.

Kondepudi, D. and Prigogine, Y. (1998). *Modern Thermodynamics: From Heat Engines to Dissipative Structures*. New York: John Wiley and Sons.

Konuma, T., Kimura, T., Matsumoto, S. et al. (2011). 'Time-resolved small-angle X-ray scattering study of the folding dynamics of barnase', *J. Mol. Biol*. 405: 1284–94.

Koshland, J. D. E., Nemethy, G. and Filmer, D. (1966). 'Comparison of experimental binding data and theoretical models in proteins containing subunits', *Biochemistry-US* 5: 365–85.

Krywka, C., Sternemann, C., Paulus, M., Tolan, M., Royer, C. and Winter, R. (2008). 'Effects of osmolytes on pressure-induced unfolding of proteins: a high pressure SAXS study', *Chem. Phys. Chem*. 9: 2809–15.

Kwok, L. W., Shcherbakova, I., Lamb, J. S. *et al.* (2006). 'Concordant exploration of the kinetics of RNA folding from global and local perspectives', *J. Mol. Biol.* 355: 282–93.

Lamb, J., Kwok, L., Qiu, X., Andresen, K., Park, H.Y. and Pollack, L. (2008). 'Reconstructing three-dimensional shape envelopes from time-resolved small angle X-ray scattering data', *J. Appl. Crystallogr.* 41: 1046–52.

Lamb, J., Zoltowski, B. D., Pabit, S. A., Crane, B. R. and Pollack, L. (2008). 'Time resolved dimerization of a PAS-LOV protein measured with photocoupled small angle X-ray scattering', *J. Am. Chem. Soc.* 130: 12226–7.

Lange, G., Mandelkow, E.-M., Jagla, A. and Mandelkow, E. (1988). 'Tubulin oligomers and microtubule oscillations', *Eur. J. Biochem.* 178: 61–9.

Lechner, M. D., Steinmeier, D. G., Vennemann, N., Ibel, K. and Oberthür, R. (1985). 'Small angle neutron scattering under high pressure', *Makromol. Chem. Rapid. Commun.* 6: 281–4.

Lee, K. K., Tsuruta, H., Hendrix, R. W., Duda, R. L. and Johnson, J. E. (2005). 'Cooperative reorganization of a 420 subunit virus capsid', *J. Mol. Biol.* 352: 723–35.

Leimkühler, M., Goldbeck, A., Lechner, M. D. and Witz, J. (2000). 'Conformational changes preceding decapsidation of bromegrass mosaic virus under hydrostatic pressure: a small-angle neutron scattering study', *J. Mol. Biol.* 296: 1295–305.

Leimkühler, M., Goldbeck, A., Lechner, M.D., Adrian, M., Michels, B. and Witz, J. (2001). 'The formation of empty shells upon pressure induced decapsidation of turnip yellow mosaic virus', *Arch. Virol.* 146: 653–67.

Lindenberg, A. M., Acremann, Y., Lowney, D. P. *et al.* (2005). 'Time-resolved measurements of the structure of water at constant density', *J. Chem. Phys.* 122: 204507.

Lorenzen, M., Riekel, C., Eichler, A. and Häussermann, D. H. (1993). 'A high pressure cell for small angle X-ray scattering', *J. Phys. I* 3(C8): 487–90.

Loupiac, C., Bonetti, M., Pin, S. and Calmettes, P. (2002). 'High-pressure effects on horse heart metmyoglobin studied by small-angle neutron scattering', *Eur. J. Biochem.* 269: 4731–7.

Lu, H. P., Xun, L. and Xie, X. (1998). 'Single-molecule enzymatic dynamics', *Science* 282: 1877–82.

Ma, H., Wan, C. and Zewail, A. H. (2006). 'Ultrafast T-jump in water: studies of conformation and reaction dynamics at the thermal limit', *J. Am. Chem. Soc.* 128: 6338–40.

Makarov, D. E. and Plaxco, K. W. (2009). 'Measuring distances within unfolded biopolymers using fluorescence resonance energy transfer: The effect of polymer chain dynamics on the observed fluorescence resonance energy transfer efficiency', *J. Chem. Phys.* 131: 085105.

Makowski, L., Bardhan, J., Gore, D. *et al.* (2011). 'WAXS studies of the structural diversity of hemoglobin in solution', *J. Mol. Biol.* 408: 909–21.

Makowski, L., Rodi, D. J., Mandava, S., Minh, D. D. L., Gore, D. B. and Fischetti, R. F. (2008). 'Molecular crowding inhibits intramolecular breathing motions in proteins', *J. Mol. Biol.* 375: 529–46.

Malmerberg, E., Omran, Z., Hub, J. S. *et al.* (2011). 'Time-resolved WAXS reveals accelerated conformational changes in iodoretinal-substituted proteorhodopsin', *Biophys. J.* 101: 1345–53.

Mandelkow, E.-M., Lange, G., Jagla, A., Spann, U. and Mandelkow, E. (1988). 'Dynamics of the microtubule oscillator: role of nucleotides and tubulin-MAP interactions', *EMBO J.* 7: 357–65.

Mandelkow, E. M., Harmsen, A., Mandelkow, E. and Bordas, J. (1980). 'X-ray kinetic studies of microtubule assembly using synchrotron radiation', *Nature* 287: 595–9.

Marmiroli, B., Grenci, G., Cacho-Nerin, F. *et al.* (2009). 'Free jet micromixer to study fast chemical reactions by small angle X-ray scattering', *Lab Chip* 9: 2063–9.

Marx, A., Jagla, A. and E., M. (1990). 'Microtubule assembly and oscillations induced by flash photolysis of caged-GTP', *Eur. Biophys. J.* 19: 1–9.

Marx, A. and Mandelkow, E. (1994). 'A model of microtubule oscillations', *Eur. Biophys. J.* 22: 405–21.

Matsudaira, P., Bordas, J. and Koch, M. H. J. (1987). 'Synchrotron X-ray diffraction studies on actin structure during polymerization', *P. Natl. Acad. Sci. USA* 84: 3151–5.

Matsui, T., Tsuruta, H. and Johnson, J. E. (2010). 'Balanced electrostatic and structural forces guide the large conformational change associated with maturation of T = 4 virus', *Biophys. J.* 98: 1337–43.

Matsumura, Y., Shinjo, M., Mahajan, A., Tsai, M.-D. and Kihara, H. (2010). 'α-helical burst on the folding pathway of FHA domains from Rad53 and Ki67', *Biochimie* 92: 1031–9.

May, R. P., Hendriks, J. and Crielaard, W. (2005). 'Real-time neutron scattering investigations of biological signal transduction dynamics', *Proceedings of the International Symposium on Research Reactor and Neutron Science, KAERI/GP-234/2005*, Daejon, Korea.

McClare, C. W. F. (1971). 'Chemical machines, Maxwell's demon and living organisms', *J. Theor. Biol.* 30: 1–34.

Meersman, F., Atilgan, C., Miles, A. J. *et al.* (2010). 'Consistent picture of the reversible thermal unfolding of hen egg-white lysozyme from experiment and molecular dynamics', *Biophys. J.* 99: 2255–63.

Meersman, F., Dobson, C.M. and Heremans, K. (2006). 'Protein unfolding, amyloid fibril formation and configurational energy landscapes under high pressure conditions', *Chem. Soc. Rev.* 35: 908–17.

Meersman, F., Smeller, L. and Heremans, K. (2005). 'Extending the pressure–temperature state diagram of myoglobin', *Helvetica Chim. Acta* 88: 546–56.

Merchant, K. A., Best, R., B., Louis, J. M., Gopich, I. V. and Eaton, W. A. (2007). 'Characterizing the unfolded states of proteins using single-molecule FRET spectroscopy and molecular simulations', *P. Natl. Acad. Sci. USA* 104: 1528–33.

Millet, I. S., Townsley, L. E., Chiti, F., Doniach, S. and Plaxco, K. W. (2002). 'Equilibrium collapse and the kinetic 'foldability' of proteins', *Biochemistry-US* 41: 321–5.

Monod, J., Wyman, J. and Changeux, J. P. (1965). 'On the nature of allosteric transitions. A plausible model', *J. Mol. Biol.* 12: 88–118.

Moody, M. F., Vachette, P., Foote, A. M., Tardieu, A. M., Koch, M. H. J. and Bordas, J. (1980). 'Stopped-flow X-ray solution scattering: the dissociation of aspartate transcarbamylase', *P. Natl. Acad. Sci. USA* 77: 4040–3.

Nagamura, T., Kurita, K., Tokikura, E. and Kihara, H. (1985). 'Stopped-flow X-ray scattering device with a slit-type mixer', *J. Biochem. Bioph. Meth.* 11: 277–86.

Nagy, G., Posselt, D., Kovacs, L. *et al.* (2011). 'Reversible membrane reorganizations during photosynthesis in vivo: revealed by small-angle neutron scattering', *Biochem J.* 436: 225–30.

Nayak, A., Sorci, M., Krueger, S. and Belfort, G. (2009). 'A universal pathway for amyloid nucleus and precursor formation for insulin', *Proteins* 74: 556–65.

Niimura, N., Minezaki, Y., Ataka, M. and Katsura, T. (1995). 'Aggregation in supersaturated lysozyme solution studied by time resolved small-angle neutron scattering', *J. Cryst. Growth* 154: 136–44.

Nishikawa, Y., Fujisawa, T., Inoko, Y. and Moritoki, M. (2001). 'Improvement of a high pressure cell with diamond windows for solution X-ray scattering of proteins', *Nucl. Instrum. Meth. A* 467: 1384–7.

Nordén, B., Wittung-Stafshede, P., Ellouze, C., Kim, H.-K., Mortensen, K. and Takahashi, M. (1998). 'Base orientation of second DNA in RecA.DNA filaments', *J. Biol. Chem.* 273: 15682–6.

Norman, A. I., Ivkov, R., Forbes, J. G. and Greer, A. C. (2005). 'The polymerization of actin: Structural changes from small angle scattering', *J. Chem. Phys.* 123: 154904.

Oliveira, C. L. P., Behrens, M. A., Pedersen, J. S., Erlacher, K., Otzen, D. and Pedersen, J. S. (2009). 'A SAXS study of glucagon fibrillation', *J. Mol. Biol.* 387: 147–61.

Oosawa, F. and Asakura, S. (1975). *Thermodynamics of the Polymerization of Protein*. London: Academic Press.

Ortore, M. G., Spinozzi, F., Mariani, P. *et al.* (2009). 'Combining structure and dynamics: non-denaturing high-pressure effect on lysozyme in solution', *J. R. Soc. Interface* 6: S619–34.

Paliwal, A., Asthagiri, D., Bossev, D. P. and Paulaitis, M. E. (2004). 'Pressure denaturation of staphylococcal nuclease studied by neutron small-angle scattering and molecular simulation', *Biophys. J.* 87: 3479–92.

Panick, G., Malessa, R., Winter, R., Rapp, G., Frye, K. J. and Royer, C. A. (1998). 'Structural characterization of the pressure-denatured state and unfolding/refolding kinetics of staphylococcal nuclease by synchrotron small-angle X-ray scattering and Fourier-transform infrared spectroscopy', *J. Mol. Biol.* 275: 389–402.

Panine, P., Finet, S., Weiss, T. M. and Narayanan, T. (2006). 'Probing fast kinetics in complex fluids by combined rapid mixing and small angle scattering', *Adv. Colloid Interfac.* 127: 9–18.

Park, H. Y., Qiu, X., Rhoades, E. *et al.* (2006). 'Achieving uniform mixing in a microfluidic device: hydrodynamic focusing prior to mixing', *Anal. Chem.* 78: 4465–73.

Park, S., Bardhan, J. P., Roux, B. and Makowski, L. (2009). 'Simulated X-ray scattering of protein solutions using explicit-solvent models', *Structure* 130: 134114.

Pérez, J., Vachette, P., Russo, D., Desmadril, M. and Durand, D. (2001). 'Heat-induced unfolding of neocarzinostatin, a small all-β protein investigated by small-angle X-ray scattering', *J. Mol. Biol.* 308: 721–43.

Phillips, C. M., Mizutani, Y. and Hochstrasser, R. M. (1995). 'Ultrafast thermally induced unfolding of RNase A', *P. Natl. Acad. Sci. USA* 92: 7292–6.

Pilling, M. J. and Seakins, P. W. (1995). *Reaction Kinetics*. Oxford: Oxford University Press.

Plaxco, K. W., Millett, I. S., Segel, D. J., Doniach, S. and Baker, D. (1999). 'Chain collapse can occur concomitantly with the rate-limiting step in protein folding', *Nature Struct. Biol.* 6: 554–6.

Pollack, L. (2011). 'Time resolved SAXS and RNA folding', *Biopolymers* 95: 543–549.

Pollack, L. and Doniach, S. (2009). 'Time-resolved X-ray scattering and RNA folding', in Herschlag, D. (ed.), *Biophysical, Chemical, and Functional Probes of RNA Structure, Interactions and Folding, Part B*. Amsterdam: Elsevier/Academic Press.

Pollack, L., Tate, M. W., Darnton, N. C. *et al.* (1999). 'Compactness of the denatured state of a fast-folding protein measured by submillisecond small-angle X-ray scattering', *P. Natl. Acad. Sci. USA* 96: 10115–7.

Pollack, L., Tate, M. W., Finnefrock, A. C. *et al.* (2001). 'Time resolved collapse of a folding protein observed with small angle X-ray scattering', *Phys. Rev. Lett.* 86: 4962–5.

Potschka, M., Koch, M. H. J., Adams, M. L. and Schuster, T. M. (1988). 'Time-resolved solution X-ray scattering of tobacco mosaic virus coat protein: kinetics and structure intermediates', *Biochemistry-US* 27: 8481–91.

Prassl, R., Pregetter, M., Amenitsch, H. *et al.* (2008). 'Low density lipoproteins as circulating fast temperature sensors', *PLoS One* 3: e4079.

Pressl, K., Kriechbaum, M., Steinhart, M. and Laggner, P. (1997). 'High pressure cell for small- and wide-angle scattering', *Rev. Sci. Instrum.* 68: 4588–92.

Pühse, M., Szweda, R. T., Ma, Y., Jeworrek, C., Winter, R. and Zorn, H. (2009). 'Marasmius scorodonius extracellular dimeric peroxidase: Exploring its temperature and pressure stability', *Biochim. Biophys. Acta* 1794: 1091–8.

Puigdomenech, J., Perez-Grau, L., Porta, J., Vega, M. C., Sicre, P. and Koch, M. H. J. (1989). 'A time-resolved synchrotron radiation X-ray solution scattering study of DNA melting', *Biopolymers* 28: 1505–14.

Ramachandran, P. L., Lovett, J. E., Carl, P. J. *et al.* (2011). 'The short-lived signaling state of the photoactive yellow protein photoreceptor revealed by combined structural probes', *J. Am. Chem. Soc.* 133: 9395–404.

Ravindra, R. and Winter, R. (2003). 'On the temperature–pressure free energy landscape of proteins', *Chem. Phys. Chem.* 4: 359–65.

Renner, W., Mandelkow, E. M., Mandelkow, E. and Bordas, J. (1983). 'Self-assembly of microtubule protein studied by time-resolved X-ray scattering using temperature jump and stopped flow', *Nucl. Instrum. Methods* 208: 535–40.

Roder, H., Maki, K. and Cheng, H. (2006). 'Early events in protein folding explored by rapid mixing methods', *Chem. Rev.* 106: 1836–61.

Roessle, M., Manakova, E., Lauer, I. *et al.* (2000). 'Time-resolved small angle scattering: kinetic and structural data from proteins in solution', *J. Appl. Crystallogr.* 33: 548–51.

Roh, J., H., Guo, L., Kilburn, J. D., Briber, R. M., Irving, T. and Woodson, S. A. (2010). 'Multistage collapse of a bacterial ribozyme observed by time-resolved small-angle X-ray scattering', *J. Am. Chem. Soc.* 132: 10150–4.

Rouget, J.-B., Schroer, M. A., Jeworrek, C. *et al.* (2010). 'Unique features of the folding landscape of a repeat protein revealed by pressure perturbation', *Biophys. J.* 98: 2712–21.

Russell, R., Millet, I. S., Tate, M. W. *et al.* (2002). 'Rapid compaction during RNA folding', *P. Natl. Acad. Sci. USA* 99: 4267–71.

Russell, R., Millett, I. S., Doniach, S. and Herschlag, D. (2000). 'Small angle X-ray scattering reveals a compact intermediate in RNA folding', *Nat. Struct. Biol.* 7: 367–70.

Russo, D., Durand, D., Calmettes, P. and Desmadril, M. (2001). 'Characterization of the denatured states distribution of neocarzinostatin by small-angle neutron scattering and differential scanning calorimetry', *Biochemistry-US* 40: 3958–66.

Sano, Y., Inoue, H., Hiragi, Y., Urakawa, H. and Kajiwara, K. (1995). 'Solution X-ray scattering study of reconstitution process of tobacco mosaic virus particle using low-temperature quenching', *Biophys. Chem.* 55: 239–45.

Sano, Y., Inoue, H. and Hiragi, Y. (1999). 'Differences of reconstitution process between tobacco mosaic virus and cucumber green mottle mosaic virus by synchrotron small angle X-ray scattering using low-temperature quenching', *J. Protein Chem.*, 18: 801–805.

Sayers, Z., Koch, M. H. J., Bordas, J. and Lindberg, U. (1985). 'Time resolved X-ray scattering study of actin polymerization from profilactin', *Europ. Journ. Biophysics* 13: 99–108.

Schlatterer, J. C., Kwok, L. W., Lamb, J. S. *et al.* (2008). 'Hinge stiffness is a barrier to RNA folding', *J. Mol. Biol.* 379: 859–70.

Schroer, M. A., Markgraf, J., Wieland, D. C. F. *et al.* (2011). 'Nonlinear pressure dependence of the interaction potential of dense protein solutions', *Phys. Rev. Letters* 106: 178102.

Schroer, M. A., Paulus, M., Jeworrek, C. *et al.* (2010). 'High-pressure SAXS study of folded and unfolded ensembles of proteins', *Biophys. J.* 99: 3430–7.

Segel, D. J., Bachmann, A., Hofrichter, J., Hodgson, K. O., Doniach, S. and Kiefhaber, T. (1999). 'Characterization of transient intermediates in lysozyme folding with time-resolved small-angle X-ray scattering', *J. Mol. Biol.* 288: 489–99.

Semisotnov, G. V., Kihara, H., Kotova, N. V. *et al.* (1996). 'Protein globularization during folding. A study by synchrotron small-angle X-ray scattering', *J. Mol. Biol.* 262: 559–74.

Shiu, Y.-J., Jeng, U.-S., Huang, Y.-S. *et al.* (2008). 'Global and local structural changes of cytochrome C and lysozyme characterized by a multigroup unfolding process', *Biophys. J.* 94: 4828–36.

Shkumatov, A. V., Chinnathambi, S., Mandelkow, E. and Svergun, D. I. (2011). 'Structural memory of natively unfolded tau protein detected by small-angle X-ray scattering', *Proteins* 79: 2122–31.

Skouri-Panet, F., Quevillon-Cheruel, S., Michiel, M., Tardieu, A. and Finet, S. (2006). 'sHSPs under temperature and pressure: The opposite behaviour of lens alpha-crystallins and yeast HSP26', *Biochim. Biophys. Acta* 1764: 372–83.

Sosnick, T. R. and Barrick, D. (2011). 'The folding of single domain proteins: Have we reached a consensus?' *Curr. Opin. Struc. Biol.* 21: 12–24.

Spann, U., Renner, W., Mandelkow, E.-M., Bordas, J. and Mandelkow, E. (1987). 'Tubulin oligomers and microtubule assembly studied by time-resolved X-ray scattering: separation of prenucleation and nucleation events', *Biochemistry-US* 26: 1123–32.

Suarez, G., Oronsky, A. L., Bordas, J. and Koch, M. H. J. (1985). 'Synchrotron radiation X-ray scattering in the early stages of in vitro collagen fibril formation', *P. Natl. Acad. Sci. USA* 82: 4693–6.

Sugiyama, T., Miyashiro, D., Akao, D. *et al.* (2009). 'Quick shear-flow alignment of biological filaments for X-ray fiber diffraction facilitated by methylcellulose', *Biophys. J.* 97: 3132–8.

Svensson, A.-K., E., Bilsel, O., Kondrashkina, E., Zitzewit, J. A. and Matthews, C. R. (2006). 'Mapping the folding free energy surface for metal-free human Cu,Zn superoxide dismutase', *J. Mol. Biol.* 364: 1084–102.

Tsukamoto, S., Yamashita, T., Yamada, Y. *et al.* (2009). 'Non-native α-helix formation is not necessary for folding of lipocalin: Comparison of burst-phase folding between tear lipocalin and β-lactoglobulin', *Proteins* 76: 226–36; and erratum, DOI: 210.1002/prot.22593.

Tsuruta, H., Brennan, S., Rek, Z., Irving, T.C., Tompkins, W.H. and Hodgson, K.O. (1998). 'A wide-bandpass multilayer monochromator for biological small-angle scattering and fiber diffraction studies', *J. Appl. Crystallogr.* 31: 672–82.

Tsuruta, H., Kihara, H., Sano, T., Amemiya, Y. and Vachette, P. (2005). 'Influence of nucleotide effectors on the kinetics of the quaternary structure transition of allosteric aspartate transcarbamylase', *J. Mol. Biol.* 348: 195–204.

Tsuruta, H., Nagamura, T., Kimura, K. *et al.* (1989). 'Stopped-flow apparatus for X-ray scattering at subzero temperature', *Rev. Sci. Instrum.* 60: 2356–8.

Tsuruta, H., Vachette, P., Sano, T. *et al.* (1994). 'Kinetics of the quaternary structure change of aspartate transcarbamylase triggered by succinate, a competitive inhibitor', *Biochemistry-US* 33: 10007–12.

Tuma, R., Tsuruta, H., French, K. H. and Prevelige, P. E. (2008). 'Detection of intermediates and kinetic control during assembly of bacteriophage P22 procapsid', *J. Mol. Biol.* 381: 1395–1406.

Uzawa, T., Akiyama, S., Kimura, T. *et al.* (2004). 'Collapse and search dynamics of apomyoglobin folding revealed by submillisecond observations of a-helical content and compactness', *P. Natl. Acad. Sci. USA* 101: 1171–6.

Uzawa, T., Kimura, T., Ishimori, K. *et al.* (2006). 'Time-resolved small-angle X-ray scattering investigation of the folding dynamics of heme oxygenase: implication of the scaling relationship for the submillisecond intermediates of protein folding', *J. Mol. Biol.* 357: 997–1008.

Váró, G. and Lanyi, J. K. (1991). 'Thermodynamics and energy coupling in the bacteriorhodopsin photocycle', *Biochemistry-US* 30: 5016–22.

Virtanen, J. J., Makowski, L., Sosnick, T. R. and Freed, K. F. (2011). 'Modeling the hydration layer around proteins: Applications to small- and wide-angle X-ray scattering', *Biophys. J.* 101: 2061–9.

Wagner, O., Zinke, J., Dancker, P., Grill, W. and Bereiter-Hahn, J. (1999). 'Viscoelastic properties of f-actin, Microtubules, f-actin/α-actinin, and f-actin/Hexokinase determined in microliter volumes with a novel nondestructive method', *Biophys. J.* 76: 2784–96.

West, J. M., Xia, J., Tsuruta, H., Guo, W., O'Day, E. M. and Kantrowitz, E. R. (2008). 'Time evolution of the quaternary structure of Escherichia coli aspartate transcarbamoylase upon reaction with the natural substrates and a slow, tight-binding inhibitor', *J. Mol. Biol.* 384: 206–18.

Westenhoff, S., Malmerberg, E., Arnlund, D. *et al.* (2010). 'Rapid readout detector captures protein time-resolved WAXS', *Nat. Methods* 7: 775–6.

Winter, R. and Dzwolak, W. (2005). 'Exploring the temperature–pressure configurational landscape of biomolecules: from lipid membranes to proteins', *Phil. Trans. R. Soc. A* 363: 537–63.

Woenckhaus, J., Köhling, R., Winter, R., Thiyagarajan, P. and Finet, S. (2000). 'High pressure-jump apparatus for kinetic studies of protein folding reactions using the small-angle synchrotron X-ray scattering technique', *Rev. Sci. Instrum.* 71: 3895–9.

Wu, Y., Kondrashkina, E., Kayatekin, C., Matthews, C. R. and Bilsel, O. (2008). 'Microsecond acquisition of heterogeneous structure in the folding of a TIM barrel protein', *P. Natl. Acad. Sci. USA* 105: 13369–72.

Xiao, Y., Partha, R., Krebs, R. and Braiman, M. (2005). 'Time-resolved FTIR spectroscopy of the photointermediates involved in fast transient H^+ release by proteorhodopsin', *J. Phys. Chem. B* 109: 634–41.

Analysis of interparticle interactions

8.1	Basic physical chemistry of interactions	259
8.2	Experimental SAS studies of protein–protein interactions	265
8.3	Structure factor calculations for proteins	266
8.4	Interactions in nucleic acids	280
References		281

Most SAS studies as described in this book are devoted to the study of the sizes and shapes of single biological molecules or large well-defined molecular complexes such as viruses or ribosomes, and for these purposes dilute solutions with no interactions between dissolved macromolecules are used. There are, however, situations where for biological, medical or biophysical reasons one may wish to study interactions between macromolecules and the processes causing aggregation. Examples are the transparency of the eye lens (Tardieu and Delaye 1988), the aggregation of amyloid proteins responsible for human pathologies such as Alzheimer's and Creutzfeld–Jakob disease, the mechanisms of macromolecular crystallisation so important for high-resolution MX (Tardieu *et al.* 2001) or problems of protein storage. The present chapter describes how information about the interactions can be obtained from SAS patterns.

8.1 Basic physical chemistry of interactions

As explained in Chapter 5 and illustrated in Fig. 8.1, the effective structure factor resulting from the interactions between particles in solution can be obtained by dividing the scattering pattern of a solution by the form factor corresponding to the scattering of an isolated particle in an ideal solution. Approaches based on the GIFT method and relying on the real space representation of the form factor by the distance distribution and of the structure factor by the pair correlation function were also proposed (Fritz and Glatter 2006, Fukasawa and Sato 2011).

In the low-q range the ratio between the scattering pattern of a concentrated solution and that of a very dilute one is affected not only by the structure factor but also by changes in contrast due to the increasing protein, salt or cosolute concentrations. In the high-q range the ratio should be independent of contrast, and any effect of the structure factor is expected to increase with protein concentration. Interestingly, accurate measurements of the WAXS pattern of protein solutions in the range

$3 < q < 20\,\text{nm}^{-1}$ for concentrations of 4–250 mg/mL revealed significant changes in the intensities relative to dilute solutions (Makowski *et al.* 2008). These effects, which depend on the structure of the proteins and on temperature, are most pronounced at concentrations below 50 mg/mL and vanish at higher concentrations. They are thus clearly not related to the structure factor and have been attributed to inhibition of intramolecular breathing motions due to molecular crowding. Similar effects have been found in the case of ligand binding or mutations and other causes of structural diversity (Makowski *et al.* 2011a, Makowski *et al.* 2011b). The topic provides an interesting area for future research, especially as it has been proposed that crowding and hydrodynamic interactions dominate macromolecular motion *in vivo* (Ando and Skolnick 2010).

The theoretical approaches used to calculate the structure factor from effective pair potentials are extensions of developments made in the framework of the theory of the structure of simple liquids and colloidal suspensions.

The interactions of biological macromolecules and colloids are usually discussed in terms of a hard core potential describing the fact that the macromolecules do not interpenetrate and of electrostatic (ionic) and Van der Waals interactions and hydrogen bonding, the last one leading to hydrophobic interactions and hydration (Israelachvili 2011).

The hard-core interactions result in an excluded volume fraction $\varphi = vc$, where v is the partial specific volume (cm^3/g) and c the concentration (g/cm^3). The Van der Waals interactions and hydrogen bonding are short range, extending at most over a few tenths of nm, whereas the electrostatic Coulombic interactions extend over up to a few nm. Some authors have also proposed hydrophilic forces (Ben-Naim 1990). Hydration and other forces depending on electrodynamic fluctuations are usually not taken into account, though they are important for ion-specific effects (Boström *et al.* 2005).

As all these interactions have the same origin and depend in the same manner on the quality of the solvent as those determining the structure and folding of individual macromolecules (Bolen and Rose 2008), they are similarly influenced by temperature, pressure and the chemical potential of the different species in solution.

The attractive depletion interactions (Asakura and Oosawa 1954) are, however, specific to solutions of colloidal particles or proteins containing non-absorbing polymers, such as polyethylene glycol (PEG). They are entropic in nature and result from the fact that polymers larger than the distance between colloidal particles are excluded from the volume between the particles. These interactions depend on the size and concentration of the polymer and of the proteins, and have a range of the order of the radius of gyration of the polymer, which is significantly larger than that of the Van der Waals interactions ((Israelachvili 2011). For a statistical mechanical explanation, see also Klein (2002).

When the long-range repulsive interactions dominate, the positions and orientations of the macromolecules become correlated and the

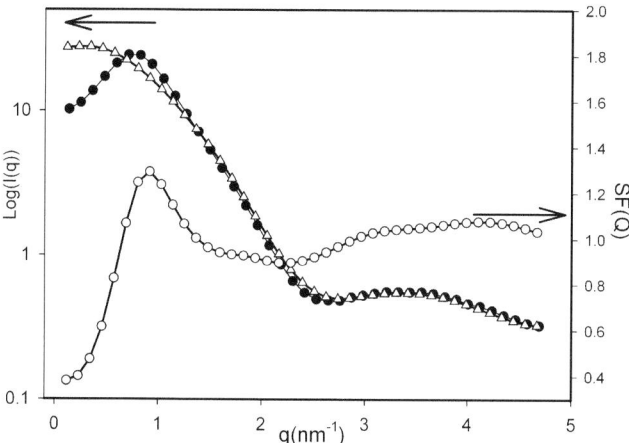

Fig. 8.1 Experimental structure factor (o) obtained by dividing the scattering pattern of a lysozyme solution (200 mg/mL •) by that of a dilute solution (5 mg/mL △). Buffer: 20 mM HEPES, pH 7.8, 20° C. Note that the scattering curves are displayed on a logarithmic scale and the structure factor on a linear scale. The symbols display one point in 50.

solutions become structured, whereas when the weaker short-range attractive forces dominate, polydisperse solutions containing monomers and oligomers or complexes and their components are formed.

There is a considerable literature on protein interactions based on static and dynamic light scattering or other methods, which is indispensable reading for those interested in the subject.

Interactions are, of course, related to the phase diagram of solutions and properties such as solubility and crystallisation.

Most SAS measurements are made in the undersaturated regime in the schematic phase diagram of a protein solution in Fig. 8.2 (Asherie 2004, Chayen 2004, Luft *et al.* 2011). In contrast, studies exploring nucleation and crystallisation cover the supersaturated regime beyond the solubility curve, where crystals and solutions are in equilibrium. In the metastable region, supersaturation is usually too low for spontaneous nucleation to occur, but liquid-liquid phase separation

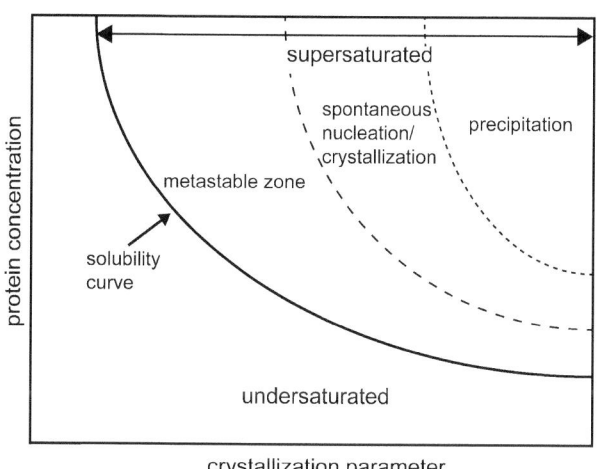

Fig. 8.2 Schematic phase diagram of a protein solution. The crystallisation parameter may correspond to salt or other precipitant (such as PEG) concentration or temperature. For details, see the text.

may occur at higher protein concentrations. In the crystallisation zone, supersaturation is sufficient for spontaneous nucleation and crystallisation, whereas at higher supersaturation disordered precipitates form in a liquid–solid phase transition. Different crystallisation methods follow different paths at different rates through these regions (Chayen 2004).

As already mentioned in Chapter 7, crystallisation and assembly are closely related phenomena requiring in most cases static as well as (slow) kinetic measurements. Whereas the solubility curve is a thermodynamically well-defined boundary, this is not the case for the other lines which also depend on kinetic factors.

In vivo intermolecular interactions may be modulated by post-synthetic modification (for example, glycation or phosphorylation), whereas in practical applications they may be modified by cross-linking the proteins with low molecular mass polymers such as polyethylene glycols (PEGylation) (Castelletto *et al.* 2008, He *et al.* 2010, Svergun *et al.* 2008).

8.1.1 Chemical potential, virial expansion and protein crystallisation

In thermodynamics, interactions appear as deviations from ideality described by the so-called virial coefficients. These also determine the conditions for crystallisation.

At constant pressure and temperature the thermodynamic properties of solutions are entirely determined by the partial molar free energies or chemical potentials corresponding to the total free energy per molecule of each of the constituents. This has two components: the self-energy of the molecule and its thermal energy. The self-energy can be regarded as the energetic cost of transferring a molecule in vacuum from infinity to the bulk solution at constant temperature and pressure. It is thus the sum of all interactions of the molecule with surrounding ones and of all changes in energy of the solvent molecules resulting from this transfer. Hence, the effective pair potential $U(r)$ describes the change in the sum of the self-energies of a pair of molecules in solution as a function of their distance (Israelachvili 2011).

If it is also a perfect solution an ideal solution is one where there is no difference in interaction energy between solute and solvent, so that the enthalpy of mixing is zero and the entropy of mixing (ΔS_m) is given by simple statistics as $\Delta S_m = -R \sum_i n_i \ln x_i$, where n_i and x_i are the numbers of molecules of type i and their mole fractions, respectively (Kondepudi and Prigogine 1998). In that case, the chemical potential at constant temperature and pressure of all components in the solution with mole fraction x_i is given by

$$\mu_i(x_i) = \mu_i^0 + RT \ln(x_i) \qquad (8.1)$$

where μ_i^0 is the chemical potential of a reference state.

For non-ideal solutions with weak interactions this expression can be replaced for the solute by eq. (8.2), based on the virial expansion in the linear approximation (that is, neglecting all terms depending on higher powers of the concentration c (mg/mL)), where M is the molar mass and the reference state corresponds to the ideal solution at $c = 1$ mg/mL, R is the gas constant and T the absolute temperature:

$$\mu = \mu^0 + RT \ln c + 2A_2 RTMc + \ldots \qquad (8.2)$$

As illustrated in Fig. 8.3, the sign of the second virial coefficient A_2 determines whether the chemical potential of the solute increases faster ($A_2 > 0$, good solvent) as a function of its concentration than in the ideal solution ($A_2 = 0$), due to solvation, excluded volume effects and/or repulsive interactions, or more slowly ($A_2 < 0$, poor solvent), due to preferential solute–solute interactions. The Gibbs–Duhem equation ($\sum_i n_i d\mu_i = 0$), where n_i is the number of moles of component i, implies that, conversely, the chemical potential for a good solvent decreases faster as a function of solute concentration than in the ideal solution, and more slowly for a poor solvent (see, for example, Van Holde *et al.* 1998). The value of A_2 is thus a measure of the quality of the solvent that can be used to establish the 'crystallisation slot' of proteins (George and Wilson 1994). For globular proteins, A_2 usually has a parabolic behaviour as a function of pH with a minimum, which can be negative in the vicinity of the isoelectric point, reflecting the fact that the molecules are no longer charged and the repulsive electrostatic interactions vanish. The well-known salting-in and salting-out phenomena are due mainly to the effects of changes in salt concentration on the electrostatic interactions, including charge screening and specific ion effects. A good understanding of the meaning of A_2 is therefore indispensable for the interpretation of scattering patterns from concentrated solutions.

It should also be kept in mind that sample preparation is a crucial step in studies of interactions and for obtaining reliable results commercial lyophilised or other protein samples, which often contain significant

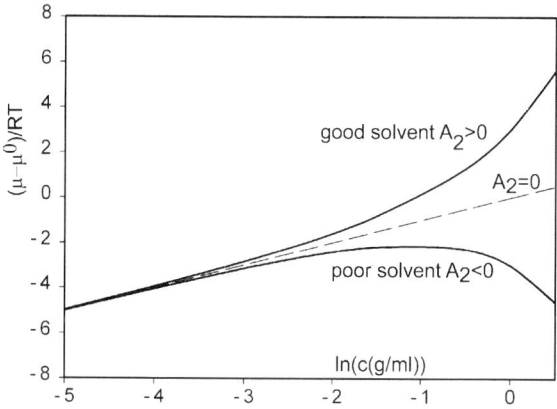

Fig. 8.3 The chemical potential of the solute increases linearly as a function of the logarithm of concentration in an ideal solution (dashed line, $A_2 = 0$), faster in a good solvent ($A_2 > 0$) and more slowly in a poor solvent ($A_2 < 0$). A MM of 50 kDa and a value of $A_2 = 3 \times 10^{-5}$ mL mol g^{-2} have been assumed here.

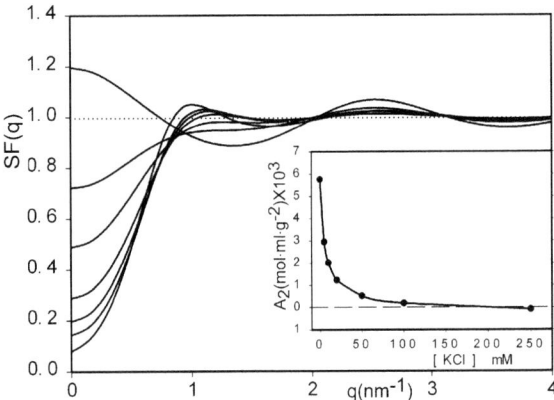

Fig. 8.4 Fits to the experimental structure factors of aqueous solutions of lysozyme (68 mg/mL) at KCl concentrations of 0, 5, 10, 20, 50, 100 and 250 mM, from bottom to top at the origin. The dotted line corresponds to an ideal solution. Insert: Dependence of A_2 on the KCl concentration. Below 200 mM KCl the interactions are repulsive ($A_2 > 0$), and above they become attractive ($A_2 < 0$) and dimerisation occurs.

amounts of salt, must be purified and deionised before further use (see, for example, Tardieu *et al.* 1999).

8.1.2 Osmotic pressure and the structure factor at $q = 0$

In the linear approximation of the virial expansion, A_2 is linked to the osmotic pressure (Π) by

$$\frac{\Pi}{cRT} = \frac{1}{M} + A_2 c + \cdots \tag{8.3}$$

The scattered light intensity ($I_L(c, 0)$) or osmotic compressibility ($\partial \Pi / \partial c$) and osmotic pressure yield simple independent measurements of the value of the structure factor at the origin (Ducruix *et al.* 1996) as

$$I_L(c, 0) \sim McS(c, 0) = (RT/M)(\partial \Pi / \partial c)^{-1} \tag{8.4}$$

and hence,

$$[S(c, 0)]^{-1} = 1 + 2MA_2 c + \cdots \tag{8.5}$$

The value of $I_L(c, 0)$, A_2 and R_g^2 can be obtained by double extrapolation to zero concentration and zero scattering angle in a so-called Zimm plot ($cI(q)^{-1}$ *versus* $q^2 + kc$), where k is an arbitrary constant to facilitate plotting (Zimm and Stockmayer 1949).

The osmotic pressure is also related to the partition function $Z(c)$ of the solution:

$$\Pi = (RT/M)Z(c) \tag{8.6}$$

In the case of non-interacting hard spheres, $Z(c) = c(1 + \varphi + \varphi^2 - \varphi^3)/(1 - \varphi)^3$ (Carnahan and Starling 1969) and $S(c, 0) = (1 - \varphi)^4/(1 + 4\varphi + 4\varphi^2 - 4\varphi^3 + \varphi^4)$ (Ducruix *et al.* 1996), where φ is the volume fraction of spheres in solution. This illustrates that the existence of a structure factor does not require interactions beyond those resulting from the excluded volume.

A comparison of the results obtained by different methods (osmotic pressure, quasi-elastic light scattering and SAS) has been made for solutions of γ-crystallin (Bonneté *et al.* 1997).

The effects of repulsive interactions on the scattering patterns and their relationship to compressibility and osmotic properties were studied very early on for spherical (Zernike and Prins 1927) and cylindrical particles (Oster and Riley 1952).

In practice, and in contrast with light scattering, it is, however, not possible to obtain $S(c, 0)$ directly from X-ray or neutron scattering data, as this requires extrapolation to the origin. Fits to the experimental structure factors over an extended range based on effective pair potentials, as shown in Fig. 8.4, give more reliable values.

8.2 Experimental SAS studies of protein–protein interactions

The influence of protein size on the interactions has been investigated in a series of studies (such as aprotinin (BPTI) (6.5 kDa) (Grouazel *et al.* 2002), lysozyme (14.8 kDa) (Ducruix *et al.* 1996), γ-crystallins (21 kDa) (Malfois *et al.* 1996), urate oxidase (128 kDa) (Vivarès and Bonneté 2004), aspartate carbamoylase (ATCase) (306 kDa) (Budayova *et al.* 1999), α-crystallins (800 kDa) (Finet and Tardieu 2001) and brome mosaic virus (4.6×10^3 kDa) (Casselyn *et al.* 2001), aiming mainly at establishing general conditions for nucleation and crystallisation; that is, finding the coexistence region between solution and crystals below the precipitation zone in the phase diagram in Fig. 8.2.

Among a number of other interesting observations, the results illustrate that the position of the first maximum in the structure factor is not simply related to the distance between macromolecules but rather to their size (Ducruix *et al.* 1996).

The formation of BPTI decamers in acidic conditions provides an interesting model for the assembly of finite structures, even if it has no obvious biological relevance (Hamiaux *et al.* 2004).

Nucleation is often accompanied by liquid–liquid phase separation detectable by the reversible appearance of liquid droplets and cloudiness in a supersaturated protein solution when the temperature is lowered below the cloud point (Grouazel *et al.* 2002) or by liquid–solid phase separation (precipitation). For BPTI the concentrated phase sediments to the bottom of the capillary, and the difference is in protein concentration only at constant salt concentration.

Independently of the pH for the smaller proteins (BPTI and lysozyme, γ-crystallins) addition of salt, which screens electrostatic interactions, suffices to induce attractive interactions, whereas for the larger ones (urate oxidase and ATCase) PEG or similar additives inducing depletion interactions are necessary to achieve this goal.

ATCase provided a good example of a large monodisperse oligomer to study qualitatively charge effects as a function of pH, the shielding by different salts and the synergetic effects of salt and PEGs of different molar mass (Budayova *et al.* 1999). The Van der Waals attraction is much reduced with this large and less compact protein compared to the smaller ones. The simultaneous occurrence of attractive and repulsive effects in some of the scattering patterns, with regions of different protein:PEG concentration ratios, is indicative of repulsive and attractive interactions with similar ranges in the condition of these experiments. This is consistent with attractive depletion interactions having a larger range than Van der Waals interactions.

A similar metastable liquid–liquid phase separation was also observed by SAXS in urate oxidase/PEG solutions, and its relationship to crystal morphology was monitored by video microscopy (Vivarès and Bonneté 2004). It was found that in the protein-rich phase the interactions were repulsive, whereas in the protein-poor phase they were attractive.

The phase diagram of brome mosaic virus as a function of concentration of different PEGs presents features similar to those observed with proteins as well as with colloids (Casselyn *et al.* 2001). As in the case of proteins there is a region of slow nucleation/crystallisation and a region of fast precipitation, but in contrast with the situation with proteins the PEG-induced precipitates are microcrystals resulting from a liquid to solid/crystal transition characteristic of colloidal systems with short-range attractions.

Despite the progress made in understanding the forces involved, crystallisation remains largely an empirical process, but the information obtained by various methods including SAS is useful for the design of high-throughput screening methods (Luft *et al.* 2011).

8.3 Structure factor calculations for proteins

To calculate structure factors from pair potentials a number of simplifications are usually made. For globular macromolecules one assumes that they are spherical and that their interactions are independent of their relative orientation, and hence can be described by a central potential $U(r)$. The force between two macromolecules is then simply $F = dU/dr$.

In order to minimise the number of formulae in the present chapter, derivation of the structure factor for spherical molecules is given in Appendix 3, together with the definitions of the direct and indirect correlation functions, the pair distribution function and the Ornike–Zernike equation. Some of the closure relationships relating the pair potential with the direct correlation function needed to solve this equation are also described.

As explained above, usually only interactions between macromolecules are considered with all other ones (such as macromolecule–solvent and solvent–solvent (and cosolvent)) compounded in the effective pair potential. Even small changes in solvent properties can have significant effects on the apparent protein–protein interactions, as illustrated by the stronger interactions and the correspondingly lower solubility of aprotinin (BPTI) in D_2O compared to H_2O—effects partly related to the differences in density between the solvents (Budayova-Spano *et al.* 2000).

One very practical consequence is that D_2O may induce aggregation in SANS experiments. It has been observed on many occasions that a protein which is soluble in an H_2O buffer aggregates in a D_2O buffer of the same composition. This phenomenon was observed well before SANS experiments on protein solutions in sedimentation studies of the protein phycocyanin (Lee and Berns 1968). This effect sometimes results from rather trivial causes; for example, pH is not the same as pD, and the appropriate correction described in Chapter 3 must be applied. In many instances it is, however, probably due to the difference in hydrophobic interactions in H_2O and D_2O, and may partly depend on kinetic effects.

8.3.1 Pair potentials

Simulations of the structure factor require an appropriate description of the effective interaction potential between pairs of macromolecules. This is a critical issue in Monte Carlo simulations (see, for example, Pedersen 2002b) as well as in the more analytical approach. Only a few representative functions frequently used in building model potentials will be considered here.

As its name implies, the hard-sphere potential corresponds to an infinite repulsion at distances (r) smaller than the diameter of the spheres—a case that would apply to billiard balls, for example.

$$
\begin{aligned}
U(r) &= \infty, \quad r < \sigma \\
&= 0, \quad\;\; r > \sigma
\end{aligned}
\tag{8.7}
$$

As already mentioned, at sufficiently high concentrations the hard-sphere interactions (excluded volume) result in a structure factor which is different from 1, even in the absence of attractive or repulsive interactions.

Most potentials are sums of attractive and repulsive components and the repulsive hard-sphere potential can easily be made to include attractive interactions as in the hard-sphere potential with square-well attraction:

$$
\begin{aligned}
U(r) &= \infty, \quad r < \sigma \\
&= -\varepsilon, \quad \sigma < r < \gamma\sigma \\
&= 0, \quad\;\; r > \gamma\sigma \qquad \text{with } \gamma \cong 1.5
\end{aligned}
\tag{8.8}
$$

Note the discontinuity at $r = \sigma$ in both cases.

Probably the best known simple function is the Lennard-Jones potential, where n = 12:

$$U(r) = 4\varepsilon\left[\left(\frac{\sigma}{r}\right)^{n} - \left(\frac{\sigma}{r}\right)^{6}\right] \qquad (8.9)$$

The value of the collision diameter σ (where the potential is zero) also defines the position of the energy minimum at $r_{min} = \sqrt[6]{2}\sigma$ and the depth of the potential is $U(r_{min}) = \varepsilon$. For atomic systems the r^{-6} term corresponds to dipole-induced dipole attractive interactions. This potential combines a repulsive (first) term and an attractive (second) term of the form $U(r) \propto r^{-n}$, where n is a positive integer. The dependence of $U(r)$ on r—in this case the value of n—determines the range of the potential as illustrated in Fig. 8.3. When $n > 3$ the interactions are short-ranged, whereas for $n \leq 3$ they are long-ranged. It can be shown that $n > 3$ for intermolecular interactions (Israelachvili 2011). The repulsive part of the Lennard-Jones potential corresponds to $n = 12$, but higher values of n are not uncommon when a softer repulsive core than the hard-sphere potential is required (Abramo *et al.* 2011).

Note that there is nothing in the Lennard-Jones potential with its longer-range (r^{-6}) attraction that prevents infinite association (such as crystallisation). To avoid this one often adds a repulsive barrier at an appropriate value of σ/r. The problem of crystallisation or its prevention is common to many areas of materials science where slightly different potentials have been developed (for a review, see Doye *et al.* 2007).

Although Coulomb interactions between two charges have an r^{-1}-dependence in vacuum and are long-range, in solution their range is limited due to screening by counterions and better described by exponential functions (Israelachvili 2011).

Such interactions can usually be described over relatively small ranges of r by a screened Coulomb potential (also known as Yukawa potential) represented by an exponential of the form $\pm |J| \exp[-(r-\sigma)/r_0]$, where J is the depth of the potential, r_0 is a range parameter, and the other symbols have the same significance as previously.

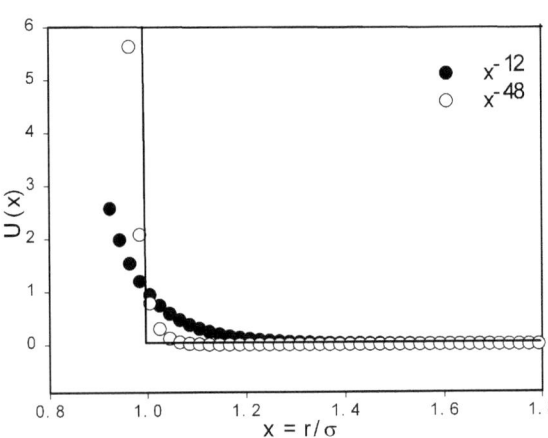

Fig. 8.5 Repulsive potentials. Hard sphere (full line), Lennard-Jones (r^{-12}) (•) and r^{-48} (○).

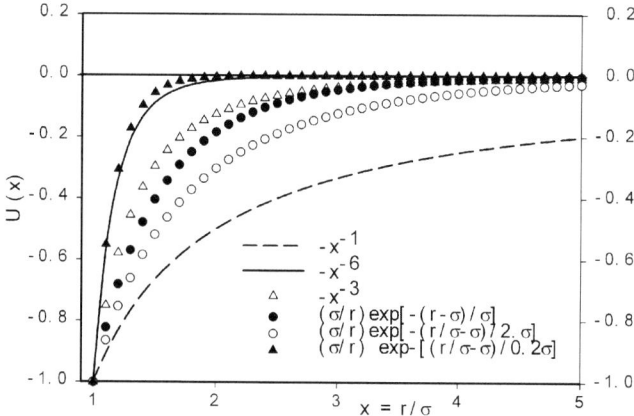

Fig. 8.6 Attractive potentials for different parameters.

For some simple cases like the hard-sphere potential or the screened Coulomb potential, formulae are available to calculate the structure factor (Pedersen 2002a).

8.3.2 The DLVO potential: repulsive and attractive interactions

Most quantitative studies of protein–protein interactions have been carried out with potentials derived from a particularly successful model in colloid science, the DLVO (Derjaguin, Landau, Verwey, Overbeek) potential consisting of a hard core ($U_{HC}(r)$), repulsive Coulomb interactions ($U_R(r)$) and attractive Van der Waals interaction ($U_A(r)$). An example of DLVO potential successfully used in a number of studies on protein–protein interactions and the effects of cosolutes within the HNC approximation (see Appendix 3) is described below (Tardieu *et al.* 1999):

$$U_{HC}(r) = +\infty \qquad\qquad\qquad r < \sigma$$

$$U_R(r) = +|J_R|\,\frac{\sigma}{r}\,\exp[-(r-\sigma)/r_{0R}] \qquad r > \sigma \qquad (8.10)$$

$$U_A(r) = -|J_A|\,\frac{\sigma}{r}\,\exp[-(r-\sigma)/r_{0A}] \qquad r > \sigma$$

For compact macromolecules with charge Z_p, the depth of the repulsive Coulomb potential is given by

$$J_R = \left(Z_p^2/\sigma\right)L_B/(1+0.5\sigma/\lambda_D)^2 \text{ and } r_{0R} = \lambda_D = \left(4\pi L_B \sum_i \rho_i Z_i^2\right)^{-1/2} \qquad (8.11)$$

$L_B = e^2/(4\pi\varepsilon_0\varepsilon_s k_B T)$ is the Bjerrum length, where e is the charge of the electron, $\varepsilon_0 = 8.854 \times 10^{12}\,C^2\,J^{-1}\,m^{-1}$, the permittivity of free space, and $\varepsilon_s \approx 80$, the relative permittivity of the solvent. The Debye length, $\lambda_D = 3/\sqrt{I}$, is related to the ionic strength $I = 0.5\sum c_i Z_i^2$, the sums running over all types of ions with number concentration ρ_i, molar concentration c_i and charge Z_i in the solution.

The attractive part of the potential is based on the Van der Waals potential for two (large) spherical colloidal particles given in its original form as (Verwey and Overbeek 1999):

$$U_A(r) = \frac{-A}{12} \left[\frac{\sigma^2}{r^2 - \sigma^2} + \frac{\sigma^2}{r^2} + 2\ln\left(\frac{r^2 - \sigma^2}{\sigma^2}\right) \right] \tag{8.12}$$

The Hamaker constant A, given in units of $k_B T$, is indirectly related to the depth of the potential and is defined by the optical properties (absorption frequencies, relative permittivity and refractive index) of the solute and the solvent. The potential, which varies as expected as r^{-6} at large distances and diverges for $r = \sigma$, is thus expected to depend only on the size of the globular proteins. It is well approximated by a short-range Yukawa potential with $J_A = -2.5\,kT$ and $r_{0A} = 0.3$ nm. Note that the temperature has a much stronger influence on the Van der Waals interactions than on the electrostatic interactions.

The different components of the effective potential for a lysozyme solution in distilled water are shown in Fig. 8.7. Note that this potential combines relatively longer-range repulsion with short-range attraction, in contrast with the Lennard-Jones potential in Figs. 8.5 and 8.6, which has short-range repulsion and longer-range attraction.

A similar approach was followed in a recent study on interactions in undersaturated and supersaturated lysozyme solutions, except that here the structure factor was calculated in the random phase approximation (RPA) based on a perturbation of the Percus-Yevick hard-sphere reference potential (see Section 5.5) with the van der Waals and Coulombic components of the DLVO potential. Additional attractive terms were also included to explain salt-specific effects and protein self-association (Narayanan and Liu 2003). Good fits were obtained at low salt concentrations (8.6 mM NaCl) using only the van der Waals and Coulombic terms in the perturbation potential, whereas at high salt concentrations (860 mM NaCl) an osmotic attractive potential due to the excluded volume effect of salt ions also had to be included. In oversaturating conditions the depth of the attractive potential was overestimated and the calculated value of A_2 incorrect, which was interpreted as being due to additional repulsive hydration interactions.

Although these potentials very successfully describe many of the commonly observed phenomena, due, for example, to protein concentration, temperature, pH and ionic strength, they fail to explain specific effects such as the Hofmeister series (Hofmeister 1888) (for a review see Lo Nostro and Ninham 2012) or the effects of addition of polymers such as polyethylene glycol (PEG), resulting in attractive depletion forces (Israelachvili 2011). This requires the introduction of additional attractive or repulsive terms and/or to modify the charge on the macromolecule at a fixed pH (Tardieu *et al.* 1999).

8.3.3 Specific ion effects

The repulsive part of the DLVO potential depends on the charge and ionic strength but not on the nature of the ions. Moreover, the attractive potential $U_A(r)$, which has a range of the order of 0.3 nm, becomes negligible for large proteins. Studies on the salt dependence of the scattering patterns are consistent with screening of electrostatic repulsion, but it has been known for more than a century that, at least for monovalent salts, the differences between the effects of cations on protein–protein interactions (or, for example, solubility) are less pronounced than those of anions. SAXS studies have shown that an additional anion-dependent attractive contribution in the form of a variable depth but constant range of the attractive term and a modified charge is necessary to explain the observations (Tardieu *et al.* 1999). This term suffices to fit the scattering curves but does not give any direct insight into its origin, which was only recently explained quantitatively on the basis of SAXS studies on several proteins (Boström *et al.* 2005, and references therein).

The additional attraction increases in the direct order of the Hofmeister series (for 0.4 M Na salts $SO_4^{2-} > H_2PO_4^- > CH_3COO^- > Cl^- > Br^- \approx NO_3^- > I^- > SCN^-$; Lo Nostro and Ninham 2012) when pH > pI and in the inverse order (that is, $SCN^- > Cl^- > CH_3COO^- \ldots$) when pH < pI. The theoretical explanation requires the introduction of non-electrostatic forces (dispersion forces related to polarisability), which can become dominant at high salt concentrations when the electrostatic ones are reduced by screening. The ion distributions for model protein surfaces are obtained by solving an ion-specific Poisson–Boltzmann equation for charge-regulated surfaces. This model predicts an increase of the double-layer repulsion in the order of the direct Hofmeister series when pH < pI and the anions are counterions, and in the reverse order for pH > pI where they are co-ions. The observed effects depend on non-linear coupling of non-electrostatic and electrostatic forces which determine the ion distributions. Further progress was made recently in this area. but has not yet been applied to the interpretation of SAS curves (Lima *et al.* 2007, Boström *et al.* 2011).

As revealed by SANS for BSA at pH 6.2–6.8 the anion dependence of A_2 follows the inverse Hofmeister series ($SCN^- > Cl^- > SO_4^{2-} \ldots$) (that is, the additional attraction follows the direct series) as expected, since pH > pI = 4.9 (Zhang *et al.* 2012).

The effect of multivalent cations, where more important specific effects can be expected, have hitherto hardly been studied (Ianeselli *et al.* 2010), though they have a strong influence on protein solubility (Bénas *et al.* 2002).

8.3.4 Small cosolutes

Small cosolutes, which destabilise proteins (such as urea or tetramethylurea) or stabilise them (such as TMAO or glycerol), or may have different effects like trifluoroethanol (TFE), also influence the effective pair potentials. The forces involved are also largely non-electrostatic, but have not yet been studied in any detail, and most results so far are phenomenological.

The effect of glycerol, sucrose, guanidinium hydrochloride and tri-fluoroethanol on the interactions of lysozyme have been investigated (Javid *et al.* 2007b) using the random phase approximation (RPA) approach of Narayanan and Liu (2003). The stabilisation of BSA at low TFE concentrations and increased attractive interactions at higher TFE concentrations have been analysed using an RPA model with a DLVO perturbation potential (Carrotta *et al.* 2009). The same approach was used in SANS experiments on preferential hydration of lysozyme in water/glycerol mixtures (Sinibaldi *et al.* 2007), where thirty-five curves obtained in different conditions (protein, H_2O/D_2O and H-glycerol/ D-glycerol concentrations) were simultaneously fitted. The attractive term in the effective potential decreases with increasing glycerol concentration, and preferential hydration was clearly detected. A similar global analysis has been performed with lysozyme–urea solutions which suggested that the charge of the protein may be modified by preferential binding of urea (Ortore *et al.* 2008). These two studies provide a direct link to recent thermodynamic investigations (Schellman 2003, Schellman 2005). Note that combined SANS/SAXS measurements are particularly well suited to investigate binding of small cosolutes to biological macromolecules in a model-free manner (Gabel *et al.* 2009).

The structure factors of lysozyme (68 mg/mL) in distilled water with salt (0–250 mM KCl) and urea (250 mM) can be well fitted following the approach of Tardieu *et al.* (1999), using a constant attractive term and a variable charge (Niebuhr and Koch 2005). In contrast, the salt series in the presence of 250 mM TMAO, which does not interact with the protein, can only be fitted using a variable attractive potential. Use of a fixed charge and variable attractive potential also improves the fit to the other data. The influence of natural cosolvent mixtures on the non-linear pressure effects in concentrated lysozyme solutions and the

Fig. 8.7 Components of the potential for a lysozyme solution (68 mg/mL) in distilled water (σ = 3.24 nm). Compare the short range of the attractive van der Waals interactions with that of the depletion potential in Fig. 8.8.

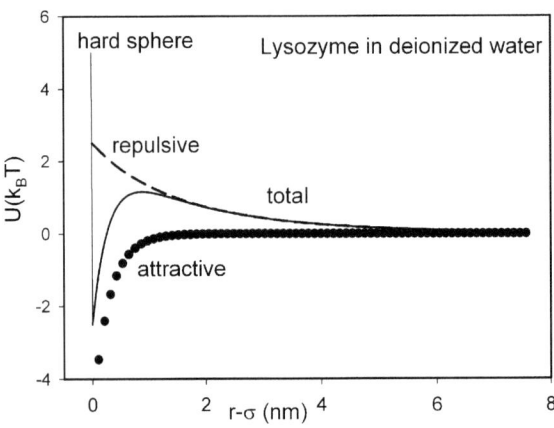

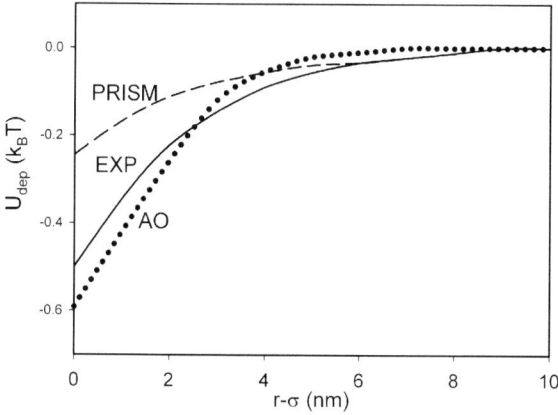

Fig. 8.8 Comparison of the depletion potential obtained by fit to SAXS data (EXP) for urate oxidase ($\sigma = 7$ nm) with 3% PEG8000 and calculated with the Asakura–Oosawa and PRISM potentials. Adapted from Vivarès *et al.* (2002b), with the permission of the European Physical Journal (<http://www.epj.org/>).

counteraction of urea and pressure by TMAO were also explained in terms of changes in the attractive potential and in the hydration shell (Schroer *et al.* 2011b).

Osmolytes may also affect the distribution of conformations in the native state of flexible proteins such as Hsp90 (Street *et al.* 2010). This complicates the study of intermolecular interactions, as each of the conformations has a different form factor and structure factor.

Given the physiological importance of osmolytes, further research in this area would be useful, especially to establish their influence on the hydration shell which in SAS appears to have a thickness of about one water molecule (Svergun *et al.* 1998), but may extend > 1 nm according to THz spectroscopy (Ebbinghaus *et al.* 2007).

8.3.5 Large cosolutes: depletion interactions

The effect of PEG on the interaction of α-crystallins (900 kDa) was modelled using a one-component model taking only protein–protein interactions into account and adding a Yukawa-type attractive depletion term to model the effect of the polymer (Finet and Tardieu 2001). Addition of salt alone fails to reach attractive conditions with such large proteins where the potential reduces to the hard-sphere and Coulombic contributions. Relatively loose molecules like α-crystallins behave like soft spheres because PEG can access cavities in the outer surface, making it more difficult to fit the structure factor over an extended range. The potential in this case has a Coulombic component and an attractive depletion component with a longer range than that characteristic of Van der Waals interactions.

The influence of PEG on the phase diagram of urate oxidase (UO) was investigated using a two-component approach with potentials for the UO–UO, PEG–PEG and UO–PEG interactions (Vivarès *et al.* 2002a). The measured effective potential, which is insensitive to protein concentration, was compared with two models of depletion interactions: the Asakura–Oosawa (AO) (Asakura and Oosawa 1954) potential and the more recent thermal PRISM (Polymer Reference Interaction Site Model) potential (Kulkarni *et al.* 2000).

The AO potential is given by:

$$U(r) = +\infty \qquad \text{for } 0 < r < \sigma = 2R$$

$$U(r) = -(4\pi/3)d^3 N_A c_p k_B T \left[1 - \frac{3r}{4d} + \frac{r^3}{16d^3} \right] \quad \text{for } \sigma \le r \le 2d \ \text{ with } d = \sigma/2 + R_{\text{pol}}$$

$$U(r) = 0 \qquad \text{for } r > 2d$$

where N_A is Avogadro's number, c_p the polymer concentration ($N_A c_p$ is the polymer number concentration), σ the diameter of the protein. R_{pol} is usually taken as the radius of gyration of the polymer, but in the study above (Vivarès *et al.* 2002a) the effective hard-sphere radius of the polymer was used. The value of $d = \sigma/2 + R_{\text{pol}}$ is an approximation to the minimum protein–polymer distance.

At low PEG concentrations, this potential, which does not take into account deviations from hard-sphere behaviour of the polymer or the protein, was found more attractive than the measured one.

A better agreement between measured and calculated potential was found with the thermal PRISM depletion potential for a polymer with degree of polymerisation N and radius of gyration R_g (Kulkarni *et al.* 2000):

$$U(r) = +\infty \qquad \text{for } 0 < r < 2R = \sigma$$

$$U(r) = -k_B T \cdot \ln \left(1 + \frac{\pi z}{3} \left(\frac{R}{r} \right) \left(\frac{R}{\sigma_s} \right) \exp[-(r - 2R)/\xi_p] \right) \quad \text{for } r \ge 2R$$

The effective statistical segment length $\sigma_s = \sqrt{\frac{6}{N}} R_g$ for $c_{\text{pol}} < c^*_{\text{pol}}$ $\propto N/R_g^3$ (c^*_{pol} is the overlap or semidilute crossover concentration), $z = N N_A c_p \sigma_s^3 / M_{wp}$ is the reduced polymer segment number density, where N_A is Avogadro's number, c_p the mass concentration of the polymer, and M_{wp} its weight average molecular weight. The correlation length for the spatial polymer concentration fluctuations is $\xi = \sigma_s / \left[\sqrt{(1 - T/T_s)} \cdot \sqrt{(12/N) + (\pi z/3)} \right]$, where T_s is the spinodal temperature for the polymer solvent separation, which depends on c_p and is approximated by the lowest critical solution temperature (that is, the lowest temperature at which PEG/water solutions display a liquid–liquid phase separation).

Figure 8.8 illustrates the fit of the two models to the experimental data for a 3% PEG 8000 solution. Note the larger range of this attractive potential compared to the Van der Waals interactions.

Monte Carlo calculations of the phase diagram of urate oxidase–PEG using a DLVO potential valid for urate oxidase without PEG and the AO model for the depletion interactions are only in qualitative agreement with the experimental results (Wenzel *et al.* 2007).

8.3.6 Effects of temperature and pressure

For α-crystallins in 150 mM phosphate buffer (pH = 6.8 > pI ≈ 5) there is hardly any influence of temperature on the scattering pattern between 10°C and 20°C, as expected when Coulombic repulsions dominate.

The short range (~0.3 nm) attractive interactions in the DLVO model explain phenomena such as the fluid–fluid phase in lysozyme or that in γ-crystallins, associated with cold cataract, below a critical temperature (Malfois *et al.* 1996).

The temperature dependence of the SAXS pattern of lysozyme (74 mg/mL) at pH = pI = 10.5, where the Coulombic interactions vanish, was measured between 7°C where phase separation occurs and 30°C, and it was found that the attractive interactions decrease with increasing temperature (Tardieu *et al.* 1999). The shape of the scattering curves as a function of temperature is salt-dependent. With salts like $NaNO_3$ a good fit could be obtained only by simultaneously increasing the depth of the attractive potential and the charge at lower temperature. The amplitude of the effect, which is well measured by the dependence of A_2 on temperature (in 0.5 M NaCl it increases from -7.1×10^{-4} at 10°C to -2.3×10^{-4} mol mL g^{-2} at 30° C), decreases in the reverse order of the Hofmeister series and is more pronounced at low ionic strength (Bonneté *et al.* 1999). In contrast, with BPTI (21 mg/mL, 350 mM KSCN, pH 4.9 < pI = 10.6) the percentage of decamers in the solution increases with temperature from about 5–50% between 5°C and 30°C (Grouazel *et al.* 2002). This is also the case with other entropy-driven assembly processes like those observed with virus capsids (such as tobacco mosaic virus protein), tubulin or actin (see Section 7.3), or amyloidogenic proteins like insulin (Javid *et al.* 2007a). Systematic studies of the effect of temperature over a broader range in fixed salt and pH conditions do not seem to be available. Things may also be different at higher temperatures where lysozyme also unfolds. Clearly, the effects depend on the extent to which the effective potential reflects protein–protein or protein–solvent interactions.

The effects of pressure on the intermolecular interactions of lysozyme (100 mg/mL) have been studied in D_2O and H_2O with not entirely consistent results. At pressures below 0.15 GPa a significant reduction in the depth of the attractive potential and a discontinuity in density of hydration water were found around 70 MPa in D_2O at an ionic strength of 30 mM (Ortore *et al.* 2009). In H_2O at non-denaturing pressures (<0.4 GPa), concentrated lysozyme solutions display a non-linear variation of the attractive potential over a wide range of concentrations and ionic strengths with a minimum around 0.15–0.2 GPa (Schroer *et al.* 2011a, Moeller *et al.* 2012), which is approximately also the pressure at which a break was observed in the anomalous component of the static structure factor of bulk water (Cunsolo *et al.* 2009). The pressure effects were taken as an indication that the pressure dependence of the attractive potential is due to changes in the water structure. Whether the differences are due to the properties of the two solvents is not known.

Note, however, that there is no consensus on the interpretation of the temperature and pressure-dependence of the SAS patterns of water (see, for example, Cunsolo *et al.* 2009, Clark *et al.* 2010, Nilsson and Pettersson 2011, and references therein).

8.3.7 Comparison between different studies

As illustrated above, structure factor calculations have hitherto been limited to a few model proteins, and the results are consistent when the same approach is used. It is more difficult to compare the results of different approximations, as illustrated by studies on BSA. Early SANS measurements on BSA solutions (3–120 mg/mL) at different pH values (5.1 and 7.0 > pI = 4.9) and LiCl concentrations (0–300 mM) (Bendedouch and Chen 1983) were fitted using a mean spherical model with an interaction potential consisting of a hard-sphere core and a screened Coulomb potential (Hayter and Penfold 1981). This is one of the earliest papers where the structure factor of a protein solution was calculated from the charge and ionic strength for pH 7.0, and an expression for the pair distribution function for an attractive interaction was introduced for pH 5.1 close to the isoelectric point where there is no detectable repulsive component. The effective charge on the protein at different pH values and salt concentrations reproduced the values expected from previous independent proton titration and Cl^--binding curves. The shape of the hydrated molecule was approximated by a prolate ellipsoid ($7 \times 2 \times 2 \, nm^3$). In later studies covering a larger q-range, an oblate ellipsoid ($1.7 \times 4.2 \times 4.2 \, nm^3$) was found to provide the best fit to the scattering pattern of dilute solutions (Zhang *et al.* 2007, Zhang *et al.* 2012). This difference in not relevant for the discussion of the structure factor, as its effects become detectable only above $q = 1.5 \, nm^{-1}$, and in recent work the form factor calculated from the human serum albumin structure in the PDB was used (Barbosa *et al.* 2010). The net protein charge (Z_p) varies in different studies, partly because of the well-known limitations of the DLVO potential at higher ionic strength (Finet and Tardieu 2001, and references therein). At pH 7 in absence of salt values $|Z_p| = 8$ (40–120 mg BSA/mL) (Bendedouch and Chen 1983), $|Z_p| = 10–13$ (20–100 mg BSA/mL) and $|Z_p| = 18–7$ (200–500 mg BSA/mL) (Zhang *et al.* 2007), $|Z_p| = 13 \pm 2$ (25 mg BSA/mL) (Barbosa *et al.* 2010), which should be compared with $|Z_p| = 12$, found by accurate potentiometric titration (Salis *et al.* 2011). Whereas the values of $|Z_p|$ in these two last studies follow the same trend between pH 5 and 9, the calculated value at pH 4.0 ($|Z_p| = 10 \pm 2$) is much lower than the potentiometric value ($|Z_p| = 20$). This suggests that, as pointed out early on (Tardieu *et al.* 1999), the concept of charge as measured by different methods, or calculated with different approximations, needs clarification.

For lysozyme the values of parameters such as the net protein charge, the depth of the potentials and the Hamaker constant found in the literature also differ for identical solution conditions (Narayanan and

Liu 2003). As an example, the particle diameters and hence the volumes also vary in the simulations of different groups. For spherical particles diameters and volumes ranging from ($\sigma = 3.24$ nm, $V = 17.8$ nm^3, $\bar{v}_{calc} = 0.703$ cm^3 g^{-1}) (Tardieu *et al.* 1999, Niebuhr and Koch 2005), ($\sigma = 3.4$ nm, $V = 20.6$ nm^3, $\bar{v}_{calc} = 0.838$ cm^3 g^{-1}) (Stradner *et al.* 2004) have been used, and for prolate ellipsoids ($\sigma_a = \sigma_b = 3.19$ nm and $\sigma_c = 4.5$ nm, $V = 21$ nm^3, $\bar{v}_{calc} = 0.854$ cm^3 g^{-1}) (Liu *et al.* 2005) and ($\sigma_a = \sigma_b = 3.19$ nm and $\sigma_c = 4.8$ nm, $V = 25$ nm^3, $\bar{v}_{calc} = 1.01$ cm^3 g^{-1}) (Shukla *et al.* 2008a). The generally accepted value for the partial specific volume of lysozyme is $\bar{v} = 0.703$ cm^3 g^{-1} (see, for example, Van Holde *et al.* 1998).

Even when the data are similar, interpretation of structure factors is by no means unique, and this has recently led to a number of controversies such as that related to the existence of an equilibrium cluster phase in concentrated lysozyme solutions (Stradner *et al.* 2004, Shukla *et al.* 2008a, Stradner *et al.* 2008, Shukla *et al.* 2008b). In the conditions of the experiments the structure factors display maxima at $q_c \approx 0.8$–1.0 nm^{-1} and $q_m \approx 2.2$ nm^{-1}, which Stradner *et al.* attributed to cluster–cluster correlations and protein–protein correlations within clusters, respectively. Maxima near $q = 0.6$ nm^{-1} and $q = 2.7$ nm^{-1} were also observed in simulations of lysozyme solutions using a two-Yukawa potential with the MSA and HNC approximations and Monte Carlo calculations (Broccio *et al.* 2006). The maximum at lower q is more pronounced when the attractive well is deeper, and was therefore also attributed to cluster formation. Shukla *et al.* observed a systematic shift of this maximum towards smaller angles with increasing concentration, and found this compatible with the calculated structure factors using a two-Yukawa potential in conditions where no aggregates are present in solution (Shukla *et al.* 2008a).

It is in general not entirely justified to attribute maxima in the structure factors to distances between specific entities, nor can one simply relate the value of $2\pi/q_{max}$ to the average distance between particles. This would imply a (concentration)$^{1/3}$-dependent shift of q_{max} to larger q-values, which is not always observed. Shifts of q_{max} are observed in strongly repulsive systems, but for hard spheres the position of q_{max}, which is related to the particle diameter, is independent of concentration (Ducruix *et al.* 1996). Polydispersity is another factor that must be taken into account, as the formation of larger aggregates is expected to shift q_{max} to smaller q-values (Klein 2002).

Poorer fits in the low-q region of the patterns of lysozyme solutions, especially at lower temperatures where the attractions are stronger, reveal the presence of aggregates—another name for clusters—but they are not the dominating species (Shukla *et al.* 2008a). A combined neutron spin echo and SANS study concluded that in the case of lysozyme monomers and dimers dominate at low concentrations, whereas at volume fractions above $\varphi = 0.15$ dynamic clusters are formed (Porcar *et al.* 2010). The preparative procedures are, however, not given in detail in this study. A similar study later suggested that the low-q

maximum in the structure factor was due to the formation of an inter-mediate range order structure, and concluded that 'there is no direct correlation between cluster formation in a solution and the existence of the cluster peak' (Liu *et al.* 2011).

Another controversy concerned the origin of a sharp peak at the origin of the structure factor attributed to long-range attraction depending on the type of anion and ion concentration (Liu *et al.* 2005), but this was shown to depend on the preparation (Stradner *et al.* 2006, Liu *et al.* 2006). At this stage it thus seems that all results can be explained in terms of a hard core potential, short-range attraction and long-range repulsion.

That equilibrium clusters can form in colloidal systems seems beyond doubt. There is also no doubt that some protein solutions have a strong tendency to form aggregates—depending on the conditions. One should, however, carefully distinguish between different mechanisms and types of aggregation, including those that may be relevant to pathology and/or practical applications (Philo and Arakawa 2009). An example of medical interest is insulin, where a deep attractive potential is necessary to explain self-assembly into fibrils when the temperature is raised (Javid *et al.* 2007a).

These controversies highlight some of the limitations of scattering methods and of current models of interactions, but some of the difficulties most probably arise from suboptimal experimental procedures (such as preparation of solutions by dilution rather than dialysis, insufficient desalting of commercial protein batches, oxidation, or use of filters, which are known to release material that scatters at low angles). Caution is also required when combining results obtained in H_2O and D_2O, as the interactions are different in these two solvents (Budayova-Spano *et al.* 2000).

The more fundamental question of whether a freshly centrifuged or chromatographed solution is at equilibrium also remains open. Clearly, kinetics plays a significant role in this area, which in practice is very important for protein storage.

Monte Carlo simulations (Pedersen 2002b) and molecular dynamics calculations are more flexible than the analytical approach, which has the advantage of providing a deeper physical insight. Moreover, they may help to statistically analyse and visualise some phenomena such as cluster formation. Attempts have been made to combine molecular dynamics calculations with isotropic potentials and spherical particles for SAS modelling (Kim *et al.* 2008) and to use patchy models of aeolotopic interactions (Gögelein *et al.* 2008), but this has not yet been applied to SAS data.

8.3.8 Very large systems

Systems leading to large-scale structures, which often occur in the food and health-care industry, are usually too opaque to obtain information about the bulk by light-scattering or microscopy, and the resolution of conventional SAS is also insufficient. Large supramolecular protein clusters can be formed by a number of proteins, and their structure also affects their catalytic activity as shown for subtilisin (Javid *et al.* 2011).

One of the most studied examples of large systems is that of milk and dairy products. Casein micelles in milk have radii of 50–100 nm and also contain calcium phosphate nanoclusters, but their internal structure is still a matter of debate (for a review, see, for example, De Kruif *et al.* 2012). The different protein constituents of casein micelles have a pI between 4.1 and 5.8. Most common dairy products, such as yoghurt or curd, are obtained by lowering the pH of milk to this range, which leads to aggregation of the micelles. Acidification leads first to shrinking of the micelles, which subsequently aggregate, but the two processes are independent (Moitzi *et al.* 2011). Interactions of micelles can be studied by osmotic stress techniques (Bouchoux *et al.* 2009) that can also be combined with SAS to obtain structure factors. There is a transition from a turbid fluid solution to a nearly transparent gel at a protein concentration around 150–250 mg/mL. Above this concentration the changes in the structure factor reveal a non-affine deformation of the internal structure of the micelles, which has led to propose a sponge model for the micelles (Bouchoux *et al.* 2010). At this stage the internal structure of the micelles seems too complex for meaningful simulations of the interactions using potentials.

Large-scale systems can be studied by USANS/USAXS, or even better, SESANS, when the dimensions exceed about 5 μm. An example of distance-distribution functions obtained by SESANS illustrates the range of sizes in common dairy products. Milk contains micelles with an average diameter of 240 nm, whereas yoghurt has aggregates of typically between about 3 and 20 micelles. Scattering experiments are performed on skimmed milk to avoid the strong scattering of large fat micelles. Modelling of such systems is more complex, as it must take into account reaction and diffusion processes affecting the slow kinetics (hours–days) of gelation. Monte Carlo simulations suggest that gelation starts with reaction-limited cluster association and evolves to diffusion-limited cluster association of sticky hard spheres on a three-dimensional

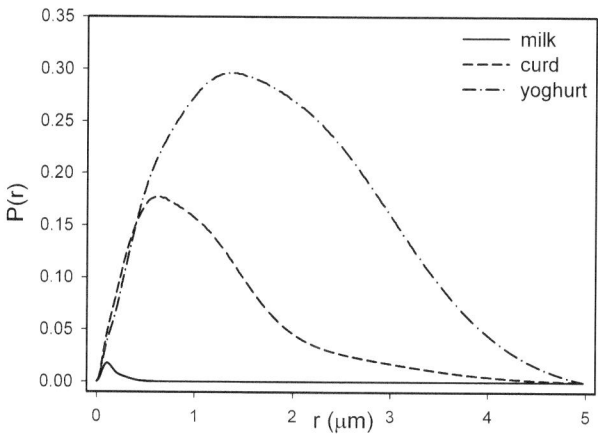

Fig. 8.9 Distance distribution function for milk, curd and yoghurt in D_2O obtained by SESANS. Note the scale on the abscissa. Adapted from Tromp and Bouwman (2007), with permission from Elsevier, © 2007.

lattice (Van Heijkamp *et al.* 2010). In contrast with the SANS work above, however, the overall size and structure of the micelles did not seem affected by gelation.

Clearly, further research will be required to obtain an integrated picture over all levels of structural hierarchy.

8.4 Interactions in nucleic acids

The interactions of nucleic acids and other linear polyelectrolytes are determined largely by the ion atmosphere around the molecules, and are therefore more difficult to describe than those in the preceding sections. To simplify matters it is usually assumed that the ions have the same number of charges. The Poisson–Boltzmann (PB) equation gives the Laplacian of the potential ($\nabla^2 \Psi = \frac{\partial^2 \Psi}{\partial x^2} + \frac{\partial^2 \Psi}{\partial y^2} + \frac{\partial^2 \Psi}{\partial z^2}$) as a function of the dielectric constant ε, the vacuum permittivity ε_0, the charge of the ions ($\pm Z$) and their number concentration in the bulk (ρ_{ion}), and taking the definition of the hyperbolic sine function into account ($\sin h(x) = (e^x - e^{-x})/2$), this yields a non-linear differential equation:

$$\nabla^2 \Psi = \frac{Ze\rho_{ion}}{\varepsilon_s \varepsilon_0} \left(e^{Ze\Psi/k_B T} - e^{-Ze\Psi/k_B T} \right) = \frac{2Ze\rho_{ion}}{\varepsilon_s \varepsilon_0} \sin h(Ze\Psi/k_B T) \quad (8.13)$$

This equation can be linearised when the electrostatic potential is small and $Ze\Psi/k_B T \ll 1$. Taylor expansion of the two exponentials then yields, with the same symbols as in eq. (8.11):

$$\nabla^2 \Psi = \frac{e^2 \sum_i \rho_i Z_i^2}{\varepsilon_s \varepsilon_0} (\Psi/k_B T) = \frac{e^2 \sum_i \rho_i Z_i^2}{\varepsilon_s \varepsilon_0 k_B T} \Psi = \Psi/\lambda_D^2 \quad (8.14)$$

ASAXS has been used to count the number of monovalent or divalent cations around RNA (Pabit *et al.* 2009) and DNA (Pabit *et al.* 2010). Low-resolution models for nucleic acids have also been used successfully in conjunction with Poisson–Boltzmann calculations (Lipfert *et al.* 2007a), and the number of ions bound compares well with those obtained with all atom models derived from crystallography. This approach was used to assess the free energy changes induced by counter-ion shielding for a glycine riboswitch, and it was shown that simple electrostatics can account for the observed data on salt-dependence of folding (Lipfert *et al.* 2007b).

Screening of double-stranded RNA (dsRNA) was found to be more efficient than for dsDNA. The topology of the RNA A-helix results in a different distribution of counterions than in the case of the B-helix of DNA (Pabit *et al.* 2009). Numerical solutions to the non-linear Poisson–Boltzmann equation gave results in good agreement with the experimental observations in terms of numbers of ions bound. More applications of SAXS to study RNA interactions with ions, including time-resolved analysis of RNA folding, can be found in recent reviews (Pollack 2011a, Pollack 2011b).

Counterion-modulated repulsion and attraction was investigated using DNA duplexes tethered by a flexible neutral link in solution, and it was concluded that counterion-induced attractive forces are not significant under physiological conditions (Bai *et al.* 2005). In this case a Yukawa potential was used to describe attraction and repulsion between phosphate groups as a function of their distance. The electrostatic energy of the random and extended low-salt ensembles was calculated using the non-linear Poisson–Boltzmann model. A critical assessment of this approach comparing SAXS results and Poisson–Boltzmann calculations indicated that the PB equation gives a satisfactory description in the case of monovalent salts, but that there are large deviations for divalent salts (Bai *et al.* 2008).

The interactions of different monovalent salts with chromatin and DNA indicate that the number of ions bound to the solute is independent of their nature, and that there are no major changes in hydration (Koch *et al.* 1987).

Weak polyelectrolytes like polysaccharides and alginates play a very important role in the food industry, where their gelation properties are often used. Modelling of such systems (Denef *et al.* 1996, Mischenko *et al.* 1996) would also require methods based on molecular models which are presently not available.

The overview above illustrates that despite the progress made over the last decades the study of interactions of biological macromolecules remains a challenging area of research both from an experimental as well as theoretical point of view. Although there are many similarities with colloids there are also significant differences, and the field could therefore gain a lot from a closer collaboration between biochemists and soft-condensed-matter experts.

References

Abramo, M. C., Caccamo, C., Calvo, M. *et al.* (2011). 'Molecular dynamics and small-angle neutron scattering of lysozyme aqueous solutions', *Phil. Mag.* 91: 2066–76.

Ando, T. and Skolnick, J. (2010). 'Crowding and hydrodynamic interactions likely dominate in vivo macromolecular motion', *P. Natl. Acad. Sci. USA* 107: 18457–62.

Asakura and Oosawa, F. (1954). 'Surface tension of high-polymer solutions', *J. Chem. Phys.* 22: 1255.

Asherie, N. (2004). 'Protein crystallization and phase diagrams', *Methods* 34: 266–72.

Bai, Y., Chu, V. B., Lipfert, J., Pande, V. S., Herschlag, D. and Doniach, S. (2008). 'Critical assessment of nucleic acid electrostatics via experimental and computational investigation of an unfolded state ensemble', *J. Am. Chem. Soc.* 130: 12334–41.

Bai, Y., Das, R., Millett, I. S., Hersschlag, D. and Doniach, S. (2005). 'Probing counterion modulated repulsion and attraction between nucleic acid duplexes in solution', *P. Natl. Acad. Sci. USA* 102: 1035–45.

Barbosa, L. R. S., Ortore, M. G., Spinozzi, F., Mariani, P., Bernstorff, S. and Itri, R. (2010). 'The importance of protein–protein interactions on the pH-induced conformational changes of bovine serum albumin: A small-angle X-ray scattering study', *Biophys. J.* 98: 147–57.

Ben-Naim, A. (1990). 'Solvent effects on protein association and protein folding', *Biopolymers* 29: 567–96.

Bénas, P., Legrand, L. and Ríès-Kautt, M. (2002). 'Strong and specific effects of cations on lysozyme chloride solubility', *Acta Crystallogr. D* 58: 1582–7.

Bendedouch, D. and Chen, S. H. (1983). 'Structure and interparticle interaction of BSA in solution studied by small-angle neutron scattering', *J. Phys. Chem.* 87: 1473–7.

Bolen, D. W. and Rose, G. D. (2008). 'Structure and energetics of the hydrogen-bonded backbone in protein folding', *Annu. Rev. Biochem.* 77: 339–62.

Bonneté, F., Finet, S. and Tardieu, A. (1999). 'Second virial coefficient: variations with lysozyme crystallization conditions', *J. Cryst. Growth* 196: 403–14.

Bonneté, F., Malfois, M., Finet, S., Tardieu, A., Lafont, S. and Veesler, S. (1997). 'Different tool to study interaction potentials in γ-crystallin solutions: relevance to crystal growth', *Acta Crystallogr. D* 53: 438–47.

Boström, M., Parsons, D. F., Salis, A., Ninham, B. W. and Monduzzi, M. (2011). 'Possible origin of the inverse and direct Hofmeister series for lysozyme at low and high salt concentration', *Langmuir* 27: 9504–11.

Boström, M., Tavares, F. W., Finet, S., Skouri-Panet, F., Tardieu, A. and Ninham, B. W. (2005). 'Why forces between proteins follow different Hofmeister series for pH above and below pI', *Biophys. Chem.* 117: 217–24.

Bouchoux, A., Cayemitte, P. E., Jardin, J., Gésan-Guizou, G. and Cabane, B. (2009). 'Casein micelle dispersions under osmotic stress', *Biophys. J.* 96: 693–706.

Bouchoux, A., Gésan-Guiziou, G., Pérez, J. and Cabane, C. (2010). 'How to squeeze a sponge: casein micelles under osmotic stress, a SAXS study', *Biophys. J.* 99: 3754–62.

Broccio, M., Costa, D., Liu, Y. and Chen, S.-H. (2006). 'The structural properties of a two-Yukawa fluid: simulation and analytical results', *J. Chem. Phys.* 124: 084501.

Budayova, M., Bonneté, F., Tardieu, A. and Vachette, P. (1999). 'Interactions in solution of a large oligomeric protein', *J. Cryst. Growth* 196: 210–19.

Budayova-Spano, M., Lafont, S., Astier, J. P., Ebel, C. and Veesler, S. (2000). 'Comparison of solubility and interactions of aprotinin (BPTI) solutions in H_2O and D_2O', *J. Cryst. Growth* 217: 311–19.

Carnahan, N. F. and Starling, K. E. (1969). 'Equation of state for nonattracting rigid spheres', *J. Chem. Phys.* 51: 635–6.

Carrotta, R., Manno, M., F.M., G. *et al.* (2009). 'Protein stability modulated by a conformational effector: effects of trifluoroethanol on bovine serum albumin', *Phys. Chem. Chem. Phys.* 11: 4007–18.

Casselyn, M., Perez, J., Tardieu, A., Vachette, P., Witz, J. and Delacroix, H. (2001). 'Spherical plant viruses: interactions in solution, phase diagrams and crystallization of brome mosaic virus', *Acta Crystallogr. D* 57: 1799–1812.

Castelletto, V., Hamley, I. W., Clifton, L. A. and Green, R. J. (2008). 'Osmotic pressure and aggregate shape in BSA/poly(ethylene glycol)-lipid/dextran solutions', *Biophys. Chem.* 134: 34–8.

Chayen, N. E. (2004). 'Turning protein crystallisation from an art into a science', *Curr. Opin. Struc. Biol.* 14: 577–83.

Clark, G. N. I., Hura, G. L., Teixeira, J., Soper, A. K. and Head-Gordon, T. (2010). 'Small-angle scattering and the structure of ambient liquid water', *P. Natl. Acad. Sci. USA* 107: 14003–7.

Cunsolo, A., Formisano, F., Ferrero, C., Bencivenga, F. and Finet, S. (2009). 'Pressure dependence of the large-scale structure of water', *J. Chem. Phys.* 131: 194501–5.

De Kruif, C. G., Huppertz, T., Urban, V. C. and Petukhov, A. V. (2012). 'Casein micelles and their internal structure', *Adv. Colloid Interfac.* 171–2: 36–52.

Denef, B., Mischenko, N., Koch, M. H. J. and Reynaers, H. (1996). 'Small-angle X-ray scattering of kappa- and iota-carrageenan in aqueous and in salt solutions', *Int. J. Biol. Macromol.* 18: 151–9.

Doye, J. P. K., Louis, A. A., Lin, I.-C. *et al.* (2007). 'Controlling crystallization and its absence: proteins, colloids and patchy models', *Phys. Chem. Chem. Phys.* 9: 2197–205.

Ducruix, A., Guilloteau, J. P., Ríès-Kautt, M. and Tardieu, A. (1996). 'Protein interactions as seen by solution X-ray scattering prior to crystallogenesis', *J. Cryst. Growth* 168: 28–39.

Ebbinghaus, S., Kim, S. J., Heyden, M. *et al.* (2007). 'An extended dynamical hydration shell around proteins', *P. Natl. Acad. Sci. USA* 104: 20749–52.

Finet, S. and Tardieu, A. (2001). 'α-crystallin interaction forces studied by small angle X-ray scattering and numerical simulations', *J. Cryst. Growth* 232: 40–9.

Fritz, G. and Glatter, O. (2006). 'Structure and interaction in dense colloidal systems: evaluation of scattering data by the generalized indirect Fourier transformation method', *J. Phys. Condens. Matter* 18: S2403–19.

Fukasawa, T. and Sato, T. (2011). 'Versatile application of indirect Fourier transformation to structure factor analysis: from X-ray diffraction of molecular liquids to small angle scattering of protein solutions', *Phys. Chem. Chem. Phys.*, 13: 3187–96.

Gabel, F., Ringkjobing Jensen, M., Zaccai, G. and Blackledge, M. (2009). 'Quantitative modelfree analysis of urea binding to unfolded ubiquitin using a combination of small angle X-ray and neutron scattering', *J. Am. Chem. Soc.* 131: 8769–71.

George, A. and Wilson, W. W. (1994). 'Predicting protein crystallization from a dilute-solution property', *Acta Crystallogr. D* 50: 361–5.

Gögelein, C., Nägele, G., Tuinier, R., Gibaud, T., Stradner, A. and Schurtenberger, P. (2008). 'A simple patchy colloid model for the phase behavior of lysozyme dispersions', *J. Chem. Phys.* 129: 085102.

Grouazel, S., Pérez, J., Astier, J.-P., Bonneté, F. and Veesler, S. (2002). 'BPTI liquid–liquid phase separation monitored by light and small-angle X-ray scattering', *Acta Crystallogr. D* 58: 1560–3.

Hamiaux, C., Pérez, J., Prangé, T., Veesler, S., Riès-Kautt, M. and Vachette, P. (2004). 'The BPTI decamer observed in acidic pH crystal forms preexists as a stable species in solution', *J. Mol. Biol.* 297: 697–712.

Hayter, J. B. and Penfold, J. (1981). 'An analytic structure factor for macroion solutions', *Mol. Phys.* 42: 109–18.

He, L., Wang, H., Garamus, V. M. *et al.* (2010). 'Analysis of monoPEGylated human galectin-2 by small-angle X-ray and neutron scattering: concentration dependence of PEG conformation in the conjugate', *Biomacromolecules* 11: 3504–10.

Hofmeister, F. (1888). 'Zur Lehre von der Wirkung der Salze. Zweite Mitteilung', *Arch. Exp. Pathol. Pharmakol.* 24: 247–60.

Ianeselli, L., Zhang, F., Skoda, M. W. F. *et al.* (2010). 'Protein–protein interactions in ovalbumin solutions studied by small-angle scattering: Effect of ionic strength and the chemical nature of cations', *J. Phys. Chem. B* 114: 3776–83.

Israelachvili, J. (2011). *Intermolecular and Surface Forces.* New York: Academic Press.

Javid, N., Voggt, K., Krywka, C., Tolan, M. and Winter, R. (2007a). 'Capturing the interaction potential of amyloidogenic proteins', *Phys. Rev. Lett.* 99: 028101–4.

Javid, N., Voggt, K., Roy, S. *et al.* (2011). 'Supramolecular structures of enzyme clusters', *J. Phys. Chem. Lett.* 2: 1395–9.

Javid, N., Vogtt, K., Krywka, C., Tolan, M. and Winter, R. (2007b). 'Protein–protein interactions in complex cosolvent solutions', *Chem. Phys. Chem.*, 8: 679–89.

Kim, S. J., Dumont, C. and Gruebele, M. (2008). 'Simulation-based fitting of protein–protein interaction potentials to SAXS experiments', *Biophys. J.* 94: 4924–31.

Klein, R. (2002). 'Interacting colloidal suspensions', in Lindner, P. and Zemb, T. (eds.), *Neutrons, X-Rays and Light: Scattering Methods Applied to Soft Condensed Matter.* Amsterdam: North Holland.

Koch, M. H. J., Vega, M. C., Sayers, Z. and Michon, A. M. (1987). 'The superstructure of chromatin and its condensation mechanism. III: Effect of monovalent and divalent cations X-ray solution scattering and hydrodynamic studies', *Eur. Biophys. J.* 14: 307–19.

Kondepudi, D. and Prigogine, I. (1998). *Modern Thermodynamics: From Heat Engines to Dissipative Structures.* New York: John Wiley & Sons.

Kulkarni, A. M., Chatterjee, A. P., Schweizer, K. S. and Zukoski, C. F. (2000). 'Effects of polyethylene glycol on protein interactions', *J. Chem. Phys.* 113: 9863–73.

Lee, J. J. and Berns, D. S. (1968). 'Protein aggregation: studies of larger aggregates of C-Phycocyanin', *Biochem. J.* 110: 465–70.

Lima, E. R. A., Biscaia, E. C. J., Boström, M., Tavares, F. W. and Prausnitz, J. M. (2007). 'Osmotic second virial coefficients and phase diagrams for aqueous proteins from a much-improved Poisson-Boltzmann equation', *J. Phys. Chem. C* 111: 16055–9.

Lipfert, J., Chu, V. B., Bai, Y., Herschlag, D. and Doniach, S. (2007a). 'Low resolution models for nucleic acids from small-angle X-ray scattering with applications to electrostatic modeling', *J. Appl. Crystallogr.* 40: s229–34.

Lipfert, J., Das, R., Chu, V. B. *et al.* (2007b). 'Structural transitions and thermodynamics of a glycine-dependent riboswitch from Vibrio cholerae', *J. Mol. Biol.* 365: 1393–1406.

Liu, Y., Fratini, E., Baglioni, P., Chen, W.-R. and Chen, S.-H. (2005). 'Effective long-range attraction between protein molecules in solutions studied by small angle neutron scattering', *Phys. Rev. Lett.* 95: 118101–4.

Liu, Y., Fratini, E., Baglioni, P., Chen, W.-R., Porcar, L. and Chen, S.-H. (2006). 'Reply', *Phys. Rev. Lett.* 96: 219801–2.

Liu, Y., Porcar, L., Chen, J. *et al.* (2011). 'Lysozyme protein solution with an intermediate range order structure', *J. Phys. Chem. B* 115: 7238–47.

Lo Nostro, P. and Ninham, B. W. (2012). 'Hofmeister phenomena: an update on ion specificity in biology', *Chem. Rev.* 112: 2286–322.

Luft, J. R., Wolfley, J. R. and Snell, E. H. (2011). 'What's in a drop? Correlating observations and outcomes to guide macromolecular crystallization experiments', *Crystal Growth Des.* 11: 651–63.

Makowski, L., Bardhan, J., Gore, D. B. *et al.* (2011a). 'WAXS studies of the structural diversity of hemoglobin in solution', *J. Mol. Biol.* 408: 909–21.

Makowski, L., Gore, D., Mandava, S. *et al.* (2011b). 'X-ray solution scattering studies of the structural diversity intrinsic to protein ensembles', *Biopolymers* 95: 531–42.

Makowski, L., Rodi, D. J., Mandava, S., Minh, D. D. L., Gore, D. B. and Fischetti, R. F. (2008). 'Molecular crowding inhibits intramolecular breathing motions in proteins', *J. Mol. Biol.* 375: 529–46.

Malfois, M., Bonneté, F., Belloni, L. and Tardieu, A. (1996). 'A model of attractive interactions to account for fluid–fluid phase separation of protein solutions', *J. Chem. Phys.* 105: 3290–300.

Mischenko, N., Denef, B., Koch, M. H. J. and Reynaers, H. (1996). 'Influence of ionic effects on the ordering and association phenomena in dilute and semidilute carrageenan solutions', *Int. J. Biol. Macromol.* 19: 185–94.

Moeller, J., Schroer, M. A., Erlkamp, M. *et al.* (2012). 'The effect of ionic strength, temperature, and pressure on the interaction potential of dense protein solutions: from nonlinear pressure response to protein crystallization', *Biophys. J.* 102: 2641–8.

Moitzi, C., Menzel, A., Schurtenberger, P. and Stradner, A. (2011). 'The pH induced sol–gel transition in skim milk revisited. A detailed study using time-resolved light and X-ray scattering experiments', *Langmuir* 27: 2195–203.

Narayanan, J. and Liu, X. Y. (2003). 'Protein interactions in undersaturated and supersaturated solutions: a study using light and X-ray scattering', *Biophys. J.* 84: 523–32.

Niebuhr, M. and Koch, M. H. J. (2005). 'Effects of urea and trimethylamine-N-oxide (TMAO) on the interactions of lysozyme in solution', *Biophys. J.* 89: 1978–83.

Nilsson, A. and Pettersson, L. G. M. (2011). 'Perspective on the structure of liquid water', *Chem. Phys.* 389: 1–34.

Ortore, M. G., Sinibaldi, R., Spinozzi, F. *et al.* (2008). 'New insights into urea action on proteins: A SANS study of the lysozyme case', *J. Phys. Chem. B* 112: 12881–7.

Ortore, M. G., Spinozzi, F., Mariani, P. *et al.* (2009). 'Combining structure and dynamics: non-denaturing high-pressure effect on lysozyme in solution', *J. R. Soc. Interface* 6: S619–34.

Oster, G. and Riley, D. P. (1952). 'Scattering from cylindrically symmetric systems', *Acta Crystallogr.* 5: 272–6.

Pabit, S. A., Meisburger, S. P., Li, L., Blose, J. M., Jones, C. D. and Pollack, L. (2010). 'Counting ions around DNA with anomalous small-angle X-ray scattering', *J. Am. Chem Soc.* 132: 16334–6.

Pabit, S. A., Qiu, X., Lamb, J. S., Li, L., Meisburger, S. P. and Pollack, L. (2009). 'Both topology and counterion distribution contribute to the more effective charge screening in dsRNA compared to dsDNA', *Nucleic Acids Res.* 37: 3887–96.

Pedersen, J. S. (2002a). 'Modelling of small-angle scattering data', in Lindner, P. and Zemb, T. (eds.), *Neutrons, X-Rays and Light: Scattering Methods Applied to Soft Condensed Matter*. Amsterdam: North Holland.

Pedersen, J. S. (2002b). 'Monte Carlo simulation techniques applied in the analysis of small angle scattering data from colloids and polymer systems', in Lindner, P. and Zemb, T. (eds.), *Neutrons, X-Rays and Light: Scattering Methods Applied to Soft Condensed Matter*. Amsterdam: North Holland.

Philo, J. S. and Arakawa, T. (2009). 'Mechanisms of Protein Aggregation', *Curr. Pharm. Biotechno.* 10: 348–51.

Pollack, L. (2011a). 'SAXS studies of ion–nucleic acid interactions', *Ann. Rev. Biophys.* 40: 225–42.

Pollack, L. (2011b). 'Time resolved SAXS and RNA folding', *Biopolymers* 95: 543–9.

Porcar, C., Falus, P., Chen, W.-C. *et al.* (2010). 'Formation of the dynamic clusters in concentrated lysozyme protein solutions', *J. Phys. Chem. Lett.* 1: 126–9.

Salis, A., Boström, M., Medda, L. *et al.* (2011). 'Measurements and theoretical interpretation of points of zero charge/potential of BSA protein', *Langmuir* 27: 11597–604.

Schellman, J. A. (2003). 'Protein stability in mixed solvents: A balance of contact interaction and excluded volume', *Biophys. J.* 85: 108–25.

Schellman, J. A. (2005). 'Destabilization and stabilization of proteins', *Quart. Rev. Biophys.* 38: 351–61.

Schroer, M., Markgraf, J., Wieland, D. C. F. *et al.* (2011a). 'Nonlinear pressure dependence of the interaction potential of dense protein solutions', *Phys. Rev. Lett.* 106: 178102.

Schroer, M. A., Zhai, Y., Wieland, D. C. F. *et al.* (2011b). 'Exploring the piezophilic behavior of natural cosolvent mixtures', *Angew. Chem. Int. Ed.* 50: 11413–6.

Shukla, A., Mylonas, E., Di Cola, E. *et al.* (2008a). 'Absence of equilibrium cluster phase in concentrated lysozyme solutions', *P. Natl. Acad. Sci. USA* 105: 5075–80.

Shukla, A., Mylonas, E., Di Cola, E. *et al.* (2008b). 'Reply to Stradner *et al.*: Equilibrium clusters are absent in concentrated lysozyme solutions', *P. Natl. Acad. Sci. USA* 105: E76.

Sinibaldi, R., Ortore, M. G., Spinozzi, F. *et al.* (2007). 'Preferential hydration of lysozyme in water/glycerol mixtures: A small-angle neutron scattering study', *J. Chem. Phys.* 126: 235101–9.

Stradner, A., Cardinaux, F. and Schurtenberger, P. (2006). 'Comment on "Effective long-range attraction between protein molecules in solution studied by small angle neutron scattering"', *Phys. Rev. Lett.* 96: 219802.

Stradner, A., Cardinaux, P., Egelhaaf, S. U. and Schurtenberger, P. (2008). 'Do equilibrium clusters exist in concentrated lysozyme solutions?', *P. Natl. Acad. Sci. USA* 105: E75.

Stradner, A., Sedgwick, H., Cardinaux, P., Poon, W. C. K., Egelhaaf, S. U. and Schurtenberger, P. (2004). 'Equilibrium cluster formation in concentrated protein solutions and colloids', *Nature* 432: 492–5.

Street, T. O., Krukenberg, K. A., Rosgen, J., Bolen, D. W. and Agard, D. A. (2010). 'Osmolyte-induced conformational changes in the Hsp90 molecular chaperone', *Protein Sci.* 19: 57–65.

Svergun, D. I., Ekström, F., Vandegriff, K. D. *et al.* (2008). 'Solution structure of poly(ethylene) glycol-conjugated hemoglobin revealed by small-angle X-ray scattering: Implications for a new oxygen therapeutic', *Biophys. J.* 94: 173–81.

Svergun, D. I., Richard, S., Koch, M. H. J., Sayers, Z., Kuprin, S. and Zaccai, G. (1998). 'Protein hydration in solution: Experimental observation by X-ray and neutron scattering', *P. Natl. Acad. Sci. USA* 95: 2267–72.

Tardieu, A. and Delaye, M. (1988). 'Eye lens proteins and transparency: from light transmission theory to solution X-ray structural analysis', *Annu. Rev. Biophys. Biophys. Chem.* 17: 47–70.

Tardieu, A., Finet, S. and Bonneté, F. (2001). 'Structure of the macromolecular solutions that generate crystals', *J. Cryst. Growth* 232: 1–9.

Tardieu, A., Le Verge, A., Malfois, M. *et al.* (1999). 'Proteins in solution: from X-ray scattering intensities to interaction potentials', *J. Cryst. Growth* 196: 193–203.

Tromp, R. H. and Bouwman, W. G. (2007). 'A novel application of neutron scattering on dairy products', *Food Hydrocolloids* 21: 154–8.

Van Heijkamp, L. F., De Schepper, I. M., Strobl, M., Tromp, R. H., Heringa, J. R. and Bouwman, W. G. (2010). 'Milk gelation studied with small angle neutron scattering techniques and Monte Carlo simulations', *J. Phys. Chem. A* 114: 2412–26.

Van Holde, K. E., Johnson, W. C. and Ho, P. S. (1998). *Principles of Physical Biochemistry*. Upper Saddle River, NJ: Prentice-Hall.

Verwey, E. J. W. and Overbeek, J. T. G. (1999). *Theory of the Stability of Lyophobic Colloids*. Mineola, NY: Dover Publications.

Vivarès, D., Belloni, L., Tardieu, A. and Bonneté, F. (2002). 'Catching the PEG-induced attractive interaction between proteins', *Eur. Phys. J. E* 9: 15–25.

Vivarès, D. and Bonneté, F. (2004). 'Liquid–liquid phase separations in urate oxidase/PEG mixtures: characterization and implications for protein crystallization', *J. Phys. Chem. B* 108: 6498–507.

Wenzel, N., Pagan, D. L. and Gunton, J. D. (2007). 'Phase diagram for a model of urate oxidase', *J. Chem. Phys.* 127: 165101–5.

Zernike, F. and Prins, J. A. (1927). 'Die Beugung von Röntgenstrahlen in Flüssigkeiten als Effekt der Molekülanordnung', *Z. Physik* 41: 184–94.

Zhang, F., Roosen-Runge, F., Skoda, M. W. A. *et al.* (2012). 'Hydration and interactions in protein solutions containing concentrated electrolytes studied by small-angle scattering', *Phys. Chem. Chem. Phys.* 14: 2483–93.

Zhang, F., Skoda, M. W. A., Jacobs, R. M. J., Martin, R. A., Martin, C. M. and Schreiber, F. (2007). 'Protein interactions studied by SAXS: Effect of ionic strength and protein concentration for BSA in aqueous solutions', *J. Phys. Chem. B* 111: 251–9.

Zimm, B. and Stockmayer, W. H. (1949). 'The dimensions of chain molecules containing branches and rings', *J. Chem. Phys.* 17: 1301–14.

9 SAS in multidisciplinary studies

9.1 Automation and
    high-throughput SAS   286

9.2 Joint use with high-resolution
    methods   291

9.3 SAS and low-resolution
    crystallography   298

9.4 Complementary biophysical
    methods   303

9.5 Bioinformatics and model
    validation   309

References   314

In this chapter we shall consider some specific aspects and recent trends in the use of solution SAS to characterise biological macromolecules. Among them, an important development is automation of SAS measurements and data analysis procedures, paving the way for large-scale studies of macromolecular solutions (Section 9.1). Most modern applications of SAS make use of multipronged approaches, with SAS being utilised together with structural and biophysical methods, and typical strategies pursued in these multidisciplinary studies will be presented (Sections 9.2–9.4). Finally, the use of SAS with bioinformatics tools and the problem of assessment and validation of three-dimensional models generated from the SAS data are considered in Section 9.5.

9.1 Automation and high-throughput SAS

In any physical or biological experiment, automation of data collection and data analysis, if possible, are substantial factors facilitating the measurements, reducing human errors and improving reliability of the results. Structural studies in biology have already entered the stage of a systemic approach where SAS is one of the major players in elucidating overall structure and kinetics of macromolecular assemblies at different levels of structural organisation, different time scales and under various environmental conditions. Automation is especially important in large-scale studies, allowing numerous samples to be measured and analysed, but the automated facilities also pave the way for unattended and remote operation.

9.1.1 Requirements for high-throughput studies

High-throughput studies are possible only if the measurements themselves are sufficiently fast and if they require small amounts of sample. In this sense, automation is most crucial for SR studies, especially on third-generation sources, where measurements are made in seconds on a few μL sample volumes. For laboratory X-ray sources and also for neutron scattering, where the measuring times are minutes to hours, automation is still a very important factor making the experiments convenient. For SANS, calibrated cuvettes can be employed, and automatic sample changers are used to position prefilled cuvettes in the beam (see Chapter 2). SANS requires much larger volumes than SAXS, making high-throughput studies more problematic, but has an advantage of non-destructiveness; that is, the same samples may in principle be reused for measurements at different conditions (such as after dialysis). For X-rays it is virtually impossible to fabricate capillaries that will have identical background scattering, and one and the same sample holder must be used and cleaned between experiments.

A SAXS measurement protocol for biological samples in solution requires repeated exposures of the specimen to the X-ray radiation, usually at different concentrations. For each data set:

- the specimen is brought to the measurement position in the sample holder;
- the beam path is cleared;
- a (usually two-dimensional) scattering data set is collected;
- the beam parameters are monitored for later normalisation;
- the beam path is blocked;
- the specimen is removed from the sample holder and the latter is cleaned out.

This sequence is repeated for all solutes and relevant solvents to be measured.

Further, each two-dimensional scattering image collected by the detector usually needs to be radially averaged to obtain a one-dimensional data set. The averaged data must be appropriately normalised, and the scattering by the pure solvent must be subtracted from the solute scattering. From the resultant difference patterns, overall parameters and, if possible, *ab initio* shapes should be calculated.

If there are only a few samples to be measured, and if the measurement cycle is sufficiently long, the entire procedure may not need automation. However, high-throughput X-ray studies on synchrotrons can be done on hundreds of samples with the full measuring cycle (loading, exposure and cleaning) within a few minutes per sample. Manual operation would make this procedure prone to errors, for example, introduced due to fatigue after long hours of continuous measurement. Moreover, manual analysis of multiple data sets may become another source of mistakes, which can be detrimental to the overall result. This calls strongly for automation of the entire measurement and data analysis protocol. Moreover, the results should be summarised in a convenient machine- and human-readable form.

9.1.2 Large-scale analysis at SR facilities

In synchrotron MX a crucial step forward was made in the large-scale analysis of proteins and macromolecular complexes when automatic sample changers and remote operation were introduced (McPhillips *et al.* 2002). This revolution took place in synchrotron SAXS several years later, and in Chapter 2 we mentioned the synchrotron SAXS stations allowing for automated, high throughput and remote operation. A necessary prerequisite for any automated and remote operation is a dedicated robotic sample changer allowing for numerous samples to be thermostatically stored and loaded in a programmed way. Such sample changers are now available at several beamlines (Round *et al.* 2008, David and Perez 2009, Classen *et al.* 2010, Pernot *et al.* 2010, Blanchet *et al.* 2012), and they have drastically changed the way the synchrotron SAXS measurements are performed. The liquid handling robots provide temperature-controlled storage for hundreds of samples which can be measured in attended, unattended or remote operation.

The sample changers provide the necessary basis for the automation of SAXS measurements, but their use must be complemented by appropriate software to allow for a fully automatic operation. Important here are not only integration and synchronised operation of the relevant hardware elements, and online data processing to automatically compute the reduced, normalised and subtracted scattering patterns. The amount of information produced by large-scale studies is so large that the data analysis steps must also be performed, whenever possible, without user intervention. The quality assessment of the data, the calculated parameters and the constructed models must be summarised further in a convenient form.

Automation of the measurements, data processing and analysis allowing for high-throughput analysis is actively pursued at different facilities. A system utilising a robotic sample changer combined with a pipeline for online processing was running at the X33 beamline of the EMBL (Hamburg, DORIS-3 storage ring) for several years (Petoukhov *et al.* 2007, Roessle *et al.* 2007, Round *et al.* 2008). This pipeline has also been implemented at dedicated solution scattering beamlines ID14-3 at the ESRF (Grenoble) and P12 at PETRA-3 (DESY, Hamburg), utilising a newer version of the sample changer (Pernot *et al.* 2010, Blanchet *et al.* 2012). At the SIBYLS beamline (LBL, Berkeley), a Blu-Ice/DCS control system, initially developed for crystallography beamlines (McPhillips *et al.* 2002), has been adapted to automate SAXS experiments (Classen *et al.* 2010), and the data analysis procedures interlinked by scripts to form a processing pipeline. The system at SIBYLS provides rapid automated sample handling, temperature and anaerobic control, and couples the automated and semi-automated data analysis with automated archiving (Hura *et al.* 2009).

9.1.3 Integration of high-throughput hardware and software

The work on automation of the measurements and data analysis is done at individual facilities, and, given the fact that many procedures are hardware-dependent, the automated pipelines are usually tailored to the given facility. A concept for the automation of solution scattering experiments that may potentially be applicable at multiple facilities has been proposed (Franke *et al.* 2012). The overall scheme of the developed acquisition and analysis software is displayed in Fig. 9.1. Here, a service called Beamline Meta Server (BMS) is introduced to mediate the communications with the individual components of the beamline (valves, motors, digital I/O signals, monitors, detector, sample changer and so on). All these services and the BMS itself are separated into a control system-specific front end and a service-specific back end. Obviously, integrated communication protocols between devices must depend on the control system of the given beamline. The BMS implemented at the high-brilliance P12 beamline of the EMBL at the PETRA-3 storage ring (DESY, Hamburg) utilises a three-fold integrated networking environment (TINE) (Bartkiewicz and Duval 2007), an accelerator control

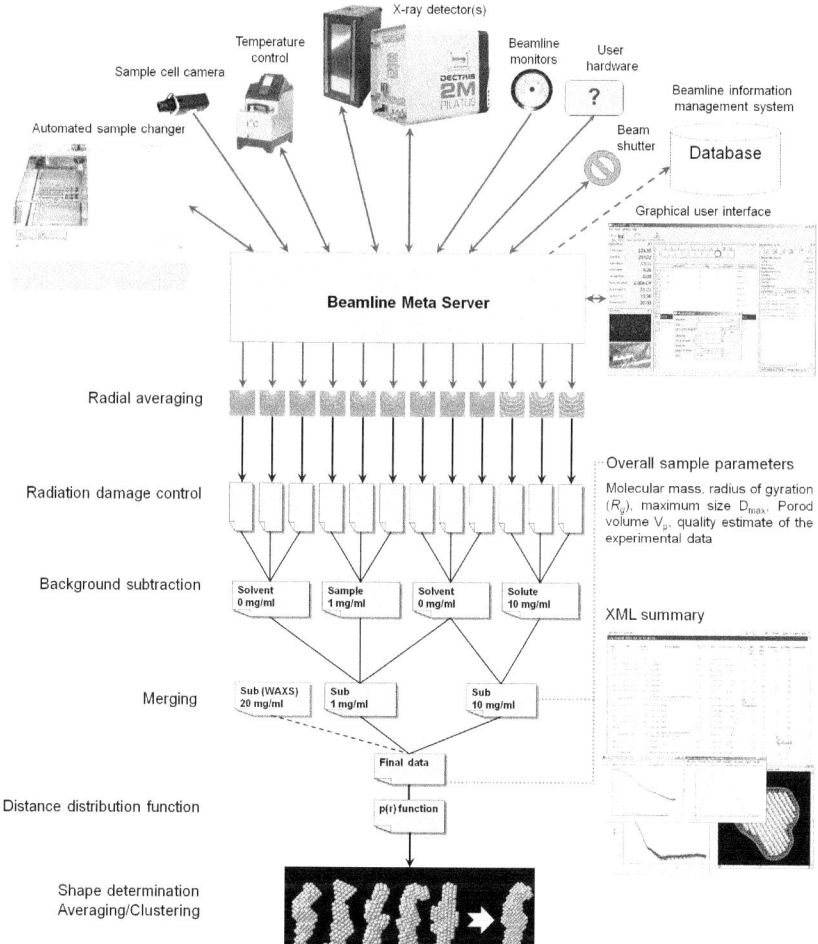

Fig. 9.1 Overall scheme of the synchrotron SAXS data acquisition, data processing and storage, following the concept of Franke *et al.* (2012).

system of DESY. Importantly, however, the front end–back end separation decouples the BMS development from the control system, making the software easily portable to other control systems by providing respective front ends for the local servers.

The control system-independent back ends coordinate the subsequent execution of user commands for all other beamline servers. The BMS is written in the Python scripting language whereby each controlled device and executable commands are encapsulated into individual classes each with a defined interface, allowing one to quickly incorporate new devices. In particular, any user device that can be operated via a Python (extension-)module can be controlled by the BMS during an experiment.

The BMS takes the beamline environment provided by appropriate monitor readings into account to assess whether a given command can safely be executed (for example, no exposure will be performed when the storage ring does not provide X-rays). The commands are not limited to driving hardware but may also run software when certain conditions are fulfilled. Command scripts are sent to the BMS and interpreted, and the command objects are queued there prior to execution. The queue can be executed not only in sequence but, taking the interdependencies between commands into account, also in parallel (for example, radial average of a collected two-dimensional image is carried out simultaneously with the cell cleaning and loading for the next measurement).

The operation of all devices, including the robotic sample changer, detector(s), temperature controls and so on, is controlled over a Graphical User Interface (GUI) to interactively select and submit the sequence of samples for the measurements. Alternatively, the sequence can be prepared by the users in advance, using an offline version of the interface. Upon submission of the queue to the BMS, all further data collection, processing and analysis are fully automated. The collected data are reduced, normalised and submitted to a processing pipeline, which performs the analysis without user intervention. The operations (depicted in the lower part of Fig. 9.1) include buffer subtraction, extrapolation to infinite dilution where appropriate, calculation of the overall parameters and pair distribution functions and *ab initio* shape reconstruction with analysis of multiple solutions. The modular automated procedures (Petoukhov *et al.* 2012) encapsulated in separated data processing tools are employed by the pipeline, which is implemented as a separate application that supervises serial or parallel execution of these modules. Each data processing step is represented by an individual component that employs one or more modularised tools to perform operations. These components are decision-making blocks that do not perform any actual data processing but instead communicate with each other by passing messages (such as when a new file becomes available for processing, or when a tool finishes). This way enables one to modify the behaviour of the pipeline to meet different requirements, for example, by including or excluding certain steps, and also to easily introduce new modules if needed. All modularised tools use an open-source library to read/write the experimental data files, allowing one to easily incorporate data at different facilities.

It is important for automated high-throughput studies that the information obtained and the history of the data analysis (including the experimental data and the computed parameters and models) be easily retrievable in human- and machine-readable forms. For a convenient hierarchical data storage and report generation, an Extensible Markup Language (XML)-based file format was employed. The XML-summary file is written and updated in the course of data collection, and remains fully consistent and readable during the entire measurement. The format of the file allows for data processing using the standard Extensible Stylesheet Language (XSL), transforming the file on the fly into a human-friendly HTML representation or into a comma-separated value table. Attempts are under way to implement sample tracking and data storage compatible with the ISPyB information management system for synchrotron MX (Delageniere *et al.* 2011). This would allow one to provide yet more efficient service and storage to the users of the large-scale facilities willing to combine scattering and crystallographic experiments.

Overall, the modular and flexible SAXS data acquisition and analysis system of Franke *et al.* (2012) can serve as a prototype of the future systems for large-scale studies, allowing for automated and remote-controlled experiments. In fact, the existence of such systems is a necessary prerequisite for the efficiency of high-throughput analysis by SAS, and one may expect that they will soon become standard components at large-scale facilities. Similar developments towards automation are also observed for laboratory X-ray sources, where hardware manufacturers such as Anton Paar, Bruker and Rigaku also seem to follow the trend towards complete automation.

9.2 Joint use with high-resolution methods

High-resolution structural methods provide extremely important complementary information for SAXS and SANS studies, as illustrated in many applications in Chapter 6. Hybrid modelling, where the available high-resolution portions are positioned based on the low-resolution solution scattering data, has become one of the most effective tools for structural biology. In this section we shall consider different aspects of the joint use of SAS with the two major high-resolution methods: MX and NMR spectroscopy.

9.2.1 Complementarity of SAS with MX and NMR

Solution scattering is highly complementary to the two high-resolution techniques, as summarised in Table 9.1. Both SAS and NMR are applied to solutions, whereas MX obviously requires well-diffracting crystals. SAS can cover a broader range of macromolecular sizes than MX and especially than NMR, which has limited applicability to large objects.

Table 9.1 Complementarity of SAS, MX and NMR.

Method/Feature	SAS	MX	NMR
Applicability	Rigid and flexible macromolecules, complexes, mixtures. Range: 10^2–10^7 Da	Rigid macromolecules, complexes. Range: 10^1–10^6 Da	Rigid and flexible macromolecules, complexes, mixtures. Range: 10^1–10^5 Da
Samples	Solutions, mM range Homogeneity required	Crystals	Solutions, mM range, isotope ^{15}N/^{13}C labelling required
Experiment/analysis	Fast (seconds to hours)	Medium (hours to days)	Slow (days to weeks)
Information provided	Overall structure	High-resolution structure	High-resolution local and overall global structure
Modelling of multidomain/subunit complexes	Most sensitive to movements	Not required/not applicable	Most sensitive to rotations (RDC) and interfaces (chemical shifts, spin labels)
Dynamic ensembles	Time averaged distributions	Single conformation captured	Time scales from spin relaxation data

For multisubunit complexes, which may be difficult to crystallise, the crystal or NMR structures of individual subunits often provide building blocks for rigid-body refinement.

There are many straightforward joint applications of SAS with MX and NMR. Solution scattering is often employed as a validation tool for the crystallographic models to verify whether the crystal structure is preserved in solution. For numerous multidomain or multisubunit structures, significant differences between crystal and solution have been reported, apparently due to the crystal packing forces, and in many cases the observed structural changes can be analysed meaningfully by rigid-body refinement (for example, Svergun *et al.* 1997, Svergun *et al.* 2000, Vestergaard *et al.* 2005, Andersen *et al.* 2006). NMR models are obtained in solution, and validation of such models by SAS is rather rare but not impossible. An example is given by a study of the Josephin domain of ataxin-3, a 182-residue protein involved in the ubiquitin/proteasome pathway. Two high-resolution models were proposed using NMR and deposited in the PDB, entries 1YZB (Nicastro *et al.* 2005) and 2AGA (Mao *et al.* 2005). The two models displayed distinctly different overall shapes, with an open cleft in the former model and a closed cleft in the latter. The SAXS data were used together with Bayesian methods to validate the models and to demonstrate that only one of them, with the open cleft, is compatible with the experimental information (Nicastro *et al.* 2006).

9.2.2 Joint applications of SAS with MX

Given the crystal structure one may rapidly identify the biologically active oligomers in solution by considering possible assemblies in the crystal. Analysis of the crystallographic contacts—using, for example, the PISA (Protein Interfaces, Surfaces and Assemblies) server (Krissinel and Henrick 2007)—may provide the assemblies, which can be sufficiently stable to be preserved in solution, and these templates can be screened against the experimental SAS data. An example is provided by a study of the C-terminus of the sarcomeric M-band protein myomesin (Pinotsis *et al.* 2008). The crystal structure revealed a tandem array of two immunoglobulin (Ig) domains My12 and My13, connected by a six-turn α-helix (Fig. 9.2). The computed scattering from the monomeric protein did not fit the experimental SAXS pattern (discrepancy χ = 3.3), and the overall parameters indicated that the protein was dimeric, in agreement with the gel filtration results. Several possible crystallographic dimers were then screened against the SAXS data (Fig. 9.2), and the best (and very good) fit with discrepancy χ = 1.1 was provided by an extended dimer mediated by My13-My13 interaction (6.68 nm^2 interface area). All other possible crystallographic dimers yielded significantly worse fits, and in particular, a compact dimer with the largest interface area (8.15 nm^2) was incompatible with the experimental data (χ = 2.8). The SAXS results therefore strongly suggested that the extended dimer in Fig. 9.2 is present in solution. This was further verified by point mutations of the residues at the dimerisation interface, impairing significantly the ability of the construct to form dimers. Further crystallographic SAXS and EM studies of a longer-construct My9-My13 fully confirmed the observed dimerisation interface and revealed a 36 nm long superhelical coil arrangement of the domains (Pinotsis *et al.* 2012). A reversible unfolding of the helical linkers observed by AFM explained the elastic properties of myomesin and its role as a ribbon, maintaining the overall structural organisation of the sarcomeric M-band in muscle.

SAXS provides perhaps the most conclusive and efficient way of validating crystallographic oligomers in solution, as illustrated by an increasing number of publications. Thus, SAXS was employed to confirm the crystal structure of RNA polymerase I-specific transcription initiation factor Rrn3, an enzyme required for RNA polymerase I transcription initiation in humans (Blattner *et al.* 2011). The 0.28-nm resolution structure of Rrn3 reveals a unique HEAT repeat fold, the asymmetric unit of the crystal containing a homodimer. The *ab initio* SAXS model yielded a shape fully consistent with the crystallographic dimer, and the scattering computed from the dimer agreed well with the experimental scattering data. Further evidence confirming the quaternary structure of Rrn3 was obtained from point mutations of the residues on the dimeric interface. Other examples are provided by validation of the extended dimeric crystal structure of a ubiquitin E3 ligase, RNF168 (Mattiroli *et al.* 2012), and of an unusually irregular protein shell of the catalytic core of an archaeal 2-oxoacid dehydrogenase multienzyme, combining cubic and dodecahedral geometrical elements (Marrott *et al.* 2012).

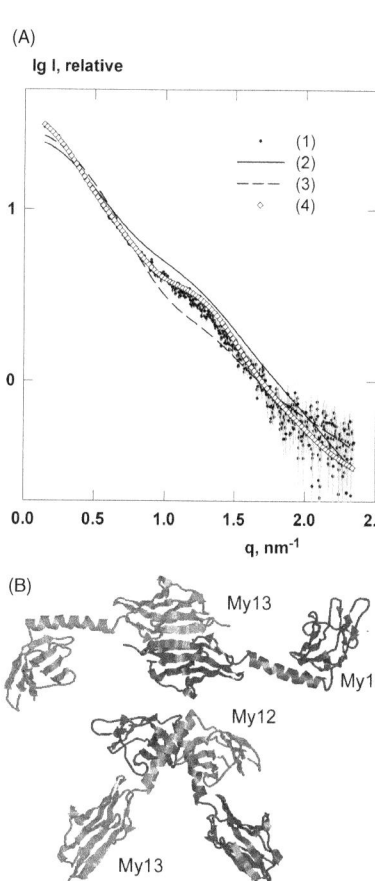

Fig. 9.2 Determination of the structural organisation of myomesin C-terminus in solution (Pinotsis *et al.* 2008). (A) experimental SAXS data from the myomesin construct in solution (1) and the computed scattering patterns from the monomer (2), and from the compact (3) and extended (4) crystallographic dimers. (B) the two crystallographic dimers (top, extended; bottom, compact).

The myomesin example in Fig. 9.2 represents a rather fortunate case, of course, where one of the crystallographic assemblies yielded an excellent fit to the experimental data, whereas all other possibilities gave poor fits. In practical applications the situation is more complicated, and very often neither of the possible oligomers gives a good fit, making the subsequent rigid body refinement necessary. Still, in many cases the scattering patterns computed from the crystallographic assemblies, and also comparison of the structure of these assemblies with the *ab initio* models, may give ideas about the possible organisation of the oligomers in solution.

Another specific area of the joint application of SAS and MX is related to the potential use of SAS-generated models for phasing of low-resolution reflections in protein crystallography (see also Section 9.3.2). The first attempts were carried out using envelopes reconstructed by the program SASHA (Hao *et al.* 1999, Hao 2001). The more detailed DR models provide even better search models for molecular replacement and phasing. The potential of such models was demonstrated in the study of a 105 kDa glycoprotein β-mannosidase from *Trichoderma reesei*, where the model obtained by GASBOR was further enhanced using available low-resolution crystallographic data phased by molecular replacement (Aparicio *et al.* 2002). The major difficulty with using low-resolution models for phasing lies in the fact that the quality of phases provided by these models is usually insufficient to use standard crystallographic programs to identify the correct position of the model within the unit cell, so that an exhaustive search must be used. A general program FSEARCH employs *ab initio* shape envelopes and bead or DR models from EM or SAXS for the determination of low-resolution phases (Hao 2006). Following the positioning of the molecular envelope or bead/dummy model in the unit cell, the low-resolution phases are extended to crystallographic resolution. This method has been used successfully for several proteins (Liu *et al.* 2003, Kollman and Quispe 2005) and continues to be developed. It was demonstrated that the GASBOR models constructed using both SAXS and WAXS data provide better chances for successful phase determination compared to those based on SAXS data only (Hong and Hao 2009).

SAS can also provide useful hints to identify suitable constructs for crystallisation, which is well illustrated by the study of the catalytic α-subunit of *E. coli* replicative DNA polymerase III (Lamers *et al.* 2006). The full-length construct containing 1,160 residues was resisting attempts at crystallisation, and the *ab initio* shape reconstructed from SAXS data by GASBOR displayed a globular core with a protuberance containing about 150 residues. This result suggested that the flexibility of a terminal part of the structure may cause the failure of the crystallisation attempts. Indeed, by cutting the C-terminal part of the protein, the construct 1-917 was successfully crystallised. The crystal structure was determined at 0.23 nm resolution, revealing a unique fold with localised similarity in the catalytic domain to DNA polymerase beta and related nucleotidyltransferases.

9.2.3 Joint applications of SAS with NMR

In rigid-body modelling against SAXS data, both MX and NMR models of individual subunits and domains are employed, and several applications have been already presented in Chapter 6. Here we stress that the fidelity of the hybrid models can be enhanced further when multiple SAS data sets are fitted simultaneously. These can be, in neutron scattering, contrast variation data or, for X-rays, scattering patterns from sub-complexes or deletion mutants. Fitting the multiple data also allows one to more reliably generate the missing portions of the structure not available as high-resolution models. One of the first comprehensive cases of modelling of a multidomain protein against multiple data sets was reported in a study of a polypyrimidine tract binding protein (PTB) (Petoukhov *et al.* 2006). This protein comprises four domains (each about 10 kDa, structures solved by NMR) interconnected by linkers. Seven SAXS curves from the full-length protein and from all possible sequential combinations of domains were fitted simultaneously, revealing the distribution of domains and tentative conformations of the linkers. The elongated shape of PTB obtained was compatible with its possible role in bridging RNA sequence motifs.

The complementarity of SAS with NMR extends well beyond providing models of domains for the rigid-body analysis. Thus, the information on the NMR chemical shifts is extremely useful in defining the interparticle interfaces and restraining the rigid-body modelling. Moreover, for multisubunit complexes or multidomain proteins, SAS provides the information on global distances and sizes, but may be somewhat less sensitive to rotations, which provide less change to the overall shape. For NMR, orientation constraints derived from the RDCs measurements in a dilute liquid crystalline media yield long-range information on the relative orientation of distant parts of the structure (for example, individual domains). The RDC data do not provide translational information, and by themselves they cannot be used to generate an unambiguous solution, though they can be very usefully combined with SAS information, as first proposed in Mattinen *et al.* (2002). Potentials for the refinement of NMR structures against SAXS data have been introduced into several structure-calculation packages; for example Xplor-NIH (Schwieters *et al.* 2003) and CNS (Brunger 2007, Gabel *et al.* 2008). The SAS data have been shown to improve significantly the accuracy of calculated structures (Grishaev *et al.* 2005, Grishaev *et al.* 2008). The extended protocols have also been shown to be useful for the determination of DNA and RNA structures by NMR (Schwieters and Clore 2007, Grishaev *et al.* 2008).

Both SAS and NMR are applicable to flexible systems, and the SAS analysis of multiple data sets from potentially flexible macromolecules is usefully complemented by NMR information. A study of organisation and flexibility of the N-terminal half of human cardiac myosin binding protein C (cMyBP-C) provides a good example (Jeffries *et al.*

2011). The construct contains five immunoglobulin domains (termed C0 to C4), and SAXS data—a total of seven scattering patterns—were collected from the full-length C0–C4 construct and from deletion mutants containing different combinations of domains (Fig. 9.3A). The modular arrangement of domains C1–C2–C3–C4 was found to be relatively fixed, while NMR relaxation studies indicated that a proline/alanine-rich linker (P/A) connecting the cardiac-specific N-terminal C0 domain to the C1 domain provides significant flexibility at the N-terminus. The structural model of cMyBP-C was constructed using the crystallographic structures of the domains (Fig. 9.3B), displaying an extended yet distinctively 'bent' modular arrangement similar to that observed for the six-Ig fragment I65-I70 from the elastic I-band fraction of the giant muscle protein titin (Von Castelmur *et al.* 2008). The flexibility of the N-terminal domain was analysed further in terms of

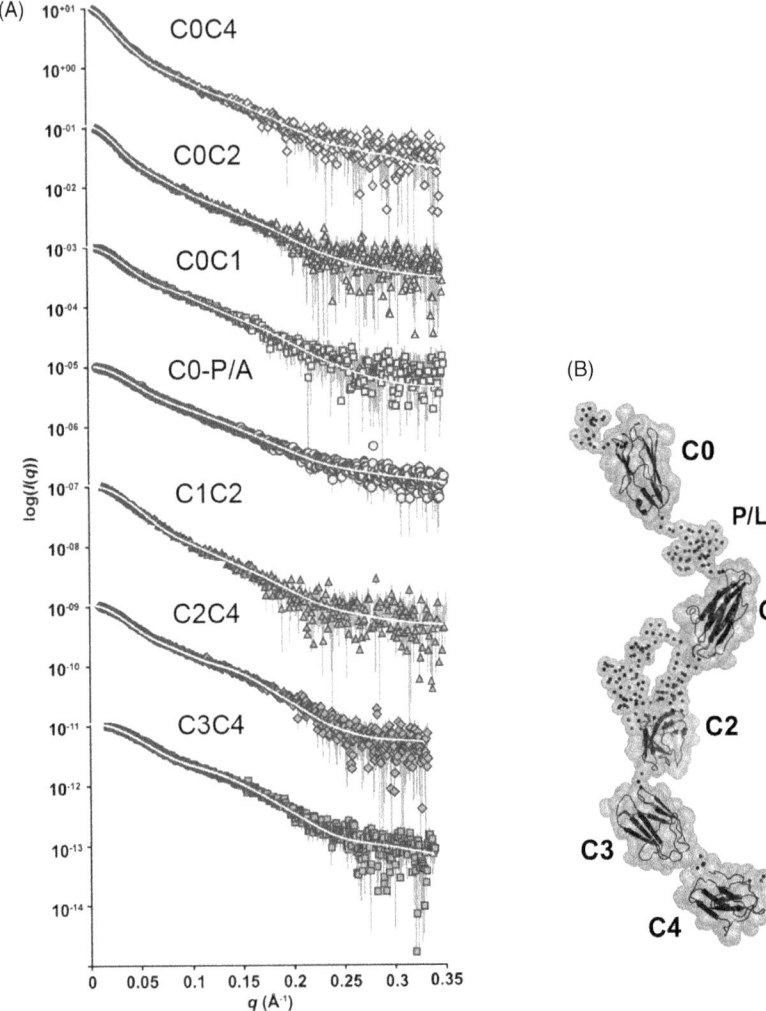

Fig. 9.3 Structural modelling of human cardiac myosin binding protein C. (A) experimental SAXS data from the cMyBP-C constructs (symbols) and the fits from the model (solid lines). The data are displaced along the logarithmic axis for clarity. (B) A representative rigid-body model of cMyBP-C. The crystal structures of the individual domains are displayed as ribbons, and the regions of unknown structure are shown as beads representing dummy residues. Adapted from Jeffries *et al.* (2011), with permission from Elsevier, © 2011.

ensemble distributions, and the results are related to the putative role of cMyBP-C as a molecular switch between actin and myosin in the muscle sarcomere.

A multipronged approach was employed successfully in the study of the tumor suppressor p53—a protein that plays a crucial role in processes such as apoptosis control and DNA repair. P53 is a homotetramer with two folded domains tethered and flanked by unstructured regions representing 37% of the whole sequence. In the modelling of p53 against the SAXS data, high-resolution structures of the folded domains from crystallography and NMR were used as rigid bodies, and the unfolded domains were represented by chains of dummy residues. The free p53 is a rather open cross-like tetrameric assembly, which collapses in the presence of DNA to tightly embrace the latter (Tidow *et al.* 2007). Interestingly, the SAXS model of the p53-DNA complex without the disordered N-terminal transactivation domains is in excellent agreement with an independent EM map of the same complex (Tidow *et al.* 2007). In a subsequent study, more detailed information about the flexibility of the isolated transactivation domain and its structure within the full length p53 complexed coactivator was obtained by joint use of SAXS and NMR (Wells *et al.* 2008).

Importantly, for flexible systems both SAS and NMR are able to provide ensemble-based description of particles, instead of a single conformation model as employed for the rigid cases. This complementarity has been used in numerous studies to comprehensively describe macromolecular ensembles (for example, Bernado *et al.* 2005, Mylonas *et al.* 2008, Blackledge 2010). An interesting concept was introduced by Bertini *et al.* (2010) to describe the configurational space of macromolecules in flexible ensembles. A maximum occurrence (MO) was defined as the maximum percent of time that a flexible macromolecule can spend in any given conformation, and a method is developed to extensively sample the conformational space and to construct MO maps from experimental data. These can be, for example, SAXS and NMR data from pseudocontact shifts and self-orientation RDCs arising from the presence of paramagnetic lanthanide ions. The method has been tested in a case study of the flexible two-domain protein calmodulin (CaM) to show that the 'closed' and 'fully extended' conformations trapped in the crystalline forms of CaM have rather small MOs of only 5% and 15%, respectively.

The joint use of scattering and NMR may provide insights not only into the structure but also the dynamics of flexible systems. A good example is given by a SAXS study of ribosomal protein L12, a two-domain protein with a 20-residue flexible linker separating the N- and C-terminal domains (Bernado *et al.* 2010). L12 forms dimers mediated by the N-terminal domains, displaying a three-lobe topology with significant flexibility, known to be critical for efficient translation. An ensemble of L12 conformations in which the structure of each domain is fixed but the domain orientations are variable was generated to reproduce the SAXS data. Simultaneously, the optimised correlation times of the reorientational eigenmodes were required to fit the ^{15}N relaxation data from NMR. The two C-terminal domains sample a large volume and extend far away from the ribosome anchor, whereby the distances between each C-terminal domain and the anchor are anticorrelated. The observed properties may promote the function of L12 to recruit translation factors and control their activity on the ribosome.

Overall, the complementarity of solution scattering to the high-resolution methods and the usefulness of their synergistic use is now widely recognised, as reflected by a rapidly increasing number of combined studies. As typical SAS measurements and their analyses require on average much less time than the MX and NMR studies, the application of SAS is becoming a routine step complementing the high-resolution studies of proteins and complexes. A good illustration of a parallel use of SAXS with MX and NMR is given by a systematic study of twenty-eight proteins taken from the targets of the Northeast Structural Genomics consortium, USA (Grant *et al.* 2011).

9.3 SAS and low-resolution crystallography

The most widespread and arguably most powerful method for determining structures at high resolution is MX, which can in the best cases produce atomic resolution, and usually provides resolution adequate to fit an atomic model, assuming certain molecular geometrical constraints. In some instances, however, in particular with some large macromolecular assemblies, high-resolution information cannot be obtained for at least part of the structure, even though highly diffracting crystals exist. It is often one particular chemical component or part of a molecule that cannot be detected by X-rays, usually due to the fact that these components are at least partially disordered in the crystal. Some examples are the nucleic acid in spherical viruses, the DNA in nucleosome core particles, the lipid in lipoproteins and detergent in crystals of detergent-solubilised membrane proteins. For X-ray crystallographers a resolution below ~4 Å is usually considered low, because the electron density obtained is insufficiently clear to fit a detailed atomic model. Here we consider mainly studies at resolutions more comparable with SAS; that is, <10 Å resolution.

Below atomic resolution the crystallographic and SAS intensities are both dominated by the contrast dependent terms in eq. (1.6). The main difference is, of course, that in the crystal the molecules are aligned, whereas in solution they are randomly oriented in solution. Hence, remembering that protein crystals invariably contain 20–80% water (on average ~40%), the scattering can be understood as arising from the contrast between the macromolecule and its surrounding aqueous environment. Indeed, the concept of contrast variation, widely used particularly in SANS but also occasionally in SAXS, was first proposed by Bragg and Perutz (1952) in their crystallographic studies of haemoglobin.

9.3.1 Low-resolution X-ray crystallography

In the early days of X-ray protein crystallography it had already been noticed that changing ammonium sulphate for water in crystals of haemoglobin changed the intensities of the lowest-angle reflections whilst leaving the higher orders unchanged (Boyes-Watson *et al.* 1947). This observation was exploited by Bragg and Perutz (1952), who measured the low-angle reflections from crystals of haemoglobin containing pure water and four different salt concentrations to determine the shape of the molecular envelope. With the development of high-resolution phasing methods there was little interest in low-resolution crystallography until the advent of studies on very large molecular complexes such as viruses and ribosomes. Harrison and Jack (1975) used phases derived from SAXS (Harrison 1969), very-low-resolution crystallography and a 2.8 nm EM reconstruction (Jack *et al.* 1975) to determine the positions of $PtCl_6$ clusters in crystals of tomato bushy stunt virus. A novel use of contrast variation in X-ray crystallography is the use of xenon as a contrast agent. Xenon (and krypton) are both more soluble in non-polar than polar solvents. Therefore, in crystals of detergent-solubilised membrane proteins they are preferentially absorbed in the detergent phase compared with the aqueous phase, thus providing a means of distinguishing the two phases. This has been demonstrated in crystals of OmpF porin (Sauer *et al.* 2002), where it could be shown that the detergent regions identified were consistent with those observed by earlier low-resolution neutron crystallography (Pebay-Peyroula *et al.* 1995).

Low-resolution X-ray crystallography and the instrumentation required for such measurements have been reviewed (Evans *et al.* 2000).

9.3.2 Low-resolution neutron crystallography

As with SAS, contrast variation is more straightforward with neutrons than with X-rays, due to the possibility of using H_2O/D_2O mixtures as contrast agents. High-resolution neutron crystallography requires crystals of $\sim 1\,mm^3$ or more, though instrumental and source improvements since around 2005 are lowering this limit. For low-resolution diffraction, due to the fact that the low-resolution data (as in SAS) is generally orders of magnitude stronger than that at high resolution, it is possible to use crystals similar to those used for high-resolution X-ray crystallography; that is, $<1\,mm^3$. In addition to the strength of the scattering, the intensity of Bragg reflections varies as the square of the wavelength, and hence long-wavelength (for example, 1 nm) neutrons strongly enhance the signal. The outstanding difficulty, as with X-rays, is the determination of the phases of the reflections—the classical 'phase problem' (Blundell and Johnson 1976)—but contrast variation makes this tractable.

It is usual for a contrast variation series to collect data from crystals containing at least four different H_2O/D_2O ratios. These are prepared by soaking the crystals for a period of weeks in the appropriate mixture to ensure the exchange of all labile protons. A small number of labile protons in the interior of the macromolecule will probably exchange on only a very long time-scale of months to years. Data are then collected by step scanning the crystal about the φ (or ω) axis of the diffractometer until a unique data set is obtained.

The crystallographic phase problem. In crystallography a molecular image can be obtained directly from the diffraction amplitudes as long as the phases are known via the well-known Fourier transform:

$$\rho(x, y, z) = \frac{1}{V} \sum_h \sum_k \sum_l \mathbf{F}(hkl) \exp[-2\pi i(hx + ky + lz)] \quad (9.1)$$

where $\mathbf{F}(hkl)$ is a complex term which can be written as

$$\mathbf{F}(hkl) = F(hkl) \exp 2\pi i \alpha_{hkl} \quad (9.2)$$

$F(hkl)$ is the structure amplitude associated with the measured intensity $I(hkl)$, and α_{hkl} the phase which is *a priori* unknown but which can be obtained experimentally by a number of methods: multiple isomorphous replacement, anomalous scattering and many variants upon these themes. For more detail see classic protein crystallography text-books such as Blundell and Johnson (1976).

If diffraction data are measured from a series of crystals where the only difference between them is that the contrast between macromolecule and solvent has been changed by modifying the H/D content of the water, then the variation of the crystallographic structure factor as a function of contrast can be expressed as:

$$\mathbf{F}(h, X) = \mathbf{F}(h, 0) + X\mathbf{F}(h, HD) \quad (9.3)$$

where h is the reciprocal lattice point h,k,l, X is the mole fraction of $[D_2O]/[D_2O]+[H_2O]$ in the crystal and $\mathbf{F}(h,HD)$ is the vector difference between the structure factor in H_2O and that in D_2O.

Multiplying by the complex conjugate we obtain the diffracted intensity:

$$I(h, 0) = F(h, 0)^2 + 2X \cos \varphi F(h, 0)F(h, HD) + F(h, HD)^2 \quad (9.4)$$

where φ is the phase angle between $\mathbf{F}(h,0)$ and $\mathbf{F}(h,HD)$.

This relationship has several important consequences for low-resolution crystallography, including the possibility of scaling together data from different contrasts and interpolating missing data (Roth *et al.* 1984). In terms of structure solution it is of fundamental importance, as it means that the phase difference, φ, between any two contrasts, of a reflection (h), may be determined except for the sign ($\pm$), if the amplitudes at three contrasts are known. Therefore, if the structure is known at any one contrast then the phases at that contrast may be calculated and then determined at any other contrast except for knowledge of the sign. In the particular case of centrosymmetric reflections where $\varphi = 0$

or π there is, of course, no ambiguity, and the phase may be calculated at any contrast (Roth 1987).

In most studies carried out to date, the structure of one component of the macromolecular complex has been determined by X-rays or could be modelled from other information. Hence structure factors calculated from the known part of the structure at a contrast where the other component is visible provide starting phases for the determination of the structure at any contrast. This is illustrated in Fig. 9.4, which demonstrates the vector relationships between structure factors at four different contrasts. The two triangles bounded by $\mathbf{F}(0)$, $\mathbf{F}(HD)$ and $\mathbf{F}(100)$ are the two possible relationships which can be constructed through knowledge of the structure factor amplitudes alone following eq. (9.1), and corresponding to the two possible signs of φ. This particular example illustrates the case of, for example, a protein/RNA complex where data would be measured at 40% D_2O where the protein is invisible, 70% D_2O where the RNA is invisible, and two other contrasts, 0% and 100% D_2O. In this case we imagine that the protein structure is known and that the RNA structure is to be determined. We may therefore calculate the phase of the structure factor in 70% D_2O, and thus determine the orientation of the phase triangle with just the ambiguity of sign corresponding to the two triangles shown. This is very closely analogous to the situation of single isomorphous replacement (Blundell and Johnson 1976) in X-ray protein crystallography. Once this (ambiguous) phase has been determined then the ambiguity may be resolved, and an approach to the true phase may be made using a number of constraints such as the invariability of the known part of the structure, solvent flattening (the assumed lack of structure in the solvent regions) or non-crystallographic symmetry averaging (Roth 1991, Roth *et al.* 1991).

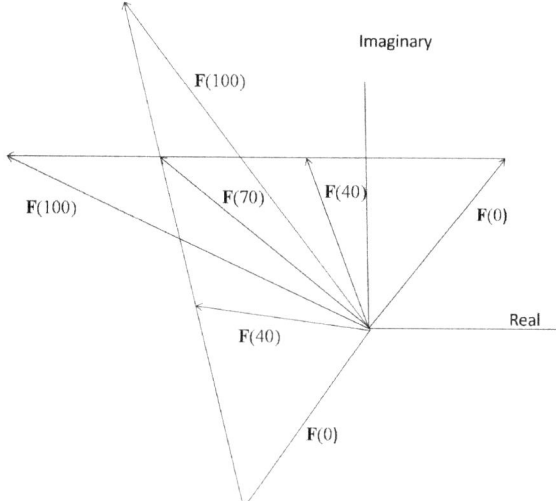

Fig. 9.4 The relationship between structure factors as a function of contrast for a hypothetical protein/RNA complex. This demonstrates that if diffraction data are measured at 0, 38, 70 and 100% $[D_2O]/[H_2O+D_2O]$ only one vector triangle, and its mirror image, can be drawn to relate the different structure factors. If, for example, the protein structure were known but the RNA structure were unknown, then the amplitude and phase at 70% D_2O mole fraction (where the RNA is invisible) could be calculated and the phase triangles uniquely oriented except for an ambiguity of sign.

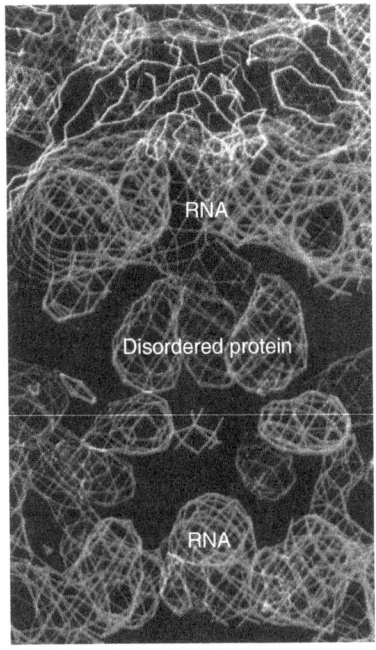

Fig. 9.5 Composite neutron scattering density map of TBSV at ~16 Å resolution from crystals soaked in 70% D_2O and 40% D_2O. The known structure of the protein capsid is represented by the polypeptide chain traced in the upper part of the figure. The density seen in the lower part of the figure represents that part of the protein which is disordered and the RNA which is all disordered, and therefore not seen in high-resolution crystallographic studies. Adapted from Timmins *et al.* (1994b), with permission from Elsevier, © 1994.

Applications Tomato Bushy Stunt Virus (TBSV). This is an icosahedral plant virus whose structure was determined by X-ray crystallography at 2.8 Å resolution (Olson *et al.* 1983). Some 25% of the protein and the nucleic acid could, however, not be located in the high-resolution structure. The radial distribution of the RNA and protein were first determined by SANS (Chauvin *et al.* 1978), but this did not provide any information on the azimuthal distribution of these components. Neutron crystallographic maps from crystals soaked in 38% and 70% D_2O helped localising the RNA and missing protein, and enabled details of the protein–nucleic acid interactions to be elucidated (Timmins *et al.* 1992). Initial phases were calculated from the known X-ray structure and applied to the neutron-structure amplitudes in 70% D_2O, and extended to other contrasts through the contrast-variation relationship. These phases were then refined using the known non-crystallographic symmetry of the particle, combined with solvent flattening, to yield a series of maps at different contrasts. Figure 9.5 shows a portion of a composite neutron-scattering map composed of density seen at 70% D_2O (RNA invisible) and density seen in 38% D_2O (protein invisible). The outer shell of protein corresponds to that seen in the X-ray maps, but the inner density and particularly the three lobes of density closely associated at the three-fold symmetry axis were invisible in the X-ray maps, most probably because of disorder. The RNA is located in the space between the two protein layers.

Lipovitellin. Lipovitellin is a lipoprotein found in the oocytes of egg-laying animals, where it sometimes forms two-dimensional crystals *in vivo*. The structure was solved using standard X-ray crystallographic techniques (Raag *et al.* 1988), but the authors were unable to locate any of the bound lipid which was known to comprise some 15% of the molecular mass. Neutron crystallographic data were measured to 12 Å resolution from crystals soaked in five different H_2O/D_2O mixtures. From these data sets the structure factors were interpolated for the molecule in 10% D_2O, at which contrast the lipid scattering is minimised. Phases calculated from the known protein structure were applied to the measured amplitudes, and this constituted a starting point for determination of the phases at contrasts where the lipid should be visible. These starting phases were refined using the known protein structure—the known proportion of lipid in the complex and solvent flattening. The resultant neutron-scattering density map showed clearly that the lipid was present as a compact condensed phase located inside the protein (Timmins *et al.* 1992).

Membrane proteins. The first crystallographic structure of a membrane protein, a bacterial photoreaction centre, appeared in 1985 (Deisenhofer *et al.* 1985), and was followed soon after by other reaction centres and porins (Allen *et al.* 1986, Weiss *et al.* 1991, Cowan *et al.* 1992). Despite these advances, however, the number of structures solved is still rather small even twenty years later, due in large part to the difficulty in obtaining crystals of these proteins. The usual method of crystallisation is to solubilise the membrane protein in a mild, non-ionic detergent, and to crystallise the protein detergent complex by methods rather similar to the crystallisation of water-soluble proteins. The crystallisation process is very poorly understood, because of low contrast and disorder, and X-ray crystallography provides no information on the organisation of the solubilising detergent in the crystal. The contrast between detergent and D_2O, however, is very high (see Section 6.4.1), and neutron diffraction is thus an ideal tool for studying the role of detergent in membrane protein crystals as well as in solution. Moreover, insofar as the detergent may mimic the lipid of the cell membrane, these studies may also yield information on lipid/protein interactions in membranes.

The first membrane proteins to be studied by neutron diffraction were the photo-reaction centres and porins (Roth *et al.* 1989, 1991, Timmins *et al.* 1994a, Pebay-Peyroula *et al.* 1995). In the photoreaction centre the detergent appears as a continuous three-dimensional structure, with detergent rings surrounding the protein and connected by short cylindrical bridges. It was suggested that the detergent forms an ellipsoidal micelle around the protein rather than a bilayer structure (Roth *et al.* 1991). OmpF porin—a trimeric protein from the purple bacterium *Rhodobacter capsulatus*—can be crystallised in two different forms—one tetragonal and one trigonal. These lead to two different forms of packing in the crystal and to two different modes of aggregation of the detergent (Pebay-Peyroula *et al.* 1995, Penel *et al.* 1998). As illustrated in Fig. 9.6, in the tetragonal form the crystal appears to be formed through detergent–detergent contacts between two lattices formed of protein and detergent. In the trigonal form the detergent forms layers, with the crystal integrity being assured by both protein–protein and protein–detergent interactions within the plane, and protein–protein interactions between planes. This demonstrates that protein crystals can form in different ways, and that even detergent–detergent contacts, which must by their nature be less rigid, can be important for crystal integrity.

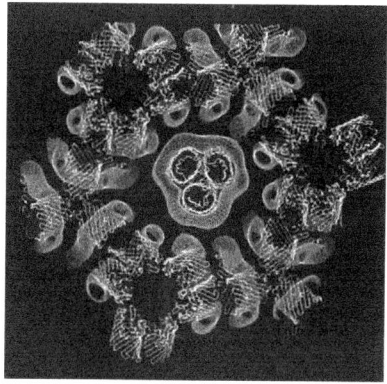

Fig. 9.6 Packing of OmpF porin and C8E4 detergent (see Table 6.1) in the tetragonal crystal form. The central porin trimer (shown as a polypeptide chain) is surrounded by a ring of detergent (hatched). Other trimers with the detergent ring seen side-on form an independent lattice which interacts with the central trimer only via detergent–detergent contacts. Reproduced from Pebay-Peyroula *et al.* (1995), with permission from Elsevier, © 1995.

9.4 Complementary biophysical methods

Many biophysical techniques are complementary to SAS, and a combined use of different techniques often leads to more reliable structural information. In this book we do not try to review systematically all biophysical techniques, but instead refer the reader to extensive classical and recent texts in the field, such as Cantor and Schimmel (1980) and Serdyuk *et al.* (2007). Here we concentrate on the techniques most frequently used in combination with SAS, and which produce the most complementary information.

9.4.1 Analytical Ultracentrifugation

AUC is a particularly useful technique to be used in conjunction with SAS for two principal reasons: (i) it allows the determination of mass, size and shape parameters similar to those determined by SAS, and (ii) it allows assessment of the aggregation state or polydispersity of samples prior to SAS experiments to verify their monodispersity. Here we present some of the basic concepts of AUC and demonstrate how the information obtained is complementary to that from SAS. More detailed information can be obtained from reviews (Lebowitz *et al.* 2002, Perkins *et al.* 2011) and books on the subject (Scott *et al.* 2005, Serdyuk *et al.* 2007).

The samples for AUC are contained in cells which are spun at high speed and the concentration of the macromolecule determined as a function of the distance from the meniscus. The concentration is measured essentially by absorption in the near UV, 258–260 nm for nucleic acids or 280 nm for proteins. Various kinds of optics are used to measure the sample concentration relative to the buffer. Interference optics can be used for samples having very low UV absorbance, and fluorescence detection allows concentrations in the ng/mL range to be detected. All AUC experiments are based on the principle that under normal gravity the movement of a macromolecule of mass up to $\sim 10^8$ cannot be measured, as the thermal energy is greater than the gravitational potential energy. The effective gravitational potential energy can be increased many-fold by placing the sample in a centrifuge and spinning it at high speed. There are two fundamentally different kinds of measurements possible in AUC: (i) sedimentation velocity (SV), which exploits a high centrifugal velocity to follow the time course of the sedimentation process, and (ii) sedimentation equilibrium (SE), by which sedimentation is carried out at low speed until a constant sedimentation profile is obtained where the diffusion balances the sedimentation. In all cases, the particle is then submitted to three forces:

$$\text{Centrifugal force: } F_{\text{sed}} = m\omega^2 r \qquad (9.5)$$

where m is the protein mass, ω is the rotor angular velocity and r is the distance from the centre of rotation.

$$\text{Buoyant force: } F_b = -m\bar{v}\rho\omega^2 r \qquad (9.6)$$

where $\bar{v}$ is the effective protein specific volume and ρ is the solvent density.

$$\text{Frictional force: } F_f = s(kT/D)\omega^2 \qquad (9.7)$$

where k is the Boltzmann constant, T is the absolute temperature, D is the diffusion coefficient and $s = v/\omega^2$ with v the absolute migration velocity.

The buoyant force (following Archimedes' principle) opposes the centrifugal (sedimentation) force as does the frictional force. The sedimentation coefficient s is measured in units of Svedberg (S) where

1S $= 10^{-13}$ sec. The result of these three forces allows one to derive the Svedberg equation:

$$\frac{s}{D} = \frac{M(1 - \bar{v}\rho)}{RT} \qquad (9.8)$$

where M is the protein molar mass and R is the gas constant, which relates the three fundamental parameters that can be accessed by sedimentation measurements: the sedimentation coefficient (obtained from the migration of the sedimentation boundary), the diffusion coefficient (obtained from the spread of the sedimentation boundary) and the molar mass.

Sedimentation velocity (SV). For SV experiments, typical concentrations depend on the protein size, but A_{280} of 1.0 is required for standard optical detection systems and 0.5–1.0 mg/mL for interference optics. A typical volume for a series of experiments at different concentrations would be 1 mL. Typical sedimentation profiles of a BSA sample (66 kDa) are shown in Fig. 9.7.

From such data the sedimentation coefficient can be determined simply by calculating the boundary displacement as a function of time. However, the complete set of sedimentation curves contains significantly more information than simply the position of the boundary, and modern analysis is most often based on the Lamm equation, which describes the evolution of the complete sedimentation profile. Several software packages, such as SEDFIT (Schuck 2000), are available for obtaining s and D and hence M from the measured SV profiles.

Sedimentation equilibrium (SE). SE measurements use lower rotor speeds than SV, such that the sedimenting species are not sedimented fully to the bottom of the cell but reach an equilibrium distribution resulting from a balance of centrifugal and diffusive forces. This equilibrium usually takes 24–48 hours to establish. For proteins of 200 kDa, measurements are typically carried out on sample volumes of

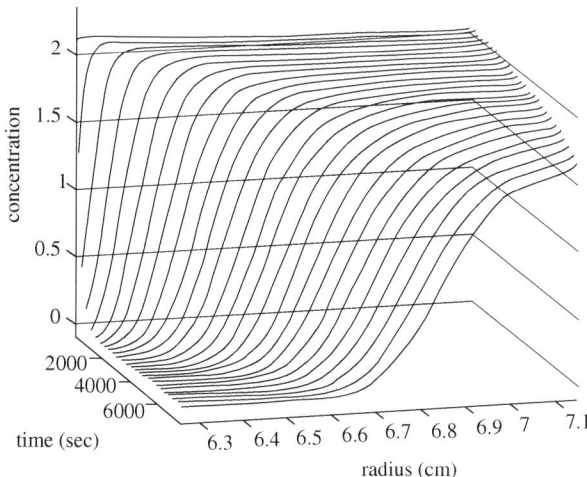

Fig. 9.7 SV data of a BSA sample. Shown are the concentration versus radius distributions at different times after start of the sedimentation at 50,000 rpm. Concentrations are in units of fringe displacement in the interference optical system, which corresponds to 0.3 mg/mL per fringe. Reproduced, with permission, from Lebowitz *et al.* (2002), © John Wiley and Sons, 2002.

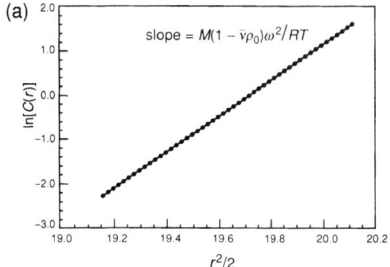

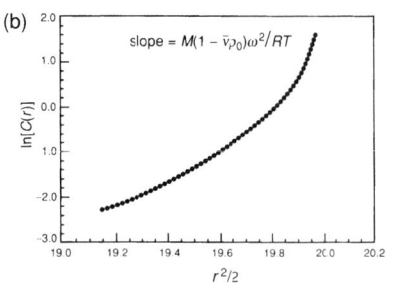

Fig. 9.8 (a) Plot of $\ln C(r)$ *vz* $r^2/2$ for a monodisperse solution of macromolecules, and (b) a similar plot for a polydisperse solution. The molecular mass is derived from the slope of the plot, $M(1 - \bar{v}\rho)\omega/RT$. Reproduced, with permission, from Serdyuk *et al.* (2007), © Cambridge University Press.

100–180 μl, and at a series of loading concentrations which may range over 1,000-fold for the same sample (Scott *et al.* 2005). At equilibrium the macromolecular concentration (C) increases exponentially towards the bottom of the sample cell, and the molecular mass can be calculated from the concentration values at different radial positions (r) in the cell expressed as (Serdyuk *et al.* 2007):

$$\frac{d\ell n(C)}{d(r^2/2)} = \frac{1}{r}C\frac{dC}{dr}\frac{\omega^2 s}{D} = \frac{M(1 - \bar{v}\rho)\omega^2}{RT} \tag{9.9}$$

The macromolecular concentration distribution between the meniscus at a and the point r obeys an exponential law called the second Svedberg equation:

$$C(r) = C(a)\exp\left[\omega^2 M(1 - \bar{v}\rho)(r^2 - a^2)/2RT\right] \tag{9.10}$$

For a monodisperse solution the plot of $\ln C(r)$ *versus* $r^2/2$ will thus be a straight line, as shown in Fig. 9.8(a), whereas for a polydisperse solution the curve will be a sum of exponentials each satisfying eq. (9.10). The higher molecular masses concentrating at the higher radii, the curve will show an upward slope with the tangential slope at each point, giving the number average molecular mass at the corresponding point in the cell.

SE curves of a solution containing an equilibrium mixture of species can be analysed in terms of the distribution of different species, as shown in Fig. 9.9, where the mutant V_L (immunoglobulin variable light chain) and domain of REI (a Bence-Jones protein) are analysed in terms of a monomer/dimer/tetramer model (Hensley 1996).

Bead modelling. Bead modelling is used in SAS through programs such as DAMMIN (Svergun 1999), as described in Chapter 4, but can also be employed to model AUC data. It is also possible to describe known structures or subunits as bead models from their known atomic coordinates. As the sedimentation of spheres of a given size and friction coefficient can be calculated exactly, one can calculate hydrodynamic

Fig. 9.9 SE data for a mutant V_L and domain of REI analysed in terms of a monomer/dimer/tetramer distribution. (b) Displays the original data (upper curve) fitted by three exponentials with residuals to the fits shown in (a). (c) Displays the distribution of each species as a function of total concentration. Reproduced from Hensley (1996), with permission from Elsevier, © 1996.

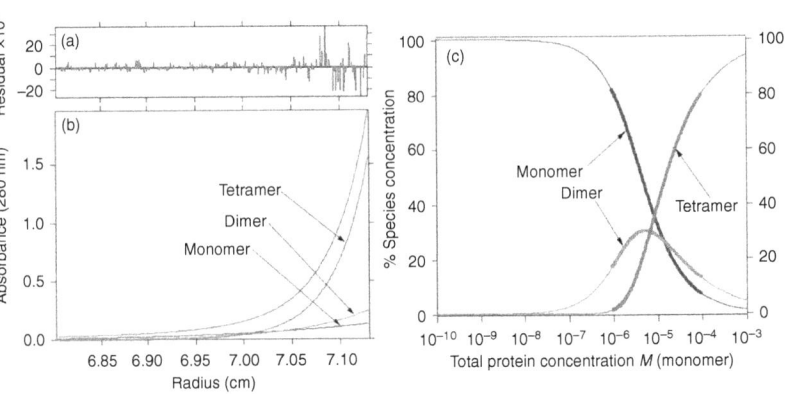

parameters of models made from such spheres (or beads). The calculation of hydrodynamic parameters from bead models has been pioneered by the group led by Garcia de la Torre, who have provided software such as HYDRO (Garcia de la Torre *et al.* 1994) to calculate parameters from anhydrous bead assemblies, and HYDROPRO (Garcia de la Torre *et al.* 2000), which calculate parameters from SAXS data, following the approach described by Ackerman *et al.* (2003). Large modular structures can in this way be modelled, and best fits obtained by first fitting to SAS data and then distinguishing between plausible models via the measured sedimentation coefficient (Perkins *et al.* 2011). Data from anhydrous models must be corrected to take into account the hydration of ~0.3 g water/g protein. This is usually done by increasing empirically the surface sphere radius by 0.31 nm (note that SAS measurements indicate that this water may be up to 10% denser than bulk water; Svergun *et al.* (1998); see Section 4.5). Another program package for bead modelling and calculation of hydrodynamic parameters is SOMO (Rai *et al.* 2005).

9.4.2 Dynamic light scattering (DLS)

DLS—also known as quasi-elastic light scattering or photo-correlation spectroscopy—is a technique which exploits the coherence of a laser beam to study the dynamical behaviour of a macromolecule in solution. A detailed description of the technique is given in Berne and Pecora (2000) and Serdyuk *et al.* (2007). For a more basic treatment, see, for example, Harding (1994b). The Brownian motion of macromolecules in solution induces a change in the local refractive index, causing a shift in wavelength and hence a fluctuation of the intensity of the light scattered from a volume element of the sample. An autocorrelator correlates intensity at time t with subsequent times $t + b\tau_s$, where b is a channel number and τ_s is a user-chosen sample time. $b\tau_s$ is referred to as the delay time. The correlation function $g^{(2)}(\tau)$ is defined as:

$$g^{(2)}(\tau) = \left[\langle I(t).I(t+\tau) \rangle / \langle I \rangle^2 \right] \qquad (9.11)$$

For a monodisperse solution of dilute macromolecules of molecular mass $\leq 10^8$ and quasi-spherical in shape, $g^{(2)}(\tau)$ is related to the translational diffusion coefficient D_m by:

$$\left[g^{(2)}(\tau) - 1 \right] = e^{-D_m q^2 \tau} \qquad (9.12)$$

D_m can thus be obtained from a plot of $\ln \left[g^{(2)}(\tau) - 1 \right]$ *versus* τ. For a polydisperse solution the correlation function is related to a sum of exponentials from the component particles, and thus a distribution of the sizes of the constituent molecular aggregates can be obtained. Commercially available instruments from companies such as Malvern (<http://www.malvern.com>), Brookhaven Instruments Corporation (<http://www.brookhaveninstruments.com>) or Wyatt Technology

(<http://www.wyatt.com>) and their associated software allow correlation functions to be measured routinely and diffusion coefficients to be calculated. From the diffusion coefficient the Stokes radius (also known as the hydrodynamic radius) of the species in solution can be calculated from the Stokes–Einstein equation:

$$R_h = \frac{k_B T}{6\pi\eta D} \tag{9.13}$$

where k_B is the Boltzmann constant, T the temperature and η the shear viscosity of the solvent. The Stokes radius R_h is defined as the radius of a hypothetical hard sphere that diffuses with the same speed as the particle under examination. It therefore includes the hydration layer, and is consequently greater than the radius calculated from, for example, the mass and specific volume of the macromolecule.

From DLS it is therefore possible to obtain a size distribution of the particles in a macromolecular solution and to determine whether the solution is sufficiently monodisperse for analysis by SAS. Solutions having a size distribution >5% are generally considered unsuitable for SAS, particularly if the aggregates present are much larger than the macromolecules being studied. DLS has also been used extensively in studies of protein crystallisation (see, for example, Bergfors 2009), as the criteria for the kind of macromolecular solution that will produce good crystals and the one which will provide good SAS data are very similar.

9.4.3 Multi-angle laser light scattering (MALLS)

Static light-scattering is essentially described by the Debye–Zimm relation (Zimm 1948):

$$\frac{Kc}{R_\theta} = \frac{1}{MP(\theta)} + 2A_2 c \tag{9.14}$$

where R_θ is the excess intensity of scattered light at angle θ, c is the sample concentration (g mL^{-1}), M is the weight-average molecular weight (molar mass), A_2 is the second virial coefficient (mL mol g^{-2}) (see Chapter 8), K is an optical parameter equal to $4\pi^2 n^2 (dn/dc)^2 / \lambda_0^4 N_A$, n is the solvent refractive index, dn/dc is the refractive index increment, N_A is Avogadro's number, λ_0 is the wavelength of the scattered light in vacuum (cm) and $P(\theta)$ is the form factor describing the angular dependence of scattered light.

From this expression it is therefore possible to obtain the z-average molecular weight and, if measurements are made at several different angles, the radius of gyration for values of Rg larger than about 10 nm. Since the late 1980s a range of commercial instruments using lasers has become available. These are able to measure light scattering simultaneously at different angles. The technique known as MALLS (for multi- or multiple-angle laser light scattering) or MALS (for multi-angle light scattering), provides a fast and accurate method of determining the weight average molecular mass of a macromolecular solution. It is particularly sensitive to high molecular weights, and is therefore an excellent tool for detecting large aggregates. MALLS is now often combined with size exclusion chromatography (SEC-MALLS) in order to separate species of different molecular weight. For a beginner's guide to MALLS, see Harding (1994a), or for more advanced texts, Serdyuk *et al.* (2007).

9.5 Bioinformatics and model validation

Data analysis in modern SAS, as most other physical experiments, relies heavily on various computational techniques. A synergistic use of SAS with various bioinformatic approaches has recently become a very useful addition to the palette of analysis tools. One of the obvious areas is validation of high-resolution models in solution. These models usually stem from MX, sometimes from NMR, but one can also rapidly validate predicted models of proteins and complexes. The programs for computing theoretical profiles from high-resolution models (see Chapter 4) are well established and freely available, and the calculations can be carried out very rapidly, allowing one to easily screen thousands of models.

9.5.1 Fold prediction and docking

Incorporation of experimental data into protein fold-prediction algorithms has proven to significantly improve their predictive capacity. Software suites such as Rosetta (Leaver-Fay *et al.* 2011) employ various NMR constraints such as chemical shifts, RDCs and pseudocontact shifts (Lange *et al.* 2012, Schmitz *et al.* 2012). The SAXS information can also be included in the Rosetta scoring function, though this option appears to be often reduced to constraining the R_g value. Other homology-modelling software, Modeller, can utilise several types of experimental restraints on the spatial structure, including SAXS data (Marti-Renom *et al.* 2000).

Several attempts have been undertaken to specifically design *ab initio* protein fold prediction algorithms assisted by SAXS data. In the approach proposed in Zheng and Doniach (2002, 2005), SAXS was employed to improve the recognition power of a threading procedure using the Dali domain library (Holm and Sander 1996). A term requiring the agreement of the scattering computed from a given model with the available SAXS data is added to the other scores (hydrophobicity, burial, contact energy and so on) to obtain the overall fitness function. About 10^4–10^5 candidate structures are generated for a given sequence by gapless threading against the set of folds in the Dali library. A set of more than seventy target proteins was screened to select the optimal weights of the individual components in the fitness function by training a neural network. Another *ab initio* approach aimed at determining the overall topology of small helical proteins was proposed by Wu *et al.* (2005). A coarse-grained Monte Carlo protocol was developed to enhance sampling of the conformational space of the protein, and a SAXS term was added to the scoring function employing knowledge-based potentials. The algorithm was able to provide structural models with overall correct topology for several examples, with the quality of reconstruction comparable to approximately 0.6 nm cryo-EM density maps.

SAS has played a limited role in fold recognition algorithms until now, which is understandable given that the technique provides overall and not high-resolution information. The use of SAS allows one to eliminate the decoys with incorrect overall shape, but with rare exceptions it is difficult to obtain much guidance about the internal structure and specifically about the fold topology within the given shape. Naturally, SAS plays a much more prominent role as a provider of information about the domain or quaternary structure, in particular, for protein–protein docking algorithms. Examples are the two recently developed methods, pyDockSAXS (Pons *et al.* 2010) and FoXSDock (Schneidman-Duhovny *et al.* 2011), which both employ global search docking programs, FTDock (Gabb *et al.* 1997) and Patch-Dock (Cheng *et al.* 2007), respectively, for orientation sampling in the complexes. The possible models are subsequently ranked by using energy-based scoring functions with the addition of the SAXS term. For both methods the success rate is nearly doubled in comparison to the docking alone, but also to purely SAXS-based rigid-body modelling. In about half of all the benchmarked cases, a near-native model was found within the top ten ranked predictions. SAXS restraints are being included in other information-driven docking software suites such as HADDOCK (Karaca *et al.* 2010), and they also play a significant role in the Integrated Modelling Platform (IMP)—an open-source software system oriented towards hybrid modelling of macromolecular assemblies (Russel *et al.* 2012). Here, the scoring function types include ones for SAXS profiles, proteomics data, EM images and density maps, NMR data, bioinformatic potentials and force fields. A comprehensive review of the existing computational methods for integrating SAXS profiles into structural modelling can be found in Schneidman-Duhovny *et al.* (2012).

9.5.2 Databases and standardisation

The bioinformatic-type approaches make extensive use of databases, and these have started to play an important role in SAS. A database DARA containing more than 10,000 computed scattering patterns from biological protein assemblies extracted from the PDB has been available since 2003 (Sokolova *et al.* 2003) at <http://dara.embl-hamburg.de/>. DARA allows for a rapid search of structural neighbours for proteins based on similarity in the SAXS and (if measured) WAXS portions of the scattering patterns. The database search returns a list of structures ranked in accordance with their agreement with the experimental scattering from the given protein. The search is very fast, such that DARA is invoked online for each and every scattering pattern coming out of the automated analysis pipeline described in Section 9.1. Revealing the 'scattering neighbours' may help in finding molecular replacement templates, and sometimes also reveals unexpected structural similarities in the overall conformations (Hura *et al.* 2009). Another open-access database, called BIOISIS <http://www.bioisis.net>, contains about fifty downloadable annotated experimental SAXS data sets, mostly collected on the SIBYLS beamline (ALS, Berkeley, USA) (Hura *et al.* 2009). The database contains also overall parameters and *ab initio* models, along with the descriptions of the experimental conditions and sequence information about the protein.

The database development becomes very important nowadays given the strong move in the biological community towards making experimental SAS data and models publicly available. In 2011 a PDB task force was established to provide recommendations on ways for incorporating SAS models into the PDB or in a related public repository. In a related move, publication standards for structural modelling of biological solution SAS data are being developed (Jacques *et al.* 2012). Depositions of the experimental data and of the derived models should allow the readers to independently assess the validity of the interpretations made by the authors.

9.5.3 Validation of SAS-based models

The move towards the deposition of the SAS models further underlines another extremely important aspect of structural studies with solution scattering. It has been indicated several times in this book that reconstruction of three-dimensional structures from the scattering data is inherently ambiguous. Given the rapidly growing number of three-dimensional models generated from SAS by various groups, the problem of overinterpretation and model validation has become rather acute. At present, there are no objective criteria to validate the models, similar to the free R-factor in crystallography (Brunger 1992). This quantity is computed over a 'test' set of reflections purposely omitted in the modelling and refinement process, and the improvement of R-free is an indication that the refinement not just fits the data but converges to a physically sensible model. It is difficult to construct a similar measure for SAS, because the amount of information is already very limited in the scattering data, and there is not much room to omit anything from the refinement. Interestingly, an R-free-like attempt was made in the first *ab initio* shape determination procedure (Svergun and Stuhrmann 1991). Here, low-resolution envelopes were computed using low-resolution data only, and the available higher-resolution data were employed to screen the possible models and select those providing best fits to higher-angle data. The approach worked reasonably well in the simulated data. In practical applications, however, higher-angle data may not be sufficiently precise, and also contain contributions from the internal structure, possibly making such screening misleading.

Clearly, any model coming out of SAS data must be consistent with the model-independent overall parameters, and this is in most cases ensured by the very fact that the model provides a good fit to the experimental scattering by the data. Another important consistency check is given by the analysis of multiple models produced by SAS, allowing assessment of their variability. As indicated in Chapter 4, convenient tools are available to analyse multiple *ab initio* reconstructions—for example, DAMAVER—which identifies the most probable model, provides the average model and quantifies the variability of the reconstruction in terms of the NSD value (see Section 4.4.2). For rigid-body modelling, DAMCLUST (Petoukhov *et al.* 2012), distributing the models into clusters with distinctly different quaternary structures, is more appropriate. Using DAMCLUST one may assess ambiguity of the quaternary structure reconstruction and test the alternative conformations, for example, by utilising the criteria employed in the docking algorithms (Cheng *et al.* 2007, Wang *et al.* 2007, Kozakov *et al.* 2008, Gajda *et al.* 2010, Karaca *et al.* 2010).

Fitting multiple scattering patterns by a single model is now possible both for *ab initio* analysis (Section 4.4) and for hybrid modelling employing high-resolution models of subunits and domains (Section 4.6). In the simultaneous fitting of multiple scattering curves from different constructs (for example, for X-rays, from a full complex and from the sub-complexes), a necessary prerequisite is the absence of conformational rearrangements within the subunits or domains in the different constructs. An ideal approach for obtaining data for the fitting of multiple curves is contrast variation in neutron scattering combined with specific perdeuteration of the samples. Here, one just needs to verify the integrity of selectively perdeuterated particles, which can be done, for example, by X-ray scattering measurements (deuteration should not change the X-ray scattering pattern). Numerous studies are presently making use of multiple curve fitting (see examples herein and in Chapter 6), and the increased information content in the data not only enhances the resolution but also improves the fidelity of the models.

The most reliable way of assessing the quality of SAS models remains *a posteriori* validation using the data from other methods. In a manner similar to the R-free approach, some pieces of information (for example, on contacts between subunits or domains) may be hidden in the refinement and the models subsequently screened against this information. Such procedures are, however, difficult to formalise, and the strategies have to be developed specifically for each individual case. Perhaps the most convincing way of verifying models of complexes is the validation of intersubunit interfaces, for example, by site-directed mutagenesis. This approach is frequently employed in MX to validate the oligomers built by crystallographic contacts. Validating SAS results in this way is less common, given that the technique usually does not provide models of sufficient detail to be checked by single-point mutations.

One of the still rather rare examples of validation of a SAXS model at a residue level is given by a study of a protein assembly relevant for the formation of an iron–sulphur cluster (Prischi *et al.* 2010). Here, a ternary complex between CyaY, a bacterial orthologue of human frataxin, desulphurase Nfs1/IscS and the scaffold protein Isu/IscU was studied. Frataxin—an essential protein of as yet unknown function—is known to interact with the two latter proteins, which are the two central components of the iron–sulphur cluster assembly. The SAXS data were collected for the individual proteins IscS, IscU, CyaY, the binary complexes IscS/IscU, IscS/CyaY and the ternary complex IcsS/CyaY/IscU (Fig. 9.10A). The shape and overall parameters computed from the scattering profile of IscS indicate that the protein is dimeric in solution, and its overall structure matches well with the available crystallographic dimer, PDB entry 1P3W (Cupp-Vickery *et al.* 2003). The two other proteins were found to be monomeric in solution. Comparison of the shapes of the binary and tertiary complexes suggested that IscU is located on the periphery and CyaY close to the centre of the ternary complex (Fig. 9.10B). Further analysis utilised the available atomic structures of the three components or their homologues. Calculations of the theoretical scattering patterns from the individual proteins confirmed the dimeric structure of IscS in solution, and also permitted one to choose

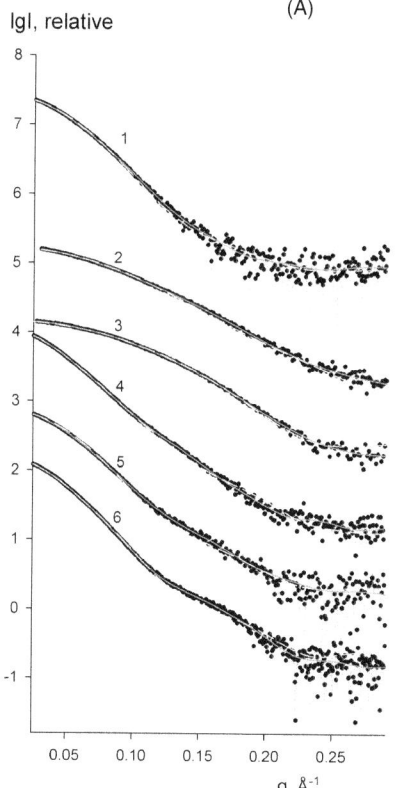

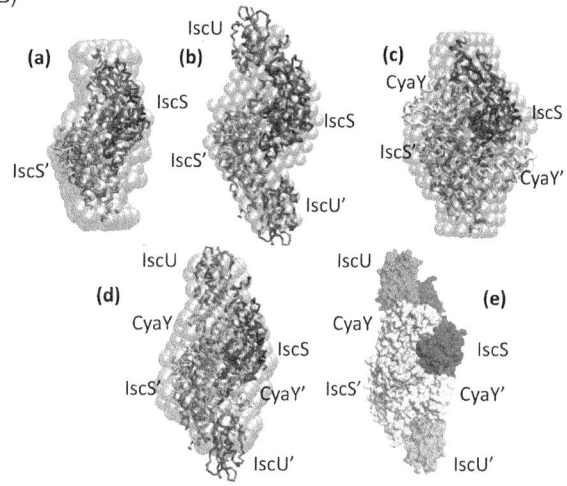

Fig. 9.10 SAXS data and models of frataxin complexes. (A) Scattering data and fits computed from the models (1–6 correspond to IscS, IscU, CyaY, IscU/IscS, Cya/IcsS and CyaY/IscS/IscU, respectively). The experimental data are represented as dots with error bars, the model scattering of typical *ab initio* models is displayed as solid lines and the scattering curves computed from high-resolution (for proteins alone) and rigid-body (for complexes) models as dashed lines. (B) The models of dimeric IscS (a), IscU/IscS (b), Cya/IcsS (c) and CyaY/IscS/IscU (d). The *ab initio* shapes are represented with grey semi-transparent spheres, the rigid-body models as C_α traces. Panel (e) displays the consensus HADDOCK-refined model in space-filling mode. Adapted from Prischi *et al.* (2010), with permission from Nature Publishing Group, © 2010.

the best model among several homology models for IscU. Then, SAS-REF was employed to construct the rigid-body model of the ternary complex fitting simultaneously the scattering data from both binary complexes (Fig. 9.10A, B). The model agrees well with the overall *ab initio* shapes, and depicts two monomeric CyaY molecules bound to the IscS dimer in a pocket between the active site and the IscS dimer interface.

To validate the SAXS model, the residues of CyaY interacting with IscS were mapped by NMR spectral perturbations in an experiment where ^2H, ^{15}N double-labelled CyaY was titrated with IscS up to a 1:1 molar ratio. Further spectral perturbations were obtained by adding IscU to the solution up to a 1:1:1 molar ratio (CyaY/IscS/IscU). In addition, the interaction surface on IscS predicted by SAXS was validated by constructing IscS mutants chosen to target residues that could potentially affect interactions with CyaY. The results of the additional experiments were fully consistent with the SAXS model positioning CyaY in a site very distinct from that occupied by IscU. The SAXS model was further refined by HADDOCK (Karaca *et al.* 2010), yielding the consensus model in Fig. 9.10B, panel (e). The final model, compatible with low-resolution SAXS data, with NMR results and the mutagenesis studies, suggests a new paradigm for understanding the role of frataxin as a regulator of IscS functions.

During the last two decades, major efforts have been invested in developing methods to build structural models based on the SAS data. Significant progress has been achieved, and the modern approaches do allow one to construct meaningful *ab initio* and hybrid models that fit the experimental SAS data. Given the inherent ambiguity of SAS data interpretation in terms of three-dimensional models, validation of these models becomes indispensable for making biologically relevant conclusions. One may expect that *a posteriori* model validation will play a major role for all future methodical developments in SAS.

References

Ackerman, C. J., Harnett, M. M., Harnett, W., Kelly, S. M., Svergun, D. I. and Byron, O. (2003). '19-Å solution structure of the filarial nematode immunomodulatory protein, ES-62', *Biophys. J.* 84: 489–500.

Allen, J. P., Feher, G., Yeates, T. O. *et al.* (1986). 'Structural homology of reaction centers from Rhodopseudomonas-Sphaeroides and Rhodopseudomonas-Viridis as determined by X-ray-diffraction', *P. Natl. Acad. Sci. USA* 83: 8589–93.

Andersen, C. B., Becker, T., Blau, M. *et al.* (2006). 'Structure of eEF3 and the mechanism of transfer RNA release from the E-site', *Nature* 443: 663–8.

Aparicio, R., Fischer, H., Scott, D. J. *et al.* (2002). 'Structural insights into the beta-mannosidase from T. Reesei obtained by synchrotron small-angle X-ray solution scattering enhanced by X-ray crystallography', *Biochemistry* 41: 9370–5.

Bartkiewicz, P. and Duval, P. (2007). 'TINE as an accelerator control system at DESY', *Measurement Science & Technology* 18: 2379–86.

Bergfors, T. M. (2009). *Protein Crystallization*. La Jolla, CA: International University Line.

Bernado, P., Blanchard, L., Timmins, P., Marion, D., Ruigrok, R. W. and Blackledge, M. (2005). 'A structural model for unfolded proteins from residual dipolar couplings and small-angle X-ray scattering', *P. Natl. Acad. Sci. USA* 102: 17002–7.

Bernado, P., Modig, K., Grela, P. *et al.* (2010). 'Structure and dynamics of ribosomal protein L12: An ensemble model based on SAXS and NMR relaxation', *Biophys. J.* 98: 2374–82.

Berne, B. J. and Pecora, R. (2000). *Dynamic Light Scattering: With Applications to Chemistry, Biology, and Physics*. Dover Publications.

Bertini, I., Giachetti, A., Luchinat, C. *et al.* (2010). 'Conformational space of flexible biological macromolecules from average data', *J. Am. Chem. Soc.* 132: 13553–8.

Blackledge, M. (2010). 'Mapping the conformational mobility of multidomain proteins', *Biophys. J.* 98: 2043–4.

Blanchet, C. E., Zozulya, A. V., Kikhney, A. G. *et al.* (2012). 'Instrumental setup for high-throughput small- and wide-angle solution scattering at the X33 beamline of EMBL Hamburg', *J. Appl. Crystallogr.* 45: 489–95.

Blattner, C., Jennebach, S., Herzog, F. *et al.* (2011). 'Molecular basis of Rrn3-regulated RNA polymerase I initiation and cell growth', *Gene. Dev.* 25: 2093–105.

Blundell, T. L. and Johnson, L. N. (1976). *Protein crystallography*, London: Academic Press.

Boyes-Watson, J., Davidson, E. and Perutz, M. F. (1947). 'An X-ray study of horse methaemoglobin .1', *Proc. R. Soc. Ser. A* 191: 83–132.

Brunger, A. T. (1992). 'Free R value: A novel statistical quantity for assessing the accuracy of crystal structures', *Nature* 355: 472–5.

Brunger, A. T. (2007). 'Version 1.2 of the crystallography and NMR system', *Nat. Protoc.* 2: 2728–33.

Cantor, C. R. and Schimmel, P. R. (1980). *Biophysical Chemistry: Part II: Techniques for the Study of Biological Structure and Function*. New York: W. H. Freeman.

Chauvin, C., Witz, J. and Jacrot, B. (1978). 'Structure of tomato bushy stunt virus: model for protein–RNA interaction', *J. Mol. Biol.* 124: 641–51.

Cheng, T. M., Blundell, T. L. and Fernandez-Recio, J. (2007). 'pyDock: Electrostatics and desolvation for effective scoring of rigid-body protein–protein docking', *Proteins* 68: 503–15.

Classen, S., Rodic, I., Holton, J., Hura, G. L., Hammel, M. and Tainer, J. A. (2010). 'Software for the high-throughput collection of SAXS data using an enhanced Blu-Ice/DCS control system', *J. Synchrot. Radiat.* 17: 774–81.

Cowan, S. W., Schirmer, T., Rummel, G. *et al.* (1992). 'Crystal-structures explain functional-properties of 2 Escherichia-coli porins', *Nature* 358: 727–33.

Cupp-Vickery, J. R., Urbina, H. and Vickery, L. E. (2003). 'Crystal structure of IscS, a cysteine desulfurase from Escherichia coli', *J. Mol. Biol.* 330: 1049–59.

David, G. and Perez, J. (2009). 'Combined sampler robot and high-performance liquid chromatography: A fully automated system for biological small-angle X-ray scattering experiments at the synchrotron SOLEIL SWING beamline', *J. Appl. Crystallogr.* 42: 892–900.

Deisenhofer, J., Epp, O., Miki, K., Huber, R. and Michel, H. (1985). 'Structure of the protein subunits in the photosynthetic reaction center of Rhodopseudomonas-Viridis at 3Å resolution', *Nature* 318: 618–24.

Delageniere, S., Brenchereau, P., Launer, L. *et al.* (2011). 'ISPyB: An information management system for synchrotron macromolecular crystallography', *Bioinformatics* 27: 3186–92.

Evans, G., Roversi, P. and Bricogne, G. (2000). 'In-house low-resolution X-ray crystallography', *Acta Crystallogr. D* 56: 1304–11.

Franke, D., Kikhney, A. G. and Svergun, D. I. (2012). 'Automated acquisition and analysis of small angle X-ray scattering data', *Nucl. Instrum. Meth. A* 689: 52–9.

Gabb, H. A., Jackson, R. M. and Sternberg, M. J. (1997). 'Modelling protein docking using shape complementarity, electrostatics and biochemical information', *J. Mol. Biol.* 272: 106–20.

Gabel, F., Simon, B., Nilges, M., Petoukhov, M., Svergun, D. and Sattler, M. (2008). 'A structure refinement protocol combining NMR residual dipolar couplings and small angle scattering restraints', *J. Biomol. NMR* 41: 199–208.

Gajda, M. J., Tuszynska, I., Kaczor, M., Bakulina, A. Y. and Bujnicki, J. M. (2010). 'FILTREST3D: Discrimination of structural models using restraints from experimental data', *Bioinformatics* 26: 2986–7.

Garcia de la Torre, J., Navarro, S. and Lopez Martinez, M. C. (1994). 'Hydrodynamic properties of a double-helical model for DNA', *Biophys. J.* 66: 1573–9.

Garcia de la Torre, J., Huertas, M. L. and Carrasco, B. (2000). 'Calculation of hydrodynamic properties of globular proteins from their atomic-level structure', *Biophys. J.* 78: 719–30.

Grant, T. D., Luft, J. R., Wolfley, J. R. *et al.* (2011). 'Small angle X-ray scattering as a complementary tool for high-throughput structural studies', *Biopolymers* 95: 517–30.

Grishaev, A., Wu, J., Trewhella, J. and Bax, A. (2005). 'Refinement of multidomain protein structures by combination of solution small-angle X-ray scattering and NMR data', *J. Am. Chem. Soc.* 127: 16621–8.

Grishaev, A., Ying, J., Canny, M. D., Pardi, A. and Bax, A. (2008). 'Solution structure of tRNAVal from refinement of homology model against residual dipolar coupling and SAXS data', *J. Biomol. NMR* 42: 99–109.

Hao, Q. (2001). 'Phasing from an envelope', *Acta Crystallogr. D* 57: 1410–4.

Hao, Q. (2006). 'Macromolecular envelope determination and envelope-based phasing', *Acta Crystallogr. D* 62: 909–14.

Hao, Q., Dodd, F. E., Grossmann, J. G. and Hasnain, S. S. (1999). '*Ab initio* phasing using molecular envelope from solution X-ray scattering', *Acta Crystallogr. D* 55: 243–6.

Harding, S. E. (1994a) 'Classical light scattering for the determination of molecular weight and gross conformation of biological macromolecules', in Jones, C., Mulloy, B. and Thomas, A. (ed.). *Methods in Molecular Biology*. New Jersey: Humana Press.

Harding, S. E. (1994b) Determination of diffusion coefficients of biological macromolecules by dynamic light scattering (DLS)', in Jones, C., Mulloy, B. and Thomas, A. (ed.), *Methods in Molecular Biology*. New Jersey: Humana Press.

Harrison, S. C. (1969). 'Structure of tomato bushy stunt virus. I. The spherically averaged electron density', *J. Mol. Biol.* 42: 457–483.

Harrison, S. C. and Jack, A. (1975). 'Structure of tomato bushy stunt virus. 3. 3-dimensional X-ray-diffraction analysis at 16Å resolution', *J. Mol. Biol.* 97: 173–91.

Hensley, P. (1996). 'Defining the structure and stability of macromolecular assemblies in solution: The re-emergence of analytical ultracentrifugation as a practical tool', *Structure* 4: 367–73.

Holm, L. and Sander, C. (1996). 'The FSSP database: Fold classification based on structure–structure alignment of proteins', *Nucleic Acids Res.* 24: 206–9.

Hong, X. and Hao, Q. (2009). 'Combining solution wide-angle X-ray scattering and crystallography: Determination of molecular envelope and heavy-atom sites', *J. Appl. Crystallogr.* 42: 259–64.

Hura, G. L., Menon, A. L., Hammel, M. *et al.* (2009). 'Robust, high-throughput solution structural analyses by small angle X-ray scattering (SAXS)', *Nat. Methods* 6: 606–12.

Jack, A., Harrison, S. C. and Crowther, R. A. (1975). 'Structure of tomato bushy stunt virus. 2. Comparison of results obtained by electron-microscopy and X-ray-diffraction', *J. Mol. Biol.* 97: 163–72.

Jacques, D. A., Guss, J. M., Svergun, D. I. and Trewhella, J. (2012). 'Publication guidelines for structural modelling of small-angle scattering data from biomolecules in solution', *Acta Crystallogr. D* 68: 620–6.

Jeffries, C. M., Lu, Y., Hynson, R. M. *et al.* (2011). 'Human cardiac myosin binding protein C: Structural flexibility within an extended modular architecture', *J. Mol. Biol.* 414: 735–48.

Karaca, E., Melquiond, A. S., de Vries, S. J., Kastritis, P. L. and Bonvin, A. M. (2010). 'Building macromolecular assemblies by information-driven docking: Introducing the HADDOCK multibody docking server', *Mol. Cell Proteomics* 9: 1784–94.

Kollman, J. M. and Quispe, J. (2005). 'The 17Å structure of the 420 kDa lobster clottable protein by single particle reconstruction from cryoelectron micrographs', *J. Struct. Biol.* 151: 306–14.

Kozakov, D., Schueler-Furman, O. and Vajda, S. (2008). 'Discrimination of near-native structures in protein–protein docking by testing the stability of local minima', *Proteins* 72: 993–1004.

Krissinel, E. and Henrick, K. (2007). 'Inference of macromolecular assemblies from crystalline state', *J. Mol. Biol.* 372: 774–97.

Lamers, M. H., Georgescu, R. E., Lee, S. G., O'Donnell, M. and Kuriyan, J. (2006). 'Crystal structure of the catalytic alpha subunit of E. Coli replicative DNA polymerase III', *Cell* 126: 881–92.

Lange, O. F., Rossi, P., Sgourakis, N. G. *et al.* (2012). 'Determination of solution structures of proteins up to 40 kDa using CS-Rosetta with sparse NMR data from deuterated samples', *P. Natl. Acad. Sci. USA* 109: 10873–8.

Leaver-Fay, A., Tyka, M., Lewis, S. M. *et al.* (2011). 'Rosetta 3: An object-oriented software suite for the simulation and design of macromolecules', *Method. Enzymol.* 487: 545–74.

Lebowitz, J., Lewis, M. S. and Schuck, P. (2002). 'Modern analytical ultracentrifugation in protein science: A tutorial review', *Protein Sci.* 11: 2067–79.

Liu, Q., Weaver, A. J., Xiang, T., Thiel, D. J. and Hao, Q. (2003). 'Low-resolution molecular replacement using a six-dimensional search', *Acta Crystallogr. D* 59: 1016–9.

Mao, Y., Senic-Matuglia, F., Di Fiore, P. P., Polo, S., Hodsdon, M. E. and De Camilli, P. (2005). 'Deubiquitinating function of ataxin-3: Insights from the solution structure of the Josephin domain', *P. Natl. Acad. Sci. USA* 102: 12700–5.

Marrott, N. L., Marshall, J. J., Svergun, D. I. *et al.* (2012). 'The catalytic core of an archaeal 2-oxoacid dehydrogenase multienzyme complex is a 42-mer protein assembly', *FEBS J.* 279: 713–23.

Marti-Renom, M. A., Stuart, A. C., Fiser, A., Sanchez, R., Melo, F. and Sali, A. (2000). 'Comparative protein structure modeling of genes and genomes', *Annu. Rev. Biophys. Biomol. Struct.* 29: 291–325.

Mattinen, M. L., Paakkonen, K., Ikonen, T. *et al.* (2002). 'Quaternary structure built from subunits combining NMR and small-angle X-ray scattering data', *Biophys. J.* 83: 1177–83.

Mattiroli, F., Vissers, Joseph h. A., Van dijk, Willem j. *et al.* (2012). 'Rnf168 ubiquitinates K13-15 on H2A/H2AX to drive DNA damage signaling', *Cell* 150: 1182–95.

McPhillips, T. M., McPhillips, S. E., Chiu, H. J. *et al.* (2002). 'Blu-ice and the distributed control system: Software for data acquisition and instrument control at macromolecular crystallography beamlines', *J. Synchrot. Radiat.* 9: 401–6.

Mylonas, E., Hascher, A., Bernado, P., Blackledge, M., Mandelkow, E. and Svergun, D. I. (2008). 'Domain conformation of tau protein studied by solution small-angle X-ray scattering', *Biochemistry* 47: 10345–53.

Nicastro, G., Habeck, M., Masino, L., Svergun, D. I. and Pastore, A. (2006). 'Structure validation of the Josephin domain of ataxin-3: Conclusive evidence for an open conformation', *J. Biomol. NMR* 36: 267–77.

Nicastro, G., Menon, R. P., Masino, L., Knowles, P. P., McDonald, N. Q. and Pastore, A. (2005). 'The solution structure of the Josephin domain of ataxin-3: Structural determinants for molecular recognition', *P. Natl. Acad. Sci. USA* 102: 10493–8.

Olson, A. J., Bricogne, G. and Harrison, S. C. (1983). 'Structure of tomato bushy stunt virus .4. The virus particle at 2.9-Å resolution', *J. Mol. Biol.* 171: 61–93.

Pebay-Peyroula, E., Garavito, R. M., Rosenbusch, J. P., Zulauf, M. and Timmins, P. A. (1995). 'Detergent structure in tetragonal crystals of OmpF porin', *Structure* 3: 1051–9.

Penel, S., Pebay-Peyroula, E., Rosenbusch, J., Rummel, G., Schirmer, T. and Timmins, P. A. (1998). 'Detergent binding in trigonal crystals of OmpF porin from Escherichia coli', *Biochimie* 80: 543–51.

Perkins, S. J., Nan, R. D., Li, K. Y., Khan, S. and Abe, Y. (2011). 'Analytical ultracentrifugation combined with X-ray and neutron scattering: Experiment and modelling', *Methods* 54: 181–99.

Pernot, P., Theveneau, P., Giraud, T. *et al.* (2010). 'New beamline dedicated to solution scattering from biological macromolecules at the ESRF', *J. Phys. Conf. Ser.* 247: 012009.

Petoukhov, M. V., Konarev, P. V., Kikhney, A. G. and Svergun, D. I. (2007). 'ATSAS 2.1: towards automated and web-supported small-angle scattering data analysis', *J. Appl. Crystallogr.* 40: S223–8.

Petoukhov, M. V., Monie, T. P., Allain, F. H., Matthews, S., Curry, S. and Svergun, D. I. (2006). 'Conformation of polypyrimidine tract binding protein in solution', *Structure* 14: 1021–7.

Petoukhov, M. V., Franke, D., Shkumatov, A. V. *et al.* (2012). 'New developments in the ATSAS program package for small-angle scattering data analysis', *J. Appl. Crystallogr.* 45: 342–50.

Pinotsis, N., Lange, S., Perriard, J. C., Svergun, D. I. and Wilmanns, M. (2008). 'Molecular basis of the C-terminal tail-to-tail assembly of the sarcomeric filament protein myomesin', *EMBO J.* 27: 253–64.

Pinotsis, N., Chatziefthimiou, S. D., Berkemeier, F. *et al.* (2012). 'Superhelical architecture of the myosin filament-linking protein myomesin with unusual elastic properties', *PLoS Biol.* 10: e1001261.

Pons, C., D'abramo, M., Svergun, D. I., Orozco, M., Bernado, P. and Fernandez-Recio, J. (2010). 'Structural characterization of protein–protein complexes by integrating computational docking with small-angle scattering data', *J. Mol. Biol.* 403: 217–30.

Prischi, F., Konarev, P. V., Iannuzzi, C. *et al.* (2010). 'Structural bases for the interaction of frataxin with the central components of iron–sulphur cluster assembly', *Nat. Commun.* 1: 95.

Raag, R., Appelt, K., Xuong, N. H. and Banaszak, L. (1988). 'Structure of the lamprey yolk lipid protein complex lipovitellin phosvitin at 2.8-Å resolution', *J. Mol. Biol.* 200: 553–69.

Rai, N., Nollmann, M., Spotorno, B., Tassara, G., Byron, O. and Rocco, M. (2005). 'SOMO (SOlution MOdeler) differences between X-ray- and NMR-derived bead models suggest a role for side chain flexibility in protein hydrodynamics', *Structure* 13: 723–34.

Roessle, M. W., Klaering, R., Ristau, U. *et al.* (2007). 'Upgrade of the small-angle X-ray scattering beamline X33 at the European Molecular Biology Laboratory, Hamburg', *J. Appl. Crystallogr.* 40: S190–4.

Roth, M. (1987). 'Best density maps in low resolution crystallography with contrast variation', *Acta Crystallogr A* 43: 780–7.

Roth, M. (1991). Crystallographic computing 5: From chemistry to biology : Papers presented at the international school on crystallographic computing held at Bischenberg, France, 29 July–5 August, 1990: International Union of Crystallography.

Roth, M., Lewit-Bentley, A. and Bentley, G. A. (1984). 'Scaling and phase-difference determination in solvent contrast variation experiments', *J. Appl. Crystallogr.* 17: 77–84.

Roth, M., Arnoux, B., Ducruix, A. and Reiss-Husson, F. (1991). 'Structure of the detergent phase and protein-detergent interactions in crystals of the wild-type (Strain-Y) Rhodobacter-Sphaeroides photochemical-reaction center', *Biochemistry* 30: 9403–13.

Roth, M., Lewit-Bentley, A., Michel, H., Deisenhofer, J., Huber, R. and Oesterhelt, D. (1989). 'Detergent structure in crystals of a bacterial photosynthetic reaction center', *Nature* 340: 659–62.

Round, A. R., Franke, D., Moritz, S. *et al.* (2008). 'Automated sample-changing robot for solution scattering experiments at the EMBL Hamburg SAXS station X33', *J. Appl. Crystallogr.* 41: 913–7.

Russel, D., Lasker, K., Webb, B. *et al.* (2012). 'Putting the pieces together: Integrative modeling platform software for structure determination of macromolecular assemblies', *PLoS Biol.* 10: e1001244.

Sauer, O., Roth, M., Schirmer, T., Rummel, G. and Kratky, C. (2002). 'Low-resolution detergent tracing in protein crystals using xenon or krypton to enhance X-ray contrast', *Acta Crystallogr. D* 58: 60–9.

Schmitz, C., Vernon, R., Otting, G., Baker, D. and Huber, T. (2012). 'Protein structure determination from pseudocontact shifts using ROSETTA', *J. Mol. Biol.* 416: 668–77.

Schneidman-Duhovny, D., Hammel, M. and Sali, A. (2011). 'Macromolecular docking restrained by a small angle X-ray scattering profile', *J. Struct. Biol.* 173: 461–71.

Schneidman-Duhovny, D., Kim, S. J. and Sali, A. (2012). 'Integrative structural modeling with small angle X-ray scattering profiles', *BMC Struct. Biol.* 12: 17.

Schuck, P. (2000). 'Size-distribution analysis of macro-molecules by sedimentation velocity ultracentrifugation and Lamm equation modeling', *Biophys. J.* 78: 1606–19.

Schwieters, C. D. and Clore, G. M. (2007). 'A physical picture of atomic motions within the Dickerson DNA dodecamer in solution derived from joint ensemble refinement against NMR and large-angle X-ray scattering data', *Biochemistry*, 46: 1152–66.

Schwieters, C. D., Kuszewski, J. J., Tjandra, N. and Clore, G. M. (2003). 'The Xplor–NIH NMR molecular structure determination package', *J. Magn. Reson.* 160: 65–73.

Scott, D., Harding, S. E., Rowe, A., Aziz, Z. and Behlke, J. (2005). *Analytical Ultracentrifugation: Techniques and Methods*. Royal Society of Chemistry.

Serdyuk, I. N., Zaccai, N. R. and Zaccai, J. (2007). *Methods in Molecular Biophysics: Structure, Dynamics, Function*. Cambridge: Cambridge University Press.

Sokolova, A. V., Volkov, V. V. and Svergun, D. I. (2003). 'Prototype of database for rapid protein classification based on solution scattering data', *J. Appl. Crystallogr.* 36: 865–8.

Svergun, D. I. (1999). 'Restoring low resolution structure of biological macromolecules from solution scattering using simulated annealing', *Biophys. J.* 76: 2879–86.

Svergun, D. I. and Stuhrmann, H. B. (1991). 'New developments in direct shape determination from small-angle scattering 1. Theory and model calculations', *Acta Crystallogr.*, A47: 736–44.

Svergun, D. I., Petoukhov, M. V., Koch, M. H. J. and Koenig, S. (2000). 'Crystal versus solution structures of thiamine diphosphate-dependent enzymes', *J. Biol. Chem.* 275: 297–302.

Svergun, D. I., Barberato, C., Koch, M. H. J., Fetler, L. and Vachette, P. (1997). 'Large differences are observed between the crystal and solution quaternary structures of allosteric aspartate transcarbamylase in the R state', *Proteins* 27: 110–17.

Svergun, D. I., Richard, S., Koch, M. H., Sayers, Z., Kuprin, S. and Zaccai, G. (1998). 'Protein hydration in solution: Experimental observation by X-ray and neutron scattering', *P. Natl. Acad. Sci. USA* 95: 2267–72.

Tidow, H., Melero, R., Mylonas, E. *et al.* (2007). 'Quaternary structures of tumor suppressor p53 and a specific p53 DNA complex', *P. Natl. Acad. Sci. USA* 104: 12324–9.

Timmins, P. A., Poliks, B. and Banaszak, L. (1992). 'The location of bound lipid in the lipovitellin complex', *Science* 257: 652–5.

Timmins, P. A., Pebay-Peyroula, E. and Welte, W. (1994a). 'Detergent organisation in solutions and in crystals of membrane proteins', *Biophys. Chem.* 53: 27–36.

Timmins, P. A., Wild, D. and Witz, J. (1994b). 'The three-dimensional distribution of RNA and protein in the interior of tomato bushy stunt virus: A neutron low-resolution single-crystal diffraction study', *Structure* 2: 1191–1201.

Vestergaard, B., Sanyal, S., Roessle, M. *et al.* (2005). 'The SAXS solution structure of RF1 differs from its crystal structure and is similar to its ribosome bound cryo-em structure', *Mol. Cell* 20: 929–38.

Von Castelmur, E., Marino, M., Svergun, D. I. *et al.* (2008). 'A regular pattern of Ig super-motifs defines segmental flexibility as the elastic mechanism of the titin chain', *P. Natl. Acad. Sci. USA* 105: 1186–91.

Wang, C., Schueler-Furman, O., Andre, I. *et al.* (2007). 'RosettaDock in CAPRI rounds 6–12', *Proteins* 69: 758–63.

Weiss, M. S., Kreusch, A., Schiltz, E. *et al.* (1991). 'The structure of porin from Rhodobacter-capsulatus at 1.8 Å resolution', *FEBS Lett.* 280: 379–82.

Wells, M., Tidow, H., Rutherford, T. J. *et al.* (2008). 'Structure of tumor suppressor p53 and its intrinsically disordered N-terminal transactivation domain', *P. Natl. Acad. Sci. USA*, 105: 5762–7.

Wu, Y., Tian, X., Lu, M., Chen, M., Wang, Q. and Ma, J. (2005). 'Folding of small helical proteins assisted by small-angle X-ray scattering profiles', *Structure* 13: 1587–97.

Zheng, W. and Doniach, S. (2002). 'Protein structure prediction constrained by solution X-ray scattering data and structural homology identification', *J. Mol. Biol.* 316: 173–87.

Zheng, W. and Doniach, S. (2005). 'Fold recognition aided by constraints from small angle X-ray scattering data', *Protein. Eng. Des. Sel.* 18: 209–19.

Zimm, B. H. (1948). 'The scattering of light and the radial distribution function of high polymer solutions', *J. Chem. Phys.* 16: 1093–9.

Conclusions and future prospects

Although small angle scattering as a technique was first developed in the 1930s it has only had an important impact in the field of structural biology since the 1970s, and yet more widely since the end of the 1990s. This has been due to a number of different reasons:

1. The development of neutron and SR sources and detectors.
2. The evolution in structural biology from a reductionist approach, which was dominated by high-resolution techniques such as MX and NMR, to structural systems biology, where objects of ever larger size and complexity determined at lower resolution are of interest to understand function.
3. The realisation that MX can at best provide snapshots of macromolecules in the rather artificial conditions of the crystal and that function is essentially determined by changes in the conformational space of the macromolecules.
4. The progress in modelling which has allowed three-dimensional molecular models to be calculated, both *ab initio* and using hybrid approaches.
5. The multidisciplinary approach exploiting SAS and other biophysical techniques in a complementary fashion.

SAS has now reached a stage where it is one of the major techniques in the armoury of the structural biologist, and one that can be used by a relative non-expert. It is, however, important to understand the range and limitations of the technique which are two of the main reasons for writing this monograph. When combined with excellent biochemistry, providing well-characterised samples, SAS is an extraordinary technique for determining the low-resolution structures and the structural flexibility of macromolecules in the size range from a few kDa to hundreds MDa in molecular mass or a few nanometres to hundreds of nanometres in dimension.

In this monograph we have attempted to illustrate both the similarities and the particularities of SAXS and SANS. SAXS benefits from a much higher flux and brilliance of the radiation, which has considerable advantages for the time-scale of measurements and the small amounts of material required, but is also associated with radiation damage. SANS brings with it all the advantages of contrast variation, allowing the visualisation of individual components of a complex but requiring longer experiments using more precious biological material.

For the future one can hope on advances both in biochemistry and molecular/cell biology and also in X-ray and neutron techniques. Biological advances will help us to produce more high quality biological material and to reconstitute an increasing number of macromolecular complexes in stable form for study. X-ray techniques are on the point of making a massive leap forward in terms of flux with the fourth-generation SR sources, the XFELs.

With the present X-ray sources one can expect that time-resolved SAXS will enter the microsecond domain by means of microfluidics. ASAXS on biologically relevant atoms and ions, not sufficiently utilised hitherto because of high X-ray absorption at lower energies, should become a useful add-on. New experimental procedures using perhaps methods based on fluctuation scattering will definitively have to be developed, especially for work on the XFELs. For SANS, new higher flux spallation sources are coming online, and may provide a significant increase in flux and signal-to-noise. Spin contrast variation has barely been exploited, and there could well be further advances in deuteration techniques via *in vitro* cell free systems.

One of the major factors, which will further boost solution SAS applications in biology, is full automation of the experiment, data processing, analysis and validation. In combination with much lower sample volumes and concentrations required on modern instruments, especially with the emerging microfluidic techniques, large-scale studies will be feasible on microgram amounts of purified material. This will broaden immensely the range of samples amenable to SAS analysis, and also bring new biological users to the technique who will profit from automated model construction without the necessity of becoming experts in the field of SAS.

Appendices

Appendix 1: Basic physics and mathematics of wave phenomena

Waves   Imagine that point **P** in the left panel of Fig. A1.1 moves counterclockwise on a circle with radius OA with a constant **angular velocity** of $-\omega$ radians per second (that is, $\omega T = 2\pi$, where T is the **period** and $\nu = 1/T$ the **frequency**). The oscillatory motion of the projection of OP on the y-axis ($\Psi(t) = $ OQ), which has a maximum value or **amplitude**, OA = A, is called **simple harmonic motion**. At any time t, $\Psi(t)$ is in general given by $\Psi(t) = A\sin(-\omega t + \alpha)$. The angle $(-\omega t + \alpha)$ is called the **phase** and α is the phase at $t = 0$ or initial phase. The velocity is $d\Psi/dt = -\omega A\cos(-\omega t + \alpha)$ and the acceleration $d^2\Psi/dt^2 = -\omega^2 A\sin(-\omega t + \alpha)$. As the origin of the circle is arbitrary, one can always reset the time so as to have $\alpha = 0$.

This simple geometric construction is useful for describing the behaviour of many physical phenomena with a periodic behaviour that does not necessarily imply any circular motion. The relationship between simple harmonic motion and wave motion can be easily understood by imagining that the tip of a pen is attached to Q and that the paper moves at constant velocity to the right, as illustrated in Fig. A1.1.

The distance between two successive points with identical state of motion is the **wavelength** λ, and the corresponding time interval is the **period** T. The phase change between such points (such as P' and P'') is 2π and $k = 2\pi/\lambda$ is the **phase constant** describing the change of phase per unit distance, whereas

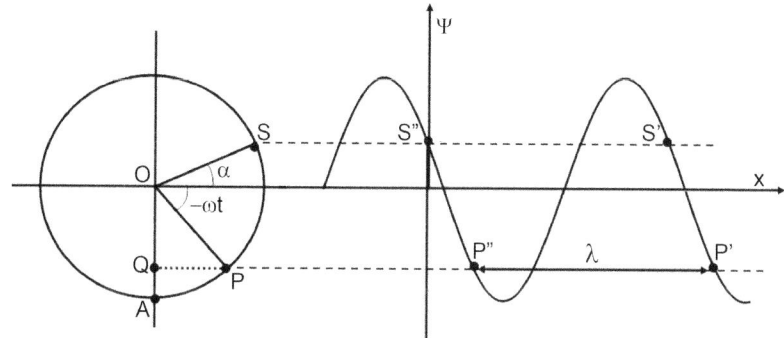

Fig. A1.1 Left: Simple harmonic motion. Right: Displacement (OQ) as a function of distance of propagation (x) at a time t. A similar graph is obtained for a fixed x as a function of time.

$v_\varphi = \lambda/T$, the **phase velocity** represents the distance travelled per unit time. Since $T = 2\pi/\omega$, $v_\varphi = \omega/k$.

The phase velocity v_φ is related to the properties of the medium through which the wave propagates, in the case of electromagnetic waves to the refractive index. In vacuum $v_\varphi = c = (\varepsilon_0\mu_0)^{-1/2}$, where ε_0 is the vacuum permittivity, μ_0 is the vacuum permeability and $c = 299,792,458\,\mathrm{ms}^{-1}$ is the velocity of light, corresponding, by definition, to a refractive index $n = c/v_\varphi = 1$. When the phase velocity in a medium depends on the frequency there is **dispersion** (for example, when white light traverses a prism). As the refractive index for X-rays is always very close to 1, dispersion can be neglected except close to absorption edges.

Waves can be **longitudinal** (such as sound waves) or **transverse** (such as electromagnetic waves), depending on whether the displacement is in the direction of propagation of the wave or perpendicular to it.

The right panel in Fig. A1.1 illustrates that for a wave propagating in the x direction in a homogeneous medium, the harmonic motion of any point x at time t is the same as at $x = 0$ at the time $t' = t - x/v_\varphi$. Since $\Psi(0,t) = A\sin(-\omega t + \alpha)$, it is clear that $\Psi(x,t) = \Psi(0,t') = A\sin(-\omega t + \omega x/v_\varphi) = A\sin(-\omega t + kx)$. With this convention the phase $\varphi = (kx - \omega t)$ at a given x decreases with time, whereas at a given time it increases with distance. Note that in the literature, other conventions for the direction of the x- or time-axis, yielding different expressions for the phase, are also used.

Solutions of the wave equation   Simple harmonic motion as in Fig. A1.1 can be represented by a differential equation, called the **wave equation**:

$$\frac{\partial^2\Psi(x,t)}{\partial t^2} = -\omega^2\frac{\partial^2\Psi(x,t)}{\partial x^2} = -\frac{k^2}{v_\varphi^2}\frac{\partial^2\Psi(x,t)}{\partial x^2} \tag{A1.1}$$

In three dimensions this becomes, with $\Psi = \Psi(x,y,z,t) = \Psi(\mathbf{r},t)$ and ∇^2 the Laplace operator:

$$\frac{\partial^2\Psi(\mathbf{r},t)}{\partial t^2} = -\frac{k^2}{v_\varphi^2}\left[\frac{\partial^2\Psi}{\partial x^2} + \frac{\partial^2\Psi}{\partial y^2} + \frac{\partial^2\Psi}{\partial z^2}\right] = -\frac{k^2}{v_\varphi^2}\nabla^2\Psi(\mathbf{r},t) \tag{A1.2}$$

As sine and cosine are solutions of eq. (A1.1) the more general form of an harmonic oscillation is $\Psi(x,t) = A\sin(kx-\omega t+\alpha)+B\cos(kx-\omega t+\alpha)$, corresponding to a wave without defined origin in space or time.

The advantage of choosing sinusoidal functions for the description of waves is that, in contrast to other possible solutions, their shape is not distorted even when they propagate through dispersive media. Also, as explained in eq. (A1.16), more complex wave shapes can be represented as sums of sinusoidal waves.

It is easy to verify using partial derivatives that any function, not necessarily sinusoidal, of the form $\Psi(x,t) = f(x - v_\varphi t)$, corresponding, as illustrated in Fig. A1.2, to a wave travelling in the x-direction, is a solution of eq. (A1.1), and that a wave travelling in the opposite direction $\Psi'(x,t) = g(x+v_\varphi t)$ is, of course, also a solution, since changing v_φ to $-v_\varphi$ gives the same result.

The most general solution to the wave equation is thus represented by the sum of two waves travelling in opposite directions:

$$\Psi(x,t) = f(x - v_\varphi t) + g(x + v_\varphi t) \tag{A1.3}$$

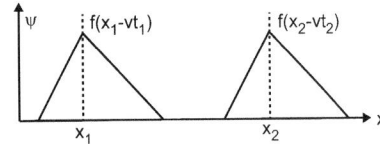

Fig. A1.2 Since $f(x_1 - vt_1) = f(x_2 - vt_2)$, $x_1 - vt_1 = x_2 - vt_2$ and $v = (x_2 - x_1)/(t_2 - t_1)$ is positive if $x_2 > x_1$ and $t_2 > t_1$ and the wave $f(x-vt)$ thus moves in the $+x$ direction which corresponds to the case of Fig. A1.1. The wave $f(x + vt)$ moves along the $-x$ direction.

The **principle of superposition** states that if $\Psi_1(x,t) = f_1(x - v_\varphi t)$ and $\Psi_2(x,t) = f_2(x - v_\varphi t)$ are solutions of the wave equation, $\Psi(x,t) = \Psi_1(x,t) + \Psi_2(x,t)$ is also a solution. This is a consequence of the fact that the wave equation is linear in $f(x,t)$ (that is, if $f_1(x - v_\varphi t)$ and $f_2(x - v_\varphi t)$ are solutions, $g(x - v_\varphi t) = a_1 f_1(x - v_\varphi t) + a_2 f_2(x - v_\varphi t)$ is also a solution).

A special case of superposition is that of two identical waves travelling in opposite directions:

$$A\sin(kx + \omega t) + A\sin(kx - \omega t) \tag{A1.4}$$

Using the relationship $\sin(a) + \sin(b) = 2\sin 1/2(a+b) \cdot \cos 1/2(a-b)$, one finds a resultant wave of the form $A\sin(kx) \cdot \cos(\omega t)$. In systems with a small number of degrees of freedom such solutions are called the (normal) **modes** of the system, whereas in continuous systems they are referred to as **standing waves**. They correspond to situations where every point x has a simple harmonic motion in time with local amplitude $A\sin(kx)$, but the wave does not propagate. X-ray standing-wave techniques play an important role in the study of surfaces and interfaces with high spatial resolution and chemical selectivity.

Solutions to the wave equation explicitly involving sinusoidal functions are somewhat impractical to handle, and one therefore usually prefers to represent them by using complex exponentials.

Complex numbers   Complex numbers $z = [x,y]$ are points in the complex plane which remaps the familiar X, Y-plane in a way that simplifies the representation of waves and signals. In the usual orthogonal coordinate system a complex number is represented as a point $[x,y]$ or as a vector $(Oz) = \mathbf{R}$ between the origin (O) and (x,y), as illustrated in Fig. A1.3. In this system the X-axis is referred to as the real axis and the Y-axis as the imaginary axis.

The complex number $z = [x,y]$ consists of a real (Re) and an imaginary (Im) part, which are both real numbers ($[x,0]$ and $[y,0]$), and the imaginary unit $i = [0,1]$, such that:

$$z = [x,0] + [0,1][y,0] = [x,y] = [x,0] + [0,y] = \text{Re}(z) + i\,\text{Im}(z) = x + iy$$
$$\text{with } [0,1][0,1] = i^2 = [-1,0] = -1$$

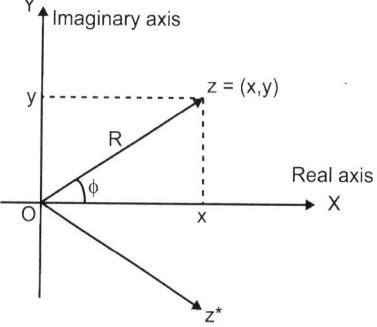

Fig. A1.3 Representation of complex numbers in an orthogonal coordinate system.

Multiplication by the imaginary unit $[0,1]$ rotates the vector $[y,0]$ which is parallel to the real axis counterclockwise by $90°$ degrees and transforms it into a vector $[0,y]$ parallel to the imaginary axis.

Addition and multiplication of complex numbers follow the rules of ordinary algebra. If $z_1 = x_1 + iy_1$ and $z_2 = x_2 + iy_2$, $z_1 + z_2 = (x_1 + x_2) + i(y_1 + y_2)$ and $z_1 z_2 = x_1 x_2 - y_1 y_2 + i(x_1 y_2 - x_2 y_2)$. The solutions ($u$) of the equations $z_1 + u = z_2$ and $z_1 u = z_2$, respectively, define subtraction and division by $z_1 \neq [0,0]$. The complex conjugate of $z = [x,y] = x + iy$ is $z^* = [x,-y] = x - iy$, so that $zz^* = x^2 + y^2$.

As illustrated in Fig. A1.3, geometrically conjugation corresponds to a reflection through the real axis. Note that when taking the conjugate of a complex expression, all numbers must be conjugated.

It is in general preferable to use the **polar form** to represent complex numbers where z is specified by its distance $R = \sqrt{x^2 + y^2}$ from the origin O (its modulus or amplitude), and the angle φ (the phase) between the horizontal and the line through O and z:

$$z = R(\cos\varphi + i\sin\varphi) = R\exp(i\varphi) \tag{A1.5}$$

The last equality (Euler's identity) and its counterpart $z^* = R(\cos\varphi - i\sin\varphi) = R\exp(-i\varphi)$ are easily proven by series expansion.

Multiplication of two complex numbers $a = R_1\exp(i\varphi_1)$ and $b = R_2\exp(i\varphi_2)$ then simplifies to

$$ab = R_1 R_2 \exp[i(\varphi_1 + \varphi_2]$$

Differentiation and integration reduce respectively to multiplication and division by i.

The solutions of the wave equation can thus be represented as:

$$\Psi(x,t) = \mathrm{Re}(\exp i[kx - \omega t + \alpha]) \text{ or } \Psi(x,t) = \mathrm{Re}(\exp(ikx)\exp(-i\omega t)\exp(i\alpha)) \tag{A1.6}$$

In the complex representation of sinusoidal waves only the real part represents a physical quantity (that is, the physical wave is the projection of the complex wave on the real axis). By an appropriate choice of origin one can make $\alpha = 0$ and drop the last exponential. In applications like X-ray scattering one is generally interested only in the time-averaged pattern, and one can ignore the factor $\exp(-i\omega t)$, since after multiplication by its complex conjugate it averages to 1 over a period (that is, $\frac{1}{T}\int_0^T dt = 1$). In contrast, in applications like electronics where one deals with time-varying signals such as voltages and currents at a given point, one ignores the spatial factor $(\exp(ikx))$ and deals only with signals of the form $V(t) = V_0 \exp{-(i\omega t)}$.

Plane waves and spherical waves The surface over which the phase of a wave is constant is called the **wavefront**. In practice, an image or a scattering pattern results from the superposition of harmonic waves with the same frequency but different phase constants emitted by several sources. Close to the sources in the **near field** the solutions to the wave equation are therefore better represented by the superposition of harmonic **spherical waves** of the form: $\Psi(r,t) = \frac{\hat{A}}{r}\sin(\omega t - kr)$, where the constant $\hat{A}$ is the **source strength** and the amplitude $\hat{A}/r$ decreases with the distance from the source as required for energy conservation. As illustrated in Fig. A1.4 this corresponds to concentric spheres of constant radius r and hence constant $\Psi(r,t)$.

When the distance between the source and the detector is large, in the **far field**, the wavefront can be approximated by a plane wave $\Psi(\mathbf{r},t) = A\sin(\omega t - \mathbf{k}\cdot\mathbf{r})$, where $\mathbf{k}\cdot\mathbf{r} = $ constant defines the equation of a plane perpendicular to the direction of propagation of the wave given by the wavevector $\mathbf{k}$.

Polarisation A travelling plane wave $E_x(z,t) = E_x\cos(\omega t - kz)$, propagating along $\mathbf{z}$ and with the transverse component of the field oscillating in the x,z plane, is said to be linearly polarised along $\mathbf{x}$.

More generally, at a fixed position z, the oscillations of the electric field of a plane wave propagating along $\mathbf{z}$ can described as the superposition of two waves—one linearly polarised along $\mathbf{x}$ and the other along $\mathbf{y}$, $\mathbf{E}(z,t) = E_x(\cos\omega t + \varphi_x)\mathbf{e}_x + E_y(\cos\omega t + \varphi_y)\mathbf{e}_y$, where $\varphi_x = k_x x$ and $\varphi_y = k_y y$. The components of the field E_x and E_y are independent, but the changes of E_x and B_y or E_y and $-B_x$ relative to z and t are the not, as they are coupled by Maxwell's equations.

If $E_y = 0$ or $E_x = E_y$ and $\varphi_y = \varphi_x$ or $\varphi_x \pm \pi$ the radiation is linearly polarised, whereas if $E_x = E_y$ and $\varphi_y = \varphi_x - \pi/2$ (that is, the x-oscillation leads the

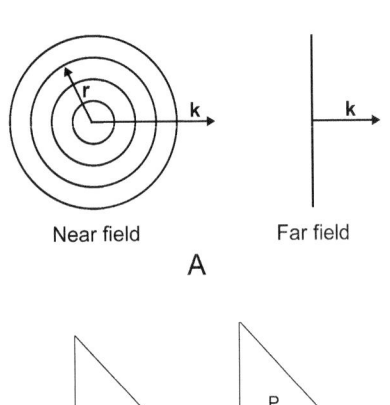

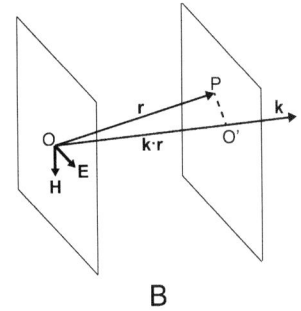

Fig. A1.4 A: In the near field, spherical harmonic waves, shown here in a plane through a diameter of the sphere, are the best representation, whereas in the far field the curvature of the wavefront can be neglected. B: Electromagnetic waves, with electric field $\mathbf{E}$ and magnetic field $\mathbf{H}$ are often represented as plane waves corresponding to planes of constant phase $\varphi = \mathbf{k}\cdot\mathbf{r}$ perpendicular to the direction of propagation of the wave.

y-oscillation by $\pi/2$) it is circularly polarised and if $E_x \neq E_y$ and $\varphi_y \neq \varphi_x$ it is elliptically polarised. The radiation from a bending magnet in a storage ring, for example, is linearly polarised in the plane of the orbit, and elliptically polarised above and below this plane. Optical elements affect the polarisation of the radiation. Although polarisation can be neglected in most case in SAS, this is not the case for measurements at larger angles or in situations where reflections occur.

Interferences   Consider two travelling waves with unit amplitude and slightly different frequencies:

$$A(x,t) = \cos(\omega_1 t - k_1 x) = \cos(a) = \mathrm{Re}\left[\exp i(\omega_1 t - k_1 x)\right] = \mathrm{Re}[\exp(ia)]$$

$$B(x,t) = \cos(\omega_2 t - k_2 x) = \cos(b) = \mathrm{Re}\left[\exp i(\omega_2 t - k_2 x)\right] = \mathrm{Re}[\exp(ib)]$$

Since $\cos(a+b) = \cos a \cos b - \sin a \sin b$ and $\cos(a-b) = \cos(a) \cdot \cos(b) + \sin(a) \cdot \sin(b)$, $\cos(a) \cdot \cos(b) = 1/2 \cos(a+b) + 1/2 \cos(a-b)$. Hence taking $\alpha = a+b$ and $\beta = a - b$ or $a = 1/2(\alpha+\beta)$, $b = 1/2(\alpha-\beta)$, one obtains $\cos(\alpha) + \cos(\beta) = 2\cos(1/2(\alpha+\beta))\cos(1/2(\alpha-\beta))$.

The sum $\Psi(x,t) = A(x,t) + B(x,t)$ is thus given by:

$$\Psi(x,t) = \exp(ia) + \exp(ib) = \exp(1/2 i(a+b))[\exp(1/2 i(a-b)) + \exp(-1/2 i(a-b))]$$

$$= \exp(1/2 i(a+b))[2\cos(1/2(a-b))]$$

The real part of this expression is: $\Psi(x,t) = \cos(1/2(a+b))[2\cos(1/2(a-b))]$.

Setting $1/2(\omega_1 + \omega_2) = \omega$, $1/2(k_1 + k_2) = k$, $1/2(\omega_1 - \omega_2) = \Delta\omega$ and $1/2(k_1 - k_2) = \Delta k$, this simplifies to:

$$\Psi(x,t) = \cos(\omega t - kx)[2\cos(\Delta\omega t - \Delta k x)] \tag{A1.7}$$

As illustrated in Fig. A1.5, the two waves beat and this defines a wave group. The amplitude function (envelope) varies slowly with maxima at $\Delta x = 2\pi/\Delta k$ (with t fixed) and the phase function varies rapidly with maxima at $\Delta x = 2\pi/k$ (with t fixed). The maxima of the amplitude propagate with the **group velocity** $v_g = \Delta\omega/\Delta k$ and the planes of constant phase with the phase velocity $v_\varphi = \omega/k$.

The group velocity is the one associated with the transport of energy. If there is dispersion (that is, v_φ depends on frequency), the group velocity and the phase velocity differ.

Real waves, which are never purely monochromatic but have a certain wavelength ($\Delta\lambda$) or frequency ($\Delta\nu$) spread, will thus not extend infinitely but

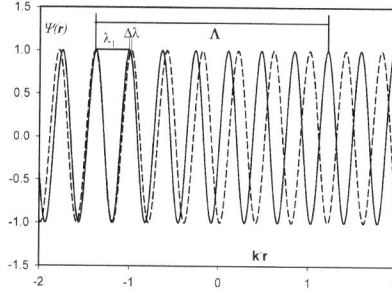
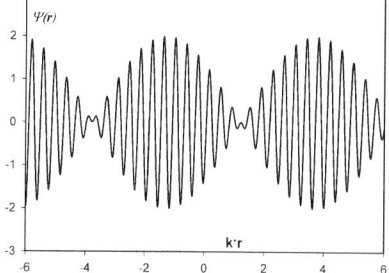

Fig. A1.5 Left: Two waves with slightly different phase factors ($\mathbf{k} \cdot \mathbf{r}$) will be out of phase by π after travelling a distance Λ corresponding to the longitudinal or temporal coherence length. Right: The superposition of the two waves leads to wave packets or a wave train.

consist of wave trains or wave packets which are temporally and spatially limited. This phenomenon determines the coherence of light sources.

Coherence   The **temporal or longitudinal coherence length** of radiation is the distance over which two waves with wavelengths λ and $\lambda + \Delta\lambda$ become out of phase by π. It describes the fact that the effective frequency range of monochromatic radiation is of the order of the reciprocal of the duration of a wave ($2\Delta\lambda$ = full width half maximum).

$$\Lambda = N\lambda_1 = \left(N - \frac{1}{2}\right)\lambda_2 \Rightarrow \Lambda \cong \frac{1}{2}\frac{\lambda^2}{\Delta\lambda} \tag{A1.8}$$

The **transverse one-sigma coherence area** corresponds to the area (S) of the sample at a distance R from the source which is coherently illuminated by a quasi-monochromatic incoherent source with horizontal and vertical source sizes σ_x and σ_y:

$$S = \lambda_0^2 R^2 / 4\pi \sigma_x \sigma_y \tag{A1.9}$$

Note that the transverse coherence does not depend on the wavelength spread but only on geometry (source size and distance between source and object).

If the source becomes very small ($\sim 20\,\mu$m) and R large (~ 50 m), as is the case with some instruments at modern synchrotron radiation sources, the samples are partially coherently illuminated over larger areas (mm^2), even if the source is not coherent.

A particularly important application of localised non-periodic wave trains or pulses, which arise from the superposition of a large number of oscillations with equal amplitudes and nearly equal phases (nearly equal $\mathbf{k}$ and ω), is the quantum-mechanical description of particles like neutrons in terms of probability amplitude waves illustrated in Fig. A1.6.

The energy of such particles with mass m is $E = \hbar\omega = h\nu$, where $\hbar = h/2\pi$, h is Planck's constant and ν the frequency, and their momentum is $\mathbf{p} = m\mathbf{v}$ is $\mathbf{p} = \hbar\mathbf{k}$. The velocity of the particle is the group velocity of the wave train: $v_g = \frac{c^2 p}{E}$, where E is the energy of the particle. For neutrons one finds that since $p = m_n v = h/\lambda$, here m_n is the mass of the neutron:

$$\lambda(\text{nm}) = \frac{h}{m_n v_g} = 396.6 \text{ m}^2\text{s}^{-1}/v_g(\text{ms}^{-1}) \tag{A1.10}$$

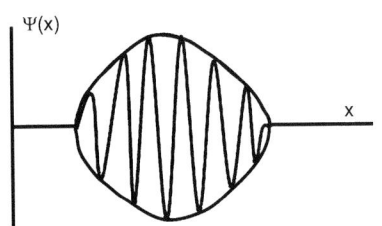

Fig. A1.6 Localised wave train or pulse representing a particle in terms of probability amplitude.

Dirac δ function   Before discussing the basic mathematics underlying the relationship between a scattering (or diffraction) pattern and a structure, it is useful to introduce the **Dirac δ function**, which plays an important role in the solution of many scattering problems. This function, $\delta(x)$, has a value $\delta(x) = \infty$ for $x = 0$ and $\delta(x) = 0$ for $x \neq 0$, with $\int_{-\infty}^{\infty}\delta(x)dx = 1$, and corresponds to an infinitely narrow and high spike at $x = 0$. Similarly, $\delta(x - a) = \infty$ for $x = a$ and $\delta(x) = 0$ for $x \neq a$ corresponds to the same spike shifted to a. It is easy to see that with this definition the integral $\int_{-\infty}^{\infty}f(x)\delta(x - a)dx = f(a)$ selects or filters out a single value from the function $f(x)$.

An alternative and perhaps more intuitive definition of the δ function is obtained by taking the limit of a Gaussian with vanishing width and constant area: $\delta(x) = \lim_{a\to\infty}\sqrt{a/\pi}\exp(ax^2)$, which also implies that $\delta(x) = \frac{\delta(x)}{|m|}$, where m is a constant.

Convolution   The convolution of two functions $f(x)$ and $g(x)$ is defined as:

$$f(x)^*g(x) = \int\limits_{-\infty}^{\infty} f(u)\,g(x-u)\,du \qquad (A1.11)$$

where u is a dummy variable running over all values of x. In some definitions the integral is multiplied by a factor of $1/\sqrt{2\pi}$. The relationship between the functions $g(u)$ and $g(x-u)$ for a given value of x is easily understood by noting that $g(-u)$ results from the inversion of $g(u)$ through the origin of the abscissa, which is equivalent to flipping $g(u)$ around the y-axis, as illustrated in Fig. A1.7.B. The function $g(x-u)$ is obtained by shifting the origin of $g(-u)$ by x along the abscissa.

An interesting case is that of convolutions involving δ functions. If $g(x) = \delta(x)$, the integral $\int_{-\infty}^{\infty} f(u)\delta(x-u)du = f(x)$. As in eq. (A1.11) this is repeated for every value of x, it is easy to see that $f(x)^*g(x) = f(x)$ (that is, the function $f(u)$ is simply transferred to the space of x). Similarly, if $g(x) = \delta(x-a)$, which is a δ function with a peak at a, $\int_{-\infty}^{\infty} f(u)\delta(x-(u-a))du = f(x+a)$, which again reproduces the function $f(x)$ but this time shifted by $-a$ along the abscissa.

In general, as illustrated in Fig. A1.7, the value of $f(x)^*g(x)$ is obtained by repeating the sequence FLIP–SHIFT–MULTIPLY–INTEGRATE for each value of x: flip $g(u)$ around the y-axis to obtain $g(-u)$, shift $g(-u)$ by x to obtain $g(x-u)$, multiply by $f(u)$ for all values of u, and integrate the product.

In the case of functions of several variables (for example, $f(u) = f(x,y,z)$ and $g(u) = g(x,y,z)$) the function $g(u)$ is inverted through a centre of symmetry at the origin to obtain $g(-u) = g(-x,-y,-z)$, and its origin is displaced by a vector $\mathbf{r}$:

$$f(\mathbf{r})^*g(\mathbf{r}) = \int\limits_{-\infty}^{\infty} f(\mathbf{u})\,g(\mathbf{r}-\mathbf{u})\,d\mathbf{u} \qquad (A1.12)$$

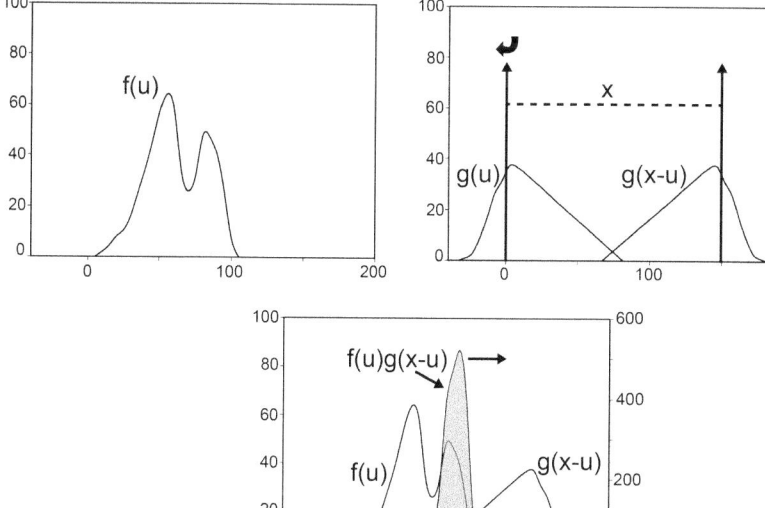

Fig. A1.7 The value of the convolution of the two functions $f(x)$ and $g(x)$ is obtained by flipping $g(u)$ around the y-axis to obtain $g(-u)$. The origin of $g(-u)$ is shifted to x (here $x = -150$) to obtain $g(x-u)$. The functions $f(u)$ and $g(x-u)$ are multiplied and the value of the convolution is obtained by integrating the result $f(x)^*g(x) = \int_{-\infty}^{\infty} f(u)g(x-u)du$ corresponding to the shaded area. These operations must be repeated for all values of x.

Fig. A1.8 Convolution of a linear array of equally spaced δ functions (lattice) with a motif yields a linear crystal.

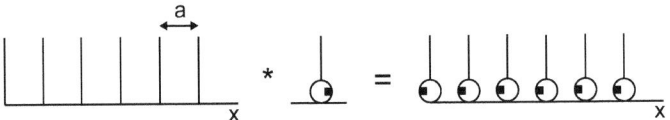

Here again the operations must be repeated for all possible vectors **r** in the range (volume) where the functions are defined.

Convolution is commutative: $f(x)*g(x) = g(x)*f(x)$, and distributive: $f(x)*(g(x) + h(x)) = f(x)*g(x) + f(x)*h(x)$.

An application of convolutions: making crystals, chain molecules and solutions

An infinite linear lattice can be represented as a sum of equally spaced δ function:

$$L(x) = \sum_{n=-\infty}^{\infty} \delta(x - na) \qquad (n \text{ integer}) \qquad \text{(A1.13a)}$$

As illustrated in Fig. A1.8, a crystal can be described as the convolution of the electron density distribution within one unit cell, $\rho(x)$, which is the motif, with the lattice, which for real crystals is a three-dimensional array of δ functions. For a linear lattice:

$$L*\rho(x) = \rho(x)*L = \int_{-\infty}^{\infty} \rho(u) \sum_{n=-\infty}^{\infty} \delta[x - (u - na)]\, du = \sum_{n=-\infty}^{\infty} \rho(x + na)$$

$$\text{(A1.13b)}$$

A similar approach is useful in many circumstances, as illustrated in Fig. A1.9. In these cases one uses a pseudo-lattice where the δ functions are no longer necessarily regularly spaced.

Correlation Correlation $f(x) \circ g(x) = f(x) * g(-x)$ is similar to convolution, except that $g(u)$ does not get flipped. Correlation is thus a repetitive sequence of SHIFT–MULTIPLY–INTEGRATE operations.

$$f(x) \circ g(x) = \int_{-\infty}^{\infty} f(u)\, g(x + u)\, du \qquad \text{(A1.14)}$$

As in the case of convolution, in some definitions the integral is multiplied by a factor of $1/\sqrt{2\pi}$. If $g(x)$ is even (that is, if $g(x) = g(-x)$), convolution and

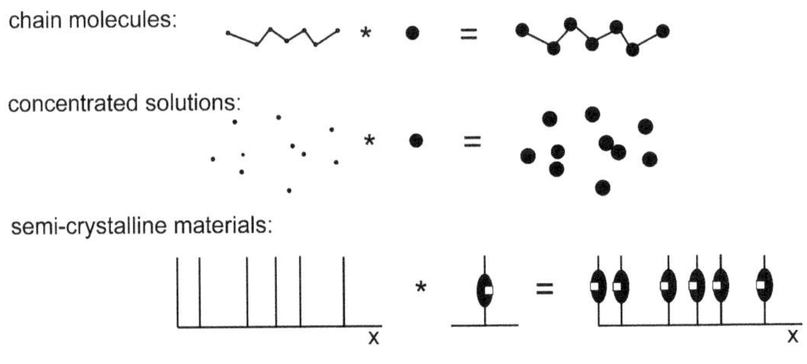

chain molecules:

concentrated solutions:

semi-crystalline materials:

Fig. A1.9 Examples of convolutions used in the description of chain molecules, concentrated solutions or semicrystalline materials.

correlation produce, of course, the same result. The most important case in the context of X-ray scattering is that of autocorrelation:

$$f(x) \circ f(x) = f(x)^* f(-x) \tag{A1.15}$$

The averaged autocorrelation function of the density distribution $\gamma(r) = \langle \rho(r)^* \rho(-r) \rangle$ is the correlation function of the particle, which is related to the distance distribution function $p(r) = r^2 \gamma(r)$.

Fourier series and Fourier transforms　Fourier series and transforms are indispensable tools for describing all kinds of signals. If the signal is periodic as in crystallography one uses Fourier series; if it is non-periodic as in scattering one uses Fourier transforms.

Any single-valued periodic function $f(x)$ which is piecewise-differentiable over the interval $[-\pi, \pi]$ can be represented as a sum of harmonic functions or **Fourier series**. One can always map the interval $[-\pi, \pi]$ to any interval $[-L, L]$ over which the function is periodic by taking $x = x' \pi / L$. Rather than using sines or cosines only with a phase shift, it is more convenient to use sums of these functions where n is integer:

$$f(x) = \frac{a_0}{2} + \sum_{n=1}^{\infty} (a_n \cos nx + b_n \sin nx) \tag{A1.16}$$

The Fourier coefficients of the Fourier series of $f(x)$, a_n and b_n can be found easily by integration (using the fact that $\cos(nx)\sin(kx) = 1/2[\sin(n+k)x - \sin(n-k)x]$ and $\sin(nx)\sin(kx) = 1/2[-\cos(n+k)x + \cos(n-k)x]$ and remembering that integrals from $-\pi$ to π of $\cos(nx)$ and $\sin(nx)$ with n integer are zero.

$$a_0 = \frac{1}{\pi} \int_{-\pi}^{\pi} f(x)\, dx \tag{A1.17}$$

$$a_k = \frac{1}{\pi} \int_{-\pi}^{\pi} f(x) \cos kx\, dx \tag{A1.18}$$

$$b_k = \frac{1}{\pi} \int_{-\pi}^{\pi} f(x) \sin kx\, dx \tag{A1.19}$$

The first term in the series, $a_0/2$ is the average of $f(x)$ over the period.

The Fourier series can also be written conveniently in complex form, taking into account that as $\cos \varphi + i \sin \varphi = \exp(i\varphi)$, and hence, $\cos \varphi = (e^{i\varphi} + e^{-i\varphi})/2$ and $\sin \varphi = (e^{i\varphi} - e^{-i\varphi})/2i = -i(e^{i\varphi} - e^{-i\varphi})/2$:

$$f(x) = \frac{a_0}{2} + \sum_{n=1}^{\infty} \left[a_n \left(\frac{e^{inx} + e^{-inx}}{2} \right) - i b_n \left(\frac{e^{inx} - e^{-inx}}{2} \right) \right] = \frac{a_0}{2} + \sum_{n=1}^{\infty} \left[\left(\frac{a_n - i b_n}{2} \right) e^{inx} + \left(\frac{a_n + i b_n}{2} \right) e^{-inx} \right]$$

$$f(x) = c_0 + \sum_{n=1}^{\infty} c_{+n}\, e^{inx} + c_{-n}\, e^{-inx} = \sum_{n=-\infty}^{\infty} c_n\, e^{inx} \tag{A1.20}$$

As a simple example of Fourier series expansion we consider the periodic function $f(x)$ with period 2π and $f(x) = -1$ for $-\pi < x < 0$ and $f(x) = 1$ for $0 \leq x \leq \pi$

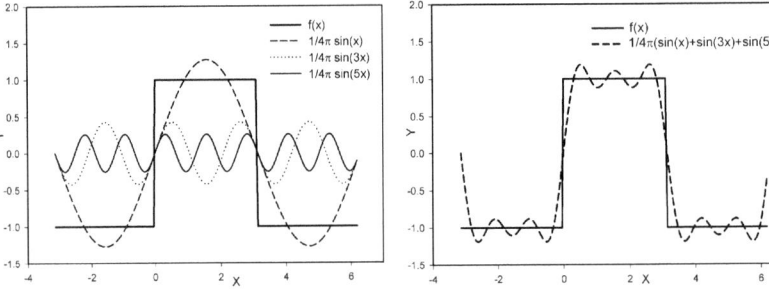

Fig. A1.10 Left: First three terms of the Fourier series of the function $f(x) = -1$ for $-\pi < x < 0$ and $f(x) = 1$ for $0 \le x \le \pi$. Right: $f(x)$ and sum of the first three terms of the Fourier series. A large number of terms is required to dampen the ripples.

in Fig. A1.10, which represents a square wave. In this case, $a_0 = 0$, because the function oscillates around 0, $a_n = 0$, because the function is odd (that is, $f(x) = -f(-x)$) and $b_n = (2/\pi n)(1 - \cos n\pi)$, which is equal to 0 for n even and $4/\pi n$ for n odd, and hence:

$$f(x) = (4/\pi)[\sin x + \sin 3x/3 + \sin 5x/5 + \dots \sin(2n+1)x/(2n+1)]$$

Whereas the Fourier series for odd functions like the one above contain only the coefficients b_n associated with the sine terms, even functions, where $f(x) = f(-x)$, contain only the coefficients a_n associated with the cosine terms. In other words, odd functions are obtained as sums of odd functions (sines) and even functions as sums of even functions (cosines). Functions which are neither odd nor even are represented as sums of odd and even functions.

Note that a truncated series like that in Fig. A1.10 provides only a low-resolution picture, since the low-index terms, which correspond to low frequencies, define the broad features, and the higher-index terms increase the detail. Fourier series are the most important mathematical tool used in the description of the periodic three-dimensional electronic densities in crystals. The description of the electron density $\rho(x, y, z)$ in the unit cell in terms of the Fourier coefficients (that is, the structure factors $F(hkl)$) are given in the International Tables of Crystallography for all space groups using expressions analogous to eq. (A1.20). For centrosymmetric crystals the electron density is an even function ($\rho(x, y, z) = \rho(\bar{x}, \bar{y}, \bar{z})$), the phases are restricted to 0 and π, and the structure factors are real numbers, whereas for non-centrosymmetric crystals the structure factors are imaginary numbers with a real and an imaginary part and the phase varies between 0 and 2π.

Fourier series are rarely used in SAS, because most objects are not periodic except for the example in the case of lipid systems. However, by extending the concept of Fourier series to the case where the harmonics vary continuously from 0 to ∞, one obtains **Fourier transforms,** which can be used to describe non-periodic structures and are the main mathematical tool in SAS.

The Fourier integral theorem states that for any function defined over $[-\infty, \infty]$ which is piecewise-differentiable and absolutely integrable (that is, for which $\int_{-\infty}^{\infty} |f(x)dx|$ converges):

$$f(x) = \int_{0}^{\infty} [A(k)\cos(kx) + B(k)\sin(kx)]dk \qquad (A1.21)$$

where $A(k) = \frac{1}{\pi} \int\limits_{-\infty}^{\infty} f(x) \cos(kx)dx$ and $B(k) = \frac{1}{\pi} \int\limits_{-\infty}^{\infty} f(x) \sin(kx)dx$. These equations are similar to eqs. (A1.16–A1.19), but with the major difference that here k is a continuous rather than an integer variable.

Using the dummy variable u, which runs over all values of x, eq. (A1.21) can be rewritten as:

$$f(x) = \int\limits_{k=0}^{\infty} \left[\frac{1}{\pi} \int\limits_{-\infty}^{\infty} f(u) \cos(ku) \cos(kx)du + \frac{1}{\pi} \int\limits_{-\infty}^{\infty} f(u) \sin(ku) \sin(kx)du \right] dk \qquad \text{(A1.22)}$$

$$= \frac{1}{\pi} \int\limits_{k=0}^{\infty} \left\{ \int\limits_{-\infty}^{\infty} f(u) [\cos(ku) \cos(kx) + \sin(ku) \sin(kx)]du \right\} dk \qquad \text{(A1.23a)}$$

As the expression between square brackets is $\cos k(x - u) = 1/2(e^{ik(x-u)} + e^{-ik(x-u)})$ one obtains, taking into account that $\int_0^\infty (e^{ik(x-u)} + e^{-ik(x-u)})dk = \int_{-\infty}^\infty e^{-ik(x-u)}dk$ and changing the limits of integration for k:

$$f(x) = \frac{1}{\pi} \int\limits_{k=0}^{\infty} \int\limits_{u=-\infty}^{\infty} f(u) \cos k(x-u)du\, dk = \frac{1}{2\pi} \int\limits_{k=-\infty}^{\infty} \left[\int\limits_{-\infty}^{\infty} f(u)e^{iku}du \right] e^{-ikx}dk \qquad \text{(A1.23b)}$$

The function $F(k) = \frac{1}{\sqrt{2\pi}} \int\limits_{-\infty}^{\infty} f(x)e^{ikx}dx$ is the **Fourier transform** of $f(x)$, which

we shall note $\Im(f(x))$, and $f(x) = \frac{1}{\sqrt{2\pi}} \int\limits_{-\infty}^{\infty} F(k)e^{-ikx}dk$, which is the **inverse**

Fourier transform of $F(k)$ (noted $\Im^{-1}(F(k))$).

In the literature, confusion often arises from the fact that there are a number of definitions of the Fourier transform differing by the factors in front of the integrals representing $f(x)$ and $F(k)$ and the sign convention for the exponential. The product of the constants in front of the integrals is always dimensionless and is equal to $1/2\pi$, and with this limitation any two arbitrary constants $[a, b]$ may be used to define a Fourier transform pair as

$$F(k) = \sqrt{\frac{|b|}{(2\pi)^{1-a}}} \int\limits_{-\infty}^{\infty} f(x)e^{ibkx}dx \text{ and } f(x) = \sqrt{\frac{|b|}{(2\pi)^{1+a}}} \int\limits_{-\infty}^{\infty} F(k)e^{-ibkx}dk \quad \text{(A1.24)}$$

Modern physics prefers $[0,1]$, classical physics used $[-1,1]$, pure mathematics uses $[1,-1]$ and signal processing $[0,-2\pi]$ (see Eric W. Weisstein, 'Fourier Transform', in *MathWorld*, A Wolfram Web Resource: <http://mathworld.wolfram.com/FourierTransform.html>).

This implies that if $f(x)$ is defined over $[-L, L]$, $F(k)$ will be defined over $[-1/2\pi L, 1/2\pi L]$ and corresponds to the definition of reciprocal space in solid-state physics, which differs by a factor 2π from the crystallographic definition.

As a simple illustration, consider the Fourier transform of the function $f(x) = 1$ for $|x| < a$ and $f(x) = 0$ for $|x| > a$, which defines a rectangular box of unit height and length a in Fig. A1.11.

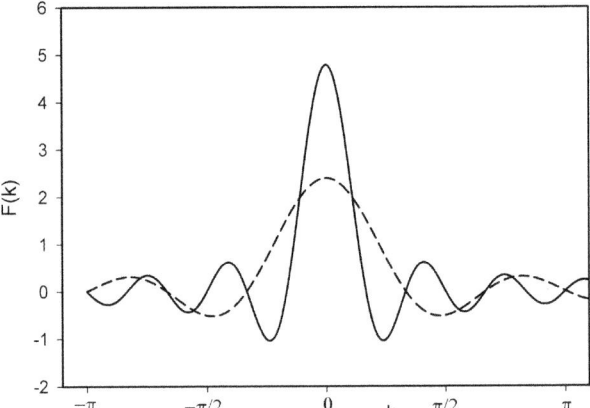

Fig. A1.11 Fourier transforms of the function $f(x) = 1$ for $|x| < a$ and $f(x) = 0$ for $|x| > a$ which defines a rectangular box for two values of $a : a = 3$ (dashed line) and $a = 6$ (full line).

$$F(k) = \frac{1}{\sqrt{2\pi}} \int_{-\infty}^{\infty} f(x)\, e^{ikx}\, du = \frac{1}{\sqrt{2\pi}} \int_{-a}^{a} (1)\, e^{ikx}\, du = \frac{1}{\sqrt{2\pi}} \left. \frac{e^{ikx}}{ik} \right|_{-a}^{a}$$

$$= \frac{1}{\sqrt{2\pi}} \left(\frac{e^{ika} - e^{-ika}}{ik} \right) = \sqrt{\frac{2}{\pi}}\, \frac{\sin ka}{k} \; , \; k \neq 0 \quad and \quad \sqrt{\frac{2}{\pi}}a \quad for \quad k = 0 \qquad (A1.25)$$

Note that the value of $F(0)$ is proportional to the area of the box, and that the width of the central maximum ($\Delta k = \pi/a$) is inversely related to the width of the box (that is, the interval over which $f(x)$ is defined, $\Delta x = 2a$). The relationship $\Delta k \Delta x = 2\pi$ is known in quantum mechanics as Heisenberg's uncertainty principle (since $p = \hbar k$, $\Delta p \Delta x = h$, Planck's constant), which arises from the representation of particles by probability amplitude waves. As time and angular velocity ($\omega = 2\pi\nu$) are also reciprocal, $\Delta\omega\Delta t = 2\pi$ or since $\Delta E = \hbar\Delta\omega$, $\Delta E\Delta t = h$) is an alternative expression of this principle.

Clearly, as $\Delta k \Delta x = 2\pi$, when the interval Δx becomes very large, $F(k)$ becomes very narrow, and when $\Delta x \to \infty$ (that is, $f(x) = 1(x)$, a function which has a constant value of 1 from $-\infty$ to $+\infty$), $F(k) \to \delta(k)$. Hence,

$$\Im(\delta(x)) = 1(k) \qquad (A1.26)$$

and

$$\Im(1(x)) = \frac{1}{\sqrt{2\pi}} \int_{-\infty}^{\infty} e^{ikx} dx = \sqrt{2\pi}\delta(k) \qquad (A1.27)$$

Table A1.1 gives a list of useful functions for small-angle scattering and their Fourier transforms.

The following properties of Fourier transforms are important:

1. The Fourier transform of a linear combination of functions is a linear combination of their transforms:

$$\Im(af(x) + bg(x)) = a\Im(f(x)) + b\Im(g(x)) \qquad (A1.28)$$

Table A1.1 A few functions which are useful in small angle scattering and their Fourier transforms.

Function ($f(x)$)	Fourier transform ($F(k)$)
$\exp(-\pi x^2)$	$\exp(-\pi k^2)$
$\cos(2\pi ax)$	$[\delta(k+a) + \delta(k-a)]/2$
$\sin(2\pi ax)$	$[\delta(k+a) - \delta(k-a)]/2$
$\prod(x,a) = 1, \quad \|x\| \leq a/2$ $\qquad\quad 0, \quad \|x\| > a/2$	$a\sin(\pi ak)/(\pi ak)$
$\Lambda(x,a) = 1 - \|x\|/a, \|x\| \leq a$ $\qquad\quad 0, \|x\| > a$	$a[\sin(\pi ak)/(\pi ak)]^2$

$\prod(x,a)$ represents a rectangle and $\prod(x,a) \cdot \prod(x,b) \cdot \prod(x,c)$ a parallelipiped.
$\Lambda(x,a)$ represents a triangle.

2. The Fourier transform of the complex conjugate is the Fourier transform of the original function inverted through the origin:

$$\Im(f^*(x)) = \Im(f(-x)) \tag{A1.29}$$

3. The Fourier transform of the product of two functions is the convolution of their Fourier transforms (multiplication theorem):

$$\Im(f(x) \cdot g(x)) = \Im(f(x)) * \Im(g(x)) \tag{A1.30}$$

4. The Fourier transform of the convolution of two functions is the product of their Fourier transforms (convolution theorem):

$$\Im(f(x) * g(x)) = \Im(f(x)) \cdot \Im(g(x)) \tag{A1.31}$$

Here is a simple proof, where u is a dummy variable.
Taking

$$F(k) = \frac{1}{\sqrt{2\pi}} \int_{-\infty}^{\infty} f(u)e^{iku}du \quad \text{and} \quad g(x-u) = \frac{1}{\sqrt{2\pi}} \int_{-\infty}^{\infty} G(k)\, e^{-ik(x-u)}\, dk$$

the Fourier transform of the convolution $f(x)^*g(x)$ is

$$H(k) = \frac{1}{\sqrt{2\pi}} \int_{-\infty}^{\infty} \left[\frac{1}{\sqrt{2\pi}} \int_{-\infty}^{\infty} f(u)g(x-u)du \right] e^{ikx}dx$$

and setting $x - u = w$, expressing the exponential as a product and separating the parts containing u and w,

$$H(k) = \frac{1}{\sqrt{2\pi}} \int_{-\infty}^{\infty} \left[\frac{1}{\sqrt{2\pi}} \int_{-\infty}^{\infty} f(u)g(w)du \right] e^{iku} e^{ikw} \frac{dx}{dw}dw = \frac{1}{\sqrt{2\pi}} \int_{-\infty}^{\infty} f(u)e^{iku}du \frac{1}{\sqrt{2\pi}} \int_{-\infty}^{\infty} g(w)e^{ikw}dw$$

The names of dummy variables like u and w are arbitrary, and they can thus also be replaced by x so that

$$H(k) = \Im(f(x)^* g(x)) = \frac{1}{\sqrt{2\pi}} \int_{-\infty}^{\infty} f(x) e^{ikx} dx \frac{1}{\sqrt{2\pi}} \int_{-\infty}^{\infty} g(x) e^{ikx} dx = F(k) \cdot G(k)$$

This proof is equally valid for the inverse transform, as only the sign of the exponential changes. The convolution theorem can be rewritten as

$$\Im(\Im^{-1}(F(k))^* \Im^{-1}(G(k))) = F(k) \cdot G(k) \Rightarrow (\Im^{-1}(F(k))^* \Im^{-1}(G(k))) = \Im^{-1}(F(k) \cdot G(k))$$

Making use of the analogy between $\Im$ and $\Im^{-1}$, one simply obtains the multiplication theorem (eq. A1.30).

5. The Fourier transform of a correlation is the product of the Fourier transform of one function with the Fourier transform of the complex conjugate of the other. This property follows directly from (2) and (4).

$$\Im(f(x) \circ g(x)) = F(k) \cdot G^*(k) \tag{A1.32}$$

6. The similarity theorem states that a change of scale by a factor a of the abscissa in real space leads to a contraction by $1/a$ of the abscissa in reciprocal space, as implied by eq. (A1.24).

$$\Im(f(ax)) = \frac{1}{\sqrt{2\pi}} \int_{-\infty}^{\infty} f(ax) e^{ikx} dx = \frac{1}{\sqrt{2\pi}} \int_{-\infty}^{\infty} f(ax) e^{i(\frac{k}{a})ax} \frac{dx}{d(ax)} d(ax) = |a|^{-1} F\left(\frac{k}{a}\right) \tag{A1.33}$$

7. If a function is shifted along the abscissa by x_0 its transform is multiplied by e^{ikx_0}.

$$\Im(f(x - x_0)) = \frac{1}{\sqrt{2\pi}} \int_{-\infty}^{\infty} f(x - x_0) e^{ikx} dx$$

$$= \frac{1}{\sqrt{2\pi}} \int_{-\infty}^{\infty} f(x - x_0) e^{ik(x - x_0)} e^{ikx_0} \frac{dx}{d(x - x_0)} d(x - x_0) = F(k) e^{ikx_0} \tag{A1.34}$$

This expression can be easily extended to three-dimensional space, and is very useful to calculate the scattering amplitudes of an assembly when the amplitudes of the individual subunits in the proper orientation are known.

8. The Fourier transform of the derivative of a function $f(x)$ is given by

$$\Im(f'(x)) = \frac{1}{\sqrt{2\pi}} \int_{-\infty}^{\infty} f'(x) e^{ikx} dx = -ikF(k) \tag{A1.35}$$

Setting $du = f'(x)dx$ and $v = e^{ikx}$ ($u = f(x)$, $dv = ike^{ikx}dx$), integration by parts ($\int v du = uv - \int u dv$) yields $\Im(f'(x)) = \frac{1}{\sqrt{2\pi}}[f(x)e^{ikx}\Big|_{-\infty}^{\infty} - ik \int_{-\infty}^{\infty} f(x)e^{ikx}dx] = -ikF(k)$.

The first term in the last equation is the product of $f(x)$ and an oscillating function, which vanishes if $\lim_{x \to \infty}(f(x)) = 0$.

9. Fourier transforms of odd and even functions

As already mentioned, an even function is one where $f(x) = f(-x)$ (for example, $\cos(x)$), and an odd function is one where $f(x) = -f(-x)$ (for example, $\sin(x)$). As in the case of the Fourier series above, the Fourier transform of an even function has only cosine terms (it is real), and $F(k)$ and $f(x)$ are cosine transforms of each other:

$$F(k) = \sqrt{\frac{2}{\pi}} \int_0^\infty f(x) \cos(kx) \, dx \qquad (A1.36)$$

Conversely, the Fourier transform of an odd function has only sine terms (it is purely imaginary), and $F(k)$ and $f(x)$ are sine transforms of each other:

$$F(k) = \sqrt{\frac{2}{\pi}} \int_0^\infty f(x) \sin(kx) \, dx \qquad (A1.37)$$

Since $1/2[f(x)+f(-x)]+1/2[f(x)-f(-x)] = E(x)+O(x)$, every function can be partitioned in even $E(x)$ and odd $O(x)$ parts, and its Fourier transform can thus also be written as a sum of Fourier transforms of these parts.

If $F(k)$ and $G(k)$ are cosine or sine transforms, **Parseval's identity** applies:

$$\int_0^\infty F(k) \, G(k) \, dk = \int_0^\infty f(x) \, g(x) \, dx \qquad (A1.38)$$

In general, however, $\int_{-\infty}^\infty F(k) \, G^*(k) \, dk = \int_{-\infty}^\infty f(x) \, g^*(x) \, dx$, and if $f(x) = g(x)$ this becomes

$$\int_{-\infty}^\infty F(k) \, F^*(k) \, dk \int_{-\infty}^\infty \left| F(k)^2 \right| dk = \int_0^\infty f(x) \, f^*(x) \, dx = \int_0^\infty \left| f^2(x) \right| dx \qquad (A1.39)$$

which states that the total energy (or intensity) in a wave is the integral of the energies (or intensities) in all its Fourier components.

For real functions such as the electron density or scattering length density one has the following useful relationships:

$$f(x) = \frac{1}{\sqrt{2\pi}} \int_{-\infty}^\infty F(k) e^{-ikx} dk \quad \text{and} \quad f(-x) = \frac{1}{\sqrt{2\pi}} \int_{-\infty}^\infty F(k) e^{ikx} dk \qquad (A1.40)$$

$$F(k) = \frac{1}{\sqrt{2\pi}} \int_{-\infty}^\infty f(x) e^{ikx} dx \quad \text{and} \quad \Im(f(-x)) = \frac{1}{\sqrt{2\pi}} \int_{-\infty}^\infty f(-x) e^{ikx} dx = F(-k) \qquad (A1.41)$$

$$F^*(k) = \frac{1}{\sqrt{2\pi}} \int_{-\infty}^\infty f(x) e^{-ikx} dx \quad \text{and} \quad F(-k) = \frac{1}{\sqrt{2\pi}} \int_{-\infty}^\infty f(x) e^{-ikx} dx \qquad (A1.42)$$

and hence,

$$F(-k) = F^*(k) = \Im(f(-x)) \qquad (A1.43)$$

The Fourier transform $\Im(f(x) * f(-x)) = F(k) \cdot F(-k)$, in agreement with the convolution theorem.

Since for a real function, $F^*(k) = F(-k)$: $F(k) \cdot F(-k) = F(k) \cdot F^*(k)$, the Fourier transform of the autocorrelation of the real function $f(x)$ is the squared modulus of its transform

$$F(k) \cdot F^*(k) = |F(k)|^2 \qquad (A1.44)$$

If $f(x)$ is the electron density or scattering length density, $F(k)$ is the scattering amplitude, and the product $F(k) \cdot F^*(k) = |F(k)|^2$ represents the scattered intensity.

Appendix 2: Spherical harmonics and their applications for SAS

Formalism of spherical harmonics   The spherical harmonics represent the angular portion of the solution to Laplace's equation (Edmonds 1957) defined as a complete set of angular functions in spherical coordinates $(r, \omega) = (r, \theta, \varphi)$. The real and imaginary parts of the spherical harmonics are trigonometric functions of two integer indices $0 \leq l < \infty$ and $-l \leq m \leq l$, expressed as

$$Y_{lm}(\theta, \varphi) = \sqrt{\frac{(2l+1)(l-m)!}{4\pi(l+m)!}} P_l^{|m|}(\cos\theta)\exp(im\varphi) \qquad (A2.1)$$

where $P_l^m(\cos\theta)$ are associated Legendre polynomials of the first kind (Abramowitz and Stegun 1964). The spherical harmonics are orthonormal; that is:

$$\int_0^\pi \int_0^{2\pi} Y_{lm}(\omega)Y_{l'm'}(\omega)d\omega = \delta_{ll'}\delta_{mm'} \qquad (A2.2)$$

where δ denotes the Kronecker δ function (Abramowitz and Stegun 1964). This orthogonality makes the spherical harmonics very useful for all operations dealing with the spherical averaging of three-dimensional functions. Similarly to Fourier series, the lower-order terms define the gross structural features (low frequencies), whereas the higher harmonics describe finer details (high frequencies); see Table A2.1.

Scattering intensity using spherical harmonics expansion   The scattering intensity from monodisperse systems comes from the spherical averaging over all orientations of the particle

$$I(\mathbf{q}) = \langle A(\mathbf{q}) \cdot A^*(\mathbf{q})\rangle_\Omega \qquad (A2.3)$$

where (q, Ω) are spherical coordinates in reciprocal space. The scattering amplitude $A(q)$ is the Fourier transform of the excess scattering length density

Table A2.1 Analytical expressions for selected spherical harmonics.

L	m	$Y_{lm}(\theta, \varphi)$
0	0	$\sqrt{\dfrac{1}{4\pi}}$
1	0	$\sqrt{\dfrac{3}{4\pi}}\cos\theta$
1	± 1	$\sqrt{\dfrac{3}{8\pi}}[\sin\theta\exp(\pm i\varphi)]$
2	0	$\sqrt{\dfrac{5}{64\pi}}[3\cos(2\theta)+1]$
2	± 1	$\dfrac{3}{2}\sqrt{\dfrac{5}{24\pi}}[\sin(2\theta)\exp(\pm i\varphi)]$
2	± 2	$\dfrac{3}{2}\sqrt{\dfrac{5}{96\pi}}[(1-\cos(2\theta))]\exp(\pm 2i\varphi)]$
Zonal, $m=0$		$\sqrt{\dfrac{(2l+1)}{4\pi}}P_l(\cos\theta)^*$
Sectorial, $m=l$		$\sqrt{\dfrac{(2l+1)}{4\pi(2l)!}}(2l-1)!!\sin^l\theta\exp(il\varphi)$

*$P_l(\cos\theta)$ are Legendre polynomials (Abramowitz and Stegun, 1964).

distribution, which can be written as a sum over the contributions from individual atoms positioned at the coordinates $\mathbf{r}_j = (r_j, \omega_j) = (r_j, \theta_j, \varphi_j)$

$$A(\mathbf{q}) = \sum_{j=1}^{N} f_j(q)\exp(i\mathbf{q}\mathbf{r}_j) \tag{A2.4}$$

Here, N is the number of atoms and $f_j(q)$ are their form factors. Using spherical harmonics in real and reciprocal space, the exponential function can be expressed as

$$\exp(i\mathbf{q}\mathbf{r}) = 4\pi \sum_{l=0}^{\infty} \sum_{m=-l}^{l} i^l j_l(qr) Y_{lm}^*(\omega) Y_{lm}(\Omega) \tag{A2.5}$$

where $j_l(sr)$ are spherical Bessel functions (Abramowitz and Stegun 1964). The scattering amplitude can therefore be represented as

$$A(\mathbf{q}) = \sum_{l=0}^{\infty} \sum_{m=-l}^{l} A_{lm}(q) Y_{lm}(\Omega) \tag{A2.6}$$

where $A_{lm}(q)$ are partial amplitudes

$$A_{lm}(q) = 4\pi i^l \sum_{j=1}^{N} f_j(q) j_l(qr_j) Y_{lm}^*(\omega_j) \tag{A2.7}$$

Substituting eq. (A2.6) into eq. (A2.3) and accounting for eq. (A2.2), the scattering intensity is a sum of individual multipole components:

$$I(q) = 2\pi^2 \sum_{l=0}^{\infty} \sum_{m=-l}^{l} |A_{lm}(q)|^2 \tag{A2.8}$$

This formalism, introduced in SAS by Harrison (1969), Stuhrmann (1970a), and Stuhrmann (1970b), is extremely useful in rapid calculation of the scattering intensities from monodisperse systems. In practical applications, the sum over l is truncated, depending on the required accuracy of the representation of the structure, by the values ranging from $L = 2$ (overall anisometry) up to $L = 50$ or 100 (precise calculation of the scattering patterns for wide angles).

Scattering intensity from a binary complex   Apart from computation of the scattering from a single particle, spherical harmonics are very useful for rapid evaluation of the scattering from complexes consisting of multiple subunits with known structure. To illustrate this, let us consider a complex of two bodies, A and B, both with known structures. Their scattering amplitudes in reference positions and orientations are denoted $A(\mathbf{q})$ and $B(\mathbf{q})$, respectively. The scattering intensity from the complex is calculated as

$$I(q) = \left\langle |A(\mathbf{q}) + B(\mathbf{q})|^2 \right\rangle_{\Omega} = I_A(q) + I_B(q) + 2 \left\langle A(\mathbf{q}) B^*(\mathbf{q}) \right\rangle_{\Omega} \tag{A2.9}$$

To construct an arbitrary complex, body A may be fixed while body B is rotated and translated into another position, C. Describing the rotation by the Euler angles α, β and γ (Edmonds 1957) and the shift by a vector $\mathbf{u} = (u_x, u_y, u_z)$, the scattering amplitude $C(\mathbf{q})$ is

$$C(\mathbf{q}) = \exp(i\mathbf{q}\mathbf{u})\Pi(\alpha\beta\gamma)[B(\mathbf{q})] \tag{A2.10}$$

where $\Pi(\alpha\beta\gamma)$ is the rotational operator. The new scattering intensity is

$$I_C(q) = I_A(q) + I_B(q) + 2 \left\langle A(\mathbf{q}) C^*(\mathbf{q}) \right\rangle_{\Omega} \tag{A2.11}$$

Note that neither translation nor rotation change the intensity from the second body, but they influence only the cross-term. Using the multipole representation of the amplitudes $A(q)$ and $C(q)$ as in eq. (A2.8) yields

$$I_C(q) = 2\pi^2 \sum_{l=0}^{\infty} \sum_{m=-l}^{l} \left(|A_{lm}(q)|^2 + |B_{lm}(q)|^2 + 2Re\left[A_{lm}(q)C_{lm}^*(q)\right] \right) \tag{A2.12}$$

Importantly, the partial amplitudes $C_{lm}(q)$ of the rotated and translated subunit B can be expressed analytically via the $B_{lm}(q)$ functions of this subunit in the reference position. Especially convenient is the description of a rotation by the Euler angles α, β and γ, for which the following equation holds:

$$B'_{lm}(q, \alpha, \beta, \gamma) = \sum_{k=-l}^{l} D^l_{mk}(\alpha, \beta, \gamma)B_{lk}(q) \tag{A2.13}$$

where $D^l_{mt}(\alpha, \beta, \gamma)$ are the elements of the finite rotation matrix (Edmonds 1957), and $B'_{lm}(q, \alpha, \beta, \gamma)$ are the partial amplitudes of the rotated particle. The equations for the shifted amplitudes are somewhat more complicated

(Svergun 1991), but the calculations are significantly simplified if the direction u coincides with the Z-axis. In this case (Svergun *et al.* 1997)

$$B'_{lm}(q, \mathbf{u}) = (-1)^m \sum_{p=0}^{\infty} j_p(qu) \sum_{k=|l-p|}^{l+p} d_{lm}(k, p) \sum_{j=-k}^{k} B_{kj}(q) \qquad (A2.14)$$

where the coefficients $d_{lm}(k,p)$ are computed using the $3j$ Wigner symbols (Edmonds 1957)

$$d_{lm}(k,p) = i^p (2p+1) \sqrt{(2l+1)(2k+1)} \begin{pmatrix} l & p & k \\ 0 & 0 & 0 \end{pmatrix} \begin{pmatrix} l & p & k \\ -m & 0 & m \end{pmatrix} \qquad (A2.15)$$

Eqs. (A2.12–A2.15) allow one to rapidly compute the scattering intensity from a complex upon translation and rotation of one subunit. Given that rotations are performed much faster than translations, it is computationally more efficient to always perform translations only along the Z-axis. This can be achieved easily by appropriately rotating the entire coordinate system prior to the translation operation such that the direction of the translation coincides with Z. The frame is rotated back after the translation is performed.

Scattering intensity from a symmetric homodimer   The computations are significantly simplified for homodimeric structures possessing P2 point symmetry. Let us denote the partial amplitudes of the monomer placed at the origin as $A_{lm}(q)$. Assuming that the two-fold axis coincides with Y, and starting from the monomer positioned at the origin, an arbitrary dimer is constructed by rotating the monomer and translating it by u_z along Z. The second monomer is then obtained rotating the first one by π over Y. This operation is described by Euler angles $\alpha = \gamma = 0$, $\beta = \pi$, which from eq. (A2.13) is simply equivalent to a complex conjugate. The scattering intensity from a homodimer is therefore expressed through the partial amplitudes of a rotated monomer $B_{lm}(q)$ (computed from $A_{lm}(q)$ using eq. (A2.13)) and its displacement along Z:

$$I_C(q, u_z) = 2 \sum_{l=0}^{\infty} \sum_{m=-l}^{l} \left| (-1)^m \sum_{p=0}^{\infty} Re[i^p j_p(qu_z)(2p+1) \right.$$

$$\left. \sum_{k=|l-p|}^{l+p} \sqrt{(2l+1)(2k+1)} \begin{pmatrix} l & p & k \\ 0 & 0 & 0 \end{pmatrix} \begin{pmatrix} l & p & k \\ -m & 0 & m \end{pmatrix} \sum_{j=-k}^{k} B_{kj}(q) \right|^2 \qquad (A2.16)$$

In this equation the number of summands is lower than for the general case, as all imaginary terms, all terms with odd m, and also all zonal harmonics (terms $l0$) with odd l are not present. Therefore, modelling of a homodimer in P2 requires fewer parameters (four instead of six) and allows more rapid computations than the modelling for the general heterodimer case.

Scattering intensity from a multibody complex   The above formalism using spherical harmonics is easily extended to a multibody complex consisting of K subunits with known structure. Denoting the scattering amplitude from each subunit in a reference position as $A^{(k)}(q)$, the scattering intensity $I_C(q)$ of the entire complex is expressed as:

$$I_C(q) = \left\langle \left| \sum_{k=1}^{K} C^{(k)}(q) \right|^2 \right\rangle_{\Omega} , \quad C^{(k)}(q) = \exp(iqr_k)\Pi(\alpha_k\beta_k\gamma_k)[A^{(k)}(q)] \quad (A2.17)$$

where the modified scattering amplitudes $C^{(k)}(q)$ of each body depend in the general case on six parameters, the vector of the shift r_k, and the Euler rotation angles α_k, β_k, and γ_k. Using spherical harmonics yields a convenient analytical representation of the intensity in the form

$$I_C(q) = 2\pi^2 \sum_{l=0}^{\infty} \sum_{m=-l}^{l} \left| \sum_{k=1}^{K} C_{lm}^{(k)}(q) \right|^2 \quad (A2.18)$$

Here, $C_{lm}^{(k)}(q)$ are the partial scattering amplitudes of the kth body, which are readily computed with eqs. (A2.13–A2.15). In the general case, the number of independent parameters for such a modelling is $6(K-1)$, as one body may be fixed without loss of generality.

Similarly to the above case of the homodimer, symmetry can significantly speed up the computations by lowering the number of independent parameters and by reducing the number of non-zero terms in eq. (A2.18). Thus, for symmetric particles having point groups Pn and Pn2, it can be assumed without loss of generality that the n-fold axis coincides with the Z-axis and the two-fold axis (in the case of Pn2 symmetry) coincides with the Y-axis, which leads to the specific selection rules for the spherical harmonics. In this case, summation in eq. (A2.18) runs only over symmetry-independent rigid bodies in the ensemble and only over m equal to zero or multiples of n and, moreover, in the case of Pn2, all zonal harmonics (terms of order $l0$) with odd l as well as all imaginary parts will vanish.

Termination effects   The spherical harmonics therefore provide a very convenient framework to rapidly compute the scattering from multibody ensembles. The main advantages are analytical equations avoiding the need of spherical averaging and the easy use of symmetry. One should make a caveat that the computation of the translation involves a termination effect due to the limited number of harmonics. This is evident from eq. (A2.14), where the summation index over the coupling terms expressed using $3j$ Wigner symbols runs, generally speaking, to infinity. Truncation of the sum in eq. (A2.14) leads to a loss of intensity at higher angles, depending on the magnitude of the shift. This can be easily understood in terms of the scattering from a spherically symmetric particle. If one positions the particle at the origin, only $L = 0$ term is present in eq. (A2.8), whereas moving the particle breaks the radial symmetry and non-zero terms appear, which grow with the magnitude of the shift. Note that the intensity from the particle computed using eq. (A2.8) remains unchanged, and only its multipole components are redistributed. Given that the rigid-body modelling is employed to analyse the overall quaternary structure, only lower-angle data (up to about 2 nm resolution) are used for the modelling. Termination effects are typically negligible for this range when using a reasonable number of terms (usually, up to $L = 15$) in the expansion in eq. (A2.8) to describe the individual bodies. Still, the rigid-body modelling algorithms always keep the entire assembly close to the origin to minimise termination effects. Note that the rotation operation in eq. (A2.13) does not have termination effects and is always precise.

Appendix 3: Interactions between spherical molecules

Equation (5.7) describes the scattering of pairs of particles, whether in direct contact or not. Whereas it is very useful for rigid-body modelling of complexes, in the case of solutions where particles have a broad distribution of relative orientations its use would lead to intractable calculations. It is therefore generally assumed that one deals with spherical particles with interactions described by a central potential. As explained below, this leads to concepts such as the pair distribution function, the direct and indirect correlation functions, and the Ornstein–Zernike relationship, which are often referred to in the literature. Different models used in studies of interactions (see Chapter 8) correspond to different approximations to the closure relation required to solve the Ornstein–Zernike relationship.

The structure factor   If the solution contains N identical spherical particles its structure can be considered as resulting from the convolution of a particle with scattering density $M(r)$ with an array of δ functions.

$$M(r) * \sum_{i=1}^{N} \delta(\mathbf{r} - r_{\mathbf{i}}) \tag{A3.1}$$

The Fourier transform of $M(r)$ is the scattering factor of the particle $f(q)$, and the scattering amplitude is the product of the Fourier transforms of an isolated particle and of the array of δ functions:

$$A(q) = f(q) \sum_{i=1}^{N} \exp(i\mathbf{q} \cdot \mathbf{r_i}) \tag{A3.2}$$

The corresponding intensity normalised to a single particle is

$$N^{-1} \left[A(q) \cdot A^*(q) \right] = I(q) = N^{-1} f(q) \sum_{i=1}^{N} \exp(i\mathbf{q} \cdot \mathbf{r_i}) \cdot f(q) \sum_{j=1}^{N} \exp(-i\mathbf{q} \cdot \mathbf{r_j})$$

$$= |f(q)|^2 \cdot N^{-1} \sum_{i=1}^{N} \sum_{j=1}^{N} \exp(i\mathbf{q} \cdot (\mathbf{r_i} - \mathbf{r_j})) \tag{A3.3}$$

The second factor in this expression is called the structure factor or static structure factor when it is necessary to differentiate it from the time-dependent dynamic structure factor. Note that the double sum yields cosine terms only, and thus represents real numbers. For an ideal solution where the particles are randomly distributed, its value is 1, and the intensity is just that due to an isolated particle. In practice, the structure factor is obtained by dividing the experimental scattering pattern of a concentrated solution by that of an infinitely dilute solution scaled to the same concentration, or possibly a scattering curve calculated from a crystallographic model.

Particle distribution functions and the pair distribution function   The description of the distribution of the centres N identical non-overlapping particles occupying fixed positions in a fixed volume V at a fixed temperature T in terms of Dirac δ functions reflects the fact that this function represents a density. For particles in solution which have no fixed positions one has to consider a time average or equivalently an average over a large set of systems with the same values of N, V and T (that is, a canonical ensemble) represented here by the angular brackets.

$$\rho^{(1)}(\mathbf{r}) \equiv \left\langle \sum_{i=1}^{N} \delta(\mathbf{r} - r_i) \right\rangle \tag{A3.4}$$

The average number density of the particles in such a system is $\rho \equiv N/V$, and the probability of finding a particle in the volume element $d\mathbf{r}$ is $\rho^{(1)}(\mathbf{r})d\mathbf{r}$, so that

$$\int_V \rho^{(1)}(\mathbf{r})d\mathbf{r} = N \tag{A3.5}$$

Similarly, one can define a two-particle distribution function

$$\rho^{(2)}(\mathbf{r}, \mathbf{r}') \equiv \left\langle \sum_{i=1}^{N} \sum_{j \neq i} \delta(\mathbf{r} - \mathbf{r_i})\delta(\mathbf{r}' - \mathbf{r_j}) \right\rangle \tag{A3.6}$$

The expression between brackets is zero except when there is a particle at $\mathbf{r} = \mathbf{r_i}$ and any of the other $N - 1$ particles at $\mathbf{r}' = \mathbf{r_j} \neq \mathbf{r_i}$. The probability of finding a particle in the volume element $d\mathbf{r}$ and another one in $d\mathbf{r}'$ independently of the position of the other particles is thus $\rho^{(2)}(\mathbf{r}, \mathbf{r}')d\mathbf{r}d\mathbf{r}'$, and integration over $\mathbf{r}'$ yields

$$d\mathbf{r} \int \rho^{(2)}(\mathbf{r}, \mathbf{r}')d\mathbf{r}' = (N - 1)\left\langle \sum_{i=1}^{N} \delta(\mathbf{r} - \mathbf{r_i}) \right\rangle d\mathbf{r} = (N - 1)\rho^{(1)}(\mathbf{r})d\mathbf{r} \tag{A3.7}$$

This indicates that the probability of simultaneously finding particles in the volume element $d\mathbf{r}$ and in a volume element $d\mathbf{r}'$ anywhere else in the volume is $(N - 1)$ times the probability of finding a particle in $d\mathbf{r}$. Intuitively this must be so, because if one particle is fixed at r (that is, $\rho^{(1)}(\mathbf{r})d\mathbf{r} = 1$), integration over $d\mathbf{r}'$ will yield the value of the number or remaining particles in the volume (that is, $N - 1$).

As intermolecular interactions influence the probability of finding a particle at $\mathbf{r}$ if there is one at $\mathbf{r}'$ and *vice versa*, the probability is not simply the product of the probabilities for $\mathbf{r}$ and $\mathbf{r}'$, $\rho^{(1)}(\mathbf{r})\rho^{(1)}(\mathbf{r}')d\mathbf{r}d\mathbf{r}'$, but rather

$$\rho^{(2)}(\mathbf{r}, \mathbf{r}')d\mathbf{r}d\mathbf{r}' \equiv \rho^{(1)}(\mathbf{r})\rho^{(1)}(\mathbf{r}')g^{(2)}(\mathbf{r}, \mathbf{r}')d\mathbf{r}d\mathbf{r}' \tag{A3.8}$$

This expression defines the pair distribution function, $g^{(2)}(\mathbf{r}, \mathbf{r}')$, which in general will depend on the distance between particles as well as on their relative orientation. As the number of particles in the volume is not altered by their uneven distribution, it can be easily understood that $g(\mathbf{r}, \mathbf{r}')$ oscillates around 1 for small values of $|\mathbf{r}-\mathbf{r}'|$ where the particles interact and tends to the limit for an ideal gas $g(r) \rightarrow 1 - 1/N = 1$ at large distances ($r \rightarrow \infty$). The function

$g(\mathbf{r}, \mathbf{r}')$ thus represents the ratio between the local density and the average or bulk density (ρ), and thus describes the deviations from a constant probability corresponding to the complete randomness of the ideal gas.

For homogeneous systems such as solutions of biological macromolecules, $\rho^{(1)}(\mathbf{r}) = \rho^{(1)}(\mathbf{r}') = \rho$, and for a fixed particle at $\mathbf{r}(\rho^{(1)}(\mathbf{r}) = 1)$ the probability of finding another particle at $\mathbf{r}'$ is $\rho^{(1)}(\mathbf{r}')g^{(2)}(\mathbf{r}, \mathbf{r}')d\mathbf{r}d\mathbf{r}' = \rho g^{(2)}(\mathbf{r}, \mathbf{r}')d\mathbf{r}d\mathbf{r}'$— an expression which also integrates to $N - 1$, and $\rho^{(2)}(\mathbf{r}, \mathbf{r}') = \rho^2 g^{(2)}(\mathbf{r})$. For homogeneous systems one can freely shift the origin of the distribution so that eq. (A3.6) can be rewritten as

$$\rho^{(2)}(\mathbf{r}, \mathbf{r}') = \left\langle \sum_{i=1}^{N}\sum_{j\neq i} \delta(\mathbf{r}' + \mathbf{r} - \mathbf{r_i})\delta(\mathbf{r}' - \mathbf{r_j}) \right\rangle \qquad (A3.9)$$

The second δ function is equal to one when $\mathbf{r}' = \mathbf{r_j}$, and the first one corresponds to a distribution where the origin has been shifted to $\mathbf{r}' = \mathbf{r_j}$, and it will thus be equal to one when $i \neq j$ and $\mathbf{r} = \mathbf{r_j} - \mathbf{r_i}$. The averaging is done by successively placing each particle j at the origin, and dividing by N. The expression between triangle brackets is integrated over $\mathbf{r}'$, and it is easily shown, taking into account that $\delta(\mathbf{r}' - \mathbf{r_j}) = 1$ if $\mathbf{r}' = \mathbf{r_j}$ and 0 otherwise, that

$$\left\langle N^{-1} \sum_{i=1}^{N}\sum_{j\neq i} \delta(\mathbf{r} + \mathbf{r_j} - \mathbf{r_i}) \right\rangle = \left\langle N^{-1} \int \sum_{i=1}^{N}\sum_{j\neq i} \delta(\mathbf{r}' + \mathbf{r} - \mathbf{r_i})\delta(\mathbf{r}' - \mathbf{r_j})d\mathbf{r}' \right\rangle$$

$$= N^{-1} \int \rho^{(2)}(\mathbf{r}' + \mathbf{r}, \mathbf{r}')d\mathbf{r}' = \rho g^{(2)}(\mathbf{r})$$

$$(A3.10)$$

$g^{(2)}(\mathbf{r}, \mathbf{r}') = g(\mathbf{r})$ with $\mathbf{r} = \mathbf{r}' - \mathbf{r}$, and using eq. (A3.7) one finds that

$$\rho \int g(\mathbf{r})d\mathbf{r} = N - 1 \qquad (A3.11)$$

Equation (A3.10) is related to the probability of finding a particle in a volume element $d\mathbf{r}$ at a distance $\mathbf{r}$ of a particle located at the origin. For isotropic solutions the result does not depend on the direction of observation but only on the distance $r = |\mathbf{r_2} - \mathbf{r_1}|$ and $g(\mathbf{r}) = g(r)$.

It is possible to define higher-order pair distribution functions involving three or more particles at different locations:

$$\rho^{(n)}(\mathbf{r_1}, \ldots \mathbf{r_n})d\mathbf{r_1} \ldots d\mathbf{r_n} = \frac{N!}{(N-n)!}$$

(probability of finding particles at $\mathbf{r_1}$ and $\mathbf{r_2}$ and $\mathbf{r_3} \ldots$ and $\mathbf{r_n}$) with

$$\int \rho^{(n)}(\mathbf{r_1}, \ldots \mathbf{r_n})d\mathbf{r_1} \ldots d\mathbf{r_n} = \frac{N!}{(N-n)!} \qquad (A3.12)$$

Equation (A3.7) is a special case of this expression. These distributions correspond to increasingly detailed descriptions of the structure, and illustrates that a pair distribution function is not uniquely related to higher-order distributions and *a fortiori* to the structure.

Confusion may sometimes arise from the fact that $g(r)$ is often referred to as the radial distribution function (RDF)—a name which is also used to refer to the average number of particles in a spherical shell of thickness ($\Delta r = r_2 - r_1$) centred on a particle at the origin:

$$\int_{r1}^{r2} \rho g(r)dr = 4\pi r^2 \rho g(r)dr = \rho R(r)dr \tag{A3.13}$$

The RDF is $R(r) \equiv 4\pi^2 g(r)$. When the integration is from $r_1 = 0$ to r_2 one obtains the average number of particles in a sphere of radius $r = r_2 - r_1$ centred on the origin particle or coordination number or number of nearest neighbours:

$$n \equiv \rho \int_0^{r1} R(r)dr \tag{A3.14}$$

Structure factor and pair correlation function   The autocorrelation function of the density, in this case the number density, can also be written in terms of Dirac δ functions:

$$G(\mathbf{r}) = N^{-1} \int \langle \rho(\mathbf{r'} + \mathbf{r})\rho(\mathbf{r'}) \rangle d\mathbf{r'} = \left\langle N^{-1} \int \sum_i^N \sum_{j\neq i} \delta(\mathbf{r'} + \mathbf{r} - \mathbf{r_i})\delta(\mathbf{r'} - \mathbf{r_j})d\mathbf{r'} \right\rangle + \delta(\mathbf{r}) \tag{A3.15}$$

The first term of this convolution is the same as eq. (A3.8), whereas the second one corresponds to the N self-terms for which $\mathbf{r_i} = \mathbf{r_j}$ (that is, $\mathbf{r} = 0$).

For an homogeneous isotropic system this simplifies to

$$G(\mathbf{r}) = \left\langle N^{-1} \sum_{i=1}^N \sum_{j\neq 1} \delta(\mathbf{r} + \mathbf{r_j} - \mathbf{r_i}) \right\rangle + \delta(\mathbf{r}) = \rho g(r) + \delta(\mathbf{r}) \tag{A3.16}$$

As described for an isolated particle in Chapter 3, the intensity—here the structure factor—is the Fourier transform of the autocorrelation function of the corresponding structure. Hence, for an homogeneous fluid the structure factor $S(\mathbf{q})$ can be written as the Fourier transform of $G(\mathbf{r})$ and hence also of $g(\mathbf{r})$.

$$S(\mathbf{q}) = N^{-1} \left\langle \sum_{i=1}^N \sum_{j=1}^N \exp(i\mathbf{q} \cdot \mathbf{r_i}) \exp(i\mathbf{q} \cdot \mathbf{r_j}) \right\rangle = 1 + N^{-1} \left\langle \sum_i^N \sum_{j\neq i} \exp(i\mathbf{q} \cdot (\mathbf{r_i} - \mathbf{r_j})) \right\rangle$$

$$= 1 + N^{-1} \left\langle \sum_i^N \sum_{j\neq i} \int \int \exp(i\mathbf{q} \cdot (\mathbf{r} - \mathbf{r'}))\delta(\mathbf{r} - \mathbf{r_i})\delta(\mathbf{r'} - \mathbf{r_j})d\mathbf{r}d\mathbf{r'} \right\rangle \tag{A3.17}$$

$$= 1 + N^{-1} \int \int \exp(i\mathbf{q} \cdot (\mathbf{r} - \mathbf{r'}))\rho^{(2)}(\mathbf{r}, \mathbf{r'})d\mathbf{r}d\mathbf{r'} = 1 + \rho \int \exp(i\mathbf{q} \cdot \mathbf{r})g(r)d\mathbf{r}$$

The properties of Fourier transforms imply that $g(\mathbf{r})$ is the (inverse) Fourier transform of $S(q)$:

$$\rho g(\mathbf{r}) = \frac{1}{(2\pi)^3} \int \exp(-i\mathbf{q} \cdot \mathbf{r})(S(\mathbf{q}) - 1)d\mathbf{q} \tag{A3.18}$$

For an isotropic system, $S(q)$ depends only on $q = |\mathbf{q}|$ and

$$S(q) = 1 + 2\pi\rho \int r^2 g(r) \int_{-1}^{1} \exp(iqr\cos\theta)d(\cos\theta)dr = 1 + 4\pi\rho \int r^2 g(r)\frac{\sin(qr)}{qr}dr \qquad (A3.19)$$

Equation (A3.17) can also be written as

$$S(q) = 1 + \rho \int \left[1 + (g(\mathbf{r}) - 1)\right]\exp(i\mathbf{q}\cdot\mathbf{r})d\mathbf{r} \qquad (A3.20)$$

Since $\int \exp(i\mathbf{q}\cdot\mathbf{r})d\mathbf{r} = \delta(\mathbf{q})$ (see Appendix 1), this yields

$$S(\mathbf{q}) = 1 + (2\pi)^3\rho\delta(\mathbf{q}) + \Im(h(\mathbf{r})) = 1 + \rho\cdot\Im(h(\mathbf{r})) \qquad (A3.21)$$

The second term in eq. (A3.21) can be neglected, as it contributes only to the forward scattering $I(0)$ and cannot be measured. The function $h(\mathbf{r}) \equiv g(\mathbf{r}) - 1$ is the (total) pair correlation function.

Correlation functions and the Ornstein–Zernike relationship   The Ornstein–Zernike relation (eq. A3.22) describes the fact that the total correlation $h(r)$ results from the direct correlation $c(r)$ and from the indirect correlation propagated through all possible paths between other particles. This leads to a multicentre integral representing a convolution. Its solution requires the graph theoretical techniques of the theory of liquids (Hansen and McDonald 1986). For homogeneous isotropic one-component systems the Ornstein–Zernike relationship is

$$h(r) = c(r) + \rho \int c(|\mathbf{r} - \mathbf{r}'|)h(\mathbf{r}')d\mathbf{r}' \qquad (A3.22)$$

Expressions for multicomponent systems are also available (Belloni 1985). Equation (A3.22) can also be written as

$$h(1,2) = c(1,2) + \rho \int c(1,3)h(2,3)dr_3 \qquad (A3.23)$$

By recursion, one finds that it represents an expansion of $h(r)$ in terms of powers of the number density:

$$h(1,2) = c(1,2) + \rho \int c(1,3)c(2,3)dr_3 + \rho^2 \iint c(1,3)c(3,4)c(4,2)dr_3dr_4 \quad (A3.24)$$

The coefficients of this expansion contain structural information about increasingly large clusters of particles and thus become increasingly difficult to evaluate.

Using the properties of Fourier transforms, eq. (A3.22) readily yields:

$$\Im(h(r)) = \frac{\Im(c(r))}{1 - \rho\Im(c(r))} \qquad (A3.25)$$

Combining eqs. (A3.21) and (A3.25) one obtains:

$$S(c,q) = 1 + \rho\cdot\Im h(r) = 1/(1 - \rho\Im c(r)) \qquad (A3.26)$$

If $S(c,q)$ is accurately known one can obtain $\Im(c(r))$ and hence also $c(r)$ by Fourier transformation, and eq. (A3.25) would yield $\Im(h(r))$ and hence also $h(r)$. $g(r)$ can be obtained from eq. (A3.18). $c(r)$ has a range comparable to that of the pair potential $U(r)$, whereas $h(r)$ has a longer range—a matter which plays an important role in the approximations to eq. (A3.24).

In principle, $S(c,q)$ and $g(r)$ can be used to evaluate thermodynamic quantities. This has been done successfully for simple liquids, but not for solutions of macromolecules because of the many simplifications involved, though general trends can be obtained reliably.

As the Ornstein–Zernike relationship contains two unknowns, a closure relationship is required to solve the equation. An exact closure relationship has been derived using the theory of liquids, where $U(r)$ is the pair potential, $\gamma(r) = h(r) - c(r)$, the indirect correlation function, and $E(r)$ a sum of multicentre integrals representing the so-called bridge diagrams in the graph representation of the Ornstein–Zernike relationship:

$$g(r) = \exp[-U(r)/kT + \gamma(r) + E(r)] \qquad (A3.27)$$

It can be understood intuitively that in the limit of infinite dilution, $g(r) = \exp[-U(r)/kT]$.

Closure relationships for the Ornstein–Zernike relationship   As the general closure relationship is too difficult to handle in practice because $E(r)$ cannot be evaluated exactly, a number of approximate closure relationships, defining the terms or classes of terms that are neglected in the expansion of $h(r)$ in terms of powers of the number density of particles in eq. (A3.24), have been developed. The usefulness of these approximations depends among others on the importance of long-range electrostatic interactions in the potential.

Equation (A3.27) leads to the following expression for the direct correlation function:

$$c(r) = \exp[-U(r)/kT + \gamma(r) + E(r)] - 1 - \gamma(r) \qquad (A3.28)$$

The simplest closure is the mean spherical approximation (MSA) obtained by neglecting $E(r)$ and expanding the exponential to the second term, which yields $c(r) = -U(r)/kT$. The MSA, which has an analytical solution (Hoye and Blum 1977), corresponds to a hard repulsive core with diameter σ and an attractive tail such that $g(r) = 0$ for $r \le \sigma$ (that is, $h(r) = -1$ and $c(r) = -U(r)/kT$ for $r > \sigma$.

If one neglects $E(r)$ and uses a series expansion up to the second term for the exponential of $\gamma(r)$:

$$c(r) + \exp[-U(r)/kT](1 + \gamma(r)) - 1 - h(r) + c(r) \qquad (A3.29)$$

Since $h(r) \equiv g(r) - 1$, this yields the Percus–Yevick approximation:

$$g(r) = \exp[-U(r)/kT](1 + \gamma(r)) \qquad (A3.30)$$

For hard spheres (see Chapter 8), for $r < \sigma$, $U(r) = \infty$ and eq. (A3.29) becomes $c(r) = -1 - h(r) + c(r)$, from which it is obvious that $h(r) = -1$, and as $h(r) \equiv g(r) - 1$, $g(r) = 0$. For $r > \sigma$, $U(r) = 0$ and eq. (A3.29) becomes $c(r) = 1 + \gamma(r) - 1 - h(r) + c(r)$, and as $\gamma(r) = h(r) - c(r)$, $c(r) = 0$.

Another useful solution of the Ornstein–Zernike equation in the Percus–Yevick approximation is given by the adhesive pair potential (Baxter 1968):

$$U(r)/kT = \infty \text{ for } 0 < r < \sigma$$
$$= \ln[(12\tau\Delta/(\sigma + \Delta)] \, \sigma \le r \le \sigma + \Delta$$
$$= 0 \quad \sigma + \Delta < r$$

The stickiness τ corresponds to a reduced temperature, and Δ is the interpenetration length. The limit of a series of calculations with variable Δ for $\Delta \rightarrow 0$ gives the same results as the hard sphere approximation (De Kruif *et al.* 1989).

The hard-sphere model is often taken as the reference system with known exact direct correlation function c_0 in the random phase approximation (RPA) (see (Hansen and McDonald 1986)). If the perturbation is not too large, $c(r) \cong c_0(r) - U(r)/kT$. The structure factor $(S(q))$ is obtained from that of the reference system $(S_0(q))$, with particle number density ρ as

$$S(q) = S_0(Q)[1 + (\rho/k_B T)S_0(Q)\Im u(r)]^{-1} \tag{A3.31}$$

For hard spheres with diameter σ and volume fraction ϕ, the structure factor is related to the first-order spherical Bessel function $j_1(q\sigma)$ by Narayanan and Liu (2003):

$$S_0(q)^{-1} = 1 - 12\phi \frac{[\phi(3 - \phi^2) - 2]}{(1 - \phi)^4} \frac{j_1(q\sigma)}{q\sigma} \tag{A3.32}$$

The perturbation potential $U(r)$ usually consists of the attractive and repulsive components of the DLVO potential (see Chapter 8), and in some cases of additional attractive terms to explain salt-specific effects and the self-association of proteins (Narayanan and Liu 2003).

Another approximation that has provided many useful results is the hypernetted chain closure (HNC). In this case one also neglects the bridge diagrams $E(r)$ and rearranges the terms in a manner similar to eq. (A3.28) to obtain

$$g(r) = \exp[-U(r)/kT + \gamma(r)] \tag{A3.33}$$

To fit a set of structure factors for a series of measurements the data and the correlation functions and their transforms are sampled on an evenly spaced grid of points i, with $1 \leq i \leq N$, with $q_i = i\Delta q$ in reciprocal space, and $r_i = i\Delta r$ in real space with a cutoff at $r = N\Delta r$.

The solution is found by iteration, starting with an initial guess for $\gamma(r)$ (for example, $\gamma_{in}(r) = 0$), and values of the potential to compute the direct correlation function $c(r)$. In the case of the HNC approximation one uses the following expression:

$$c(r) = h(r) - \gamma(r) = (g(r) - 1) - \gamma(r) = \exp(-U(r)/kT + \gamma(r)) - 1 - \gamma(r) \tag{A3.34}$$

Fourier transformation of $c(r)$ yields $\Im c(r)$, and one can use eqs. (A3.25) and (A3.26) to calculate $\Im h(r)$, $S(c, q)$ and $\Im \gamma_{out}(r)$, and then $\gamma_{out}(r)$ by inverse Fourier transformation. The new approximation $\gamma_{out}(r)$ is then used as input for the next iteration, repeating the process until convergence is achieved (that is, $\left\{ \frac{1}{N} \sum_{i=1}^{N} [\gamma_{out}(ri) - \gamma_{in}(r(i)]^2 \right\} < \varepsilon_{max}$ (Belloni 1993)). Various numerical techniques to speed up convergence have been described.

Appendix 4: Web resources

General
Small Angle Scattering Commission of the International Union of Crystallography
<http://www.iucr.org/resources/commissions/small-angle-scattering>

The Collective Action for Nomadic Small Angle Scatterers (canSAS)
<http://www.cansas.org/>

Information about synchrotron and free electron laser light source facilities
<http://www.lightsources.org>

Biological small angle scattering on Wikipedia
<http://en.wikipedia.org/wiki/Biological_small-angle_scattering>

List of SAXS synchrotron stations and laboratory cameras
<http://en.wikipedia.org/wiki/Small-angle_X-ray_scattering>

A list of neutron sources
<http://www.ncnr.nist.gov/nsources.html>

World Directory of SANS Instruments
<http://www.ill.eu/instruments-support/instruments-groups/groups/lss/more/world-directory-of-sans-instruments/>

Program packages and databases
A forum on biological SAS programs
<http://www.saxier.org/forum/>

ATSAS page
<http://www.embl-hamburg.de/biosaxs/software.html>

MULCH modules for the computation of contrast variation data
<http://smb-research.mmb.usyd.edu.au/NCVWeb/>

The Small Angle Scattering ToolBox
<http://sastbx.als.lbl.gov/wiki/index.php/Main_Page>

Integrative Modelling Platform
<http://salilab.org/imp/>

Docking software HADDOCK
<http://www.nmr.chem.uu.nl/haddock/>

Protein Data Bank (PDB), the major repository of high-resolution macromolecular structures
<http://www.rcsb.org/>

BIOISIS database of models and scattering data
<http://www.bioisis.net/>

DARA database for the search of structural neighbours
<http://dara.embl-hamburg.de/>

Online SAS resources
ATSAS online
<http://www.embl-hamburg.de/biosaxs/atsas-online/>
Online versions of DAMMIN, DAMMIF, GASBOR, MONSA, CRYSOL, SASREF, EOM, and DANESSA.

Servers for computation of SAXS profiles:
<http://modbase.compbio.ucsf.edu/foxs/index.html> (program FOXS)
<http://spin.niddk.nih.gov/bax/nmrserver/saxs1/> (program AXES)
<http://lorentz.immstr.pasteur.fr/aquasaxs.php> (program AQUASAXS)

Molecular mass estimation from Porod volume (MoW server)
<http://www.if.sc.usp.br/~saxs/>

Visualisation tools for low-resolution SAS models
Rasmol
<http://rasmol.org/>

VMD
<http://www.ks.uiuc.edu/Research/vmd/>

SITUS
<http://situs.biomachina.org/tutorial_saxs.html>

Other methods
Computation of hydrodynamic parameters, program HYDROPRO
<http://leonardo.inf.um.es/macromol/programs/hydropro/hydropro.htm>

Computation of hydrodynamic parameters, program US-SOMO
<http://www.sas.uthscsa.edu/>

NMR and SAXS tools on a GRID
<http://www.wenmr.eu/>

Bioinformatics tools
Protein structure prediction
<http://en.wikipedia.org/wiki/List_of_protein_structure_prediction_software>

RNA structure prediction
<http://en.wikipedia.org/wiki/List_of_RNA_structure_prediction_software>

PISA (generation of possible biological assemblies from crystallographic contacts)
<http://www.ebi.ac.uk/msd-srv/prot_int/pistart.html>

References

Abramowitz, M. and Stegun, I. A. (1964). *Handbook of Mathematical Functions with Formulas, Graphs, and Mathematical Tables*. Washington: US Government Printing Office.

Baxter, R. J. (1968). 'Percus–Yevick equation for hard spheres with surface adhesion.' *J. Chem. Phys*. 49: 2770–4.

Belloni, L. (1985). 'A hypernetted chain study of highly asymmetrical polyelectrolytes.' *Chem. Phys*. 99: 43–4.

Belloni, L. (1993). 'Inability of the hypernetted chain integral equation to give a spinodal line', *J. Chem. Phys*. 98: 8080–95.

De Kruif, C. G., Rouw, P. W., Briels, W. J., Duits, M. H. G., Vrij, A., and May, R. P. (1989). 'Adhesive hard-sphere colloidal dispersions. A small-angle neutron-scattering study of stickiness and the structure factor', *Langmuir*, 5: 422–8.

Edmonds, A. R. (1957). *Angular Momentum in Quantum Mechanics*. Princeton, NJ: Princeton University Press.

Hansen, J.-P. and McDonald, I. R. (1986). *Theory of Simple Liquids*, San Diego, CA: Academic Press.

Harrison, S. C. (1969). 'Structure of tomato bushy stunt virus. I. The spherically averaged electron density', *J. Mol. Biol.* 42: 457–83.

Hoye, J. S. and Blum, L. (1977). 'Solution of Yukawa closure of Ornstein-Zernike equation', *J. Statist. Phys.* 16: 399–413.

Narayanan, J. and Liu, X. Y. (2003). 'Protein interactions in undersaturated and supersaturated solutions: a study using light and X-ray scattering', *Biophys. J.* 84: 523–32.

Stuhrmann, H. B. (1970a). 'Ein neues Verfahren zur Bestimmung der Oberflaechenform und der inneren Struktur von geloesten globularen Proteinen aus Roentgenkleinwinkelmessungen', *Zeitschr. Physik. Chem. Neue Folge* 72: 177–98.

Stuhrmann, H. B. (1970b). 'Interpretation of small-angle scattering of dilute solutions and gases. A representation of the structures related to a one-particle scattering functions', *Acta Crystallogr.* A26: 297–306.

Svergun, D. I. (1991). 'Mathematical methods in small-angle scattering data analysis', *J. Appl. Crystallogr.* 24: 485–92.

Svergun, D. I., Volkov, V. V., Kozin, M. B., Stuhrmann, H. B., Barberato, C., and Koch, M. H. J. (1997). 'Shape determination from solution scattering of biopolymers', *J. Appl. Crystallogr.* 30: 798–802.

Abbreviations list

AFM	Atomic force microscopy
AD(T)P	Adenosine di(tri)phosphate
AUC	Analytical ultracentrifugation
BPTI	Bovine pancreatic trypsin inhibitor
BMS	Beamline Meta Server
BSA	Bovine serum albumin
BSSA	Boltzmann simplex simulated annealing
CCD	Charge-coupled device (X-ray detector)
CD	Circular dichroism
CMC	Critical micellar concentration
DAM	Dummy atom model
DCM	Double-crystal monochromator
DLS	Dynamic light scattering
DLVO	Derjaguin, Landau, Verwey, Overbeek (potential)
DNA	Deoxyribonucleic acid
DR	Dummy residue
dsRNA	Double-stranded ribonucleic acid
DTT	Dithiothreitol
EM	Electron microscopy
EOM	Ensemble optimization method
FRET	Fluorescence resonance energy transfer
FTIR	Fourier transform infrared (spectroscopy)
FWHM	Full width at half maximum
GA	Genetic algorithm
(G)IFT	(Generalized) indirect Fourier transform
GnHCl	Guanidine hydrochloride
HPLC	High-performance liquid chromatography
IDP	Intrinsically disordered protein
KB	Kirkpatrick–Baez (X-ray mirror)
LDAO	Lauryldimethylamine-oxide
L(H)DL	low(high-)density lipoprotein
MA(L)LS	Multi-angle (laser) light scattering
MD	Molecular dynamics
MM	Molecular mass
MX	Macromolecular crystallography
NSD	Normalised spatial discrepancy
NMR	Nuclear magnetic resonance

PDB   Protein Data Bank
PEG   Polyethylene glycol
PDDF   Pair-distance distribution function
RDC   Residual dipolar coupling
RNA   Ribonucleic acid
RPA   Random phase approximation
SA   Simulated annealing
SAS   Small angle scattering
SASE   Self amplified spontaneous emission
SAX(N)S   Small angle X-ray (neutron) scattering
SESANS   Spin echo small angle neutron scattering
SR   Synchrotron radiation
SVD   Singular value decomposition
TFE   trifluoroethanol
TMAO   trimethylammoniume N-oxide
TOF   time of flight
USAX(N)S   Ultra-Small angle X-ray (neutron) scattering
(X)FEL   (X-ray) Free Electron Laser

List of most important symbols and variables

λ	radiation wavelength
2θ	scattering angle
q	$4\pi \sin(\theta)/\lambda$, momentum transfer, in other publications called also Q, s, or h
$d\sigma/d\Omega\ (q)$	differential scattering cross-section
$I(q)$	scattering intensity (for dilute monodisperse systems, particle form factor)
$S(q)$	structure factor
ρ	scattering-length density
$\rho_p(\mathbf{r})$	scattering-length density distribution of the particle
ρ_s	scattering-length density of the solvent
$\Delta\rho(\mathbf{r}) = \rho_p(\mathbf{r}) - \rho_s$	excess scattering-length density distribution
$\Delta\rho = <\Delta\rho(\mathbf{r})> -\rho_s$	contrast
R_g	radius of gyration
D_{max}	maximum particle size
$p(r)$	pair distance distribution function, also called distance distribution
T	transmission
V_p	particle volume (Porod volume)
$\bar{v}_p$	partial specific volume of a macromolecule
(r, ω)	spherical coordinates in real space
(q, Ω)	spherical coordinates in reciprocal space

Index of computer programs

A
AQUASAXS 127, 351
AXES 126, 351

B
BUNCH 132–134, 185, 211

C
CORAL 133–134
CREDO 132
CRYSOL 126, 160, 248
CRYSON 126

D
DALAI_GA 115, 120, 174
DAMAVER 121, 312
DAMCLUST 131, 312
DAMMIN 118–124, 174–177, 186, 202–205, 306
DAMMIF 120–124

E
EM2DAM 127

F
FSEARCH 294
FOXS 127, 351
FOXSDOCK 130, 310

G
GASBOR 124, 134, 158, 174–179, 294
GIFT 99, 165
GNOM 99, 154–155

H
HADDOCK 310–314
HYDRO 127, 307
HYDROPRO 178, 307

I
ITP 99

M
MASSHA 129
MONSA 120, 176–177, 190
MULCH 351

O
OLIGOMER 156

P
PISA 293

S
SASHA 114, 119, 124, 294
SASMODEL 120
SASREF 130–131, 158, 184, 186, 314
SAXS3D 120
SEDFIT 305
SOMO 307, 351
SUPCOMB 121

Index

A

Ab initio analysis 112–124, 182
Absolute measurements 70, 85
Absorption:
  edge 26
  neutrons 69
  X-rays 26, 69
Aggregation 73, 82, 85, 102, 259, 267, 304
Analytical ultracentrifugation 2, 65, 174, 179, 185, 303–307

B

Background 2–3, 56, 69
  incoherent 5, 18, 25, 68, 135
  subtraction 70–73, 75, 78
Bacteriophage 6, 247
Basic scattering functions 22
Bead modelling 115–122, 171, 306
Bessel functions 94, 100, 109, 339
Bioinformatics 309
Brightness (spectral) 27

C

Calibration sample 85–87
Characteristic function (see correlation function, particle)
Clusters:
  heavy atom 139, 299
  in solution 277
  models 131, 312, 347
Collimation (slit, block, point) 43
  Bonse-Hart 58
Contrast 3, 21
  variation 5, 22
Convolution 19–20, 31, 97, 329
Correlation function:
  direct 343
  dynamic light scattering 307
  indirect 343
  particle 95
  cross section 100
  interparticle 164
  pair 95, 347
  thickness 100
Crystallography:
  macromolecular 46–50, 131, 159, 182, 208, 211, 287, 291–298, 309, 312, 350
  neutron 17, 299
  X-ray 1, 128, 136, 186–192, 299–303

D

Debye formula 96, 122
Delta function (Dirac delta function) 328
Depletion interaction 260, 273
Detectors:
  neutron 53, 55
  pixel 19, 86
  X-ray 44, 46, 48–49
Deuteration 24, 134
DLVO potential 269
Differential scattering cross section 15, 85
Distance distribution function 83, 93, 331
DNA damage 7, 206
Dummy atom model 115–124
Dummy residue 122, 171, 297
Dynamic light scattering 307

E

Electron microscopy 2, 119, 127, 130, 176, 184–188, 232, 294
Ensemble optimization 160–163, 210–211
Errors (propagation, random, systematic) 79–84

F

Field orientation (electric/magnetic) 229
Flexible systems 159–163, 205, 295–297
Fluctuations 21–22, 87, 248, 271, 285
Focusing 41–49
  Kirkpatrick Baez 48, 50–51
  hydrodynamic 227
Folding proteins 6, 67, 107, 157–158, 220, 239
  RNA 243
Form factor 4, 20, 95–96, 99, 122, 154, 166, 259, 276, 308, 339
Forward scattering 82–83, 103–105, 156, 347
Fourier transform 15–16, 19, 21, 33, 93, 97–98, 111, 143, 300, 331, 347
  indirect 84, 97, 99–100, 259
Free electron laser 27, 35–37, 220

G

Guinier 3
  approximation 106, 143
  plot 82, 206, 240
  zone (range) 77, 82

H

Hankel transform 100, 109
H-D exchange 5, 22–24, 26, 68, 103–104, 126, 138–139, 205, 300
Heavy-atom labels 23, 134, 136, 138–139
High pressure 224–225, 233–237
High throughput studies 286–291
Homogeneous particles 101, 110, 112
Hybrid methods 7, 128–134, 291, 295, 310, 320

I

Ill-posed problems 98, 154
Information content 31, 110–111, 120, 122, 312
Interactions 152–166
  extrapolation to zero concentration 66, 164
  protein-protein 4, 74, 85, 95, 163–166, 237, 259–280
  nucleic acids 280–281
Interference 2, 4, 15, 18, 20, 35, 91, 136–141, 152, 164, 166, 203, 305, 327
Invariants 106, 121
Isomorphous replacement 138, 140, 145, 300–301
Isotopic replacement (see H-D exchange)

K

Kratky:
  plot 106–108, 159, 178, 209, 240, 243
  camera 44–45

L

Lagrange multiplier 99
Largest dimension see Maximum particle size
Light Scattering
  dynamic 65, 165, 205, 261
  multi-angle 51, 65, 308
Light triggered processes 228, 247
Lipoproteins 198, 302
  High density 204–205
  Low density 204, 223

M

Macromolecular crystallography (see crystallography, macromolecular)
Maximum particle size 4, 76, 83–84, 96–99, 115, 236, 249
Membrane proteins 67, 198–204, 298–303
Micelles 67, 154, 199–204, 303
    critical micellar concentration 67, 199
Mixing stopped flow 225–237
    continuous flow 226–230
Mixtures 91, 292, 306
    equilibrium 4, 152–166, 192
    non-equilibrium 158
    oligomeric 5, 152–166, 192–198, 222
Molecular mass 4, 82
    determination 15, 103, 305
Momentum transfer 2, 14, 25
Monochromatization
    neutrons (velocity selector) 40
    X-rays 46–50
Multipole expansion 108–114, 124–128, 340–342
Muscle 51, 173–174, 223, 293, 296–297

N

NMR 2, 24
    application examples 182–184, 194–195, 208, 211–212
    joint use with SAS 128–135, 291–292, 295–298, 309–310
    ensemble analysis 160, 163
Nucleic acids 22–24, 95, 105, 115, 124, 135, 180, 185, 243, 280, 302
Nucleosome 186, 298

O

Oligomeric mixtures see Mixtures
Ornstein-Zernike formula 164, 343, 347–348

P

Pair potential 165, 260, 262, 265–267, 348
Particle:
    homogeneous 82, 85, 93, 96, 101, 110
    lamellar 100
    rodlike 100, 108, 165
    spherical 20, 154–155, 164–166, 277–278, 343
    surface 106–107, 109
    volume 19, 82, 102, 106, 156
Penalty function 117, 119, 120, 123, 132
Pink beam:
    definition 51
    applications 238, 241, 244
Porod:
    asymptotic 79, 106, 112
    invariant 106
    plot 107
Protein
    folding 67
    intrinsically disordered 5, 153, 209

Q

Quaternary structure 114, 125, 128, 131, 136, 180–185, 195, 236, 247, 293, 310, 312

R

Radiation damage 5, 45, 66–67, 72–74, 226–227, 243, 246, 248, 321
Radius of gyration 4, 77, 82, 94, 101–103, 156
Resolution
    instrumental 53, 55–56
    spatial 21
    time 49, 51
Ribosome
    glassy 141
    triangulation 136
        30S triangulation 189
        50S shape 188
        50S/70S models 189–190
Rigid body refinement 7–8, 128–133, 163, 179, 182–185, 295, 310, 312, 342
Residual dipolar coupling 163, 208, 292, 295, 297, 309

S

Salts 5, 22–23, 67, 72, 163, 193, 238, 259, 261, 263–266, 270–281
Sample concentration 66, 106, 141, 304, 308
    size 53
    volume 66, 95, 141, 224–227, 286–305
Sampling theorem 98, 111–112, 120
Scattering
    amplitude 15–16, 19, 109, 116, 120, 123, 125, 128, 163, 336, 338–343
    anomalous 17, 26, 35, 300
    cross section 15, 19, 76, 85, 86, 88, 103
    density 2–5, 19–22, 61, 87, 103, 106, 108–109, 125–126, 200, 302, 343
    excess 3, 102, 104
    effective atomic 125
    elastic 2, 13–17, 25, 221
    incoherent 5, 18, 25, 61, 68–70, 76, 79, 87
    inelastic 14, 76
    intensity 5, 19–20, 83, 93–94, 106–107, 109–113, 116, 125, 131, 137, 340
    length 15–26
    light 51, 65, 205, 231–232, 246, 261, 265, 278, 307–308
    shape 22, 109, 122, 181
    simple bodies 94
    spin 18, 190
    wide-angle 21, 44, 73, 340
Shannon theorem (see sampling theorem)
Shape determination 47, 114–115, 118, 121, 171–181, 188, 311
Simulated annealing 99, 117, 124
Singular value decomposition 157–158, 196–197, 222, 231, 240, 244, 248
Size distribution 153–154, 166, 308
Solvent:
    bound 123, 125–126
    density 22, 125, 304

Sources:
    neutron 37–38
    reactor 38–40
    spallation source 40–43
    X-ray 29
    laboratory sources 29–32
    synchrotron radiation
        bending magnet 33
        free electron laser (FEL) 35–37
        undulator 33–35
        wiggler 33–35
Solution
    monodisperse 4, 65, 91, 93, 165, 306–307
    polydisperse 261, 306–307
    ideal 259, 262–264, 343
    perfect 262
Spectral brightness see Brightness
Spherical harmonics 6, 108–109, 114–120, 128–129, 132, 188, 338–342
Splines 81, 97–98, 143
Structure
    factor 4, 20, 95, 99, 163–166, 260–261, 264, 266, 343
    prediction 351
Stuhrmann plot 205
    definition 103
Symmetry 114
    axial 6
    constraints, 119, 124, 130, 133
    ensemble generation 163
    practical applications 179–180, 182–183, 186, 193–194, 202–203, 207
    non-crystallographic, 302
    homodimer 114, 341
Synchrotron radiation see Sources
Systems
    binary 106
    fibrillar 100
    lamellar 100
    monodisperse 93–125
    polydisperse 152–166

T

Temperature:
    jump 223, 229–232
    scan 223, 230–231
Time-resolved measurements 7, 35, 228, 247
Time structure 27–28, 33
Triangulation 134–145, 188–189

U

Uniqueness 21, 109, 114, 116, 122, 277

V

Virial coefficients 262–264, 308

W

Wigner coefficients 113, 129, 341–342
Wavelength 13, 323

Z

Zimm plot 82, 101, 264